Archives of Virology

Supplementum 5

O. W. Barnett (ed.)
Potyvirus Taxonomy

Springer-Verlag Wien New York

Dr. Ortus W. Barnett, Jr.
Department of Plant Pathology and Physiology
Clemson University, Clemson, South Carolina, U.S.A.

With 57 Figures

Library of Congress Cataloging-in-Publication Data

Potyvirus taxonomy / O. W. Barnett (ed.)
      p.   cm.   —   (Archives of virology.   Supplementum.   ISSN
   0939-1983 ; 5)
      Includes bibliographical references and index

      ISBN-13:978-3-211-82353-8   e-ISBN-13: 978-3-7091-6920-9
      DOI: 10.1007/978-3-7091-6920-9

   1. Potyviruses.   2. Potyviruses—Nomenclature.   I. Barnett, O. W.
   (Ortus W.), 1939–   . II. Series.
   QR411.P68   1992
   567'.6483—dc20   92-25228
                                                 CIP

ISSN 0939-1983
ISBN-13:978-3-211-82353-8 Springer-Verlag Wien New York
e-ISBN-13: 978-3-7091-6920-9 Springer-Verlag New York Wien

Preface

The Potyvirus Study Group of the Plant Virus Subcommittee of the International Committee on Taxonomy of Viruses found that discussing the many ramifications of potyvirus nomenclature and classification by mail made for slow progress due to the size of the group and the diversity of ideas among study group members. To expedite these deliberations, a workshop on Potyvirus Taxonomy was held at the Biologische Bundesanstalt (BBA) in Braunschweig, Federal Republic of Germany on 2–4 September 1990. Rudolf Casper of the BBA and Gunter Adam of the Deutsche Sammlung von Mikroorganismen und Zellkulturen (DSMI) hosted the workshop which was sponsored by the Monsanto Agricultural Company and the Deutsche Sammlung von Mikroorganismen und Zellkulturen.

Persons interested in potyviruses or virus taxonomy were contacted about this workshop. News of the workshop spread and eventually there were 65 participants from 14 countries. Presentations by participants covered many aspects of the field of potyvirology including a general overview of potyviruses, serological relationships, nucleic acid sequences, biological properties, the pathogenesis-crop perspective, and specific situations in which taxonomic problems occur. Concepts of taxonomy also were discussed in detail.

My initial objective for the workshop was to develop a set of criteria for use in determining if a potyvirus isolate should be designated as a strain of a described potyvirus or as a new potyvirus. However, the workshop participants decided that enough information was not available to reach a consensus on this issue. Instead, after reviewing potyvirology and changes in the Rules of Nomenclature of Viruses which were made at the International Congress of Virology immediately preceding the workshop [see *Archives of Virology* Supplementum 2 (1991)], the workshop participants voted to establish the plant virus family *Potyviridae* and approved three genera and one possible genus. The approved genera were the genus *Potyvirus* composed of the aphid-borne viruses with potato virus-Y as the type species, the genus *Bymovirus* composed of the fungal-transmitted viruses with barley yellow mosaic virus as the type species, and the genus *Rymovirus* composed of the mite-borne viruses with ryegrass mosaic virus as the type species. The possible genus *Ipomovirus* included the whitefly-borne virus, sweet potato mild mosaic virus [Arch. Virol. 118: 139–141 (1991)].

This volume includes much of the material discussed at the workshop but it is also a synthesis of ideas and concepts from the workshop discussions and individuals interested in potyvirus taxonomy. Many chapters start with a recapitulation of potyviruses but each evolves to make its unique point(s) stressed by the author. Rather than edit this volume to avoid all duplication, the duplication was left to emphasize areas which many authors agree upon. Some chapters are short, some are longer, but each makes its own contribution to the taxonomy of potyviruses.

Because this volume deals with virus taxonomy and is a project of the Potyvirus Study Group, taxonomic terminology must be utilized properly. Since the proposal for the family *Potyviridae* has been approved by the Plant Virus Subcommittee of ICTV, I thought it best to begin using family, genera, and species terminology in this issue. When used in the formal taxonomic sense, the first letter of virus family and genus are capitalized and the terms are printed in italics; the name of the taxon precedes the term. The virus species terms are always printed in lower case script even if part of the name is derived from a botanical Latin binomial. For example, ... family *Potyviridae* and genus *Potyvirus*; and for species, tobacco etch virus, johnsongrass mosaic virus, datura shoestring virus, or Colombian datura virus would be correct (Colombian is capitalized because it is the name of a country). To avoid confusion, vernacular usage of 'potyvirus' without the taxon should refer only to a virus in the genus *Potyvirus*; the family designation must be retained when referring to species of the family. The "List of proposed standard acronyms for plant viruses and viroids" [Arch. Virol. 120: 151–164 (1991)] also is used in this issue (see pp. 9ff). Some of the acronyms for species of *Potyviridae* seem awkward and use of family, genus, and species terminology had to be integrated with the historical usage of group. This issue is a trial in a time of rapid evolution of virus terminology. It was sometimes difficult to know when the historical terminology was proper and when to utilize the new language. I hope this issue will help plant virologists as they evolve toward usage of the family, genus, and species terminology.

The workshop and this compilation could not have occurred without the support of the Potyvirus Study Group. For 1991–1993 the group consists of G. Adam, O.W. Barnett, A.A. Brunt, J. Dijkstra, W.G. Daugherty, J.R. Edwardson, R. Goldbach, J. Hammond, J.H. Hill, R. Jordan, S. Kashiwazaki, K. Makkouk, F. Morales, S.T. Ohki, D. Purcifull, E. Shikata, D.D. Shukla, and I. Uyeda.

The authors of this volume responded to the challenge to think about viruses from the taxonomic viewpoint and produced their ideas in writing in timely fashion. I am indebted to each for their cooperation.

Dr. F.A. Murphy, President of ICTV, and Dr. C.H. Calisher, Special Issues Editor of *Archives of Virology*, were strong supporters of the topic of *Potyvirus Taxonomy* being published as a special issue of the *Archives of Virology*. I thank Faye Nicholson for typing the contributions in a uniform manner and M.T. Zimmerman for art work.

January 1992   O. W. Barnett

Contents

Overview of potyviruses and taxonomy

Serology and antigenic relationships

Virus biology and variation

Virus relationships – BCMV subgroup

Summary

Listed in Current Contents

Overview of potyviruses and taxonomy

Arch Virol (1992) [Suppl 5]: 3–16

The general properties of potyviruses

A. A. Brunt

Microbiology and Crop Protection Department, Horticulture Research International,
Littlehampton, England

Summary. The criteria used during the past three decades for including viruses in the potyvirus group are briefly discussed and evaluated. The biological and physico-chemical properties of the viruses transmitted by aphids, mites, whiteflies, or the fungus *Polymyxa graminis* are reviewed, and the taxonomic value of their molecular properties in regrouping the viruses into four groups or genera within the family *Potyviridae* is discussed.

Introduction

The classification of anisometric plant viruses primarily on the morphology and size of their particles was first proposed over 30 years ago by Brandes and Wetter [8]. One of the six groups originally proposed, then designated the potato virus Y group, contained potato virus Y (PVY) and 15 similar aphid-borne viruses with flexuous filamentous particles 720–770 nm long. The taxonomic value of particle characteristics, together with associated biological and physico-chemical properties, was approved in 1971 by the International Committee on Virus Nomenclature (which, in 1973, was renamed the International Committee on Taxonomy of Viruses; ICTV) with the recognition of 16 groups, including one with PVY as its type member [29].

During the three decades or so since the first recognition of the group, its membership has gradually increased (Table 1). However, the actual size of the group is uncertain because new members are frequently described, others originally considered distinct viruses are now recognized as strains or synonyms of another virus, and the status of some viruses is still being vigorously debated. Whatever its size, the group is generally recognized to be the largest and most important of the 35 groups of plant viruses currently recognized by ICTV [10, 75].

During the past 30 years, additional taxonomic criteria for group membership have been introduced, and the initial criteria extended and improved. Notably among the latter was the acceptance by ICTV of the acronym of the type member as the group name (potyvirus) and the inclusion as possible

Table 1. Membership of the potyvirus group 1959–1991

Year	Aphid-borne members		Others[b]	Total	References
	definitive	possible[a]			
1959	16			16	8
1964	19			19	6
1971	13	13		26	29
1976	34	11		45	22
1979	35	38		73	44
1982	48	55	11	114	45
1985	37	105		142	24
1989	68	87	11	166	49
1991	73	72	12	157	23

[a] Includes some viruses for which aphid transmissibility had yet to be demonstrated
[b] Includes viruses transmitted by, or possibly by, mites, whiteflies *(Bemisica tabaci)* or the fungus *Polymyxa graminis*

members those viruses which, although they had some properties characteristic of the group, needed to be further characterized [29].

The presence in infected plants of cytoplasmic inclusions, seen in transverse section as pinwheels, was later recognized as a common feature of potyviruses; inclusion formation thereafter became an important taxonomic criterion [19,22]. Filamentous viruses transmitted by mites, whiteflies *(Bemisia tabaci)* or the fungus *Polymyxa graminis* (Plasmodiophoromycetes) also were shown to induce pinwheel formation, and thus to be possible members of the group [43]. The qualifying range of particle lengths for group membership was extended (680–900 nm) to permit the inclusion of aphid-borne viruses which, in the presence of divalent metallic ions such as Mg^{++} [11, 26], or the sap of some plant species [13], are mostly straight and c. 850 nm long [22]. Reduction of the qualifying length to 680 nm also permitted the inclusion of mite-transmitted viruses (particles mostly c. 700 nm) and its increase to 900 nm the inclusion of the whitefly-transmitted sweet potato mild mottle virus (particles mostly 850–900 nm long); the viruses transmitted by non-aphid vectors were thereafter included in the potyvirus group, but placed in subgroups as possible members [45].

For over 25 years, the basic physico-chemical properties of particles (sedimentation coefficient, buoyant density, type and size of genomic nucleic acid, size of capsid protein, etc.) have provided accessory and supportive criteria [26, 44, 45]. More recently, however, modern analytical procedures have permitted the molecular properties of the component proteins and nucleic acids to be determined and used taxonomically [14, 37, 38, 53, 75]. As an introduction to the following contributions on the taxonomy and possible regrouping of potyviruses, the general properties of these viruses will be reviewed.

Biological properties

Natural transmission

The majority of members and possible members of the group are transmitted with various degrees of efficiency by aphids in a non-persistent manner. Some viruses such as turnip mosaic, beet mosaic, and PVY are vectored by numerous aphid species, whereas other viruses are transmitted by a single or few species [39]; aphid transmissibility has yet to be demonstrated for some possible members. Several species of aphid vectors (especially *Aphis*, *Myzus*, and *Macrosiphum* species) are often implicated in the widespread occurrence of potyviruses in a number of crops; this is probably attributable to their fecundity, polyphagy, and mobility in immature crops [51]. Although little is known about the molecular mechanisms of aphid transmissibility, transmission is known to be dependent on the presence within infected plants of a virus-coded helper component protein of M_r 53–58k [25, 56, 57, 68] and also, with some viruses at least, the composition of the capsid protein [3, 55].

Because aphid-borne potyviruses are transmitted in the non-persistent manner, it has long been thought that aphids were capable of spreading viruses only over relatively short distances; however, it is now recognized that viruliferous aphids can retain maize dwarf mosaic virus during long distance dispersal and migration, and thus are capable of initiating epidemics of this and possibly other viruses at great distances from the primary foci of infection [5, 78].

Members of other invertebrate taxa have transmitted some aphid-borne potyviruses under experimental conditions; the leafminer *Liriomyza sativae* is an inefficient vector of watermelon mosaic 2, papaya ringspot, and celery mosaic viruses [79] and the red spider mite of pea seed-borne mosaic virus (R. Khetapal and Y. Maury, pers. comm.). Recently, maize dwarf mosaic virus was reported to be transmitted by the spores of corn rust [74].

Some of the aphid-borne viruses, especially those such as bean yellow mosaic, bean common mosaic, peanut mottle, pea seed-borne mosaic, peanut stripe, and soybean mosaic viruses, are seed-borne in some leguminous hosts [35]; seed-borne infection by a few other viruses such as lettuce mosaic and maize dwarf mosaic can provide important foci of infection in non-leguminous crops [33].

Five viruses, barley yellow mosaic, barley mild mosaic, oat mosaic, rice necrosis mosaic, and wheat yellow mosaic (this latter virus is serologically very closely related to the wheat spindle streak mosaic virus occurring in North America [66] and the two viruses are thus probably synonyms [69]) are transmitted from infected to healthy plants within zoospores of the fungus *Polymyxa graminis*, and can survive for several years within its resting spores; resting spores, whether wind-blown or as surface contaminants of seeds, might provide a mode of long distance spread of the viruses [1]. These viruses are currently recognized by ICTV as possible members of the potyvirus group [45, 49].

Five viruses, agropyron mosaic, oat necrotic mottle, ryegrass mosaic, spartina mottle, and wheat streak mosaic, three of which are known to be transmitted by eriophyid mites (*Abacarus hystrix* or *Aceria tulipae*), have long been recognized as members of a separate subgroup of possible potyviruses [45, 49]; however, four others (brome streak, hordeum mosaic, onion mite-borne latent, and shallot mite-borne latent viruses) should also now be included in the group. Although little is known about the virus-vector relationships of mite-transmitted viruses , wheat streak mosaic virus is probably transmitted by mites in the persistent manner [54].

Only one virus of another potyvirus subgroup (sweet potato mild mottle virus) has long been known to be transmitted in the non-persistent manner by the whitefly *Bemisia tabaci* [34]. However, sweet potato yellow dwarf virus is also now thought to be a possible member of this subgroup.

Natural host range

Many aphid-borne potyviruses have narrow host ranges but some, such as turnip mosaic, bean yellow mosaic, and watermelon mosaic 2, infect a wide range of plant species [33].

The fungal-transmitted viruses, like their vector, are mainly confined to the Graminae. For similar reasons, viruses transmitted by, or possibly transmitted by, mites also have restricted host ranges [65]. Conversely, sweet potato mild mottle, although known to occur naturally only in sweet potato, has an extensive experimental host range [34].

Cytopathology

Potyviruses characteristically induce the intracellular formation of three-dimensional cytoplasmic inclusions within infected plants. These are seen as pinwheels in transverse section and as bundles in longitudinal section [18, 21]. These inclusions consist of a central tubule with 5–15 plates or lamellae attached; the lamellae consist of a single virus-coded protein of M_r 66–74,000 having a lattice with a periodicity of c. 5 nm [9, 43]. Inclusions are generally described as being cylindrical or conical in shape, but computer analysis of serial sections of those of wheat streak mosaic virus suggest that the inclusions of this virus at least are ellipsoid hyperboloids [48].

Pinwheel inclusions appear intracellularly within 48 h of infection and, although initially in contact with the plasmalemma and with tubules apparently aligned with plasmodesmata, they are later found scattered throughout the cytoplasm [12].

There are several characteristic structures of cytoplasmic inclusions, although inclusions are apparently constant for individual viruses in different hosts. The inclusion lamellae of those induced by some viruses may be rolled to form scrolls, those of others stacked in flat layers to form laminated aggregates and those of others are found as both scrolls and laminated aggregates [21]; some viruses induce the formation of pinwheels which have predominantly short curved laminated aggregates [20]. Potyviruses may be put in subdivisions according to the type of inclusion they induce [19, 20].

Inclusion protein, although its function is uncertain, demonstrates ATPase and helicase activities and is thus possibly involved in virus replication [17].

Some aphid-borne potyviruses, notably tobacco etch virus, induce the formation of crystalline nuclear inclusions which consist of equimolar concentrations of two virus-encoded proteins of M_r respectively, c. 49k (a polyprotein proteinase and VPg) and c. 58k (probably an RNA-dependent RNA polymerase) [40].

Some aphid-borne potyviruses such as peanut mottle, papaya ringspot, and tobacco vein mottling induce the formation of non-crystalline amorphous inclusions [19]; these consist of one protein of M_r 53–58k which is serologically related to, and possibly an aggregated form of, helper component protein [4, 15, 30]; the helper proteins of unrelated viruses are serologically distinct [67].

The particles of potyviruses are usually found scattered or in loose aggregates in the cytoplasm, but occasionally occur in monolayers adjacent to the tonoplast, in cytoplasmic bridges traversing vacuoles, or within plasmodesmata [43].

Properties of virus particles

Morphology and size

Potyviruses have slightly flexuous filamentous particles 11–15 nm in diameter; those transmitted by mites, aphids, or whiteflies are, respectively, mostly c. 700, 750 and 900 nm long. The particle lengths of some viruses, as noted previously, are markedly affected by divalent metallic ions [26]. The particles of most fungal-transmitted potyviruses are of two predominant lengths of c. 275 and 550 nm [49].

Physical properties

Particles of monopartite potyviruses usually sediment as a single component with a sedimentation coefficient ($S^o_{20,w}$) mostly c. 150–160 S, buoyant densities in caesium chloride at 20–25 °C of 1.325–1.335 g/cm³, and extinction coefficients of 2.4–2.9/(mg/ml)/cm at 260 nm. Particles of fungal-transmitted viruses have a buoyant density of 1.29 g/cm³ [70].

Chemical composition

Potyvirus particles contain c. 95% protein and c. 5% nucleic acid [33].

The capsid proteins of potyviruses consist of a single polypeptide usually of M_r 32–36,000 [33, 34, 70], although that of wheat streak mosaic virus is c. 46,000 [42]. The coat proteins each contain c. 300 amino acids. The pitch of the protein helix is 3.3–3.4 nm, and there are c. 7.7 subunits per turn [72, 73]. The polymerized protein of PVY reassembles with viral RNA into short filaments but alone into long flexuous stacked discs or rings [46, 47].

Monopartite potyviruses that have been characterized contain a single stranded positive sense RNA genome of M_r 3.0–3.5×10^6 (8.8–10.25 kb) which is polyadenylated (20–160 adenosines) at its 3′ terminus and with a virus protein (VPg) covalently linked to the 5′ terminus [27, 28, 52, 53, 64]. The genomes of potyviruses encode, or have the potential to encode, for eight proteins, the positions of which have been mapped for some viruses [17, 31, 32, 58]. Sequencing of potyviral RNAs indicates that each has a single long open reading frame [2, 16]. In vitro translation studies indicate that the genomes are translated as one or more polyproteins which undergo post-translational cleavage to produce functional virus proteins [30, 59, 71, 76].

Unlike monopartite potyviruses, the fungal-transmitted viruses have a bipartite genome [70]. RNA 1, present in the longer particles of barley yellow mosaic virus, has an M_r of 2.6×10^6 (7.6 kb) and RNA 2, contained within the shorter particles, has a size of 1.5×10^6 (3.6 kb); both have recently been fully sequenced [14, 37, 38].

The molecular properties and functions of the component proteins and genomic RNAs of potyviruses are active topics of research [17, 32, 60–63, 74, 75]. The molecular characteristics and their taxonomic value are given in other contributions to this volume.

Discussion

The viruses currently classified within the potyvirus group have some properties in common; however, those transmitted by different types of vector also have other properties by which they can be differentiated. Modern analytical procedures, although providing information on molecular properties which largely substantiate earlier taxonomic judgments, also provide data for distinguishing similar viruses and virus taxa [75]. It is now an appropriate time, therefore, to consider the possible reclassification of these filamentous viruses. One suggestion that has the support of many plant virologists is to create the family *Potyviridae*, within which can be classified the potyvirus, bymovirus, rymovirus and ipomovirus groups or genera which contain, respectively, viruses transmitted by aphids, fungus, mites, or whiteflies (Table 2). The majority of viruses which induce the formation of pinwheel inclusions in plants can be readily placed in one of these four genera. However, maclura mosaic and narcissus latent viruses induce pinwheel inclusions but differ in having filamentous particles c. 650 nm long and a capsid protein of M_r 46 kDa [41, 50]; the former is currently included in the potyvirus group and the latter in the carlavirus group. Their taxonomic status is thus uncertain, although they may well be members of another, as yet undefined, genus within the *Potyviridae* [50].

Table 2. Viruses included in the family *Potyviridae* as modified by the Potyvirus Study Group from the 5th Report of the International Committee on Taxonomy of Viruses

Acronym	Virus name
Potyvirus – members	
AlMV	alstroemeria mosaic
AmLMV	amaranthus leaf mottle
ArjMV	araujia mosaic
ALV	artichoke latent
AV1	asparagus 1
BCMV	bean common mosaic (73, 337)
BYMV	bean yellow mosaic (= crocus tomasinianus) (40)
BtMV	beet mosaic (53)
BiMoV	bidens mottle (161)
BlCMV	blackeye cowpea mosaic (= Azuki bean mosaic [AZMV]) (305)
CdMV	cardamom mosaic
CVMV	carnation vein mottle (78)
CTLV	carrot thin leaf (218)
CeMV	celery mosaic (50)
ClYVV	clover yellow vein (= pea necrosis) (131)
CSV	cocksfoot streak (59)
CDV	Colombian datura
ComMV	commelina mosaic
CABMV	cowpea aphid-borne mosaic (134)
CGVBV	cowpea green vein banding
DsMV	dasheen mosaic (191)
DSTV	datura shoestring
DeMV	dendrobium mosaic
GarMV	garlic mosaic
GSMV	gloriosa stripe mosaic
GEV	groundnut eyespot
GGMV	guinea grass mosaic (190)
HVY	helenium virus Y
HMV	henbane mosaic (95)
HiMV	hippeastrum mosaic (117)
IFMV	iris fulva mosaic (310)
IMMV	iris mild mosaic (116, 324)
ISMV	iris severe mosaic (= bearded iris mosaic) (147, 338)
JGMV	johnsongrass mosaic
LYSV	leek yellow stripe (240)
LMV	lettuce mosaic (9)
MDMV	maize dwarf mosaic

Table 2 (continued)

Acronym	Virus name
NDV	narcissus degeneration
NYSV	narcissus yellow stripe (76)
NoMV	nothoscordum mosaic
OrMV	ornithogalum mosaic
OYDV	onion yellow dwarf (158)
PRSV	papaya ringspot (= watermelon mosaic 1) (63, 84, 292)
ParMV	parsnip mosaic (91)
PWV	passionfruit woodiness (122)
PSbMV	pea seed-borne mosaic (146)
PeMoV	peanut mottle (141)
PStV	peanut stripe (= peanut mild mottle, peanut chlorotic ring mottle)
PeSMV	pepper severe mosaic
PVMV	pepper veinal mottle (104)
PTV	Peru tomato mosaic (255)
PPV	plum pox (70)
PkMV	pokeweed mosaic (97)
PVA	potato A (54)
PVV	potato V (316)
PVY	potato Y
SrMV	sorghum mosaic
SbMV	soybean mosaic (93)
SVY	statice Y
SCMV	sugarcane mosaic (88, 341)
SPFMV	sweet potato feathery mottle (= sweet potato russet crack, sweet potato A)
TamMV	tamarillo mosaic
TeMV	telfairia mosaic
TEV	tobacco etch (55, 258)
TVMV	tobacco vein mottling (325)
TCBV	tulip chlorotic blotch
TBV	tulip breaking (71)
TuMV	turnip mosaic (8)
WMV2	watermelon mosaic 2 (63, 293)
WVMV	wisteria vein mosaic
YMV	yam mosaic (314) (= dioscorea green banding)
ZYFV	zucchini yellow fleck
ZYMV	zucchini yellow mosaic (282)

Possible members

aneilema[b]
anthroxanthum mosaic[a]
aquilegia[a, b]
arracacha Y
asystasia gangetica mottle[a]
bidens mosaic
bryonia mottle
canary reed mosaic
canavalia maritima mosaic
carrot mosaic

Table 2 (continued)

Acronym	Virus name
	cassava brown streak-associated
	cassia yellow spot
	celery yellow mosaic
	chickpea bushy dwarf
	chickpea filiform
	chilli veinal mottle
	clitoria yellow mosaic
	cowpea rugose mosaic
	crinum mosaic[a]
	Croatian clover[b]
	cypripedium calceolus[a]
	daphne Y
	datura 437
	datura distortion mosaic
	datura mosaic[a]
	datura necrosis
	desmodium mosaic
	dioscorea alata ring mottle
	dioscorea trifida[b]
	dock mottling mosaic
	eggplant green mosaic
	eggplant severe mottle
	euphorbia ringspot
	ficus carica[b]
	freesia mosaic
	garlic yellow streak
	guar symptomless[a]
	habenaria mosaic
	holcus streak[a]
	Hungarian datura innoxia[a]
	hyacinth mosaic[a]
	Indian pepper mottle
	isachne mosaic[a]
	kennedya Y
	lily mild mottle
	maclura mosaic (239) (particle length and coat protein MW are atypical)
	malva vein clearing
	marigold mottle
	melilotus mosaic
	mungbean mosaic[a]
	mungbean mottle
	narcissus late season yellows (=jonquil mild mosaic)
	nerine[a, b]
	palm mosaic[a]
	papaya leaf distortion
	patchouli mottle
	passionfruit ringspot
	peanut green mosaic

Table 2 (continued)

Acronym	Virus name
	pea mosaic
	pecteilis mosaic
	pepper mild mosaic
	pepper mottle (253)
	perilla mottle
	plantain 7
	pleioblastus mosaic
	populus[a]
	primula mosaic
	primula mottle
	ranunculus mottle
	sesame yellow mosaic
	sunflower mosaic[a]
	sweet potato latent
	sweet potato vein mosaic
	sword bean distortion mosaic
	teasel mosaic
	telfairia mosaic
	tobacco vein banding mosaic
	tobacco wilt
	tradescantia/Zebrina[b]
	tropaeolum 1
	tropaeolum 2
	ullucus mosaic
	vallota mosaic
	vanilla mosaic
	vanilla necrosis
	white bryony mosaic
	wild potato mosaic
	zoysia mosaic

Rymovirus – members

RGMV	ryegrass mosaic
AgMV	agropyron mosaic
HoMV	hordeum mosaic
ONMV	oat necrotic mottle
WSMV	wheat streak mosaic

Possible members

BrSV	brome streak virus
SpMV	spartina mottle

Bymovirus – members

BaMMV	barley mild mosaic
BaYMV	barley yellow mosaic (143)
OMV	oat mosaic (145)
RNMV	rice necrosis mosaic (172)

Table 2 (continued)

Acronym	Virus name
WSSMV	wheat spindle streak mosaic (167)
WYMV	wheat yellow mosaic (latter 2 names are synonyms)

Ipomovirus – member

| SPMMV | sweet potato mild mottle (162) |

Possible member

| | sweet potato yellow dwarf |

This tentative list of viruses of the *Potyviridae* is a compilation being modified for the Sixth Report of the ICTV. Bob Milne, Istituto di Fitovirologia Applicata, has reconciled for the Plant Virus Sub-Committee of ICTV the list of acronyms with that of Hull et al. [34a]. Milne has further suggested that, to avoid possible future confusion, acronyms should not be published for tentative members. Numbers in parentheses are those of CMI/AAB Descriptions of Plant Viruses

[a] Aphid transmission not confirmed
[b] Name inadequate but denotes species in which a potyvirus has been reported

References

1. Adams MJ (1991) Transmission of plant viruses by fungi. Ann Appl Biol 118: 479–492
2. Allison RF, Johnston RE, Dougherty WG (1986) The nucleotide sequence of the coding region of tobacco etch virus genomic RNA: evidence for the synthesis of a single polyprotein. Virology 154: 9–20
3. Allison RF, Dougherty WG, Parks TD, Willis L, Johnston RE, Kelly M, Armstrong FB (1985) Biochemical analysis of the capsid protein gene and capsid protein of tobacco etch virus: N-terminal amino acids are located on the virion's surface. Virology 147: 309–316
4. Baunoch DA, Das P, Hari V (1990) Potato virus Y helper component protein is associated with amorphous inclusions. J Gen Virol 71: 2479–2482
5. Berger PH, Zeyen RJ, Groth JV (1987) Aphid retention of maize dwarf mosaic virus (potyvirus): epidemiological implications. Ann Appl Biol 111: 337–344
6. Brandes J (1964) Identifizierung von gestreckten pflanzenpathogenen Viren auf morphologischer Grundlage. Mitt Biolog Bundesanst Land Forstwirtsch Berlin-Dahlem 110: 5–130
7. Brandes J, Bercks R (1965) Gross morphology and serology as a basis for classification of elongated plant viruses. Adv Virus Res 11: 1–24
8. Brandes J, Wetter C (1959) Classification of elongated plant viruses on the basis of particle morphology. Virology 8: 99–115
9. Brakke MK, Ball EM, Hsu YH, Langenberg WG (1987) Wheat streak mosaic virus cylindrical inclusion body protein. J Gen Virol 68: 281–287
10. Brown F (1989) The classification and nomenclature of viruses: summary of results of meetings of the International Committee on Taxonomy of Viruses in Edmonton, Canada, 1987. Intervirology 30: 181–186
11. Brunt AA, Kenten RH (1971) Pepper veinal mottle virus – a new member of the potato virus Y group from peppers (*Capsicum annum* L. and *C. frutescens* L.) in Ghana. Ann Appl Biol 69: 235–243
12. Christie RG, Edwardson JR (1977) Light and electron microscopy of plant virus inclusions. Fla Agric Exp Stat Monogr Ser, no 9

13. Chamberlain JA, Catherall PL (1977) Electron microscopy of cocksfoot streak virus and its differentiation from ryegrass mosaic virus in naturally infected *Dactylis glomerata* plants. Ann Appl Biol 85: 105–112

14. Davidson AH, Prols M, Schell J, Steinbiss HH (1991) The nucleotide sequence of RNA 2 of barley yellow mosaic virus. J Gen Virol 72: 989–993

15. De Mejia MVG, Hiebert E, Purcifull DE, Thornbury DW, Pirone TP (1985) Identification of potyviral amorphous inclusion protein as a nonstructural virus-specific protein related to helper component. Virology 142: 34–43

16. Domier LL, Shaw JG, Rhoads E (1987) Potyviral proteins share amino acid sequence homology with picorna-, como-, and caulimoviral proteins. Virology 158: 20–27

17. Dougherty WG, Carrington JC (1988) Expression and function of potyviral gene products. Annu Rev Phytopathol 26: 123–143

18. Edwardson JR (1966) Electron microscopy of cytoplasmic inclusions in cells infected with rod-shaped viruses. Amer J Bot 53: 359–364

19. Edwardson JR (1974) Some properties of the potato virus Y-group. Fla Agric Exp Stat Monogr Ser, no 4

20. Edwardson JR, Christie RG, Ko NJ (1984) Potyvirus cylindrical inclusions – subdivision IV. Phytopathology 74: 1111–1114

21. Edwardson J, Purcifull DE, Christie RG (1968) Structure of cytoplasmic inclusions in plants infected with rod-shaped viruses. Virology 34: 250–263

22. Fenner F (1976) Classification and nomenclature of viruses. Second report of the International Committee on Taxonomy of Viruses. Intervirology 7: 1–116

23. Francki RIB, Fauquet CM, Knudson DL, Brown F (eds) (1991) Classification and nomenclature of viruses. Fifth report of the International Committee on Taxonomy of Viruses. Springer, Wien New York (Arch Virol [Suppl 2])

24. Francki RIB, Milne RG, Hatta T (1985) Atlas of plant viruses, vol 2. CRC Press, Boca Raton, pp 183–217

25. Govier DA, Kassanis B (1974) Evidence that a component other than the virus particle is needed for aphid transmission of potato virus Y. Virology 57: 285–286

26. Govier DA, Woods RD (1971) Changes induced by magnesium ions in the morphology of some plant viruses with filamentous particles. J Gen Virol 13: 127–132

27. Hari V (1981) The RNA of tobacco etch virus: further characterization and detection of protein linked to RNA. Virology 112: 391–399

28. Hari V, Siegel A, Rozek C, Timberlake WE (1979) The RNA of tobacco etch virus contains poly(A). Virology 92: 568–571

29. Harrison BD, Finch JT, Gibbs AJ, Hollings M, Shepherd RJ, Valenta V, Wetter C (1971) Sixteen groups of plant viruses. Virology 45: 356–363

30. Hellmann GM, Thornbury DW, Hiebert E, Shaw JG, Pirone TP, Rhoads RE (1983) Cell-free translation of tobacco vein mottling virus RNA. II. Immunoprecipitation of products by antisera to cylindrical inclusion, nuclear inclusion and helper component proteins. Virology 124: 434–444

31. Hellmann GM, Shaw JG, Rhoads RE (1985) On the origin of the helper component of tobacco vein mottling virus: translational initiation near the 5′ terminus of the viral RNA and termination by UAG codons. Virology 143: 23–34

32. Hiebert E, Dougherty WG (1988) Organization and expression of viral genomes. In: Milne RG (ed) The plant viruses, vol 4, the filamentous plant viruses. Plenum, New York, pp 159–178

33. Hollings M, Brunt AA (1981) Potyviruses. In: Kurstak E (ed) Handbook of plant virus infections: comparative diagnosis. Elsevier/North Holland, Amsterdam, pp 731–807

34. Hollings M, Stone OM, Bock KR (1976) Purification and properties of sweet potato mild mottle, a whitefly-borne virus from sweet potato (*Ipomoea batatas*) in East Africa. Ann Appl Biol 82: 511–528

34a. Hull R, Milne RG, Van Regenmortel MHV (1991) A list of proposed standard acronyms for plant viruses and viroids. Arch Virol 120: 151–164

35. Kaiser WJ (1987) Testing and production of healthy plant germplasm. Techn Bull 2, Inst Seed Pathol Dev Countries, Copenhagen, Denmark, pp 16–20

36. Kashiwazaki S, Hayano Y, Minobe Y, Omura T, Hibino H, Tsuchizaki T (1989) Nucleotide sequence of the capsid protein gene of barley yellow mosaic virus. J Gen Virol 70: 3015–3023

37. Kashiwazaki S, Minobe Y, Hibino H (1991) Nucleotide sequence of barley yellow mosaic virus RNA 2. J Gen Virol 72: 995–999

38. Kashiwazaki S, Minobe Y, Omura T, Hibino H (1990) Nucleotide sequence of barley yellow mosaic virus RNA 1: a close evolutionary relationship with potyviruses. J Gen Virol 71: 2781–2790

39. Kennedy JS, Day MF, Eastop VF (1962) A conspectus of aphids as vectors of plant viruses. Commonwealth Institute of Entomology, London

40. Knuhtsen H, Hiebert E, Purcifull DE (1974) Partial purification and some properties of tobacco etch virus induced intranuclear inclusions. Virology 61: 200–209

41. Koenig R, Plese N (1981) Maclura mosaic virus. CMI/AAB Descriptions of Plant Viruses, no 239

42. Lane LC, Skopp R (1983) The coat protein of wheat streak mosaic virus. Phytopathology 73: 791

43. Lesemann DE (1988) Cytopathology. In: Milne RG (ed) The plant viruses, vol 4, the filamentous plant viruses. Plenum, New York, pp 179–235

44. Matthews REF (1979) Classification and nomenclature of viruses. Third report of the International Committee on Taxonomy of Viruses. Intervirology 12: 131–296

45. Matthews REF (1982) Classification and nomenclature of viruses. Fourth report of the International Committee on Taxonomy of Viruses. Intervirology 17: 1–199

46. McDonald JG, Bancroft JB (1977) Assembly studies on potato virus Y and its coat protein. J Gen Virol 35: 251–263

47. McDonald JG, Beveridge TJ, Bancroft JB (1976) Self-assembly of protein from a flexuous virus. Virology 69: 327–331

48. Mernaugh RL, Gardner WS, Yocom KL (1980) Three-dimensional structure of pinwheel inclusions as determined by analytic geometry. Virology 106: 273–281

49. Milne RG (1988) Taxonomy of rod-shaped filamentous viruses. In: Milne RG (ed) The plant viruses, vol 4, the filamentous plant viruses. Plenum, New York, pp 3–50

50. Mowat WP, Dawson S, Duncan GH, Robinson DJ (1991) Narcissus latent, a virus with filamentous particles and a novel combination of properties. Ann Appl Biol 119: 31–46

51. Murant AF, Raccah B, Pirone TP (1988) Transmission by vectors. In: Milne RG (ed) The plant viruses, vol 4, the filamentous plant viruses. Plenum, New York, pp 237–273

52. Niblett CL, Calvert LA, Stark DM, Lommel SA, Beachy RN (1988) Characterization of the coat protein gene of wheat streak mosaic virus. Phytopathology 78: 1561

53. Niblett CI, Zagula KR, Calvert LA, Kendall TL, Stark DM, Smith CE, Beachy RN, Lommel SA (1991) cDNA cloning and nucleotide sequence of the wheat streak mosaic virus capsid protein gene. J Gen Virol 72: 499–504

54. Paliwal YC (1980) Relationship of wheat streak mosaic and barley stripe mosaic viruses to vector and nonvector eriophyid mites. Arch Virol 63: 123–134

55. Pirone TP, Thornbury DW (1983) Role of virion and helper component in regulating aphid transmission of tobacco etch virus. Phytopathology 73: 872–875

56. Pirone TP, Thornbury DW (1984) The involvement of a helper component in nonpersistent transmission of plant viruses by aphids. Microbiol Sci 1: 191–199

57. Sako N, Ogata K (1981) Different helper factors associated with aphid transmission of some potyviruses. Virology 112: 762–765

58. Shahabuddin M, Shaw JG, Rhoads RE (1988) Mapping of the tobacco vein mottling virus VPg cistron. Virology 163: 635–637
59. Shields SA, Wilson TMA (1987) Cell-free translation of turnip mosaic virus RNA. J Gen Virol 68: 169–180
60. Shukla DD, Ward CW (1988) Amino acid sequence homology of coat proteins as a basis for identification and classification of the potyvirus group. J Gen Virol 69: 2703–2710
61. Shukla DD, Ward CW (1989) Structure of potyvirus coat proteins and its application in the taxonomy of the potyvirus group. Adv Virus Res 36: 273–314
62. Shukla DD, Ward CW (1989) Identification and classification of potyviruses on the basis of coat protein sequence data and serology. Arch Virol 106: 171–200
63. Shukla DD, Inglis AS, McKern NM, Gough KH (1986) Coat protein of potyviruses. 2. Amino acid sequence of the coat protein of potato virus Y. Virology 152: 118–125
64. Siaw MFE, Shahabuddin M, Ballard S, Shaw JG, Rhoads RE (1985) Identification of a protein covalently linked to the 5′ terminus of tobacco vein mottling virus RNA. Virology 142: 134–143
65. Slykhuis JT (1955) *Aceria tulipae* Keifer (Acarina: Eriophyidae) in relation to the spread of wheat streak mosaic. Phytopathology 45: 116–128
66. Slykhuis JT (1976) Wheat spindle streak mosaic virus. CMI/AAB Descriptions of Plant Viruses, no 167
67. Thornbury DW, Pirone TP (1983) Helper components of two potyviruses are serologically distinct. Virology 125: 487–490
68. Thornbury DW, Hellmann GM, Rhoads RE, Pirone TP (1985) Purification and characterization of potyvirus helper component. Virology 144: 260–267
69. Usugi T, Saito Y (1979) Relationship between wheat yellow mosaic virus and wheat spindle streak mosaic virus. Ann Phytopathol Soc Jpn 45: 581–585
70. Usugi T, Kashiwazaki S, Omura T, Tsuchizaki T (1989) Some properties of nucleic acids and coat proteins of soil-borne filamentous viruses. Ann Phytopathol Soc Jpn 55: 26–31
71. Vance VB, Beachy RN (1984) Translation of soybean mosaic virus RNA *in vitro*: evidence of protein processing. Virology 132: 271–281
72. Varma A, Gibbs AJ, Woods RD, Finch JT (1968) Some observations on the structure of the filamentous particles of several plant viruses. J Gen Virol 2: 107–114
73. Veerisetty V (1978) Relationships among structural parameters of virions of helical symmetry. Virology 84: 523–529
74. von Wechmar, MB, Chauhan R, Enox E (1992) Fungal transmission of a potyvirus: uredospores of *Puccinia sorghi* transmit maize dwarf mosaic virus. In: Barnett OW (ed) Potyvirus taxonomy. Springer, Wien New York, pp 239–250 (Arch Virol [Suppl] 5)
75. Ward CW, Shukla DD (1991) Taxonomy of potyviruses: current problems and some solutions. Intervirology 32: 269–296
76. Yeh S-D, Gonsalves D (1985) Translation of papaya ringspot virus RNA *in vitro*: detection of a possible polyprotein that is processed for capsid protein, cylindrical-inclusion protein, and amorphous-inclusion protein. Virology 143: 260–271
77. Zagula KR, Kendall TL, Lommel SA (1990) Wheat streak mosaic virus genomic RNA shares sequence homology with potyviral cylindrical inclusion cistrons. Phytopathology 80: 1036
78. Zeyen RJ, Berger PH (1990) Is the concept of short retention times for aphid-borne nonpersistent plant viruses sound? Phytopathology 80: 769–771
79. Zitter TA, Tsai JH (1977) Transmission of three potyviruses by the leafminer *Liriomyza sativae* (Diptera: Agromyzidae). Plant Dis Rep 61: 1025-1028

Author's address: A.A. Brunt, Horticulture Research International, Worthing Road, Littlehampton, BN17 6LP, England.

Arch Virol (1992) [Suppl 5]: 17–23

Application of genome sequence information in potyvirus taxonomy: an overview

C.D. Atreya

Department of Plant Pathology, University of Kentucky, Lexington, Kentucky, U.S.A.

Summary. The application of protein and nucleic acid sequence analysis in evolutionary and phylogenetic studies is well established. Available sequence information for the 5′ untranslated region of potyviruses including the fungus-transmitted barley yellow mosaic virus (BaYMV) RNA-1 suggests that a 12-nucleotide conserved sequence, the "potybox" is unique to this group. Various non-structural proteins of potyviruses share considerable "signature" sequence homology across a broad spectrum of unrelated viruses, which makes their value limited to "supergroup" or "superfamily" identity. However, in potyviruses, the coat-protein N-terminal sequences and 3′ noncoding regions are variable among viruses, but similar among strains of the same virus. This suggests that these sequences may be an accurate marker of genetic related-ness. Until complete genome sequences from a large number of potyviruses become available and their value in systematics is tested, coat protein and 3′ noncoding regions remain as the choice of taxonomic indicators. The reason being, that cloning and sequencing of the coat-protein gene and 3′ noncoding regions are less complicated and time consuming and the sequences show significant differences among the virus species within the family *Potyviridae*.

Introduction

The purpose of a taxonomy for any group of organisms is to distinguish members of that group from those of other groups as well as to distinguish among members within the group. If the purpose is to deduce evolutionary relationships, the parameters chosen will be different from those to be used for diagnostics and vice versa. The most desirable form of classification should include a combination of both and should ultimately aid the practitioner, for example, a pathologist in the field if the taxonomy is for a group of pathogens or parasites.

The techniques and concepts of molecular biology provide a new rationale for the study of biological macromolecules, nucleic acids in particular, since phenotypes are reflections of genetic material. The application of protein and

Table 1. Potyvirus proteins and their functions (modified from [28])

Protein	Size (kDa)	Function
"34k" protein	28–34	putative movement protein, protease[a]
Helper component (HC), or HC-Pro	50–56	aphid transmission, processes HC/"42k" junction
"42k" protein	29–42	unknown
Cylindrical inclusion protein (CI)	c. 70	helicase, nucleotide binding protein, replication
"6k" protein	6	unknown
NIa, 49k proteinase	49–52	polyprotein processing except HC/"42k" junction, VPg
NIb, 58k protein	56–58	replication, "polymerase"
Coat protein	29–37	RNA packaging, aphid transmission

[a] Verchot et al. [30a]

nucleic acid sequence analysis in evolutionary and phylogenetic studies is well established, and the concept of molecular evolution has emerged as a new field of study. The presence of conserved amino acid "signature" sequences which are associated with similar functions in bacterial, plant, and animal viruses as well as in cellular proteins are well recognized [2, 15, 21]. For example, similarities in genetic organization and expression of evolutionarily conserved sequences between plant and animal viruses led to the creation of the "alpha-like" and "picorna-like" virus supergroups which include plant viruses [14]. Any classification based on genome sequences is appropriate and attempts in this direction are to be encouraged, provided they enhance the understanding of that particular group per se.

Since approximately one-third of all known plant viruses are in the family *Potyviridae*, and there are frequent reports of new members of this family, these viruses comprise one of the most important and dynamic of the plant virus groups. Virus morphology, formation of cylindrical inclusion bodies in the host cell, and genome organization are some of the characteristics used to include members in this family which includes aphid-, fungus-, and mite-transmitted members [25].

Excellent reviews are available on potyvirus gene organization, function of their products, and use of their sequences in taxonomy [12, 28, 31]. The gene

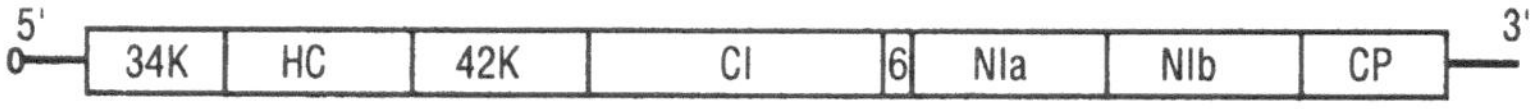

Fig. 1. Gene map of a potyvirus, tobacco vein mottling virus (TVMV). Open boxes indicate positions of the known and putative proteins in the polyprotein. Open circle at 5′ terminus represents the genome-linked protein. Refer to Table 1 for proteins, their sizes and functions

map of a potyvirus (Fig. 1) and the function of gene products is given (Table 1) for readers not familiar with these viruses. In this contribution, some of the prospects and concerns of using non-coding regions, non-structural and coat protein amino acid sequences, and/or their nucleotide sequences as taxonomic indicators are discussed.

Prospects

Role of 5' non-coding and non-structural protein sequence information in taxonomy

Although a large family, the genomic RNA of only five viruses in the *Potyviridae* has been completely sequenced [1, 10, 20, 24, 26]. Sequence information is crucial in understanding the general genetic organization of potyviruses. However, no taxonomic value can be given to this general information other than to establish a group identity. Sequence information in the 5' untranslated region of a few species including the fungus-transmitted barley yellow mosaic virus (BaYMV) RNA 1 suggests that a 12-nucleotide conserved sequence (TCAACACAACAT), the "potybox" motif, is unique to this family [1, 10, 18, 20, 24, 26, 31]. This sequence might play a critical role in packaging or translation [20]. A group specific sequence, the "tymobox" shown to be shared by most of the tymoviruses, has proven to be useful in diagnostics [8].

Sequences of some of the non-structural proteins of potyviruses such as the cylindrical inclusion protein (CI) which contains a helicase motif [22] and the nuclear inclusion (NIb) protein which is a putative replicase are highly conserved within the *Potyviridae* and also share homology in "nucleotide-binding" and replicase signature domains, respectively, with other groups. When conserved "polymerase" signature sequences are used in hierarchical clustering of plant RNA viruses, potyviruses fall under the "picorna-like" group that includes como-, nepo-, and picorna viruses; however, similar analysis using "nucleotide-binding" signature sequences present in the CI protein reveals the homogeneity within the *Potyviridae* that separates these viruses from the other groups [6]. This clearly demonstrates that sequence information can be meaningful in taxonomy, provided the right "signatures" are used with caution.

The NIa proteinase (large nuclear inclusion protein) which cleaves the potyviral polyprotein at the 42k/CI/NIa/NIb/CP junctions is a cysteine-type proteinase, structurally related to trypsin-like serine proteinases [5]. The C-terminal half of NIa is required for proteolysis [12]. In two potyviruses, tobacco etch virus (TEV) and tobacco vein mottling virus (TVMV), this domain is homologous with cowpea mosaic virus 24k proteinase and 3C proteinase of poliovirus [11]. Another non-structural protein, the helper component (HC-Pro), performs a function in the transmission of potyviruses by their aphid vectors; aphid transmission is one of the characteristics of the genus

Potyvirus [25]. Significant sequence homology was observed between the HC gene of the potyvirus and its counterpart, gene II, of the aphid transmissible cauliflower mosaic virus [11].

These various non-structural proteins are involved in critical but general functions across a broad spectrum of unrelated viruses. The sequences of these proteins are homologous because of their common function. This makes a majority of these sequences less important as taxonomic indicators other than to establish a "supergroup" or "superfamily" identity [6, 21]. The "potybox" motif may prove valuable for diagnostic purposes and as another characteristic of the family *Potyviridae*.

Role of coat protein and 3' non-coding sequence information in potyvirus taxonomy

Aphid-transmitted potyvirus genus

Amino acid and nucleotide sequences of the coat protein (CP) and its cistron, as well as 3' nontranslated sequences, are available for a large number of potyviruses. Unlike nucleotide or amino acid sequences of nonstructural proteins of the potyviruses, CP sequences share only a few homologous sequences with those of other groups of viruses [9].

Although nucleotide and amino acid sequences for the "core" of the CP are conserved among potyviruses, the CP N-terminal sequences and 3' noncoding regions are variable among potyviruses. The termini of the coat proteins are present on the surface of the virion and are the major virus-specific antigenic determinants [29]. Critical amino acid residues involved in virus transmission by aphids [3, 4] also are located near the N terminus of the CP. Removal of the projecting termini from the virus core alters aphid transmissibility [27]. Antibodies against the virus core recognize a broad spectrum of potyviruses and are thus useful in genus or even family identity [16]. On the other hand, antibodies against the CP N-terminal domain, which is virus specific, can be useful in potyvirus strain identity [16, 31].

The non-coding region at the 3' end of the potyviral genome differs considerably in length for different potyviruses and shows no significant homology among distinct potyviruses, but in related strains it is similar in length and sequence [13]. A comparison of several strains of distinct potyviruses with respect to their 3' terminal nucleotide sequence reveals a degree of homology between strains in the range of 83–99%, whereas between viruses the homology is only 39–53%, suggesting that these sequences may be an accurate marker of genetic similarity [13].

Fungus- and mite-transmitted genera

Usugi et al. [30] proposed that soil-borne viruses transmitted by the fungus *Polymyxa graminis* be excluded from the "possible" potyvirus group [25] to form the new group, now genus *Bymovirus* (type member, BaYMV). This proposal is based on their bipartite particle morphology and the absence of serological relationships with definitive members of the genus *Potyvirus*. Alignment of the BaYMV coat-protein sequence with that of five distinct potyviruses reveals only 20–25% identity [17]. Nonetheless, complete sequence information for BaYMV RNA 1 reveals that similarities do exist between potyviruses and bymoviruses in their genetic organization, suggesting that these viruses have a common ancestry [18]. The observation that a single open reading frame in RNA 2 of BaYMV potentially codes for a 94k protein and has a putative second proteinase domain homologous to the potyvirus HCPro also supports their close genetic relationship [7, 19]. However, the bipartite genome of BaYMV does not merely represent a split potyvirus genome; the 5' terminal region of potyviruses and RNA-2 of the BaYMV differ significantly in their genetic organization.

The mite-transmitted (genus *Rymovirus*) wheat streak mosaic virus (WSMV) capsid gene has recently been sequenced and shows only 18–24% homology with that of aphid-transmitted potyviruses [23]. So far sequence information on members of the whitefly-transmitted possible genus *Ipomovirus* is not available.

Conclusions

Ideally taxonomy should reflect evolutionary trends of a particular group of organisms, and evolutionary trends normally are reflected in genome organization and sequence similarity within a group. However, obligate parasites such as plant viruses have an intimate relationship with their hosts and the relationships between host and parasite are often manifested as disease symptoms in the host. Since this "phenotype" is a concern for plant pathologists, and a major element in the current classification of viruses and other plant pathogens, classification in the *Potyviridae* should also reflect the "virus–host" concept rather than only properties of the virus.

The available sequence information on potyviruses suggests that it can be used as "family-specific" and "species-specific" indicators [31]. However, all this information is derived from the virus alone with no consideration given to accommodate virus–host relationships. In other words, symptomatology and host range are not taken into account in this approach to classification in the *Potyviridae*.

There are indications that point mutations in the potyviral genome may cause drastic changes in the virus phenotype, such as symptomatology (Atreya et al., unpubl. data) and even aphid transmissibility [3, 4]. Such subtle changes in the genome would not alter the sequence homology relationships but obviously have a profound effect on the virus phenotype. Also there is some concern that the coat protein gene represents only 10% of the potyviral genome yet is given a disproportionate place in taxonomy.

Until complete genome sequences from a large number of the *Potyviridae* species become available and their value in systematics is tested, coat protein and 3' termini noncoding regions remain the choice of taxonomic indicators for the simple reason that cloning and sequencing of the 3' termini are less complicated and time consuming, and they show significant differences among the species.

Acknowledgements

I wish to thank Drs. T. P. Pirone and J. G. Shaw for critical reading of this manuscript and their helpful suggestions.

References

1. Allison RF, Johnston RE, Dougherty WG (1986) The nucleotide sequence of the coding region of tobacco etch virus genomic RNA evidence for the synthesis of a polyprotein. Virology 154: 9–20
2. Argos P (1988) A sequence motif in many polymerases. Nucleic Acids Res 16: 9909–9919
3. Atreya CD, Raccah B, Pirone TP (1990) A point mutation in the coat protein abolishes aphid transmissibility of a potyvirus. Virology 178: 161–165
4. Atreya PL, Atreya CD, Pirone TP (1991) Amino acid substitutions in the coat protein result in loss of insect transmissibility of a plant virus. Proc Natl Acad Sci USA 88: 7887–7891
5. Bazan JF, Fletterick RJ (1988) Viral cystein proteinases are homologous to the trypsin-like family of serine proteinases: structural and functional implications. Proc Natl Acad Sci USA 85: 7872–7876
6. Candresse T, Morch MD, Dunez J (1990) Multiple alignment and hierarchial clustering of conserved amino acid sequences in the replication associated proteins of plant RNA viruses. Res Virol 141: 315–329
7. Davidson AD, Prols M, Schell J, Steinbiss HH (1991) The nucleotide sequence of RNA 2 of barley yellow mosaic virus. J Gen Virol 72: 989–993
8. Ding S, Howe J, Keese P, Mackenzie A, Meek D, Osorio-Keese M, Scotnicki M, Srifah P, Torronen M, Gibbs A (1990) The tymobox, a sequence shared by most tymoviruses: its use in molecular studies of tymoviruses. Nucleic Acids Res 18: 1181–1187
9. Dolia VV, Boyko VP, Agranousky AA, Koonin EV (1991) Phylogeny of capsid proteins of rod-shaped and filamentous RNA plant viruses: two families with distinct patterns of sequence and probably structure conservation. Virology 184: 79–86
10. Domier LL, Franklin KM, Shahabuddin M, Hellman GM, Overmeyer JH, Hiremath ST, Siaw MEE, Lomonossoff GP, Shaw JG, Rhoads RE (1986) The nucleotide sequence of tobacco vein mottling virus. Nucleic Acids Res 14: 5417–5430
11. Domier LL, Shaw JG, Rhoads RE (1987) Potyvirus proteins share amino acid sequence homology with picorna-, como-, and caulimoviral proteins. Virology 158: 20–27
12. Dougherty WG, Carrington JC (1988) Expression and function of potyviral gene products. Annu Rev Phytopathol 26: 123–143
13. Frenkel MJ, Ward CW, Shukla DD (1989) The use of 3' non-coding nucleotide sequences in the taxonomy of potyviruses: application to watermelon mosaic virus 2 and soybean mosaic virus-N. J Gen Virol 70: 2775–2783
14. Goldbach R (1986) Molecular evolution of plant RNA viruses. Annu Rev Phytopathol 24: 289–310

15. Gorbalenya AE, Blinov VM, Donchenko AP, Koonin EV (1988) An NTP-binding motif is the most conserved sequence in a highly divergent monophyletic group of proteins involved in positive strand RNA viral replication. J Mol Evol 28: 258–268

16. Jordan R (1989) Mapping of potyvirus-specific and group-common antigenic determinants with monoclonal antibodies by Western-blot analysis and coat protein amino acid sequence comparisons. Phytopathology 79: 1157 (Abstr)

17. Kashiwazaki S, Hayano Y, Minobe Y, Omura T, Hibino H, Tsuchizaki T (1989) Nucleotide sequence of the capsid protein gene of barley yellow mosaic virus. J Gen Virol 70 : 3015–3023

18. Kashiwazaki S, Minobe Y, Omura T, Hibino H (1990) Nucleotide sequence of barley yellow mosaic virus RNA 1: a close evolutionary relationship with potyviruses. J Gen Virol 71: 2781–2790

19. Kashiwazaki S, Minobe Y, Hibino H (1991) Nucleotide sequence of barley yellow mosaic virus RNA 2. J Gen Virol 72: 995–999

20. Lain S, Riechmann JL, Garcia JA (1989) The complete nucleotide sequence of plum pox potyvirus. Virus Res 13: 157–172

21. Lain S, Riechmann JL, Martin MT, Garcia JA (1989) Homologous potyvirus and flavivirus proteins belonging to a superfamily of helicase-like proteins. Gene 82: 357–362

22. Lain S, Riechmann JL, Garcia JA (1990) RNA helicase: a novel activity associated with a protein encoded by a positive strand RNA virus. Nucleic Acids Res 18: 7003–7006

23. Niblett CL, Zagula KR, Calvert LA, Kendall TL, Stark DM, Smith CE, Beachy RN, Lommel SA (1991) cDNA cloning and nucleotide sequence of the wheat streak mosaic virus capsid protein gene. J Gen Virol 72: 499–504

24. Maiss E, Timpe U, Brisske A, Jelkmann W, Casper R, Himmler G, Matlanovich D, Katinger HWD (1989) The complete nucleotide sequence of plum pox virus RNA. J Gen Virol 70: 513–524

25. Matthews REF (1982) Classification and nomenclature of viruses. Fourth report of the International Committee on Taxonomy of Viruses. Intervirology 17: 1–199

26. Robaglia C, Durand-Tardif M, Tronchet M, Boudazin G, Astier-Manifacier S, Casse-Delbart F (1989) Nucleotide sequence of potato virus Y (N strain) genomic RNA. J Gen Virol 70: 935–947

27. Salomon R, Raccah B (1990) The role of the N terminus of potyvirus coat protein in aphid transmission. In: VIIIth International Congress for Virology, Berlin, August 26–31, 1990, Abstract P83-7

28. Shaw JG, Hunt AG, Pirone TP, Rhoads RE (1990) Organization and expression of potyviral genes. In: Pirone TP, Shaw JG (eds) Viral genes and pathogenesis. Springer, Berlin Heidelberg New York Tokyo, pp 107–123

29. Shukla DD, Strike PM, Tracy SL, Gough KH, Ward CW (1988) The N and C termini of the coat proteins of potyviruses are surface located and the N-terminus contains the major virus-specific epitopes. J Gen Virol 69: 1497–1508

30. Usugi T, Kashiwazaki S, Omura T, Tsuchizaki T (1989) Some properties of nucleic acids and coat proteins of soil-borne filamentous viruses. Ann Phytopathol Soc Jpn 55: 26–31

30a. Verchot J, Koonin EV, Carrington JC (1991) The 35 kDa protein from the N terminus of the potyviral polyprotein functions as a third virus-encoded proteinase. Virology 185: 527–535

31. Ward CW, Shukla DD (1991) Taxonomy of potyviruses: current problems and some solutions. Intervirology 32: 269–296

Author's address: C.D. Atreya, Department of Plant Pathology, University of Kentucky, Lexington, KY 40546, U.S.A.

Arch Virol (1992) [Suppl 5]: 25–30

Inclusion bodies

J. R. Edwardson

Department of Agronomy, University of Florida, Gainesville,
Florida, U.S.A.

Summary. All viruses in the family *Potyviridae* which have been studied cytologically (currently 111) induce cylindrical inclusions in host cytoplasm. These inclusions are controlled by portions of the virus genome, therefore, viruses which induce them are related. Viruses in other groups do not induce this type of inclusion. Cylindrical inclusions have come to be recognized as one of the main characteristics of the family *Potyviridae*. They are used in diagnosis of diseases induced by these viruses. For diagnostic purposes the family can be separated into four subdivisions on the basis of differences in cylindrical inclusion morphologies. Assigning viruses to subdivisions assists in virus identification at the specific and in some instances at the strain level.

Introduction

All potyviruses which have been studied cytologically have been observed to induce proteinaceous cytoplasmic cylindrical inclusions. Cytological examinations of infections induced by viruses in the other 33 plant virus groups have revealed inclusions, but none of them of the cylindrical type. Cylindrical inclusions are unique to infections by members of the *Potyviridae* and have an important taxonomic value. Any virus inducing cylindrical inclusions can be readily classified into this family. There are different types of cylindrical inclusions, and many potyviruses also induce other kinds of cytoplasmic inclusions as well as nuclear inclusions. Differences in inclusions are often used to identify potyviruses at the specific level and in some cases at the strain level.

Potyvirus cytoplasmic inclusions

In 1885 Molisch [20] was, if not the first, certainly one of the first to observe virus-induced inclusions. Since then there have been some misconceptions about the constituents of inclusions, probably the most widespread of these is the idea that inclusions consist largely of virus particles [1]. While inclusions induced by viruses in many groups do consist of virus particles, most of the inclusions associated with infections by members of the *Potyviridae* do not. Cylindrical inclusions consist of more-or-less curved proteinaceous sheets with inner edges converging and outer edges diverging to produce pinwheels in cross-section and bundles in longitudinal section. Attached to this central part of the cylindrical inclusions are proteinaceous sheets in the form of scrolls and/ or laminated aggregates. All components exhibit striations with c. 5 nm periodicity. The presence of cylindrical inclusions was proposed as indicating infections with potyviruses in 1966 [11]. Since then these inclusions have been recognized as a main characteristic of the family *Potyviridae* [16, 17, 19]. Cylindrical inclusions are useful in classification and in diagnosis of diseases induced by members of the *Potyviridae* because they are consistently induced, and their morphologies are not influenced by the environment, including that of the host plant. A strain of a virus induces the same type of cylindrical inclusion in different hosts. At present there are four types of cylindrical inclusions recognized [12, 14]: type 1, consisting of pinwheels and scrolls; type 2, consisting of pinwheels and laminated aggregates; type 3, pinwheels with scrolls and laminated aggregates; type 4, pinwheels with scrolls and short usually curved laminated aggregates.

Viruses inducing these types of inclusions have been separated into subdivisions: type 1 cylindrical inclusions are characteristic of 35 potyviruses assigned to subdivision I; type-2 inclusions are characteristic of 44 viruses assigned to subdivision II; type-3 inclusions are characteristic of 14 viruses assigned to subdivision III; type-4 inclusions are characteristic of 18 viruses assigned to subdivision IV. Currently there are 28 viruses of the *Potyviridae* for which the types of cylindrical inclusions induced are yet to be determined [13]. The cylindrical inclusions are unstained in Azure A (a nucleoprotein stain), and they stain green in the Calcomine Orange-Luxol Brilliant Green combination (OG) (a protein stain) [5, 6]. The subdivisions were constructed in an attempt to assist in reducing the number of comparisons and tests required for identification of a member of this family. The subdivisions are useful for this endeavor. Their usefulness is limited to some extent by (*i*) the necessity to consider the viruses for which types of cylindrical inclusions are unknown [13], and (*ii*) the capacity of different strains of some potyviruses such as those of potato virus-Y, soybean mosaic, sugarcane mosaic, and watermelon mosaic virus-2 to induce different types of cylindrical inclusions [13, 14]. It is to be expected that as cytological studies of group members progress, difficulties arising from (*i*) will decrease markedly, unfortunately difficulties associated with (*ii*) may increase.

The induction of different types of cylindrical inclusions by different strains of the same virus is to be expected since genomes change through such procedures as mutations, deletions, and recombinations. Cylindrical inclusions are controlled by portions of the virus genome: individual viruses induce their characteristic cylindrical inclusions in different hosts, and different viruses induce their characteristic cylindrical inclusions in the same host [11, 12]; cylindrical inclusion protein is serologically unrelated to host and to capsid protein [21]; studies of tobacco etch and pepper mottle virus RNAs in in vitro translations have demonstrated control of capsid and cylindrical inclusion proteins by different portions of the viral genomes [9, 10]. Cylindrical inclusions have obvious taxonomic value. The subdivisions were not designed to be subgroups and they have no apparent relationship to other taxonomic breakdowns. The different subdivisions contain some viruses with different

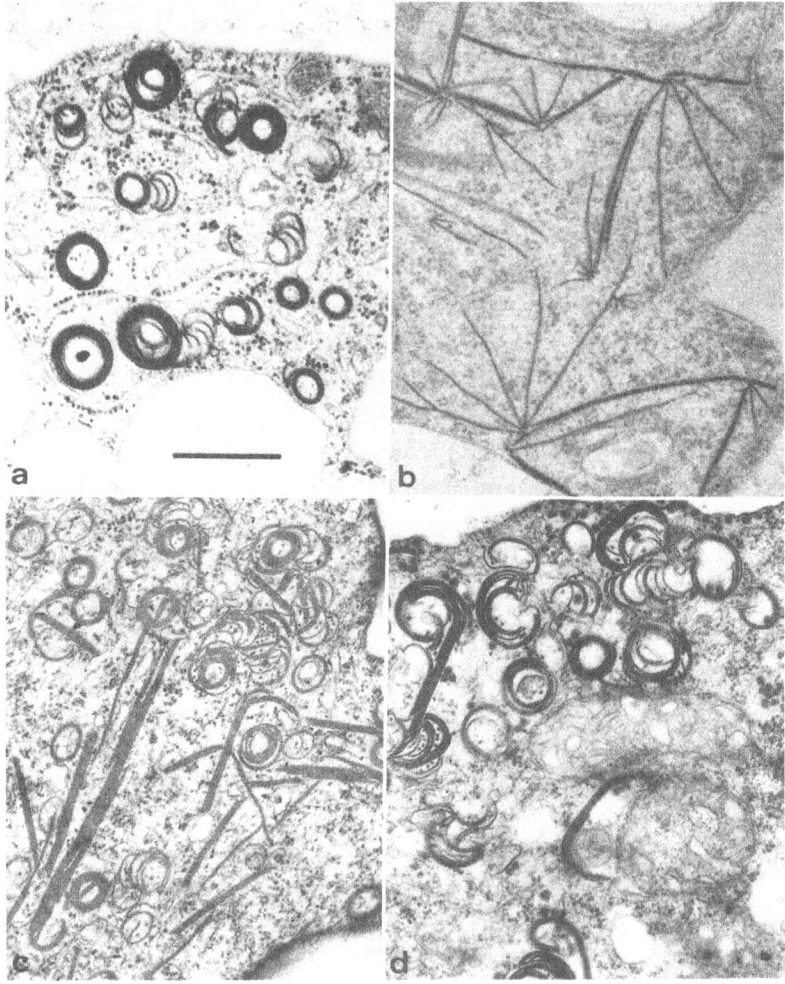

Fig. 1. The potyviruses are unified into a family by cytoplasmic cylindrical inclusions (pinwheels). Each potyvirus induces a single type of cylindrical inclusion. Examples of the four recognized types of these cytoplasmic proteinaceous inclusions are presented in electron micrographs. **a** Type 1, blackeye cowpea mosaic virus; **b** type 2, lettuce mosaic virus; **c** type 3, watermelon mosaic virus-2 (Florida isolate); **d** type 4, potato virus Y.
Bar: ca. 500 nm

types of vectors, and the capsid proteins of some viruses in different subdivisions exhibit serological relatedness. Perhaps comparisons of serological reactions of cylindrical inclusion proteins within and between subdivisions will eventually yield information useful to taxonomy.

Depending on the age of infection, the appearance of cylindrical inclusions induced by a single potyvirus such as pepper mottle can range from inconspicuous (at the plasmalemma) in an early stage of infection, to very long and "sharp" in extremely conspicuous masses (unassociated with the plasmalemma) in advanced stages of infection. In leaf-dip preparations the laminated aggregate components of cylindrical inclusions induced by certain viruses possess strikingly different morphologies such as the triangles of tobacco etch virus, the bidens mottle virus-induced rectangles with one corner clipped, and the stair-step configurations induced by pokeweed mosaic virus (S. R. Christie, unpubl.). Differences in the shapes of laminated aggregates induced by other potyviruses no doubt exist. Differences in these inclusions as they appear in negatively stained extracts or partially purified preparations may eventually be used to identify some potyviruses within the subdivisions. A function has been proposed for these intriguing cytoplasmic inclusion proteins; they have been reported to possess helicase activity [18].

Cytoplasmic inclusions of another kind are induced only by some members of the genus *Potyvirus*. These are the irregularly shaped (amorphous) inclusions which occur in cells infected with papaya ringspot virus (strain W), pepper mottle, pepper veinal mottle, potato virus-Y (strains inducing type 4 cylindrical inclusions), and turnip mosaic virus (strains inducing type 4 cylindrical inclusions). With the exception of papaya ringspot virus (W), which is a member of subdivision I, the viruses reported to induce irregular inclusions are all members of subdivision IV. These inclusions stain green in the OG combination (indicating protein) and red to magenta in Azure A (indicating the presence of ribonucleoprotein). In thin-sections the irregularly shaped inclusions induced by potato virus-Y appear to consist of straight tubules which are more widely spaced than the convoluted tubules in pepper mottle virus-induced inclusions [5]. Irregular cytoplasmic inclusions have been demonstrated to contain the helper factor for potyvirus aphid transmission [7].

Crystal-containing microbodies occur in healthy tissues of both monocotyledonous and dicotyledonous plants. The cubical crystals are constructed from host protein which is colorless in Azure A and is bright green in the OG combination. Their numbers are greatly increased in many virus infections, particularly those of the potyviruses. These crystal-containing microbodies may participate in the formation of massive cytoplasmic inclusions [5, 12], but often they occur in large aggregates which contain few if any other host constituents or other inclusions induced by the virus.

Cytoplasmic virus-induced paracrystalline inclusions have been used to separate viruses within subdivisions: subdivision I, celery mosaic, subdivision II, plum pox, and subdivision III, henbane mosaic virus [5,12].

Nuclear inclusions of the *Potyviridae*

Several potyviruses induce nuclear inclusions, for example, bean yellow mosaic virus and tobacco etch virus crystalline inclusions, blackeye cowpea mosaic virus and gloriosa stripe mosaic virus fibrous inclusions, zucchini yellow fleck virus fimbriate inclusions, and beet mosaic virus and potato virus-A lacunose globular inclusions. Some of the potyvirus-induced nuclear inclusions (NI) have been demonstrated to contain two proteins, the large NI (54 kDa) and the small NI (49 kDa) [10]. The large NI is a putative RNA dependent RNA polymerase [8], while the small NI protein has been reported to be a protease [4]. Nuclear inclusions can be used to distinguish viruses within subdivisions. It is of interest to note that only viruses in the cytoplasmic inclusion subdivision I have been reported to induce fibrous proteinaceous nuclear inclusions [13].

All tobacco etch virus (TEV) strains induce type 2 cylindrical inclusions (pinwheels and laminated aggregates) but several different TEV strains induce different types of crystalline proteinaceous nuclear inclusion. The mild etch strain induces bipyramidal inclusions, the severe etch strain induces thin truncated 4-sided pyramids, and the Madison isolate induces octahedral inclusions. Several TEV strains can be identified on the bases of differences in the nuclear inclusions they induce [13].

Conclusions

Recently Brakke [2] stated that "The potential usefulness of inclusion bodies for identification of viruses has been repeatedly pointed out, and repeatedly ignored by most of us." However, there are indications of increased interest in applying inclusions to taxonomy and diagnosis. In the ICNV report of 1971 inclusions were considered to be a main characteristic of one group [22], while in the 1982 ICTV report inclusions were listed as main characteristics in 23 groups [19]. Differences in inclusion morphologies, staining reactions, intracellular locations, and associations with specific tissues have been used to diagnose virus infections at the group level for 18 [3] or 19 plant virus groups [15].

Cylindrical inclusions indicate the presence of a member of the family *Potyviridae* in infected tissues. Information on the types of cylindrical inclusions combined with the appearance of other types of cytoplasmic and/or nuclear inclusions can be used in some cases to identify specific viruses and in some instances strains.

References

1. Bawden FC (1964) Symptomatology: changes within infected plants. In: Plant viruses and virus diseases, 4th edn. The Ronald Press, New York, pp 48–67
2. Brakke MK (1988) Perspectives on progress in plant virology. Annu Rev Phytopathol 26: 331–350

3. Brunt A, Crabtree K, Gibbs A (1990) Viruses of tropical plants. CAB. International, Wallingford, UK
4. Carrington JG, Dougherty WG (1987) Small nuclear inclusion protein encoded by a plant potyvirus genome is a protease. J Virol 61: 2540–2548
5. Christie RG, Edwardson JR (1977) Light and electron microscopy of plant virus inclusions. Fla Agric Exp Stat Monogr Ser, no 9
6. Christie RG, Edwardson JR (1986) Light microscopic techniques for detection of plant virus inclusions. Plant Dis 70: 273–279
7. DeMejia MVG, Hiebert E, Purcifull DE, Thornbury DW, Pirone TP (1985) Identification of potyviral amorphous inclusion protein as a nonstructural, virus-specific protein related to helper component. Virology 142: 34–43
8. Dougherty WG, Carrington JG (1988) Expression and function of potyviral gene products. Annu Rev Phytopathol 26: 123–143
9. Dougherty WG, Hiebert E (1980) Translation of potyvirus RNA in a rabbit reticulocyte lysate: reaction conditions and identification of capsid protein as one of the products of in vitro translation of tobacco etch and pepper mottle viral RNA. Virology 101: 466–474
10. Dougherty WG, Hiebert E (1980) Translation of potyvirus RNA in a rabbit reticulocyte lysate: identification of nuclear inclusions proteins as products of tobacco etch virus RNA translation and cylindrical inclusion protein as a product of the potyvirus genome. Virology 104: 174–182
11. Edwardson JR (1966) Electron microscopy of cytoplasmic inclusions in cells infected with rod shaped viruses. Amer J Bot 53: 359–364
12. Edwardson JR (1974) Some properties of the potato virus Y-group. Fla Agric Exp Stat Monogr Ser, no 4
13. Edwardson JR, Christie RG (1991) The potyviruses. Fla Agric Exp Stat Monogr Ser, no 16
14. Edwardson JR, Christie RG, Ko NJ (1984) Potyvirus cylindrical inclusions – subdivision IV. Phytopathology 74: 1111–1114
15. Edwardson JR, Christie RG, Purcifull DE, Petersen MA (1992) Cytological effects in diagnosis, light and electron microscopy. In: Matthews REF (ed) Diagnosis of plant virus diseases. CRC Press, Boca Raton (in press)
16. Fenner FF (1976) Classification and nomenclature of viruses. Intervirology 7: 4–115
17. Francki RIB, Fauquet CM, Knudsen DL, Brown F (eds) (1991) Classification and nomenclature of viruses. Fifth report of the International Committee on Taxonomy of Viruses. Springer, Wien New York (Arch Virol [Suppl] 2)
18. Lain S, Riechmann JL, Martin MT, Garcia JA (1989) Homologous potyvirus and flavivirus proteins belonging to a superfamily of helicase-like proteins. Gene 82: 357–362
19. Matthews REF (1982) Classification and nomenclature of viruses. Intervirology 17: 4–199
20. Molisch H (1885) Über merkwürdig geformte Proteinkörper in den Zweigen von *Epiphyllum*. Ber Deutsch Bot Ges 3: 195–202
21. Purcifull DE, Hiebert E, McDonald JG (1973) Immunospecificity of cytoplasmic inclusions induced by viruses in the potato Y group. Virology 55: 275–279
22. Wildy P (1971) Classification and nomenclature of viruses. First report of the International Committee of Nomenclature of Viruses. Monogr Virol 5: 1–81

Author's address: J. R. Edwardson, Agronomy Department, University of Florida, PVL Building 164, Gainesville, FL 32611, U.S.A.

Arch Virol (1992) [Suppl 5]: 31–46

Potyviruses, chaos or order?

L. Bos

Research Institute for Plant Protection (IPO-DLO), Wageningen, The Netherlands

Summary. At first potyviruses were easily distinguished by biological and serological properties because only a few were known and information on their host ranges was limited. The first evidence of serological cross reaction between two of these viruses was reported in 1951 and was further corroborated for three obviously distinct members of the group in 1960. In 1968 attention was drawn to the fact that some legume and non-legume potyviruses have much wider host ranges than previously known and that within the potyvirus group there is as much biological variation within viruses and overlap between viruses as there is in serology. The concept of continuity within the group was soon supported by others and became known as the "continuum hypothesis." Results with highly sensitive serological methods using polyclonal antisera were conflicting, and nucleic acid hybridization techniques did not unambiguously discriminate between potyviruses.

Recent results, obtained with antibodies directed toward epitopes located in the N-termini of the coat proteins of potyviruses, suggest that there are ways to more definitely group strains of one potyvirus and distinguish them from other potyviruses. However, there are exceptions to this rule, as in the case of bean yellow mosaic virus and clover yellow vein virus which are clearly distinct in host range, inclusion bodies, and migration velocity of coat protein, but which still react with antibodies to the N-terminal epitopes of one virus. So the question remains of whether coat-protein properties, especially the serological reactivity of N-termini, which do not alter overall virus integrity when lost, sufficiently represent the genome of a pathogenic virus entity as a single criterion for classification.

Introduction

Nature manifests itself in a seemingly chaotic abundance of forms of life. Since the publication of Linnaeus' classical *Systema Naturae* (1735) and his later *Species Plantarum* (1753), we have known there is no such chaos, but that all organisms can be arranged, according to degrees of similarities in form and structure, into a taxonomic system comprising hierarchical classes such as species, genera, and families. Such a classification is the basis for reliable nomenclature, and thus for communication, and for storage and retrieval of information on the taxonomic entities, so that everyone knows what we are dealing with and are talking about. It is also supposed to be a natural system in that it should represent phylogeny, indicating how evolution may have taken place.

As long as lack of information on intrinsic virus properties persisted, Linnaean principles were hard to apply to viruses [16], but virus taxonomy still is in its infancy and under continuing discussion. Plant viruses, in particular, are hard to describe as organisms. They do not live their own life, and there are no indications of sexuality. That is why the species issue for "viruses" has led to vigorous debate [42 , 61–63, 85] and there is persistent reluctance to denote them with Latin binomials [61]. In the laboratory, viruses can be exclusively described in terms of physico-chemistry, and this is now the predominant domain of their study by virologists. Viruses, like organisms, share genetic continuity and evolve [15]. They show high rates of mutation, and each mutant, if it occurs under selective conditions, may develop into a new variant or strain and possibly into a new "virus." This explains the tremendous variability of viruses and the continuing problem of virus delimitation and of distinguishing between viruses.

First semblance of order

During the "mosaic age" [44], when little was known about the viruses themselves, it was gradually shown that different infectious entities were responsible for the different mosaics in beans, lettuce, potatoes, soybeans, sugar beets, etc. Increasing information on host specialization led to the postulation of the existence of diseases with different sources of infection [83], suggesting the existence of distinct viruses. This was soon supported by the reactions of differential test plant species and other biological characteristics of viruses determined in expressed plant sap [52, 53]. In the 1930's this was further corroborated by early serology [38], and later by particle morphology with the advent of electron microscopy in plant virology in 1939.

In 1959, a system for classifying viruses with elongate particles gave plant virus taxonomy its real start [23]. It was the first system based on intrinsic virus properties such as particle morphology and size, which were soon thereafter used in combination with serological affinities [22]. The discrimination between morphological groups, such as those now called clostero-, poty-, potex-, carla-, tobamo- and tobraviruses, did not create much of a problem. Within morphological groups, however, erratic discrimination still haunts plant virologists when identifying viruses; that is, when diagnosing known viruses and describing new ones. The underlying recurrent question is, what is the taxonomic meaning of the word "virus"?

Potyviruses

The potyviruses often serve as an example of the difficulties encountered in virus classification and identification. This contribution reflects some of my own experiences with potyviruses, particularly those of legumes, i.e., the subgroup of viruses around bean yellow mosaic virus (BYMV) [19], including clover yellow vein virus (ClYVV) [49]. For technical details on potyviruses, reference is made to a number of excellent reviews [33, 47, 48, 64, 73–75].

Potyviruses are the largest and economically most important group of plant viruses. In 1988 the group included 79 definitive members (most are aphid-transmitted, but some with fungus, mite, and whitefly vectors were tentatively included) and 87 possible members [64]. They are characterized by flexuous, filamentous particles, c. 680–900 nm long. All group members induce characteristic cytoplasmic inclusions in their hosts, and some induce intranuclear inclusions as well. As long as only a few potyviruses were known, and advanced techniques had not yet been applied to them, those that had been studied during the 1920's and early 1930's were apparently distinct, especially in host range and serology. Classical examples are bean common mosaic virus (BCMV), bean yellow mosaic virus (BYMV), beet mosaic virus (BtMV), lettuce mosaic virus (LMV), onion yellow dwarf virus (OYDV), potato virus Y (PVY), soybean mosaic virus (SbMV), sugarcane mosaic virus (SCMV), tobacco etch virus (TEV), tulip breaking virus (TBV), and turnip mosaic virus (TuMV). These viruses obviously infected different crops, and their respective antisera were thought to be specific.

Fading borderlines; continuum hypothesis

In 1951, Beemster and van der Want [10] in Wageningen, The Netherlands, were the first to detect a serological relationship between two distinct potyviruses, BCMV and BYMV, in reciprocal tests with antisera to both viruses. This was confirmed nine years later in Braunschweig, Germany, by Bercks [11], who also detected a distant serological relationship with high-titered antisera among BYMV, BtMV, and PVY, viruses infecting totally different crops and differing slightly in "normal lengths" of their particles [12, 13]. Such distant serological relationships soon were detected in other morphological groups [14].

During the late 1960's, I was studying a number of new virus diseases of legumes, viz. a vein mosaic of *Wisteria floribunda*, a severe necrosis of pea (*Pisum sativum*), and a leafroll mosaic of the same, caused by potyviruses. The viruses concerned had several features in common with BYMV [19]. During that decade a number of virus isolates differing to a similar extent from BYMV and BCMV, or having an intermediate position, had been described as separate viruses (Table 1). A non-legume potyvirus, LMV, had already been found in a legume crop (pea) under natural conditions [3], and BtMV [68] and water-

Table 1. Viruses, described up to 1970, closely related serologically to bean yellow mosaic virus (BYMV) and bean common mosaic virus (BCMV) (from [18])

Bean western mosaic virus	Skotland and Burke (1961)
Red clover necrosis virus	Zaumeyer and Goth (1963)
Passion fruit woodiness virus	Taylor and Kimble (1964)
	Teakle and Wildermuth (1967)
Cowpea aphid-borne mosaic virus	Brandes (1964)
	Lovisolo and Conti (1966)
Clover yellow vein virus	Hollings and Nariani (1965)
	Gibbs et al. (1966)
Pea seed-borne mosaic virus	Inouye (1967)
Lupin mottle virus	Hull (1968)
Peanut mottle virus	Kuhn (1965)
	Schmidt and Schmelzer (1966)

melon mosaic virus (likely watermelon mosaic virus 2, WMV-2) [50] were found in pea crops and TuMV in groundnut (*Arachis hypogaea*) [51]. Several of these viruses already were known to be interrelated serologically and in other respects, but the nature and degree of these relationships remained obscure. This hampered characterization of my "viruses" as new ones, as there was continuing discussion on whether pea mosaic virus (PMosV) [29], which was non- or poorly pathogenic to bean (*Phaseolus vulgaris*), should be considered a separate virus or a strain of BYMV [11, 36, 81].

When comparing my *Wisteria* virus, pea necrosis virus, and pea leafroll mosaic virus with BYMV, PMosV, ClYVV, cowpea aphid-borne mosaic virus (CABMV), BtMV and LMV on a range of legume and non-legume crop and test species, all viruses had much wider host ranges than hitherto known. The non-legume potyviruses infected quite a number of legume species, and the legume potyviruses also infected many non-legumes. BYMV, for example, infected 17 out of 20 non-legumes tested. In fact, the virus increasingly occurs as a pathogen of non-legume crops such as Iridaceae (*Freesia*, *Gladiolus*, *Crocosmia*, *Crocus*, *Iris*, *Montbretia*, and *Tritonia* [19, 27, 40, 70]), and members of three other plant families (*Lisianthus russelianus* [60], *Passiflora caerulea* [66], and *Proboscidea jussieui* [67]). Seed transmission, long thought to be unique to BCMV, SbMV, and LMV, occurred with BYMV in lupins and to a lesser extent in broad bean, pea, and sweet clover [19]. The serological interrelationships among different potyviruses thus were corroborated by considerable overlap and similarities in biological properties. Nevertheless, the three new isolates could be distinguished readily in several test plants and by quantitative serology. Consequently, they were described as distinct viruses. The *Wisteria* virus, simultaneously reported from Italy [25], was named *Wisteria* vein-mosaic virus, and the two viruses of pea: pea necrosis virus and pea leafroll mosaic virus [18].

During the First International Phytopathological Congress in London in 1968, the above results were presented and attention was drawn to "the problem of variation within the potato virus Y group" [17]. The results were further documented later [18] and attention was drawn to the fact that border-lines, biologically as well as otherwise, presumed to exist among members within morphological groups, seemed to fade (*i*) as more viruses are studied in detail, (*ii*) as more related strains of these viruses are distinguished, and (*iii*) as more related viruses are described. It was stressed that "finally, all criteria for drawing borderlines, for delimiting taxonomic units within the morphological groups are failing." Within groups a continuum of forms appeared to exist. Finally, it was concluded that borderlines must be drawn arbitrarily when identifying or characterizing plant viruses [18].

Further studies in our laboratory in the early 1970's allowed the biological differentiation of bean mosaic, pea yellow mosaic, and pea necrosis strains of BYMV [20], the discrimination between the pea necrosis strain of BYMV and the pea necrosis virus [9], and the demonstration that the latter was ClYVV [21]. Through international comparison, pea leafroll mosaic virus was found to be one of the isolates of pea seed-borne mosaic virus which was described during the late 1960's more or less simultaneously with varying names in different parts of the world [41]. Cross-protection tests and electron microscopy were not of much help for distinguishing between the different potyviruses. Subjectivity of the serological and biological criteria was demonstrated in the U.S.A. by Jones and Diachun [54], who on the basis of such observations treated ClYVV and PMosV as serotypes or strains of BYMV.

As a result, Hollings and Brunt [48] in their 1981 potyvirus review conclude that "there are many conflicting reports and interpretations, and it is evident that no simple pattern of serological relationships among this group can be envisaged, nor are serological relationships obviously correlated with other biological characteristics." "Nearly all accepted potyviruses have been shown to be serologically related to at least one other potyvirus" and "different strains of one potyvirus may be as distantly related (serologically) to one another as they are to other potyviruses." Finally, while referring to Bos [18] they conclude "that in a number of instances a continuum of variants or strains exists" and that they "would agree with Bos's suggestion that the extreme variation of viruses may make it impossible to define a species concept for them, and this is nowhere more apparent than in the potyvirus group." Harrison [42], in his 1985 discussion about the usefulness and limitations of the species concept for plant viruses, feels "forced to agree with Bos [18] and Hollings and Brunt [48], all having long experience with potyviruses, that the greater the number of isolates studied the more evident it becomes that sharply defined borderlines separating individual potyviruses cannot be drawn; in several instances a continuum of variants or strains exists." Finally, Milne [64], in his more recent 1988 review of rod-shaped filamentous viruses, states that "no one has come up with a workable definition of what is a potyvirus 'virus' as distinct from a strain or pathotype; hence our appeal to the ghost of Linnaeus. The status of 'virus' is usually arrived at by a process of consensus, but with potyviruses the seemingly continuous range of variation and the large number of isolates involved make the delineation of 'viruses' an unrewarding task."

The problem is clearly reflected in the "Guidelines for the Identification and Characterization of Plant Viruses," prepared by a study group in response to a suggestion by the Plant Virus Subcommittee of the International Committee on Taxonomy of Viruses [39]. In these guidelines it is advised to consider a virus isolate to be a new virus if (*i*) the serological differentiation index (SDI – the number of twofold dilution steps separating the homologous and heterologous titers) is greater than three and the new isolate differs from similar viruses in natural and artificial host ranges and symptomatology and usually also in other properties, (*ii*) the SDI is only between one and three, but the natural and artificial host ranges are very dissimilar, or few or no common hosts are found, and there are also pronounced differences in other properties, or (*iii*) no serological cross reactivity is found with similar established viruses.

Continuing efforts towards distinction

The advent of highly sensitive serological techniques, and their applicability for rapid detection and routine diagnosis, further boosted interest in serological approaches during the late 1970's and the 1980's. Hollings and Brunt [48] summarized that "there is a great mass of accumulated information on the serological relationships between various potyviruses," but that the reports and interpretations were often conflicting. Despite intensified research, results persisted in discord. They depended greatly on whether intact or degraded virus particles were used for antiserum production and/or serological reaction, and differed according to the antiserum used [65]. Antisera to dissociated coat proteins had a much broader reactivity than those prepared to intact virus particles [72, 77] and antisera to coat protein prepared by different procedures often differed in specificity [46]. Antisera resulting from bleedings at different stages of immunization of the same animal differed greatly in reactivity (as for potexviruses) [58]. Antisera from early bleedings contained more specific antibodies, whereas those from later bleedings contained increasing amounts of cross-reacting antibodies [77, 80, 86]. Hence, the conclusion that degradation of potyvirus protein takes place during purification, possibly continues during immunization, and considerable caution is needed in the interpretation of serological results [65].

Information on particle structure and genomes of potyviruses has dramatically increased [33, 74, 75], as has information on the expression and function of potyviral gene products [30]. The filamentous particles of the aphid-transmitted potyviruses are now known to contain one molecule of ssRNA of a molecular weight of $3.0-3.5 \times 10^6$, encapsidated in a protein coat of up to 2000 subunits of a single capsid polypeptide of $30-37 \times 10^3$. The RNA molecule

(Fig. 1 in [5]) is approximately 10,000 nucleotides in length, and codes for at least six proteins formed as one large polyprotein, which is later proteolytically cleaved. The capsid protein is on the 3' end of the genome. This protein is of special interest, because it is the site of serological reactivity and plays a role in vector transmission specificity. Heterogeneity of coat-protein size is a common feature of potyviruses due to degradation during purification and storage. Limited proteolysis, as with trypsin, leads to a change in amino acid composition and loss of some serological determinants (as with TEV and pepper mottle virus) [45].

Nucleotide sequence, determined for a number of potyviruses, has attracted special taxonomic attention in studying relationships as it more completely represents the entire viral genome than any other virus character. Cloned special fragments of cDNA are useful for specific virus detection through molecular hybridization [37]. Although highly sensitive for discrimination between closely related isolates of BYMV [2], molecular hybridization analysis does not establish definitive, unambiguous hierarchical relationships among BYMV, ClYVV, and PMosV [1, 2], but molecular hybridization in conjunction with biological and serological information suggests that the three viruses are distinct and form an evolutionary continuum in relation to each other [8, 71].

N-terminal part of coat protein: a breakthrough

When studying serological relationships, intact virus particles behave differently from dissociated protein subunits obtained by sonication or pyrrolidine treatment [65]. Partial degradation of coat protein during purification and its effect on coat-protein size determination, amino acid composition, and serology have also been recognized as problems when studying the relationships between BYMV and a number of other potyviruses [65]. In recent years interest has drifted back to viral coat proteins. Their amino acid composition, especially sequence, partially explains serological behavior. Similarities in amino acid composition are more marked among viruses which appear to be closely related serologically, and their determination was the basis for proposing a BYMV subgroup, comprising BYMV, PMosV, and sweet pea mosaic virus [65, 69].

Information on coat-protein amino acid sequences provides better insight into coat-protein structure and function, including serological reactivity. By 1989, complete coat-protein amino acid sequences were known for 25 strains of 11 distinct potyviruses [74, 75]. Their coat proteins vary considerably in size (263–330 amino acids) because of differences in lengths of their N-termini. The N-terminal regions of the coat proteins of different potyviruses vary in sequence, whereas sequences in the C-terminal three quarters of the coat proteins are more homologous. In contrast, strains of the same virus have coat proteins of the same length with highly homologous N-terminal sequences.

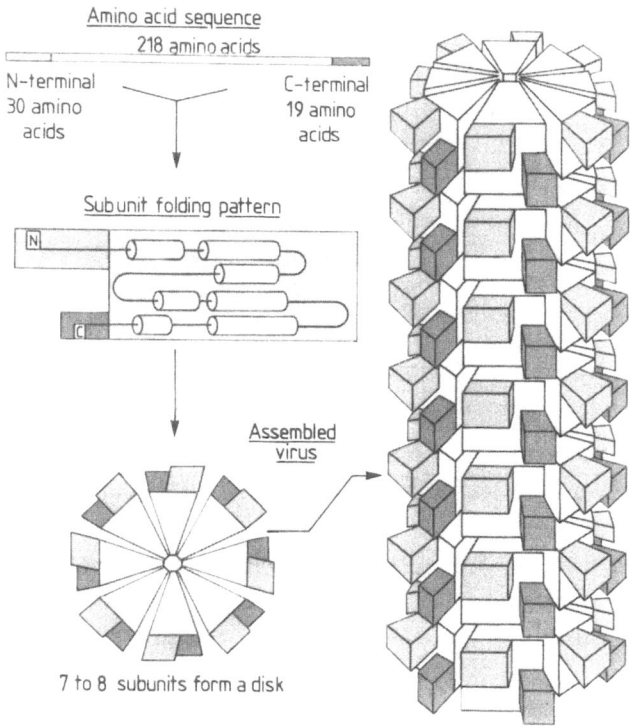

Fig. 1. Schematic drawing showing the linear amino acid sequence of the coat protein subunit, the subunit folding pattern, the surface location of N- and C-termini, and the assembly of the particle of potato virus Y. (After Shukla and Ward [75], with permission)

Analysis of possible pairings of the complete amino acid sequences of 17 strains of eight potyviruses revealed a sequence homology of 38–71% (average 54%) among different viruses and of 90–99% (average 95%) among strains of one virus. This bimodal distribution (Fig. 1 in [87]) is not consistent with the continuum hypothesis and has opened perspectives for novel approaches in taxonomy.

The N- and C-terminal regions of potyvirus coat proteins are now known to be exposed on the surface of the virus particle (Fig. 1) [4, 31, 75, 77]. Mild proteolysis of virus particles with trypsin removes 30 or more of the amino acids from the N-terminal end depending on the virus, and 18–20 amino acids from the C-terminal end. This treatment leaves fully assembled, morphologically intact, and infectious virus particles with protein cores of 215–218 amino acids [77]. The function of the surface-exposed termini is incompletely known, but there is increasing evidence that they play a key role in aphid transmissibility [43].

The viral coat proteins can be characterized by high-performance liquid-chromatographic peptide profiling [76], but serological techniques employing their virus-specific N-terminal parts also have proved to be of major interest. Shukla et al. [79] have developed a simple affinity-chromatographic procedure to isolate virus-specific antibodies from polyclonal antisera produced against intact potyvirus particles. The method involves (*i*) removal of the virus-specific N-terminal protein parts from the particles of a potyvirus using lysyl endopeptidase, (*ii*) binding the truncated coat protein to cyanogen bromide-activated Sepharose in a chromatography column, (*iii*) passing antisera to different potyviruses through the column, (*iv*) collecting antibodies that did not bind to the column and that thus are directed to the N-terminal part of the coat proteins. Employing this technique, Shukla et al. [80] showed that 17 potyvirus isolates infecting Gramineae from the U.S.A. and Australia were not all closely related strains of SCMV as previously believed, but represented four distinct potyviruses, namely johnsongrass mosaic virus (JGMV), maize dwarf mosaic virus (MDMV), SCMV, and sorghum mosaic virus. Antibodies raised against the dissociated core part of the coat protein of JGMV recognized all 15 different aphid-transmitted as well as mite- and whitefly-transmitted potyviruses studied [78, 79], further supporting inclusion of the latter into the potyvirus group. Monoclonal antibodies (MAbs), that react with a single antigenic determinant (epitope) but with broad reactivity, are now known to be directed towards epitopes in the core part of the coat protein. A broad spectrum MAb PTY1 (now commercially available) reacted with all 33 different potyviruses tested. Some other monoclonal antibodies reacted with isolates within the BYMV subgroup (BYMV, PMosV, ClYVV), others with most of the strains of BYMV tested, and others with at least one strain of BYMV [56].

Loss of antiserum specificity with age of a virus preparation is now ascribed to partial protein degradation, especially loss of N-terminal ends containing the virus-specific epitopes, during purification and storage [77]. The use of one preparation of purified virus, stored for successive immunizations, may help explain the loss of specificity with later bleedings [58, 86].

Further complications

The problem of variation within the subgroup of legume potyviruses recurred with recent attempts to identify a number of BYMV-like virus isolates from faba bean (*Vicia faba*) from West Asia and North Africa. Our earlier research had revealed the existence of strains of BYMV which were highly different pathologically, biological overlap of BYMV with ClYVV, and the existence of strains of BYMV or of a separate virus especially adapted to faba bean [20]. We therefore compared representatives of these new faba bean isolates with isolates earlier identified as BYMV and ClYVV, for their biological and serological properties, and also employed virus-specific antibodies directed towards N-terminal parts of coat proteins [32]. ClYVV isolates were easily distinguished from isolates of BYMV by their wider host range among non-legumes and by the striking enlargement of the nucleolus in infected plants, as well as by the slower electrophoretic migration of their coat protein.

Unfractionated antibodies to BYMV reacted with all isolates of BYMV and ClYVV, and, more weakly, with BCMV and even OYDV. When using antibodies specific to the N-terminus of BYMV, no reaction occurred with the latter two viruses, but all other isolates, including those of ClYVV, clearly reacted. We have no doubt that BYMV and ClYVV are two separate viruses because they clearly differ biologically, as in their pathological effects on major hosts, and in artificial and natural host ranges, which result in different ecological potentials. This conclusion is in full agreement with the "identification guidelines" proposed by Hamilton et al. [39]. According to Shukla's hypothesis based on N-terminal specific antibodies, however, our isolates from faba beans should be considered strains of one virus, i.e., BYMV including ClYVV.

The unexpected reaction of ClYVV isolates with antibodies directed to the N-termini of BYMV coat protein suggests that the two different viruses share epitopes in the N-terminal region. A possible underlying identity in amino acid sequence (four residues) in the surface-exposed N-terminal region was indeed found by Shukla et al. [77]. This homologous region was confirmed by Uyeda et al. [82] who found a 39–47% homology between ClYVV and two strains of BYMV. Likewise, the virus pairs JGMV/watermelon mosaic virus and MDMV/tobacco etch virus also have common sequences in their N-terminal coat protein region [75]. Such exceptions to the Shukla theory are new evidence of molecular-biological overlap between distinct potyviruses.

Conclusions

This more or less historical overview concentrates on the subgroup of legume-infecting potyviruses related to BYMV, and on the variation within this cluster. It has not yet touched upon the uncertain taxonomic status of a range of legume potyviruses, such as azuki bean mosaic virus, blackeye cowpea mosaic virus, and cowpea aphid-borne mosaic virus [28, 59], the necrosis-inducing strains of BCMV, peanut mottle virus, and peanut stripe virus, all related in one way or another to BYMV and BCMV. Reference has been made to a similar situation with respect to potyviruses infecting Gramineae. Identical problems resulting from virus variation appear to exist within clusters of potyviruses infecting plant species belonging to other plant families, such as those of Cucurbitaceae (WMV-1 and -2, zucchini yellow mosaic virus, etc.), Liliiflorae (TBV, *Nerine* viruses, etc.), and Solanaceae (PVY, TEV, tobacco veinal mottle virus, etc.). Within the genus *Allium* alone, non-described potyviruses more or less related to OYDV and leek yellow stripe virus seem to exist (Van Dijk and Bos, unpubl. data). The economic importance of several of these viruses necessitates better distinction among them and more rapid, efficient and reliable methods for diagnosis and detection in the certification of plant propagation material. Discrimination between the "viruses" therefore remains topical and is of more than mere academic interest. But what are the prospects?

With time, more exceptions to the Shukla hypothesis may emerge, further damping the present high spirits. In nature, black and white borderlines seem to be non-existent, and viruses may not be exempt from this rule. Thanks to molecular biology, there is progress in our understanding, but natural variation keeps haunting taxonomists. Viruses obviously evolve with time. BYMV [57], PMosV and ClYVV [7] change in symptom type and infectivity to particular hosts after long-term passage through certain hosts. This state of flux could explain potyvirus variation with time and adaptation to special hosts or groups of hosts as a consequence of agronomic isolation. This is especially true when the viruses are perpetuated in certain crops through continued vegetative propagation, such as with ornamental bulbs, *Allium* species, and potatoes. Agronomic isolation also applies to potyviruses perpetuated through seed, as for bean (BCMV), cowpea (blackeye cowpea mosaic virus and/or cowpea aphid-borne mosaic virus), groundnut (peanut mottle virus and/or peanut stripe virus), lettuce (LMV), pea (pea seed-borne mosaic virus), and others. Such viruses may well have originated from a common ancestor through divergence when accidentally arriving in an uncommon host and further mutating under this new selection pressure.

A basic problem remains of whether coat-protein properties represent the entire genome of a pathogenic entity sufficiently to be used as a single criterion for classification. The coat protein represents only 10% or less of the total genetic information of a virus, or, more precisely, with serological reactivity residing in part of the amino acid configuration, only some 2% of the total viral genetic information [84]. From the human perspective, viruses derive their significance from being pathological entities, and their biological characters cannot be ignored. On the contrary, these are of paramount importance. The final question therefore is whether serological reactivity, particularly when determined with monoclonal antibodies, can be used as a reliable parameter of a "virus" as a taxonomic unit and thus for reliable diagnosis. For instance, a panel of selected MAbs reacted with eight out of nine strains of BYMV [56], and would seem to be very reliable for BYMV detection, if the ninth strain had not been included in the test. The broadly reactive MAb PTY1 [56], advocated and already commercially distributed as a general tool for potyvirus detection, does not react with papaya ringspot virus (PRSV-W) and peanut mottle virus [55], and with *Hippeastrum* mosaic virus, two *Nerine* potyviruses and a potyvirus of *Gloriosa* [26]. So, negative reaction with this MAb does not exclude potyvirus group affinity of the virus investigated. The use of monoclonal antibodies directed to a single coat-protein epitope therefore entails risks. A change in biological characters of PMosV and ClYVV through host passage, not affecting serological reactivity [7], indicates independent inheritance.

Coat-protein properties may represent certain biological virus characters, such as vector transmissibility, and amino acid configurations involved may be directly recognizable serologically [31]. The major part of the viral genome, but for the 3' end of the RNA (coding for the N-terminal part of the coat protein), is claimed to be conserved; that is, to be little changed during evolution and to be largely similar for all potyviruses [73–75]. In fact, several domains are similar even among viruses of distinct groups, including non-plant viruses, and this allowed the proposal to establish a supergroup of picorna-like viruses including poliovirus, como-, nepo-, and potyviruses. In addition to divergence, interviral recombination of gene sets in mixed infections may have been an underlying evolutionary mechanism [34, 35]. However, the viruses with "conserved" major 5' end parts of their genomes may still be far from identical in these parts. Single point mutations which alter biological and ecological behavior [6] already are known. Despite serological relationships among the large nuclear inclusion (NI) proteins of BYMV, ClYVV and TEV, there are differences, and the antisera to the small NI proteins of BYMV and ClYVV do not react with those of TEV [24]. Apart from acting as enzymes in virus multiplication, the possible involvement of nuclear inclusion protein in pathogenesis remains unknown. Retention of most potyvirus activity and integrity after loss of N-termini of the coat-protein molecule [77] (except for aphid transmissibility [43]) seems incompatible with the idea that virus identity is fully represented by amino acid sequences in the N-termini.

The use of N-terminal-directed antibodies has proved useful for further elucidating relationships among potyviruses. Their exclusive use for identification seems unjustified since in biology any type of classification is based on a combination of characters, and a "virus" really is more than mere coat protein. For non-plant viruses "the application of serological tests to designate a virus species also is not sufficient per se" [15]. Reality, even in the virus world, thus remains complicated and dynamic, and with virological classification difficult to throughly grasp.

References

1. Abu-Samah N, Randles JW (1981) A comparison of the nucleotide sequence homologies of three isolates of bean yellow mosaic virus and their relationship to other potyviruses. Virology 110: 436–444
2. Abu-Samah N, Randles JW (1983) A comparison of Australian bean yellow mosaic virus isolates using molecular hybridisation analysis. Ann Appl Biol 103: 97–107
3. Ainsworth GC, Ogilvie L (1939) Lettuce mosaic. Ann Appl Biol 26: 279–297
4. Allison RF, Dougherty WG, Parks TD, Willis L, Johnston RE, Kelly M, Armstrong FB (1985) Biochemical analysis of the capsid protein gene and capsid protein of tobacco etch virus: N-terminal amino acids are located on the virion's surface. Virology 147: 309–316
5. Atreya CD (1992) Application of genome sequence information in potyvirus taxonomy: an overview. In: Barnett OW (ed) Potyvirus taxonomy. Springer, Wien New York, pp 17–24 (Arch Virol [Suppl] 5)
6. Atreya CD, Raccah B, Pirone TP (1990) A point mutation in the coat protein abolishes aphid transmissibility of a potyvirus. Virology 178: 161–165

7. Barnett OW, Burrows PM, McLaughlin MR, Scott SW, Baum RH (1985) Differentiation of potyviruses of the bean yellow mosaic subgroup. Acta Hortic 164: 209–216

8. Barnett OW, Randles JW, Burrows PM (1987) Relationships among Australian and North American isolates of the bean yellow mosaic potyvirus subgroup. Phytopathology 77: 791–799

9. Beczner L, Maat DZ, Bos L (1976) The relationships between pea necrosis virus and bean yellow mosaic virus. Neth J Plant Pathol 82: 41–50

10. Beemster ABR, van der Want JPH (1951) Serological investigations on the *Phaseolus* viruses 1 and 2. Antonie van Leeuwenhoek J Microbiol Ser 17: 15–26

11. Bercks R (1960) Serologische Untersuchungen zur Differenzierung von Isolaten des *Phaseolus*-Virus 2 und ihrer Verwandtschaft mit *Phaseolus*-Virus 1. Phytopathol Z 39: 120–128

12. Bercks R (1960) Serological relationships between beet mosaic virus, potato virus Y, and bean yellow mosaic virus. Virology 12: 311–313

13. Bercks R (1961) Serologische Verwandtschaft zwischen Kartoffel-Y-Virus, Rüben-mosaik-Virus und *Phaseolus*-Virus 2. Phytopathol Z 40: 357–365

14. Bercks R (1962) Der Stand der serologischen Verwandtschaftsforschung bei pflanzen-pathogenen Viren. Zentralbl Bakteriol Parasitenkd Infektionskrank Hyg II Abt 116: 1–12

15. Bishop DHL (1985) The genetic basis for describing viruses as species. Intervirology 24: 79–93

16. Bos L (1963) Linnaeus en de plantevirologie. Vakbl Biol 43: 113-128 (with English summary: Linnaeus and plant virology. Meded Inst Plztk Onderz Wageningen 321: 1–14)

17. Bos L (1968) The problem of variation within the potato virus Y group with special reference to bean yellow mosaic virus. In: Abstracts 3rd Meeting IWGLV, July 26, 1968, London, UK

18. Bos L (1970) The identification of three new viruses isolated from *Wisteria* and *Pisum* in The Netherlands, and the problem of variation within the potato virus Y group. Neth J Plant Pathol 76: 8–46

19. Bos L (1974) Bean yellow mosaic virus. CMI/AAB Descriptions of Plant Viruses, no 40

20. Bos L, Kowalska Cz, Maat DZ (1974) The identification of bean mosaic, bean yellow mosaic and pea necrosis strains of bean yellow mosaic virus. Neth J Plant Pathol 80: 173–191

21. Bos L, Lindsten K, Maat DZ (1977) Similarity of clover yellow vein virus and pea necrosis virus. Neth J Plant Pathol 83: 97–108

22. Brandes J, Bercks R (1965) Gross morphology and serology as a basis for classification of elongated plant viruses. Adv Virus Res 11: 1–24

23. Brandes J, Wetter C (1959) Classification of elongated plant viruses on the basis of particle morphology. Virology 8: 99–115

24. Chang C-A, Hiebert E, Purcifull DE (1988) Purification, characterization, and immunological analysis of nuclear inclusions induced by bean yellow mosaic and clover yellow vein potyviruses. Phytopathology 78: 1266–1275

25. Conti M, Lovisolo O (1969) Observations on a virus isolated from *Wisteria floribunda* DC in Italy. Riv Patol Veget 4: 115–132

26. Derks AFLM (1992) Some unusual serological reactions among potyviruses. In: Barnett OW (ed) Potyvirus taxonomy. Springer, Wien New York, pp 77–79 (Arch Virol [Suppl] 5)

27. Derks AFLM, Vink-van den Abeele JL, Muller PJ (1980) Bean yellow mosaic virus in some Iridaceous plants. Acta Hortic 110: 31–37

28. Dijkstra J, Bos L, Bouwmeester HJ, Hadiastono T, Lohuis H (1987) Identification of blackeye cowpea mosaic virus from germplasm of yard-long bean and from soybean, and the relationships between blackeye cowpea mosaic virus and cowpea aphid-borne mosaic virus. Neth J Plant Pathol 93: 115–133

29. Doolittle SP, Jones FR (1925) The mosaic disease in the garden pea and other legumes. Phytopathology 15: 763–771
30. Dougherty WG, Carrington JC (1988) Expression and function of potyviral gene products. Annu Rev Phytopathol 26: 123–143
31. Dougherty WG, Willis L, Johnston RE (1985) Topographic analysis of tobacco etch virus capsid protein epitopes. Virology 144: 66–72
32. Fortass M, Bos L, Goldbach RW (1991) Identification of potyvirus isolates from faba bean (*Vicia faba* L.) and the relationships between bean yellow mosaic virus and clover yellow vein virus. Arch Virol 118: 87–100
33. Francki RIB, Milne RG, Hatta T (1985) Potyvirus group. Atlas of plant viruses, vol 2. CRC Press, Boca Raton, pp 183–217
34. Goldbach R (1986) Molecular evolution of plant RNA viruses. Annu Rev Phytopathol 24: 289–310
35. Goldbach R, Wellink J (1988) Evolution of plus-strand RNA viruses. Intervirology 29: 260–267
36. Goodchild DJ (1956) Relationships of legume viruses in Australia. II. Serological relationships of bean yellow mosaic virus and pea mosaic virus. Aust J Biol Sci 9: 231–237
37. Gould AR, Symons RH (1983) A molecular biological approach to relationships among viruses. Annu Rev Phytopathol 21: 179–199
38. Gratia A (1934) Identification sérologique et classification des virus des plantes. Distinction entre l'antigène mosaïque et l'antigène végétal. C R Soc Biol Belge 115: 1239–1241
39. Hamilton RI, Edwardson JR, Francki RIB, Hsu HT, Hull R, Koenig R, Milne RG (1981) Guidelines for the identification and characterization of plant viruses. J Gen Virol 54: 223–241
40. Hammond J, Hammond RW (1989) Molecular cloning, sequencing and expression in *Escherichia coli* of the bean yellow mosaic virus coat protein gene. J Gen Virol 70: 1961–1974
41. Hampton RO, Mink G, Bos L, Inouye T, Musil M, Hagedorn D (1981) Host differentiation and serological homology of pea seed-borne mosaic virus isolates. Neth J Plant Pathol 87: 1–10
42. Harrison BD (1985) Usefulness and limitations of the species concept for plant viruses. Intervirology 24: 71–78
43. Harrison BD, Robinson DJ (1988) Molecular variation in vector-borne plant viruses: epidemiological significance. Philos Trans R Soc Lond [Biol] 321: 447–462
44. Heald FD (1926) Manual of plant diseases. McGraw-Hill, New York
45. Hiebert E, Tremaine J, Ronald WP (1984) The effect of limited proteolysis on the amino acid composition of five potyviruses and on the serological reaction and peptide maps of tobacco etch virus capsid protein. Phytopathology 74: 411–416
46. Hill JH, Benner HI (1980) Properties of soybean mosaic virus and its isolated protein. Phytopathol Z 97: 272–281
47. Hollings M, Brunt AA (1981) Potyvirus group. CMI/AAB Descriptions of Plant Viruses, no 245
48. Hollings M, Brunt AA (1981) Potyviruses. In: Kurstak E (ed) Handbook of plant virus infections: comparative diagnosis. Elsevier/North-Holland, Amsterdam, pp 731–807
49. Hollings M, Stone OM (1974) Clover yellow vein virus. CMI/AAB Descriptions of Plant Viruses, no 131
50. Inouye T (1964) A virus disease of pea caused by watermelon mosaic virus. Ber Ohara Inst Landwirtsch Biol 12: 133–143
51. Inouye T, Inouye N (1964) A virus disease of peanut caused by a strain of turnip mosaic virus. Rep Ohara Inst Agric Bot Okayama Univ 50: 51–60

52. Johnson J (1927) The classification of plant viruses. Agric Exp Stat Univ Wisconsin Res Bull 76
53. Johnson J, Grant TJ (1932) The properties of plant viruses from different host species. Phytopathology 22: 741–757
54. Jones RT, Diachun S (1977) Serologically and biologically distinct bean yellow mosaic virus strains. Phytopathology 67: 831–838
55. Jordan R (1992) Potyviruses, monoclonal antibodies and antigenic sites. In: Barnett OW (ed) Potyvirus taxonomy. Springer, Wien New York, pp 81–95 (Arch Virol [Suppl] 5)
56. Jordan R, Hammond J (1991) Comparison and differentiation of potyvirus isolates and identification of strain-, virus-, subgroup-specific and potyvirus group-common epitopes using monoclonal antibodies. J Gen Virol 72: 25–36
57. Koenig R (1976) Transmission experiments with an isolate of bean yellow mosaic virus from *Gladiolus nanus* – activation of the infectivity for *Vicia faba* by a previous passage on this host. Acta Hortic 59: 39–47
58. Koenig R, Bercks R (1968) Änderungen im heterologen Reaktionsvermögen von Antiseren gegen Vertreter der potato virus X-Gruppe im Laufe des Immunisierungsprozesses. Phytopathol Z 61: 382–398
59. Lana AF, Lohuis H, Bos L, Dijkstra J (1988) Relationships among strains of bean common mosaic virus and blackeye cowpea mosaic virus – members of the potyvirus group. Ann Appl Biol 113: 493–505
60. Lisa V, Dellavalle G (1987) Bean yellow mosaic virus in *Lisianthus russelianus*. Plant Pathol 36: 214–215
61. Milne R (1984) The species problem in plant virology. Microbiol Sci 1: 113–117
62. Milne RG (1985) Alternatives to the species concept for virus taxonomy. Intervirology 24: 94–98
63. Milne RG (1988) Species concept should not be universally applied to virus taxonomy – but what to do instead? Intervirology 29: 254–259
64. Milne RG (1988) Taxonomy of the rod-shaped filamentous viruses. In: Milne RG (ed) The plant viruses, vol 4, the filamentous plant viruses. Plenum, New York, pp 3–50
65. Moghal SM, Francki RIB (1976) Towards a system for the identification and classification of potyviruses. I. Serology and amino acid composition of six distinct viruses. Virology 73: 350–362
66. Plese N, Wrischer M (1984) A mixed infection of *Passiflora caerulea* L. with two viruses. Acta Bot Croat 43: 1–6
67. Provvidenti R, Schroeder WT (1972) Natural occurrence of bean yellow mosaic virus in *Proboscidea jussiuei*. Plant Dis Rep 56: 548–550
68. Quantz L (1958) Ein Beitrag zur Kenntnis der Erbsenvirosen in Deutschland. Nachrichtenbl Deutsch Pflanzenschutzd Stuttg 10: 65–70
69. Randles JW, Davies C, Gibbs AJ, Hatta T (1980) Amino acid composition of capsid protein as a taxonomic criterion for classifying the atypical S strain of bean yellow mosaic virus. Aust J Biol 33: 245–254
70. Russo M, Martelli GP, Cresti M, Ciampolina F (1979) Bean yellow mosaic virus in saffron. Phytopathol Medit 18: 189–191
71. Reddick BB, Barnett OW (1983) A comparison of three potyviruses by direct hybridization analysis. Phytopathology 73: 1506–1510
72. Shepard JF, Secor GA, Purcifull DE (1974) Immunochemical cross-reactivity between the dissociated capsid proteins of the PVY group of plant viruses. Virology 58: 464–475
73. Shukla DD, Ward CW (1988) Amino acid sequence homology of coat proteins as a basis for identification and classification of the potyvirus group. J Gen Virol 69: 2703–2710

74. Shukla DD, Ward CW (1989) Identification and classification of potyviruses on the basis of coat protein sequence data and serology. Arch Virol 106: 171–200
75. Shukla DD, Ward CW (1989) Structure of potyvirus coat proteins and its application in the taxonomy of the potyvirus group. Adv Virus Res 36: 273–314
76. Shukla DD, McKern NM, Cough KH, Tracy SL, Letho SG (1988) Differentiation of potyviruses and their strains by high performance liquid chromatographic peptide profiling of coat proteins. J Gen Virol 69: 493–502
77. Shukla DD, Strike PH, Tracy SL, Gough KH, Ward CC (1988) The N and C termini of the coat proteins of potyviruses are surface-located and the N terminus contains the major virus-specific epitopes. J Gen Virol 69: 1497–1508
78. Shukla DD, Ford RE, Tosic M, Jilka J, Ward CW (1989) Possible members of the potyvirus group transmitted by mites or whiteflies share epitopes with aphid-transmitted definitive members of the group. Arch Virol 105: 143–151
79. Shukla DD, Jilka J, Tosic M, Ford RE (1989) A novel approach to the serology of potyviruses involving affinity-purified polyclonal antibodies directed towards virus specific N termini of coat proteins. J Gen Virol 70: 13–23
80. Shukla DD, Tosic M, Jilka J, Ford RE, Toler RW, Langham MAC (1989) Taxonomy of potyviruses infecting maize, sorghum, and sugarcane in Australia and the United States as determined by reactivities of polyclonal antibodies directed towards virus-specific N-termini of coat proteins. Phytopathology 79: 223–229
81. Taylor RH, Smith PR (1968) The relationship between bean yellow mosaic virus and pea mosaic virus. Aust J Biol Sci 21: 429–437
82. Uyeda I, Takahashi T, Shikata E (1991) Relatedness of the nucleotide sequence of the 3'-terminal region of clover yellow vein potyvirus RNA to bean yellow mosaic potyvirus RNA. Intervirology 32: 234–245
83. Van der Meulen JGJ (1928) Voorloopig onderzoek naar de specialisatie en de infectiebronnen der mozaiekziekten van landbouwgewassen. Tijdschr Plantenz 34: 155–176
84. Van Regenmortel MHV (1982) Serology and immunochemistry of plant viruses. Academic Press, New York
85. Van Regenmortel MHV (1990) Virus species, a much overlooked but essential concept in virus classification. Intervirology 31: 241–254
86. Van Regenmortel MHV, von Wechmar MB (1970) A reexamination of the serological relationship between tobacco mosaic virus and cucumber virus 4. Virology 41: 330–338
87. Ward, CW, McKern NM, Frenkel MJ, Shukla DD (1992) Sequence data as the major criterion for potyvirus classification. In: Barnett OW (ed) Potyvirus taxonomy. Springer, Wien New York, pp 283–297 (Arch Virol [Suppl] 5)

Author's address: L. Bos, Research Institute for Plant Protection (IPO-DLO), Binnenhaven 12, P.O. Box 9060, NL-6700 GW Wageningen, The Netherlands.

Arch Virol (1992) [Suppl 5]: 47–53

What is a virus?

M. H. V. Van Regenmortel

Institut de Biologie Moleculaire et Cellulaire, Strasbourg, France

Summary. The earlier reluctance of some plant virologists to use the term "virus species" has been overcome and the species has now been accepted as the basic unit in virus classification. A virus species is a polythetic class of viruses that constitutes a replicating lineage and occupies a particular ecological niche. Because of the polythetic nature of virus species, there is no single property, such as a particular level of genome homology, that could be used as the sole criterion for delineating individual virus species.

Introduction

In order to answer the question "What is a virus?" it is essential first to distinguish several meanings of the word virus. First, it may refer to the virus particles that can be seen in an electron micrograph or that are responsible for the infection of a particular host. In this case it refers to a concrete object located somewhere in space and time. In the jargon of logicians such a concrete object is a spatiotemporally localized entity.

Second, the word virus may refer to a concrete collective entity, usually labeled a taxon, corresponding to the sum total of all the particles in a replicating lineage that existed in the past or will exist in the future. In the contemporary jargon of taxonomists, the taxon potato virus Y (PVY) is "a chunk of the genealogical nexus" comprising all virus particles of the past that had PVY features together with similar particles that will exist in the future. Although all such particles distributed over millions of years are certainly concrete objects, their continued genetic evolution implies that at some stage in time the particles may have been or will become so different as not to correspond to what we call today PVY. It can thus be seen that what is crucial is the degree of similarity or difference between evolving objects that is still compatible with their being considered in some way the same "thing." Usually there will be no difficulty in recognizing a mutant virus as a mere variant of a type virus, but it will be more difficult to decide if a distant strain differing in many properties should be considered a separate virus. This means that the definition of the viral taxon as a population of concrete objects linked by common descent does not provide an easy criterion for species demarcation.

A third meaning of the word, virus, is that of a particular class of particles defined in terms of certain properties. In this case, virus means a class, i.e., an abstract concept devoid of any spatiotemporal location [21]. Like the concept, "dog," virus in this case is an abstract category and thus purely a conceptual construction. The abstract notion of "dog" is derived from the observation of certain types of animals and of the features they have in common. Unlike a particular dog, the concept of dog cannot be owned nor can it be heard to bark in the night. In the same way, the taxonomic category of potyvirus or of PVY is an abstract concept that cannot be seen in the microscope and cannot infect plants.

Failure to distinguish between these three meanings of virus would be somewhat similar to confusing (*i*) a piece of solid gold, (*ii*) the total amount of gold present in the universe, and (*iii*) the element gold defined by its atomic number 79 (a class). However, this analogy is not totally accurate since there is a fundamental difference between the universal class of gold atoms, all of which are identical, and a class of biological objects such as viruses. Like all biological entities that possess the ability to reproduce themselves or to replicate, viruses are endowed with an intrinsic variability derived from the error-prone process of nucleic acid duplication. It is this built-in variability which allows biological systems to become adapted through selection and which in the end guarantees their survival. A population of RNA viruses derived from a single particle consists of individuals that differ at many nucleotide positions. Under equilibrium conditions in a particular environment where the climatic conditions, the host, and vectors remain unchanged, the same wild-type sequence of an RNA virus will predominate. In effect, the class of identical gold atoms can be defined in terms of properties that are collectively necessary and sufficient for membership in the class, i.e., its atomic number; in contrast, a class of viruses cannot be defined in this way.

In the remainder of this chapter, discussion will be limited mainly to the third meaning of the term, virus, namely that of a class or category of infectious agents recognizable by a number of characters.

Plant virus descriptions

For more than 20 years, B. D. Harrison and A. F. Murant, acting as editors for the Association of Applied Biologists (AAB) and the Commonwealth Mycological Institute, have produced a series of *Descriptions of Plant Viruses,* which at present comprise about 400 different viruses. These Descriptions represent a semi-official catalogue of what plant virologists regard as separate viruses. The same viruses appear as separate entities in the official classification published in the reports of the International Committee on Taxonomy of Viruses. The way in which the list of Descriptions has grown over the years has been recounted by Matthews [8] as follows: "When a new virus isolate is described in the literature, the two editors, using common sense guidelines developed by themselves, decide whether it is a new virus or merely a strain of a previously described virus. When they invite a contributor to write a description for an isolate they consider to be a new "virus," they are in effect unofficially delineating a new species of virus."

It is noteworthy that the editors of the Descriptions never explained what they meant by "virus" nor how they decided which cluster of viral isolates and strains corresponded to a separate taxon and deserved a description. Each virus for which a description was produced was given a taxonomic status by means of inverted commas. The symbol "virus" was meant to convey the message that one was referring to a separate taxon or class and not merely to a separate strain or isolate. In the case of potyviruses, a total of 54 different "viruses" had been delineated in this manner by the end of 1989, and these were brought together in another ill-defined taxonomic category known as a group. As pointed out by Matthews [8, 9] the two labels "virus" and group are nothing but semantic alternatives to the classical taxonomic categories of species and genus. However, many plant virologists have argued forcibly against the use of the terms species and genus, partly because they believed that such a usage would lead to pressure to replace vernacular English names by latinized binomials [4, 11–13]. It was also argued that the species concept was inappropriate for entities such as viruses that do not reproduce sexually.

As discussed elsewhere [21, 22] the arguments advanced against the use of the term species in virology were based on an obsolete concept of biological species defined by gene pools and reproductive isolation and applicable only to sexually reproducing organisms. However, other definitions of species exist, some of which are applicable to asexual organisms and to clonally reproducing entities such as viruses.

The justification for using the species concept in virology is that viruses are biological entities and not simply chemicals [2, 5, 21]. Viruses possess genes, replicate, evolve, and occupy particular ecological niches. The basic unit in all biological classifications is the species and there seems to be no cogent reason for not using the term in virology.

What is a virus species?

It is remarkable that more than 100 years after the publication of Darwin's *On the Origin of Species,* biologists have not yet coined a universally accepted definition of species [10, 15]. Several species concepts have been proposed, such as morphological species, biological species, and evolutionary species but the most useful idea is that introduced by Beckner [1], namely the concept of polythetic species. The nature of a polythetic class can be illustrated by the following example taken from Sattler [16]. Suppose a group of individuals is defined by a set of five properties, f1 to f5. If these properties are distributed in the way shown in Table 1, a typical polythetic class is obtained. In this class (*i*) each individual possesses a large number but not all of the properties of the set, (*ii*) each property of the set is possessed by most but not all individuals, and (*iii*) no property is possessed by all individuals.

Whereas a classical monotypic class is defined by a single property or by a set of properties that are both necessary and sufficient for membership, the members of a polythetic class need not have a single defining property in common. There is thus no need to look for an elusive single property that will define the species.

Properties and classes are related abstract entities. Whatever is said about a thing is seen as ascribing a property to it, or equivalently, to assign the thing to a universal class [14]. If a virus has a positive strand RNA genome, it is automatically a member of the class of positive strand RNA viruses. In addition, the same virus could also be a member of other universal classes defined for instance by icosahedral shape or a certain genome structure. It is important to realize, however, that although viruses can be members of a variety of universal classes where each such class is defined by a single property both necessary and sufficient for class membership, the grouping known as a virus species is not a universal class definable by a single property. Properties that belong to all the members of a virus family (such as positive strand RNA genomes and non-enveloped, isometric particle morphology in the case of the *Picornaviridae*) are properties that define the family (a universal class) and they cannot be used to define or differentiate individual species within the family.

The concept of polythetic species is very useful in biology because it makes it possible to accommodate certain individuals that lack one or another character considered typical of the class. It is thus also particularly suited for dealing with viruses that always possess considerable intrinsic variability [20]. Another advantage of the polythetic species concept is that it does not attempt to impose rigid demarcation lines on fuzzy boundaries. A useful analogy is the perception of individual colors by the human mind from a continuous spectrum of electromagnetic waves [22].

Table 1. Distribution of five properties, f1–f5, among five members of a polythetic class [16]

Individual	Properties of individual				
1		f2	f3	f4	f5
2	f1		f3	f4	f5
3	f1	f2		f4	f5
4	f1	f2	f3		f5
5	f1	f2	f3	f4	

Intuitively, virus species are often viewed as a collection of strains that are so similar that it is useful to give them a single virus name. Similarly, a virus strain could be defined as a collection of isolates that are so similar that it is useful to give them a single strain name. However, such circular definitions amount to saying that the whole consists of the sum of its parts and are of little use since they do not specify what distinguishes a virus strain from a virus species. Other definitions of virus species have been proposed [5, 6, 21] but have not become generally accepted. Finally, at a recent meeting of the Executive Committee of ICTV held in Atlanta in April 1991, the following definition of virus species was approved for inclusion in the Sixth ICTV Report: "A virus species is a polythetic class of viruses that constitutes a replicating lineage and occupies a particular ecological niche." This definition had been developed [21, 22] to stress the biological nature of viruses but avoided any reference to gene pools since these are absent in the case of many clonally reproducing viruses. The definition also incorporated the notions of genome and biological variability inherent in replicating lineage and of niche occupation, crucial for maintaining the identity of obligate parasites.

Although the above definition of virus species clarifies the meaning of the term it does not provide rules for deciding where the line should be drawn between any two species. The definition of the abstract category of virus species should not be confused with a list of diagnostic properties needed to identify the members of a particular species taxon [3]. The definition of a concept or a class does not provide the means for recognizing concrete objects. The atomic number 79 does not help in recognizing a piece of metal as being gold. It is also logically impossible for a concrete object such as a virus particle to be a component part of an entity of a different logical type such as the abstract concept of virus species [22].

How can members of a virus species be identified?

The following properties are of diagnostic value for delineating virus species: genome characteristics (sequence, recombination potential), level of antigenic cross-reactivity, host range and response, tropism, and vectors [7]. However, in view of the polythetic nature of virus species, it should be stressed that a single diagnostic property such as level of genome homology or extent of antigenic relationship will always fail to establish membership of a particular virus species. Species identity is maintained through interaction of environmental and host factors with the intrinsic plasticity of viral genomes. A single character cannot embody the resulting multifactorial complexity of the stabilizing selection that ensures successful adaptation in a particular ecological niche.

For many years attempts were made to delineate individual potyviruses on the basis only of comparative biological properties and serological relationships. The resulting taxonomy was unsatisfactory because different strains of one potyvirus frequently seemed to form a continuum with the strains of another virus, preventing the drawing of clear-cut borderlines between individual species [18, 19]. However, as the coat-protein sequences of many potyviruses became established, it was recognized that many potyviruses could be divided into species and strains on the basis of the degree of coat-protein sequence homology. Distinct potyvirus species usually exhibit sequence homologies of 40–70% whereas strains of individual potyviruses have homologies of 90–99% [17]. These data fully substantiate the truism that there is continuity within the boundaries of species and discontinuity between individual species. The separation between different strains of a potyvirus is a matter of practical convenience, for instance for enshrining symptomatological or host range differences relevant to a pathologist. However, such differences may result from a single point mutation and they should not be given taxonomic importance.

When sequence information was added to biological and serological data, a number of erroneous assignments of particular potyvirus strains could be corrected. For instance, soybean mosaic virus strain N was identified as a strain of watermelon mosaic virus 2, while several so-called strains of maize dwarf mosaic virus were recognized to be strains of johnsongrass mosaic virus [18]. Occasional oddities of nomenclature are unavoidable as illustrated, for instance, by the "watermelon" strain of papaya ringspot virus which does not infect papaya. Although sequence information has been of crucial importance for recognizing the species identity of many potyviruses, it would nevertheless be wrong to rely on a particular level of genome homology as the sole criterion for delineating individual species. Two potyviruses showing 85% coat-protein sequence homology is a case in point, since they could be considered two different species or two strains of the same species.

Virus species are polythetic classes and searching for an elusive, ultimate single defining character for class membership is doomed to failure.

References

1. Beckner M (1959) The biological way of thought. Columbia University Press, New York
2. Bishop DHL (1985) The genetic basis for describing viruses as species. Intervirology 24: 79–93
3. Ghiselin MT (1984) "Definition," "character," and other equivocal terms. Syst Zool 33: 104–110
4. Harrison BD (1985) Usefulness and limitations of the species concept for plant viruses. Intervirology 24: 71–78
5. Kingsbury DW (1985) Species classification problems in virus taxonomy. Intervirology 24: 62–70
6. Matthews REF (1982) Classification and nomenclature of viruses. Fourth report of the International Committee on Taxonomy of Viruses. Intervirology 17: 1–200
7. Matthews REF (ed) (1983) A critical appraisal of viral taxonomy. CRC Press, Boca Raton
8. Matthews REF (1985) Viral taxonomy. Microbiol Sci 2: 74–75

9. Matthews REF (1985) Viral taxonomy for the nonvirologist. Annu Rev Microbiol 39: 451–474
10. Mayr E (1988) Toward a new philosophy of biology. Belknap Press of Harvard University Press, Cambridge
11. Milne RG (1985) Alternatives to the species concept for virus taxonomy. Intervirology 24: 94–98
12. Milne RG (1988) Species concept should not be universally applied to virus taxonomy – but what to do instead? Intervirology 29: 254–259
13. Murant AF (1985) Taxonomy and nomenclature of viruses. Microbiol Sci 2: 218–220
14. Quine WV (1987) Quiddities. Belknap Press of Harvard University Press, pp 22–24
15. Rosenberg A (1985) The structure of biological science. Cambridge University Press, Cambridge
16. Sattler R (1986) Bio-philosophy. Analytic and holistic perspectives. Springer, Berlin Heidelberg New York Tokyo
17. Shukla DD, Ward CW (1988) Amino acid sequence homology of coat proteins as a basis for identification and classification of the potyvirus group. J Gen Virol 69: 2703–2710
18. Shukla DD, Ward CW (1989) Structure of potyvirus coat proteins and its application in the taxonomy of the potyvirus group. Adv Virus Res 36: 273–314
19. Shukla DD, Ward CW (1989) Identification and classification of potyviruses on the basis of coat protein sequence data and serology. Arch Virol 106: 171–200
20. Strauss JH, Strauss EG (1988) Evolution of RNA viruses. Annu Rev Microbiol 42: 657–683
21. Van Regenmortel MHV (1989) Applying the species concept to plant viruses. Arch Virol 104: 1–17
22. Van Regenmortel MHV (1990) Virus species, a much overlooked but essential concept in virus classification. Intervirology 31: 241–254

Author's address: M. H. V. Van Regenmortel, Institut de Biologie Moléculaire et Cellulaire, 15 rue René Descartes, F-67084 Strasbourg Cedex, France.

Serology and antigenic relationships

Arch Virol (1992) [Suppl 5]: 57–69

Serology of potyviruses:
current problems and some solutions

D. D. Shukla[1], **R. Lauricella**[2], and **C. W. Ward**[1]

[1] CSIRO, Division of Biomolecular Engineering, Parkville, Victoria, Australia
[2] Chiron Mimotopes Pty Ltd., Clayton, Victoria, Australia

Summary. The serological relationships among members of the family *Potyviridae* are extremely complex and inconsistent. Variable cross-reactivity of polyclonal antisera, unexpected paired relationships between distinct viruses, and lack of cross-reactions between some strains are the major problems associated with the serology of potyviruses. Recent biochemical and immunochemical investigations of coat proteins have established the molecular basis for potyvirus serology and provided explanations for most of the problems with serology of potyviruses. Information from these studies has also formed the basis for the development of several novel approaches to the accurate detection and identification of potyviruses. However, even these novel approaches are not without drawbacks and some of them cannot be applied easily in plant virus laboratories, since they require prior sequence information and facilities for peptide synthesis. These findings suggest that serology is an imperfect criterion for the identification and classification of potyviruses.

Introduction

Serology is a very useful criterion for the identification and classification of several plant virus groups [50], notably tobamoviruses [19] and tymoviruses [28]. However, it has proved most unsatisfactory when applied to the family *Potyviridae* [36]. The difficulties with the serology of members of this family are not due to problems associated with the serological techniques, but are due to inherent complexities associated with potyvirus coat proteins and particles [36–38]. In fact, sensitivity of the recent serological techniques, such as enzyme-linked immunosorbent assay (ELISA) and Western blotting, has reached a level where plant viruses can be detected even in a single infected seed or a single virus-carrying insect [38]. In spite of these developments in serological techniques, some members of the family *Potyviridae* still cannot be identified accurately by the methods currently available.

The serological relationships among members of the genus *Potyvirus* are extremely complex and inconsistent. It has been suggested that there is no simple pattern of antigenic relationship among members in the genus and serological relationships often do not correlate with biological properties [21, 24, 25, 31]. In this respect potyviruses are perhaps exceptions to the rule that viruses which are antigenically related also share most of their other properties [12]. Hollings and Brunt [24] note that, although most genus members are serologically related to at least one other member in the group and in many cases to several others, the expected serological relationships between many connected pairs have not been observed. For example, bean yellow mosaic virus (BYMV) is serologically related to lettuce mosaic virus (LMV) and bean common mosaic virus (BCMV), but no serological relationship has been observed between LMV and BCMV. Hollings and Brunt [24] also state that different strains of one potyvirus may be as distantly related serologically to one another as they are to other potyviruses. On the basis of such observations it has been suggested that unless considerable caution is used in the interpretation of serological data, serology may be a misleading approach for tracing relationships among potyviruses [12, 31].

Structure of potyvirus coat proteins and particles

Since serology reflects protein structure, the complexity in serology of potyviruses can only be resolved by a thorough understanding of the variation in their coat-protein structure. The long, flexuous, rod-shaped particles of potyviruses (680–900 nm long and 11 nm wide) consist of approximately 2000 copies of a single protein species of molecular weight ranging from 30k to 37k, and one copy of positive-sense single stranded RNA of molecular weight $3.0–3.5 \times 10^6$ [24, 25]. Recent biochemical and immunochemical investigations of particles, coat proteins, and overlapping synthetic octapeptides corresponding to entire coat-protein sequences of potyviruses have demonstrated that: (*i*) distinct members, in general, possess a coat-protein sequence identity of 38 to 71% with major differences in length and sequence of their N termini, whereas strains of individual viruses exhibit a sequence homology of greater than 90% and have N-terminal sequences that are very similar [34–36, 53]; (*ii*) the N and C termini of coat proteins appear to be located on the particle surface, as they can be removed from intact particles by mild enzyme treatment and their removal does not affect the infectivity by mechanical inoculation or morphology of virus particles [45]; (*iii*) the degradation of virus particles, known to occur during purification and storage [13], involves gradual removal of the N and C termini of their coat proteins [45]; (*iv*) the N terminus constitutes the most immunodominant region of potyvirus particles [45, 48]; and (*v*) virus-specific epitopes usually are located in the N terminus, whereas potyvirus group-specific epitopes are contained in the highly homologous core region (devoid of N and C termini) of coat proteins [40, 42, 45, 48].

Problems with potyvirus serology

The problems associated with the serology of potyviruses are mainly threefold: (*i*) variable cross-reactivity of potyvirus antisera, (*ii*) unexpected and inconsistent paired relationships between distinct potyviruses, and (*iii*) lack of cross-reactions between strains.

Variable cross-reactivity of potyvirus antisera

It is well known that antisera prepared against a potyvirus in one laboratory may differ considerably in their specificities from antisera to the same virus raised in another laboratory. Substantial variation in specificity was observed when 11 potyvirus antisera produced in different laboratories were tested with 12 distinct potyviruses [42]. A majority of the antisera reacted to some degree with all or most of the potyviruses tested whereas two antisera reacted only with their homologous viruses. Such variation in the specificity of the antisera may be due to two factors. First, the state of the purified virus preparations used for immunization may have contributed to this situation. It is known that the N and C termini of coat proteins of potyviruses are degraded, during purification and storage, by enzymes of plant or microbiol origin which co-sediment with the virus particles [45]. The usual practice in different laboratories is to use the same preparation of purified virus for successive immunizations. Since the N terminus contains the virus-specific epitopes, its removal from virus particles in situ would gradually result in virus particles containing only non-specific core epitopes. Second, the immunization procedure may have had an influence. There is considerable variation in the literature on the number, interval, and route of injections and the amount of antigen administered when producing an antiserum to plant viruses [50]. Although there is little reliable information available concerning the relative merit of different immunization procedures, these are very likely to affect the reactivities of the antibodies produced. Large differences in the reactivities of antisera taken at different stages of immunization of the same animal have been reported [27]; antisera from early bleedings contain virus-specific antibodies whereas cross-reactive antibodies begin to appear in later stages of immunization [51]. Our investigations of potyviruses have given similar results [47, 48]. Furthermore, antibodies raised against the dissociated coat protein core of a potyvirus were found to recognize all 15 different aphid-transmitted potyviruses as well as mite-transmitted rymoviruses and whitefly-transmitted ipomoviruses (possible genus) [40, 42]. These results suggest that most of the contradictory information on serological relationships among potyviruses is due to the presence in antisera of variable proportions of cross-reacting antibodies that are targeted to the highly homologous core regions of the potyvirus coat proteins.

Unexpected and inconsistent paired relationships between distinct potyviruses

The second major problem with serology of potyviruses is the presence of unexpected and inconsistent paired serological relationships between viruses which on other grounds are regarded as distinct potyviruses. Such relationships occur between biologically similar as well as distinct potyviruses. Examples of biologically distinct potyviruses forming unexpected paired relationships are johnsongrass mosaic virus (JGMV)/watermelon mosaic virus 2 (WMV 2) [45] and the B strain of maize dwarf mosaic virus (MDMV-B)/tobacco etch virus (TEV-HAT) [42]. The first virus in each pair (JGMV or MDMV-B) is known to infect only monocotyledonous plant species whereas the second virus in each pair (WMV 2 or TEV) infects only dicotyledonous plant species. The viruses in each of these two pairs also are transmitted by different aphid vectors. The pairing between potato virus Y (PVY) and TEV and between BYMV and clover yellow vein virus (ClYVV) (unpubl. obs.) are examples of biologically similar, but distinct viruses being involved in unexpected paired serological relationships. The viruses in each of these two pairs have overlapping host ranges, produce similar symptoms in several hosts and have common aphid vectors. The eight viruses involved in the above unexpected paired relationships are considered distinct potyviruses on the basis of amino acid sequences of their coat proteins [53].

The serological reactions obtained with these paired viruses are often similar in strength to the homologous reactions [42, 45]. The serological relationships can be either uni- or bi-directional. For example, an antiserum to JGMV reacted strongly with WMV 2 [45] but a WMV 2 antiserum did not react with JGMV (unpubl. obs.). Similarly, an antiserum to PVY reacted with TEV but a TEV antiserum did not react with PVY (unpubl. obs.). In contrast, antisera to MDMV-B and TEV strongly cross reacted with each other [42].

The other well known but unexplained example of paired relationship is between BYMV, LMV and BCMV. BYMV antisera cross react with LMV and BCMV but no serological relationship has been observed between LMV and BCMV [24]. Examination of the past literature on reactivities of potyvirus antisera is likely to reveal many more examples of unexpected paired relationships between distinct potyviruses [13, 24, 50, 52], suggesting that the problem of paired serological relationship in the family *Potyviridae* is more acute than is currently believed. There are two main reasons why many more such relationships have not been reported. First, workers concerned with diagnosis of potyviruses generally test their viruses with antisera to related potyviruses and not with antisera to potyviruses that are very different biologically. For example, a potyvirus isolated from a dicotyledonous plant species would generally be tested with antisera to potyviruses infecting dicots only, and in the majority of cases only with antisera to those potyviruses which infect plants of the same or related families of the host of the new isolate. Such testings will not reveal paired relationships such as JGMV/WMV 2 or MDMV-B/TEV because serological relationships between these biologically very different potyviruses would not be suspected. Second, the close serological relationships observed between biologically similar potyviruses are generally considered normal serological behaviour expected from viruses which share many other properties. However, careful re-examination of such relationships would reveal that many of them are due to unexpected paired relationships. For example, several of the 15 or so potyviruses infecting legumes [43] appear to form a "continuum" on the basis of their biological and serological properties [5, 9, 29]. Some of them (BYMV, ClYVV) cannot be clearly distinguished by monospecific antisera [3], monoclonal antibodies [34] or antibodies directed to N termini of coat proteins [11].

Examination of the JGMV and WMV 2 relationship revealed that the epitope for this paired relationship was located in the surface-exposed amino-terminal region of the coat protein since removal of this region abolished their relationship when examined in Western blotting [45]. Preliminary immunochemical analysis of overlapping, synthetic octapeptides which correspond to the surface-exposed amino-terminal regions of JGMV and WMV 2 suggests that the epitope for this paired relationship consists of the first eight amino-terminal residues (unpubl. obs.). Examination of these sequences in the two viruses [36] shows that only three of the eight residues (1, 2, and 4) are identical. However, work of Geysen and co-workers [15–17] has shown that only a few amino acid residues in an epitope (generally five to seven residues long) are key contact residues; other residues can be substituted without any apparent effect on antibody binding.

Similarly, the epitope for the MDMV-B and TEV relationship was also found to be located in the amino-terminal region of the coat proteins since antibodies directed towards amino-terminal regions of the two coat proteins gave strong cross-reactivities [42]. A comparison of sequences of MDMV-B [14] and TEV-HAT [1] showed that the first eight amino-terminal amino acid residues in the two coat proteins are identical and this sequence may be responsible for the paired relationship between these two viruses [36, 37, 42]. MDMV-B is a strain of sugarcane mosaic virus (SCMV) [47] and seven of the first eight (2 to 8) amino-terminal amino acid residues of SCMV-SC [14] also are identical to that of TEV-HAT. It will be interesting to see if SCMV-SC also forms an unexpected paired relationship with TEV-HAT.

The epitope for other unexpected paired relationships may also reside in this N-terminal region of the coat proteins. In a close analysis of the N-terminal sequences of distinct potyviruses, Shukla and Ward [36] observed that limited sequence homology can be found if the N-terminal ends are aligned and major gaps produced. Such examinations revealed that other potyviruses also have common sequences in the N-terminal region as found with the JGMV/WMV 2 and MDMV-B/TEV-HAT pairs. For example, in BYMV and CYVV six of the first eight amino-terminal amino acid residues are identical [49], and these residues may be responsible for the paired serological relationship between these two biologically similar potyviruses.

Lack of cross-reactions between some strains

The third major problem in potyvirus serology is posed by viruses such as pepper mottle virus (PepMoV) and PVY which appear to be strains of the same virus on the basis of the structure of their coat proteins [46], yet show only a distant or no serological relationship [32, 33]. Comparisons of coat-protein sequences show that PepMoV displays amino acid sequence identity of 91 to 93% with the strains of PVY [53], and the sequence differences in the surface-exposed, immunodominant amino terminal 32 amino acid region in the coat proteins of PepMoV and PVY strains are very similar in number and location to the sequence differences between the PVY strains alone [46].

Purcifull et al. [33] and Nelson and Wheeler [32] investigated the serological relationship between PepMoV and PVY using SDS-gel immunodiffusion tests and found negligible cross-reactions. When affinity-purified antibodies, targeted to amino termini of the coat proteins [42] of PepMoV and PVY, were reacted with coat proteins of the two viruses in Western blotting, no cross-reactions with the antibodies were observed (unpubl. results), supporting the findings of these other workers [32, 33]. In contrast, the same affinity purified antibodies gave strong cross-reactivities between PepMoV and PVY coat proteins when tested in antigen-coated, indirect ELISA (unpubl. results). These contradictory reactivities of the same antibodies in different serological tests is unexplained at this stage.

A similar failed cross-reaction was observed recently with a strain of passionfruit woodiness virus (PWV) [20]. Strains of PWV are closely related serologically and exhibit amino acid sequence identities of 96 to 99% in their coat proteins [44]. However, when the newly isolated K strain of the virus (PWV-K) was compared with two previously known strains, PWV-TB and PWV-M, in Western blotting using an antiserum to a common strain of the virus, PWV-K reacted very weakly while PWV-TB and PWV-M gave strong reactions [20]. This antiserum previously was shown to contain antibodies mainly to the amino terminus and very little, if any, to the core region of the coat protein as it did not recognize any of the other 11 potyviruses tested [42]. These findings suggest that the surface-exposed amino-terminal region of PWV-K is very different from the other two strains. Comparison of amino acid sequences show that PWV-K has a sequence identity of only 53% in the amino-terminal 43 amino acid residues and 92% in the core region with PWV-TB, PWV-S, and PWM-M [20]. These results suggest that the low serological reactivity of PWV-K is due to the significant sequence variation in the amino terminus of its coat protein.

Systematic epitope mapping of overlapping synthetic peptide fragments [48] corresponding to the amino-terminal amino acid sequences of PepMoV,

PVY, and the PWV strains may help to explain the low serological cross-reactions between these viruses.

Some solutions to the problems of serology

It is now well established that the surface-exposed amino terminus is the only large region in the potyvirus coat protein that is variable, virus-specific, and contains virus-specific epitopes. Therefore, antibodies targeted to the amino-terminal region should recognize a virus and its strains, whereas those directed to the homologous core region should recognize other potyviruses [40 ,42, 45, 47, 48]. Many of the problems in serology of potyviruses can be resolved if antibodies directed toward the amino-terminal region of the coat proteins, instead of those targeted to the whole coat protein (normal potyvirus polyclonal antisera), are used in serological tests. Such virus-specific antibodies can be obtained using the following approaches.

Use of early bleed antisera

Early bleed antisera to potyviruses predominantly contain antibodies to amino termini of coat proteins [47, 48]. Although such antisera may have low titers, they can be used successfully to detect most strains of a potyvirus in ELISA and Western blotting. Two early bleed antisera to JGMV obtained after three and five intravenous injections (1 mg virus/injection) contained antibodies only to epitopes located in the surface-exposed amino-terminal region and reacted with strains of the virus originating in Australia and the United States but not with five other potyviruses tested [48]. Thus, it seems that as few as three intravenous injections of animals with freshly purified viral preparations at a fortnightly interval are enough to generate virus-specific antibodies to potyviruses.

Affinity purification of polyclonal antisera on core protein-bound columns

Since polyclonal antisera to intact particles of potyviruses contain antibodies to epitopes located in the entire coat protein, Shukla et al. [42] developed a simple chromatographic procedure to obtain virus-specific antibodies from polyclonal antisera by selective removal of the core-targeted antibodies. The method involved: (i) removal of the surface-located amino-terminal region of the coat protein from particles of one potyvirus using lysyl endopeptidase, (ii) coupling the truncated coat protein core to cyanogen bromide-activated Sepharose, (iii) passing antisera to different potyviruses through the column. Antibodies that did not bind to the column were directed to the N terminus of coat proteins and were highly specific. This approach showed that 17 potyvirus strains infecting maize, sorghum and sugarcane in Australia and the U.S.A. were not all closely related strains of SCMV as previously believed, but represented four distinct potyviruses [47]. This classification of the SCMV subgroup has now been confirmed by several other biological and molecular approaches including reactivities of differential sorghum and oat cultivars; peptide profiling and amino acid sequences of coat proteins; and nucleotide sequences and molecular hybridization of the 3' non-coding regions of the viral genomes [39].

For most potyviruses and their strains, polyclonal antibodies to amino termini isolated from crude antisera using affinity chromatography will be sufficient to establish their exact identity. However, this approach will not work if the isolates being identified are involved in unexpected paired serological relationships, since epitopes for such relationships are also located in the amino-terminal region of the coat proteins [42, 45].

Affinity purification of polyclonal antisera on peptide-bound columns

Systematic immunochemical analysis of overlapping, synthetic peptide fragments corresponding to sequences of the potyvirus coat proteins [48] has helped establish the exact locations and sequences of virus-specific epitopes of several potyviruses in our laboratory (unpubl. results). Using this information, we have developed a method to isolate specific antibodies from crude antisera prepared against JGMV, WMV2 and soybean mosaic virus [18, 30]. The method involved: (i) coupling of synthetic peptide fragments corresponding to virus-specific epitopes to Thiopropyl-Sepharose 6B, (ii) packing the peptide-coupled gel in a column for FPLC, (iii) passing rabbit and sheep polyclonal antisera (raised against intact particles or dissociated coat proteins of potyviruses) over the pre-equilibrated column, (iv) elution and evaluation of the bound antibodies against a panel of potyviruses. The eluted antibodies reacted only with their homologous viruses in ELISA and Western blotting. This procedure appears to be potentially useful for selecting virus-specific antibodies from polyclonal sera displaying unexpected paired serological relationships.

Selection of virus-specific monoclonal antibodies

Monoclonal antibodies have now been produced against several potyviruses [7, 8, 10, 22, 23, 26, 34, 48]. Immunization of mice with intact potyvirus particles generates monoclonal antibodies of different specificities. Some recognize only the homologous viruses and their strains while others react with two, three, four, and in some cases with most or all potyviruses [26, 48]. Since cloning and selecting lines with desired reactivities is often a laborious procedure and constitutes the major limitation to the rapid establishment of lines producing antibodies useful for diagnostic purposes, we have developed a simple method to identify virus-specific antibodies. The method involves screening monoclonal antibodies against native and truncated (minus N terminus) potyvirus coat proteins in Western blotting. Monoclonals directed to the N terminus will only recognize the native coat protein band, whereas those directed to the coat-protein core will react with native as well as core-protein bands (D. R. Hewish et al., unpubl.).

Generation of virus-specific monoclonal antibodies by inducing tolerance to the coat-protein core region

Benjamin and Waldman [4] were able to selectively block unwanted monoclonal antibody responses. The immune system of mice can be made tolerant of certain protein antigens by administering these antigens during a brief pulse of treatment with a monoclonal antibody directed to the L3T4 molecule on helper T lymphocytes. This technique has the potential to form the basis for a generalized means of tolerance induction and appears ideally suited for potyviruses, where the mice can be made tolerant with the coat-protein core in the presence of anti-L3T4 monoclonal antibodies and then immunized with the intact virus particles to specifically obtain virus-specific antibodies targeted to the N terminus of the coat protein. Recently, this procedure was used successfully in our laboratory to obtain virus-specific antibodies to BYMV potyvirus [54].

Generation of virus-specific monoclonal antibodies by peptide-mediated electrofusion

Peptide-mediated electrofusion is an attractive procedure for obtaining specific monoclonal antibodies to proteins. The technique gives increased fusion efficiency and high selectivity [56], and seems a valuable approach for generating virus-specific antibodies to potyviruses since coat-protein sequences of many potyviruses and their strains are known [53]. The peptide-mediated electrofusion approach was used by us recently to generate virus-specific monoclonal antibodies to PVY [52]. The method involved immunization of mice with PVY native coat protein; tagging B-cells specific for defined N-terminal epitopes by incubation with the corresponding biotinylated peptide-carrier protein conjugate; bridging these tagged B-cells to biotinylated myeloma cells with streptavidin; electrofusion of the streptavidin-linked cell complexes; and screening the hybrids with a panel of potyviruses in ELISA. The 11 monoclonals, generated against a peptide corresponding to the N-terminal 30 amino acids of the PVY coat protein [41] using this procedure, were all found to react only with PVY and not with other potyviruses tested [55].

Synthetic peptide-generated antibodies

Synthetic peptides corresponding to amino acid sequences of proteins have been used to generate diagnostic reagents for virus diseases of animals [2]. It is well established that antibodies to synthetic peptides corresponding to linear segments of a protein will frequently react with the homologous sequence in the native folded protein provided the region is exposed on the surface and exhibits a repertoire of conformation similar to the immunizing peptide [6]. The synthetic peptide approach may be an attractive proposition for the production of virus-specific diagnostic reagents for potyviruses, since the virus-specific epitopes are located on the surface-exposed amino termini of coat proteins [45].

Conclusions

From the foregoing discussion it is clear that the serological relationships among distinct members of the potyvirus group are complex and inconsistent. This complexity is not due to the problems associated with the serological techniques but are due to the inherent problems associated with the structure of potyvirus particles and their coat proteins. Variable cross-reactivity of poly-clonal antisera, unexpected and inconsistent paired relationships between distinct viruses, and lack of cross-reactions between some strains are the major problems associated with serology of potyviruses. Recent biochemical and immunochemical analyses of potyvirus particles, coat proteins, and synthetic peptide fragments corresponding to coat-protein sequences have shown that the variable cross-reactivity of potyvirus antisera is due to the presence of variable proportions of core-targeted antibodies which result from the state of purified viral preparations and type of immunization schedules used. The structural information has also provided a basis for the development of several novel serological approaches for the production of antibodies of defined specificity. These are: (*i*) use of early bleed antisera; (*ii*) removal of cross-re-acting antibodies on core-protein columns; (*iii*) selective recovery of specific polyclonal antibodies on peptide-bound columns; (*iv*) screening virus-specific monoclonal antibodies by Western blotting; (*v*) selective tolerance induction to cross-reacting core-epitopes; (*vi*) epitope-specific, peptide-mediated electro-fusion; and (*vii*) synthetic peptide generated antibodies. As discussed, use of these novel approaches is not without problems. Amino-terminal serology will not work with viruses involved in paired serological relationships unless the antibodies responsible are selectively removed. Moreover, some of these newer approaches are not easily applied since they require prior sequence information and facilities for peptide synthesis and are unlikely to be used routinely in plant virus laboratories. These findings suggest that serology, while useful for detection, is an imperfect criterion for the identification and classification of potyviruses.

References

1. Allison RF, Johnston RE, Dougherty WG (1986) The nucleotide sequence of the cod-ing region of tobacco etch virus genomic RNA: evidence for the synthesis of a single polyprotein. Virology 154: 9–20

2. Arnon R (1986) Synthetic peptides as the basis for future vaccines. Trends Biochem Sci 11: 521–524

3. Barnett OW, Randles JW, Burrows PM (1987) Relationships among Australian and North American isolates of the bean yellow mosaic potyvirus subgroup. Phytopathology 77: 791–799

4. Benjamin RJ, Waldman NH (1986) Induction of tolerance by monoclonal antibody therapy. Nature 320: 449–551

5. Bos L (1970) The identification of three new viruses isolated from *Wisteria* and *Pisum* in the Netherlands and the problem of variation within the potato virus Y group. Neth J Plant Pathol 76: 8–46

6. Cariepy J, Maitzner TM, Schoolnik GK (1986) Peptide antisera as sequence-specific probes of protein conformational transitions: calmodulin exhibits calcium dependent changes in antigenicity. Proc Natl Acad Sci USA 83: 8888–8892

7. Culvar JN, Sherwood JL (1988) Detection of peanut stripe virus in peanut seed by an indirect enzyme-linked immunosorbent assay using a monoclonal antibody. Plant Dis 72: 676–679

8. Diaco R, Hill JH, Durand DP (1986) Purification of soybean mosaic virus by affinity chromatography using monoclonal antibodies. J Gen Virol 67: 345–351

9. Dijkstra J, Bos L, Bouwmeester HJ, Hadiastono T, Lohuis H (1987) Identification of blackeye cowpea mosaic virus from germplasm of yard-long bean and from soybean, and the relationships between blackeye cowpea mosaic virus and cowpea aphid-borne mosaic virus. Neth J Plant Pathol 93: 115–133

10. Dougherty WG, Willis L, Johnston RF (1985) Topographic analysis of tobacco etch virus capsid protein epitopes. Virology 144: 66–73

11. Fortass M, Bos L, Goldbach RW (1991) Identification of potyvirus isolates from faba beans (*Vicia faba* L.), and the relationships between bean yellow mosaic virus and clover yellow vein virus. Arch Virol 118: 87–100

12. Francki RIB (1983) Current problems in plant virus taxonomy. In: Matthews REF (ed) A critical appraisal of viral taxonomy. CRC Press, Boca Raton, pp 63–104

13. Francki RIB, Milne RG, Hatta T (1985) Atlas of plant viruses, vol 2. CRC Press, Boca Raton

14. Frenkel MJ, Jilka JM, McKern NM, Strike PM, Clark Jr JM, Shukla DD, Ward CW (1991) Unexpected sequence diversity in the amino terminal ends of the coat proteins of strains of sugarcane mosaic virus. J Gen Virol 72: 237–242

15. Geysen HM, Mason TJ, Rodda SJ (1988) Cognitive features of continuous antigenic determinants. J Mol Recog 1: 32–41

16. Geysen HM, Meloen RH, Barteling SJ (1984) Use of peptide synthesis to probe viral antigens for epitopes to a resolution of single amino acids. Proc Natl Acad Sci USA 81: 3998–4002

17. Geysen HM, Rodda SJ, Mason TJ, Tribbick G, Schoofs PG (1987) Strategies for epitope analysis using peptide synthesis. J Immunol Methods 102: 259–274

18. Geysen HM, Shukla DD, Lauricella R, Plompen S, Tribbick G, Ward CW (1990) Synthetic peptides in the diagnosis of plant viruses. In: Abstracts of the VIIIth International Congress for Virology, Berlin, August 26–31, 1990, abstract W32-002

19. Gibbs AJ (1977) Tobamovirus group. CMI/AAB Descriptions of Plant Viruses, no 184

20. Gough KH, Shukla DD (1992) Major sequence variation in the N-terminal region of the capsid protein of a severe strain of passionfruit woodiness virus. Arch Virol 124: 389–396

21. Harrison BD (1985) Usefulness and limitations of the species concept for plant viruses. Intervirology 25: 71–78

22. Hill EK, Hill JH, Durand DP (1984) Production of monoclonal antibodies to viruses in the potyvirus group: use in radioimmunoassay. J Gen Virol 65: 525–532

23. Himmler G, Brix U, Steinkellner H, Laimer M, Mattanovich D, Kattinger HWD (1988) Early screening of anti-plum pox virus monoclonal antibodies with different epitope specificities by means of gold-labelled immunosorbent electron microscopy. J Virol Methods 22: 351–358

24. Hollings M, Brunt AA (1981) Potyviruses. In: Kurstak E (ed) Handbook of plant virus infections: comparative diagnosis. Elsevier/North- Holland, Amsterdam, pp 731–807

25. Hollings M, Brunt AA (1981) Potyvirus group. CMI/AAB Description of Plant Viruses, no 245

26. Jordan R, Hammond J (1991) Comparison and differentiation of potyvirus isolates and identification of strain-, virus-, subgroup-specific and potyvirus group-common epitopes using monoclonal antibodies. J Gen Virol 72: 25–36

27. Koenig R, Bercks R (1968) Änderungen in heterologen Reaktionsvermögen von Antiseren gegen Vertreter der potato virus X-Gruppe im Laufe des Immunosierungsprozesses. Phytopathol Z 61: 382–392

28. Koenig R, Lesemann DE (1979) Tymovirus group. CMI/AAB Description of Plant Viruses, no 214

29. Lana AF, Lohuis H, Bos L, Dijkstra J (1988) Relationships among strains of bean common mosaic virus and blackeye cowpea mosaic virus-members of the potyvirus group. Ann Appl Biol 113: 493–505

30. Lauricella R, Shukla DD, Plompen S, Morgan P, Stanton D, Harrison M, McKerral M (1990) Production of affinity purified potyvirus-specific antisera using peptides shown to bind virus-specific antibodies. In: Abstracts of the Fifteenth Annual Lorne Conference on Protein Structure and Function, Lorne, Australia, February 11–15, 1990

31. Moghal SM, Francki RIB (1976) Towards a system for the identification and classification of potyviruses. I. Serology and amino acid composition of six distinct viruses. Virology 73: 350–362

32. Nelson MR, Wheeler RE (1978) Biological and serological characterization and separation of potyviruses that infect peppers. Phytopathology 68: 979–984

33. Purcifull DE, Zitter TA, Hiebert E (1975) Morphology, host range and serological relationships of pepper mottle virus. Phytopathology 65: 559–562

34. Scott SW, McLaughlin MR, Ainsworth AJ (1989) Monoclonal antibodies produced to bean yellow mosaic virus, clover yellow vein virus, and pea mosaic virus which cross-react among the three viruses. Arch Virol 108: 161–167

35. Shukla DD, Ward CW (1988) Amino acid sequence homology of coat proteins as a basis for identification and classification of the potyvirus group. J Gen Virol 69: 2703–2710

36. Shukla DD, Ward CW (1989) Structure of potyvirus coat proteins and its application in the taxonomy of the potyvirus group. Adv Virus Res 36: 273–314

37. Shukla DD, Ward CW (1989) Identification and classification of potyviruses on the basis of coat protein sequence data and serology. Arch Virol 106: 171–200

38. Shukla DD, Frenkel MJ, McKern NM, Ward CW (1991) Immunological and molecular approaches to the diagnosis of viruses infecting horticultural crops. In: Prakash J, Peirik RLM (eds) Horticulture – new technologies and applications. Kluwer, Dordrecht, pp 311–319

39. Shukla DD, Frenkel MJ, McKern NM, Ward CW, Jilka J, Tosic M, Ford RE (1992) Present status of the sugarcane mosaic subgroup of potyviruses. In: Barnett OW (ed) Potyvirus taxonomy. Springer, Wien New York, pp 363–373 (Arch Virol [Suppl] 5)

40. Shukla DD, Ford RE, Tosic M, Jilka J, Ward CW (1989) Possible members of the potyvirus group transmitted by mites or whiteflies share epitopes with aphid-transmitted definitive members of the group. Arch Virol 105: 143–151

41. Shukla DD, Inglis AS, McKern NM, Gough KH (1986) Coat protein of potyviruses. 2. Amino acid sequence of coat protein of potato virus Y. Virology 152: 118–125

42. Shukla DD, Jilka J, Tosic M, Ford RE (1989) A novel approach to the serology of potyviruses involving affintity purified polyclonal antibodies directed towards virus-specific N termini of coat proteins. J Gen Virol 70: 13–23

43. Shukla DD, McKern NM, Barnett OW, Ward CW (1990) Identification and classification of potyviruses infecting tropical legumes. In: Abstracts of the VIIIth International Congress of Virology, Berlin, August 26–31, 1991, abstract W87-008

44. Shukla DD, McKern NM, Ward CW (1988) Coat protein of potyviruses. 5. Symptomatology, serology and coat protein amino acid sequences of three strains of passionfruit woodiness virus. Arch Virol 102: 221–232

45. Shukla DD, Strike PM, Tracy SL, Gough KH, Ward CW (1988) The N and C termini of the coat proteins of potyviruses are surface located and the N terminus contains the major virus-specific epitopes. J Gen Virol 69: 1497–1508

46. Shukla DD, Thomas JE, McKern NM, Tracy SL, Ward CW (1988) Coat protein of potyviruses. 4. Comparison of biological properties, serological relationships and coat protein amino acid sequence of four strains of potato virus Y. Arch Virol 102: 207–219

47. Shukla DD, Tosic M, Ford RE, Jilka J, Toler RW, Langham MAC (1989) Taxonomy of potyviruses infecting maize, sorghum and sugarcane in Australia and the United States as determined by reactivities of polyclonal antibodies directed towards virus-specific N-termini of coat proteins. Phytopathology 79: 223–229

48. Shukla DD, Tribbick G, Mason TJ, Hewish DR, Geysen HM, Ward CW (1989) Localisation of virus-specific and group-specific epitopes of plant potyviruses by systematic immunochemical analysis of overlapping peptide fragments. Proc Natl Acad Sci USA 86: 8192–9196

49. Tracy SL, Frenkel MJ, Gough KH, Hanna PJ, Shukla DD (1992) Bean yellow mosaic, clover yellow vein and pea mosaic are distinct potyviruses: evidence from coat protein gene sequences and molecular hybridization involving 3'-non-coding regions. Arch Virol 122: 249–261

50. Van Regenmortel MHV (1982) Serology and immunochemistry of plant viruses. Academic Press, New York

51. Van Regenmortel MHV, von Wechmar B (1970) A reexamination of the serological relationship between tobacco mosaic virus and cucumber virus 4. Virology 41: 330–338

52. Walkey DGA, Webb MJB (1984) The use of simple electron microscope serology procedure to observe relationships of seven potyviruses. Phytopathol Z 110: 319–327

53. Ward CW, Shukla DD (1991) Taxonomy of potyviruses: current problems and some solutions. Intervirology 32: 269–296

54. Werkmeister JA, Shukla DD (1991) Selection of polyclonal antibodies to potyvirus-specific N terminus of coat proteins by induction of tolerance with monoclonal antibody. J Virol Methods 34: 71–79

55. Werkmeister JA, Tebb TA, Kirkpatrick A, Shukla DD (1991) The use of peptide-mediated electrofusion to select monoclonal antibodies against specific and homologous regions of the potyvirus coat protein. J Immunol Methods 143: 151–157

56. Wojchowski DM, Sytkowski AJ (1986) Hybridoma production by simplified avidin-mediated electrofusion. J Immunol Methods 90: 173–177

Authors' adress: D. D. Shukla, CSIRO, Division of Biomolecular Engineering, 343 Royal Parade, Parkville, Vic. 3052, Australia.

Arch Virol (1992) [Suppl 5]: 71–74

Polyclonal reference antisera may be useful for the differentiation of potyvirus species

J. Richter

Institut für Phytopathologie Aschersleben, Biologische Zentralanstalt, Aschersleben,
Federal Republic of Germany

Summary. Attention is directed to the usefulness of polyclonal antisera for the identification of potyviruses on the species level. A scheme for the preparation of such antisera is given.

*

Identification and classification of potyviruses involve some important theoretical and practical aspects. A central problem of practical importance is the need to decide rapidly and accurately whether two isolates are different virus species (i.e., separate viruses) or strains of the same species. Results from recent structural and immunochemical studies of potyviruses show that a distinction between strains and species usually is possible by serological means [2]. Simple methods which enable identification of separate species are desirable for routine diagnosis.

In our experience, direct double antibody sandwich (DAS) ELISA can be used to identify distinctive virus species. When 19 different potyvirus species in crude sap of infected plants (Table 1) or in a partially purified form (Table 2) were incubated with their corresponding antisera only the homologous combinations showed a positive reaction.

Negative cross-reactions were consistently obtained with all heterologous combinations. High-affinity antisera are especially useful for distinguishing virus species. Some prerequisites should be taken into consideration when preparing such antisera. Immunizations should be performed with low doses (µg range) of freshly prepared, highly purified virus preparations. Furthermore, it is essential to take advantage of the "memory" effect and to use adjuvants (e.g., Freund's adjuvants) to obtain a "depot" effect. A scheme for the preparation of antisera useful for the identification of potyvirus species is given in Fig. 1.

From the results in Tables 1 and 2 it may be concluded that antisera prepared in this manner have a high affinity and react very specifically. However,

Table 1. Absorbance in DAS-ELISA with potyvirus antisera and potyvirus antigens in crude sap of infected plants

Virus	PVY[a]	PVA	PVV	HMV	WMV2	TuMV	PRSV	LMV	BYMV	BCMV	PSBMV	MDMV	BtMV	SPMMV	CeMV	PPV
PVY	0.99															
PVA		2.87														
PVV			2.10													
HMV				1.44												
WMV2					0.67											
TuMV						2.08				<0.1						
PRSV							1.90									
LMV								1.52								
BYMV									0.70							
BCMV										1.84						
PSbMV				<0.1							1.58					
MDMV												2.12				
BtMV													2.28			
SPMMV														1.96		
CeMV															0.51	
PPV																0.54

[a] Abbreviations: AV1 asparagus virus 1; BtMV beet mosaic virus; BCMV bean common mosaic virus; BYMV bean yellow mosaic virus; CeMV celery mosaic virus; HMV henbane mosaic virus; LMV lettuce mosaic virus; MDMV maize dwarf mosaic virus; OYDV onion yellow dwarf virus; PRSV papaya ringspot virus; PSbMV pea seed-borne mosaic virus; PPV plum pox virus; PVA, PVV, PVY potato viruses A, V, Y; SbMV soybean mosaic virus; SPMMV sweet potato mild mottle virus; TuMV turnip mosaic virus; WMV2 watermelon mosaic virus 2

Table. 2. Absorbance in DAS-ELISA with potyvirus antisera and partially purified potyvirus antigens

Virus	Antiserum									
	PVY[a]	PVA	PVV	AV1	SbMV	BtMV	OYDV	LMV	TuMV	BYMV
PVY	1.58									
PVA		1.60								
PVV			1.83			<0.1				
AV1				0.62						
SbMV					1.48					
BtMV						0.24				
OYDV							1.8			
LMV			<0.1					1.9		
TuMV									0.4	
BYMV										1.5

[a] Acronyms as in Table 1

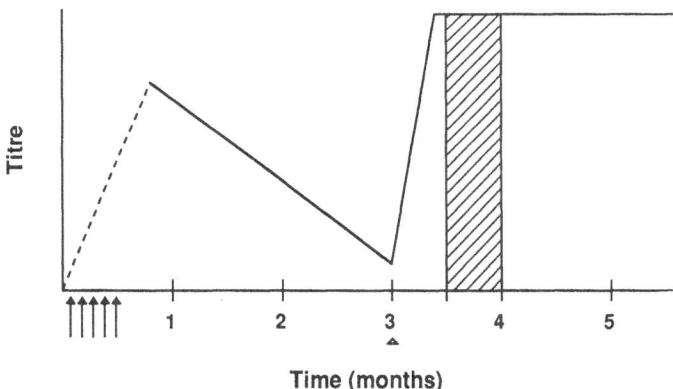

Fig. 1. A model for preparation of high-affinity antisera against potyviruses. Arrows indicate the first injection series (iv), increasing amounts (60–120 µg/rabbit). The arrowhead indicates booster injection (im) of 200 µg. The hatched area represents the time span for bleeding

based on these data, it is not justified to propose a generalized statement that there will be no cross reaction in DAS-ELISA between potyvirus antisera and heterologous potyviruses. In recent experiments cross-reactions occurred among blackeye cowpea mosaic, peanut stripe, and bean common mosaic viruses. This indicates a close serological relationship among these viruses and agrees with coat-protein peptide profiles which are so similar that McKern et al. [1] proposed that blackeye cowpea mosaic and peanut stripe viruses be considered strains of the same virus. Antisera like those prepared by our procedure may be useful as reference antisera. In this context we propose to establish potyvirus reference laboratories in different parts of the world. These laboratories should prepare and characterize potyvirus antisera for distribution as reference antisera as well as virus isolates to allow proper identification of unknown virus isolates and differentiation of potyvirus species.

References

1. McKern NM, Shukla DD, Barnett OW, Vetten HJ, Dijkstra J, Whittaker LW, Ward CW (1991) Coat protein properties suggest that azuki bean mosaic virus, blackeye cowpea mosaic virus, peanut stripe virus, and three isolates from soybean are all strains of the same virus. Intervirology 33: 121–134
2. Shukla DD, Ward CW (1989) Identification and classification of potyviruses on the basis of coat protein sequence data and serology. Arch Virol 106: 171–200

Author's address: J. Richter, Biologische Zentralanstalt, Institut für Phytopathologie Aschersleben, Theodor-Roemer-Weg 4, Postfach 162, D-O-4320 Aschersleben, Federal Republic of Germany

Arch Virol (1992) [Suppl 5]: 75–76

Proteolytic cleavage of the N-terminal region of potyvirus coat protein and its relation to host recovery and vector transmission

R. Salomon

Department of Virology, ARO, Volcani Center, Bet Dagan, Israel

Summary. The proteolytic cleavage of potyvirus coat N-terminal region is discussed with relation to host recovery and aphid transmission. The relationship of this coat protein region and the HC, may affect the host range and should be considered in this virus group classification.

*

The N-terminal region of sweet potato feathery mottle virus (SPFMV) was removed by a host protease found in *Ipomoea nil*-infected plants. These plants show attenuation, with complete recovery from viral symptoms in the young leaves from five to six leaves upward. A protease was found in leaves with a normal appearance from infected plants and no viral particles or coat protein could be detected in them; on the other hand, in uninfected plants this proteolytic activity could not be detected [1]. This protease from *I. nil* partially cleaved the N-terminal region of PVY but was completely inactive on ZYMV. Following these results the possibility of a site essential for the viral movement in the N-terminal region was raised. To test the assumption of the N-terminal region involvement in aphid transmission, PVY and ZYMV were trypsin treated, mixed with soybean trypsin inhibitor, and membrane fed to aphids with semi-purified helper component (HC). The trypsin treatment diminished aphid transmission while mechanical inoculation was hardly affected. A mechanical infection dilution curve supported this finding [2].

Recently a similar proteolytic activity was recovered from gladioli corms [3]. This protease cleaved the N-terminal region from the coats of bean yellow mosaic virus (BYMV), zucchini yellow mosaic virus (ZYMV), potato virus Y (PVY), and maize dwarf mosaic virus (MDMV). Thus, the gladiolus corm protease is probably non-specific. MDMV recovered from corn or sorghum remains intact for more than three months, unlike many other potyviruses recovered from their specific hosts; for example, sweet potato feathery mottle virus (SPFMV), from *Ipomoea* spp., ZYMV from zucchini, PVY from tobacco, and BYMV from gladioli corms. MDMV represents a somewhat exceptional situation among potyviruses with its extreme variability. Among the Israeli MDMV strains of group A a difference in size of the N-terminal region was found, with antigenic variability. However, following proteolytic cleavage the specific serological variability was abolished. Furthermore the specific antigenic properties of the N-terminal region are preserved even after dena-turation of the coat protein. The serological specificity of the N-terminal region of potyvirus coat protein, probably reflects the biological heterogeneity of these viruses. This heterogeneity is manifested in host specificity and may be linked to the N-terminal region involvement in the intracellular movement of the viruses [4].

Therefore the specific composition and biological behavior of the N-terminal region of potyvirus coat protein should be considered in classification of these viruses along with the specific antigenic activity of this region. Integrity of the coat protein subunit is essential for proper classification of potyviruses and the nature of the partial cleavage of the N-terminal region is important for this group's taxonomy.

References

1. Salomon R (1989) Partial cleavage of sweet potato feathery mottle virus coat protein subunit by an enzyme in extracts of infected symptomless leaves. J Gen Virol 70: 1943–1949
2. Salomon R, Raccah B (1990) The role of the N-terminus of potyvirus coat in aphid transmission. In: Abstracts of the VIIIth International Congress of Virology, Berlin, August 26–31, 1990, P 83-007
3. Stein A, Salomon R, Cohen J Loebenstein G (1986) Detection and characterization of bean yellow mosaic virus in corms of *Gladiolus grandiflours*. Ann App Biol 109: 147–154
4. Salomon R (1989) A possible mechanism for exclusion of plant viruses from embry-onic and meristemic tissues. Res Virol 140: 453–455

Author's address: R. Salomon, Department of Virology, ARO, Volcani Center, Bet Dagan 50-250, Israel.

Arch Virol (1992) [Suppl 5]: 77–79

Some unusual serological reactions among potyviruses

A. F. L. M. Derks

Bulb Research Centre, Lisse, The Netherlands

Summary. The determination of relationships among and the identification of potyviruses with polyclonal antibodies does not always lead to a proper conclusion. The potyvirus specific monoclonal PTY 1 does not detect all potyviruses tested.

Polyclonal antibodies

Polyclonal antibodies are widely used in the identification of viruses and in general are a very useful tool. However, for proper identification of potyviruses other methods may have to be used in addition to serology. Two examples are given as illustrations.

Tulip breaking virus (TBV) in tulips and lilies

Two antisera were prepared to TBV: one against an isolate from the tulip cv. Jack Laan (JL) and another one to an isolate from tulip cv. Texas Flame (TF). TBV can be detected in both tulips and lilies with both antisera, although better results were obtained with the JL antiserum for testing lilies and with the TF antiserum for testing tulips [2].

Hybridization with cDNA probes did not show these relationships between the isolates from tulips and lilies. A cDNA probe prepared to TBV from *Lilium longiflorum* hybridized specifically with three biologically different TBV isolates from lily and not with two TBV isolates from tulips (JL and TF). On the other hand, a cDNA probe to TBV from tulip hybridized with most tulip isolates and not with the lily isolates [1, 4]. On the basis of these cDNA hybridization results, TBV from lily and from tulip should be considered as different viruses.

Bean yellow mosaic virus (BYMV)
and freesia mosaic virus (FreMV) in freesia

A serological relationship between the two potyviruses BYMV and FreMV has never been detected by the microprecipitin or DAS-ELISA tests. However, when a heterologous coating antibody is used in DAS-ELISA a positive reaction is obtained; for instance, when an FreMV coating is used, followed by sap with BYMV and subsequently with a BYMV conjugate. In controls with normal serum for coating no color development occurred which indicated that no BYMV was absorbed directly to the ELISA plate. The serological relationship found between the two viruses in this way was partly confirmed in ISEM (immunosorbent electron microscopy) by trapping the viruses with the homologous antisera: BYMV particles were decorated by FreMV antisera, but FreMV particles were not decorated by BYMV antiserum.

Monoclonal antibodies

The potyvirus specific monoclonal PTY 1 [3] was used as prescribed by Agdia, but with a longer substrate incubation of 3 h at 37 °C plus overnight at room temperature.

Results were: (i) With some isolates of tulip breaking virus high absorption values were obtained but with other isolates very low values occurred, although by electronmicroscopy about the same numbers of virus particles were present in the samples tested. Differences among isolates of one virus were even more pronounced with BYMV: some isolates were not detected, while others gave rise to high absorption values. (ii) Many potyviruses in bulbous crops were detected, even some poorly described isolates such as those in *Polianthes*, *Stenomesson* and *Eucomis*. (iii) Four (tentative) potyviruses, viz. hippeastrum mosaic virus, two *Nerine* potyviruses and a potyvirus from *Gloriosa*, were not detected with the monoclonal PTY 1 [4]. No indications were obtained for an inhibitory effect of leaf sap constituents of *Nerine*. The *Nerine* potyviruses were not detected when the ELISA plates were coated first with γ-globulins of the homologous antisera or by using other sap dilutions. In immunoblotting the *Nerine* potyviruses did not react with the monoclonal, but reacted well with the homologous polyclonal antibodies. Hippeastrum mosaic virus was not detected in leaf sap of *Hippeastrum* or when *Hyoscyamus niger* was the source plant. (iv) The monoclonal did not react with sap of healthy plants, two carlaviruses, or one potexvirus.

Conclusions

Molecular hybridization with nucleic acid probes and serological reactions with polyclonal antisera do not lead to the same conclusions. More information is needed to confirm this statement. For instance, the lily and tulip isolates could actually be different viruses, but have common antigenic determinants. The cross reaction of FreMV antiserum with BYMV could result from antigenic sites with low affinity for the antibodies.

The unusual monoclonal antibody reactions described here show the need for use of more than a single monoclonal preparation for detection of potyviruses, even for isolates of a single virus. To be useful in potyvirus taxonomy, monoclonal antibodies need to be thoroughly characterized with a broad range of viruses.

Unexpected results such as these need to be examined in detail at the molecular level before drawing conclusions related to taxonomy.

References

1. Boonekamp PM, Asjes CJ, Derks AFLM, van Doorn J, Franssen JM, van der Linde PCG, van der Vlugt CIM, Bol JF, van Gemen B, Linthorst HJM, Memelink J, van Schadewijk AR (1990) New technologies for the detection and identification of pathogens in bulbous crops with immunological and molecular hybridization techniques. Acta Hortic 266: 483–490
2. Derks AFLM, Vink-van den Abeele JL, van Schadewijk AR (1982) Purification of tulip breaking virus and production of antisera for use in ELISA. Neth J Plant Pathol 88: 87–98
3. Jordan R, Hammond J (1991) Comparison and differentiation of potyvirus isolates and identification of strain-, virus-, subgroup-specific and potyvirus group-common epitopes using monoclonal antibodies. J Gen Virol 72: 25–36
4. Langeveld SA, Dore J-M, Memelink J, Derks AFLM, van der Vlugt CIM, Asjes CJ, Bol JF (1991) Identification of potyviruses using the polymerase chain reaction with degenerate primers. J Gen Virol 72: 1531–1541

Author's address: A. F. L. M. Derks, Bulb Research Centre, Vennestraat 22, Postbus 85, NL-2160 AB Lisse, The Netherlands.

Arch Virol (1992) [Suppl 5]: 81–95

Potyviruses, monoclonal antibodies, and antigenic sites

R. Jordan

United States Department of Agriculture, Agricultural Research Service, Plant Sciences Institute, Florist and Nursery Crops Laboratory, Beltsville, Maryland, U.S.A.

Summary. Virus-specific and cross-reactive monoclonal antibodies have been produced to at least 19 different aphid-transmitted potyviruses. This report summarizes the development of these monoclonal antibodies as well as presents information on the delineation of the virus-specific and group-common epitopes defined by these monoclonal antibodies. Virus-specific and group-common antigenic determinants were mapped by a variety of techniques, including analysis of antigen : antibody reactivity patterns, determination of N-terminal vs. trypsin-resistant core peptide-specificity, immunoanalysis of overlapping synthetic peptides, and immunoanalysis of bacterially expressed coat-protein gene products. Of those monoclonal antibodies that have been examined, monoclonal antibody-defined virus-specific epitopes are virion surface-located within the 30+ amino acid amino terminus, whereas the group-common epitopes are found in the trypsin-resistant core protein not usually located on the virion surface, as has been shown previously with certain polyclonal antibodies. New information is presented on the analysis of bean yellow mosaic virus amino terminal epitopes as well as on the identification of amino terminal antigenic determinants shared between strains of bean yellow mosaic virus and pepper mottle virus. A recommendation on the evaluation and use of a panel of potyvirus broad-spectrum reacting monoclonals as reference monoclonal antibodies for the detection and classification of aphid-transmitted potyviruses is also presented.

Introduction

The relative merits in potyvirus taxonomy of such molecular parameters as nucleic acid hybridization, gene sequence data, coat-protein sequence data, or high-pressure liquid chromatography peptide profiles, and of such phenotypic characteristics as particle morphology, host range, symptomatology, cross-protection, cytoplasmic inclusion morphology, and serology have been reviewed [39, 44, 45, 52]. Two major requirements for taxonomy include group-specific criteria and criteria that discriminate among distinct viruses and strains [54]. Coat-protein serology has been shown to be a very useful criterion for the identification and classification of most plant virus groups. This can now be used in potyvirus taxonomy with the advent of polyclonal and monoclonal antibodies that are specific to amino terminal domains or trypsin-resistant core proteins [1, 11, 28, 30, 40, 42, 52]. The overall scope of this report is to present information on the development and utility of monoclonal antibodies that define virus-specific and group-common epitopes.

The introduction of hybridoma technology has provided methods for the production of homogeneous and biochemically defined immunological reagents (monoclonal antibodies) of identical specificity, produced by a single cell line and directed against a unique epitope of the immunizing antigen. Hybridoma-produced monoclonal antibodies (MAbs) react with a single antigenic determinant (epitope), and can therefore have very unique and extremely discriminatory abilities. MAbs produced to specific potyvirus members have been highly virus-specific, if not strain-specific. However, many MAbs have shown various degrees of cross-reactivity with closely related potyviruses as well as with other less-related potyviruses.

The specific aims of this report are to briefly review the development of MAbs to potyviruses and the delineation of the virus-specific and group-common epitopes defined by these MAbs, to present new information on the determination of bean yellow mosaic virus-specific amino terminal coat-protein epitopes, and to present a recommendation on the use of reference MAbs for the identification and taxonomic determination of aphid-transmitted potyviruses.

Potyviral monoclonal antibodies

The potyviruses comprise the largest and economically most important genus of plant viruses and affect a wide range of crop plants. It is no wonder then that the majority of MAbs that have been produced to plant viruses to date have been to potyviruses.

Virus-specific and cross-reactive MAbs have been produced against at least 19 different potyviruses. Virus-specific MAbs generated to potato virus Y (PVY) by Gugerli and Fries [16] reacted with common epitopes of 26 isolates belonging to the tobacco veinal necrosis (N), common (O) and stipple streak (C) strains of PVY and did not react with any other potyviruses [12, 16]. Seven of the eight MAbs produced by Hill et al. [23] to lettuce mosaic (LMV), soybean mosaic (SbMV), or to two strains of maize dwarf mosaic virus (MDMV) were strain-specific. The other, LMV-generated, MAb had very weak cross-reaction with MDMV and SbMV when used in radioimmunoassay. Later [26], MDMV-specific MAbs to either MDMV-A and/or MDMV-B were also produced. Dougherty et al. [11] prepared ten MAbs to the capsid protein of tobacco etch virus (TEV). Three of these were TEV-specific, while the remaining seven were cross-reactive with a variety of different potyviruses including TEV, PVY, MDMV, tobacco vein mottling virus (TVMV), pepper mottle virus (PepMoV), and/or watermelon mosaic virus 2 (WMV 2). Two of nine MAbs prepared to tulip breaking virus (TBV) by Hsu et al. [25] were TBV strain-specific, whereas the other seven (including MAb TBV 7) cross reacted to at least ten other potyviruses, including bean yellow mosaic virus (BYMV), iris severe mosaic virus (ISMV), and iris mild mosaic virus (IMMV) [21,25; Jordan and Hammond, unpubl.]. MAbs prepared separately against BYMV, clover yellow vein virus (ClYVV), and pea mosaic virus (PMosV) [36], three members of the BYMV subgroup of potyviruses [5], cross reacted among the three viruses. Using as immunogen an admixture of native and denatured virions of seven different potyviruses (including BYMV, PVY, TEV, ISMV, IMMV) a panel of 30 virus-specific and broad-spectrum MAbs were developed by Jordan and Hammond [29, 30]. Fourteen of the MAbs recognize epitopes found only on strains of BYMV or BYMV-subgroup isolates, whereas the remaining 16 MAbs react with one or more of the other 43 potyvirus isolates tested. MAb PTY 1 reacted with all 55 potyvirus isolates (representing at least 33 different and distinct aphid-transmitted potyviruses) in the initial study [29, 30], and has since been shown to react with at least 135 potyvirus isolates (representing 48 distinct potyviruses) (see below, and Table 1). Six MAbs produced by Baker et al. [2] to a Florida isolate of papaya ringspot virus type-W (PRSV-W) were able to recognize isolates of PRSV-W, PRSV-P (papaya strain), and PRSV-T (a unique strain from Guadelope) but not zucchini yellow mosaic virus (ZYMV) or zucchini yellow fleck virus (ZYFV). One of the MAbs did cross react with a WMV 2 isolate. Other potyvirus-specific MAbs also have been produced to isolates of bean common mosaic virus [35, 51], peanut mottle virus (PeMoV) [37], peanut stripe virus (PStV) [8], plum pox virus [24], potato virus A [6], PVY [6, 54], TBV [6], WMV 2 [53], and ZYMV [53]. Due to their uniquely specific and/or differential reactivities, most of the above MAbs could be useful reagents for the identification and classification of potyviruses. However, as will be illustrated below, the extent of a MAb's specificity must be thoroughly examined before it is used as a taxonomic reagent.

Monoclonal antibodies also have been produced to non-structural potyviral proteins. MAbs produced by Slade et al. [46] to nuclear inclusions of TEV were shown to react with the 49 kDa proteinase or the 58 kDa putative RNA-dependent RNA polymerase. Their usefulness in the classification of potyviruses has not yet been evaluated. The 49 kDa proteinase-specific MAbs were useful in delimitating the VPg and proteinase domains of the 49 kDa protein; one of them also was able to inhibit the self-processing reaction of the 49 kDa proteinase [10, 46]. MAbs produced to the amorphous inclusion of PRSV-W [3] had broad-spectrum reactivity with 9 of 14 different potyviruses, including WMV 2, ZYMV, ZYFV, PStV, PeMoV, and SbMV [3; C. A. Baker, pers. comm.]. MAbs produced by Jordan et al. [31] to purified preparations of PVY and TVMV helper components (HC) were either specific to each of the viral HCs or showed cross-reactivity with either or both PVY and TVMV and with BYMV, PepMoV, ZYMV, and/or TEV sap extracts. MAbs specific to PRSV-W cylindrical inclusion protein have also been produced by Baker and Purcifull [4]. The potential of many of the above MAbs for use in potyvirus taxonomy is high, but not yet demonstrated.

Mapping virus-specific and group-common antigenic determinants

Well characterized monoclonal antibodies that define virus-specific and potyvirus group-common epitopes could begin to meet two of the requirements for potyvirus taxonomy; that is, reagents that discriminate among and between viruses and strains, and those that are group-specific (i.e., group cross-reactive). Before any MAb, or set of MAbs, can be used for potyvirus classification its antigen and/or epitope specificity must be determined. A few of the methods that permit delineation of MAb-defined epitopes include, antibody: antigen reactivity patterns [11, 30], determinination of the N-terminal vs. trypsin-resistant core (TRC) peptide-specificity [1, 27, 28, 42], immunoanalysis of overlapping synthetic peptides [15, 43], and immunoanalysis of bacterially expressed coat-protein gene products [19, 33] or of cell-free translation products from transcripts derived from plasmid cDNA [10]. Recent examples of using these technologies in the determination of MAb-defined virus-specific and group-common epitopes follow.

Antibody:antigen reactivity patterns

Antigens (or antibodies) that exhibit similar or identical reactivity patterns with a group or panel of test antibodies (or antigens) can be grouped together and, further, can be differentiated from those groups of antigens (or antibodies) that show markedly different reactivity patterns.

When evaluated against four strains of TEV and five other potyviruses, the nine TEV-derived MAbs of Dougherty et al. [11] defined five epitopes; two TEV-specific and three group-common. MAb TBV 7 [25] defines a group-common epitope [21, 25; Jordan and Hammond, unpubl.]. Twenty-five distinct epitopes were defined by the 30 PTY MAbs developed by Jordan and Hammond [30]; 10 BYMV-specific, 3 BYMV subgroup-specific, and 12 group-common epitopes. Unfortunately, no obvious correlations have been observed between these group-common coat-protein epitopes and virus host ranges, nor with vector specificity [17, 30].

N-terminal vs. TRC specificity

The coat-protein sequences of at least 42 strains of 20 distinct potyviruses have been determined [39, 52]. As researched and reviewed by Shukla and workers [39, 44, 45, 52], sequence comparisons and biochemical analysis show that the 30 to 95 amino acid N-terminus of the coat protein of distinct potyviruses is the only large region in the coat protein that is unique to a potyvirus [44, 45], is immunodominant [42, 43], contains virus-specific epitopes [40, 42], and is located on the virion surface [42]. The C-terminal two-thirds of the coat protein, on the other hand, has high sequence identity among distinct potyviruses [44, 45], and contains group-common epitopes [39, 44, 45].

Before the above polyclonal serological information had been compiled, Allison et al. [1] had shown that a TEV-specific MAb [11] reacted with a surface-located N-terminal epitope and that a potyvirus group cross-reactive MAb [11] detected an internal epitope on the trypsin-resistant core protein. In another study, using antigen competition analysis and endoproteinase-Arg (another enzyme that cleaves the N-terminal 39 amino acids from intact BYMV GDD [18])-treated BYMV in Western blot analysis, Jordan [27, 28] has demonstrated that the BYMV-specific epitopes are also surface-located within the N-terminus, and that the potyvirus group-common epitopes are non-surface within the endoproteinase-Arg-resistant core (=TRC) protein. This data corroborates that generated with polyclonal antisera [40, 42] and with the TEV MAbs [1, 11].

Mapping of coat protein epitopes expressed in bacteria

At least two prerequisites for epitope mapping via gene expression in bacteria are the availability of recombinant DNA clones containing all and partial relevant coding information, and that the test antibody recognizes a sequential epitope [33]. Different strategies have been used to generate gene fragments which, after insertion in an expression vector, direct the synthesis of an antigenic expression product [33]. These approaches include the construction of a series of deletion clones with Bal 31, exonuclease III, or restriction enzymes.

Epitopes defined by BYMV-specific and group cross-reactive PTY MAbs [28, 30] were mapped using bacterially expressed BYMV GDD coat protein [18]. Using carboxyterminal deletion mutants prepared by Bal31 deletions of a BYMV GDD cDNA, the virus-specific and group-common epitopes were mapped to at least six distinct domains (Fig. 1), some to within 10 to 20 amino acids [19; Hammond and Jordan, manuscript in prep.]. These results corroborated the N-terminal vs. TRC epitope specificity defined by Western blot analysis; i.e., the N-terminal virus-specific epitopes map to a 65 amino acid bacterially expressed N-terminal protein domain, and the TRC group-common epitopes map to one of five different bacterially expressed TRC protein domains.

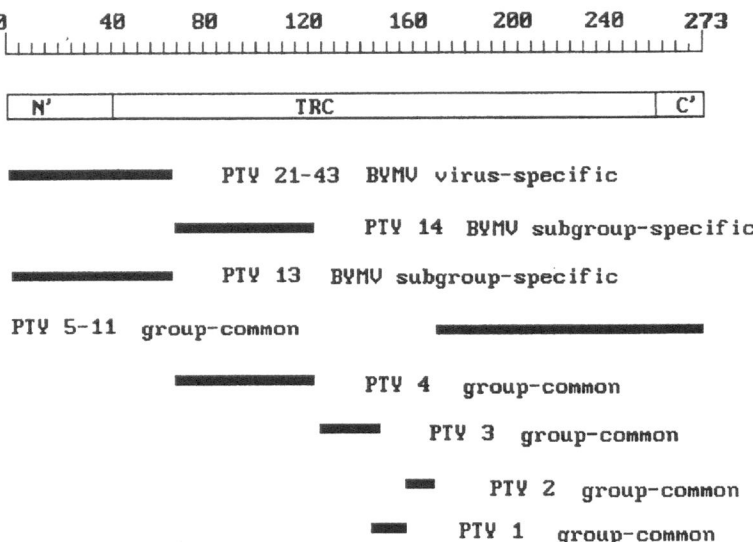

Fig. 1. Schematic representation of the expressed coat protein of BYMV GDD [18] and the predicted protein domains recognized by BYMV virus-specific MAbs (PTY 21–43), BYMV subgroup-specific MAbs (PTY 13, 14), and potyvirus group cross-reactive MAbs (PTY 1–11) [30] as determined by immunoanalysis of bacterially expressed coat-protein deletion mutants [19; Hammond and Jordan, manuscript in prep.]. The BYMV GDD coat protein is presented as the open box with the amino acid number 'ruler' located above. The predicted N-terminal and C-terminal trypsin-sensitive cleavage sites are indicated by vertical lines in the open box. The predicted domains recognized by the MAbs are represented by the solid horizontal lines below the full length coat protein. The respective MAbs recognizing each domain are listed to the right or left of each domain

Fine mapping of N-terminal sequences

In this section, the methodology and results from more recent experiments designed to more precisely map the actual amino acid sequence defined by five unique BYMV-specific PTY MAbs [30] will be presented. MAb PTY 43 defines an N-terminal, virion surface-located epitope present on only one strain (BYMV GDD) of 11 strains of BYMV [30; Jordan, this report] and on white lupin mosaic virus, a potential new member of the BYMV subgroup [22]. MAb PTY 37 defines an N-terminal, surface-located epitope present on only five of 11 BYMV strains [30; this report]. MAbs PTY 21, 30, and 33 define N-terminal, surface-located epitopes present on 7 to 9 of 11 strains of BYMV [30; this report] and two isolates of PepMoV (PepMoV NC165, from J. Moyer [31]; and PepMoV C from California [50]). The fine mapping of these MAbs was determined by immunoanalysis of overlapping octapeptides and/or by direct comparison of the amino terminal sequences of immuno-reactive and non-immunoreactive potyviruses.

Epitope analysis using overlapping octapeptides

Ninety-two overlapping octapeptides (overlapping the last five of the previous eight amino acids) covering the 273 aa BYMV GDD coat-protein sequence [18] were synthesized on polyethylene pins using the "epitope mapping kit", per manufacturer's instructions (Cambridge Research Biochemicals), essentially as described [15, 43]. ELISA procedures using the pin-bound octapeptides with MAbs PTY 37 and 43 (as tissue culture supernatants [30]) were as described [15, 30, 43].

MAb PTY 43 reacted strongly ($A_{405} = 1.4$) with peptide #10 (SEGQSVRQ) and weakly ($A_{405} = 0.4$) with peptide #11 (QSVRQIVP). MAb 37 reacted strongly ($A_{405} = 1.2$) only with peptide #11 (QSVRQIVP). The amino terminal sequence location of both peptides are identified in Fig. 2. Of the four BYMV subgroup isolates whose N-terminal sequences are illustrated in Fig. 2A, BYMV GDD is the only one detected by MAbs PTY 43 and 37. The BYMV GDD amino terminal sequence is also the only published potyvirus coat-protein sequence that contains the amino acid sequences identified by immunoanalysis of the octapeptides (data not shown).

N-terminal amino acid sequence comparisions

Shukla et al. [44, 45, 52] have demonstrated that coat-protein amino acid sequence data can be used to identify and differentiate distinct potyviruses and their strains. The coat-protein amino acid sequence data for three strains of BYMV (D, GDD, and CS) and one strain of ClYVV [49], all members of the BYMV subgroup, have been determined [7, 18, 47, 49] and compared [49, 52; Jordan, data not shown]. Analysis of the data for the three BYMV isolates indicates that they have high sequence identity (88–90%) and are strains of BYMV, whereas ClYVV 30 has much lower sequence identity with the BYMVs, suggesting that it should be a distinct virus.

When just the 39 amino acid N-terminal domains of these four viruses are aligned (without the introduction of gaps) and compared (GDD, CS, and 30 [49]) (GDD, CS, D, and 30; Fig. 2A), the homology between the BYMVs and ClYVV is between 44 to 51%, whereas the homology between/among the three BYMV isolates is 67% (D:CS, GDD:CS) to 85% (GDD:D). Stretches of three to six amino acids homologous between/among the four BYMV-subgroup isolates could account for much of the serological cross-reactivity observed between/among biologically distinguishable BYMV-subgroup isolates [5, 36], especially when antibodies directed to the N-termini are used [13].

Seven of the ten BYMV-specific PTY MAbs [30] were able to recognize their respective epitopes on BYMV GDD and BYMV D, but not on either BYMV CS or ClYVV 30 [unpubl.]. Because there is such high amino acid homology between GDD and D, potential epitopes shared only between GDD

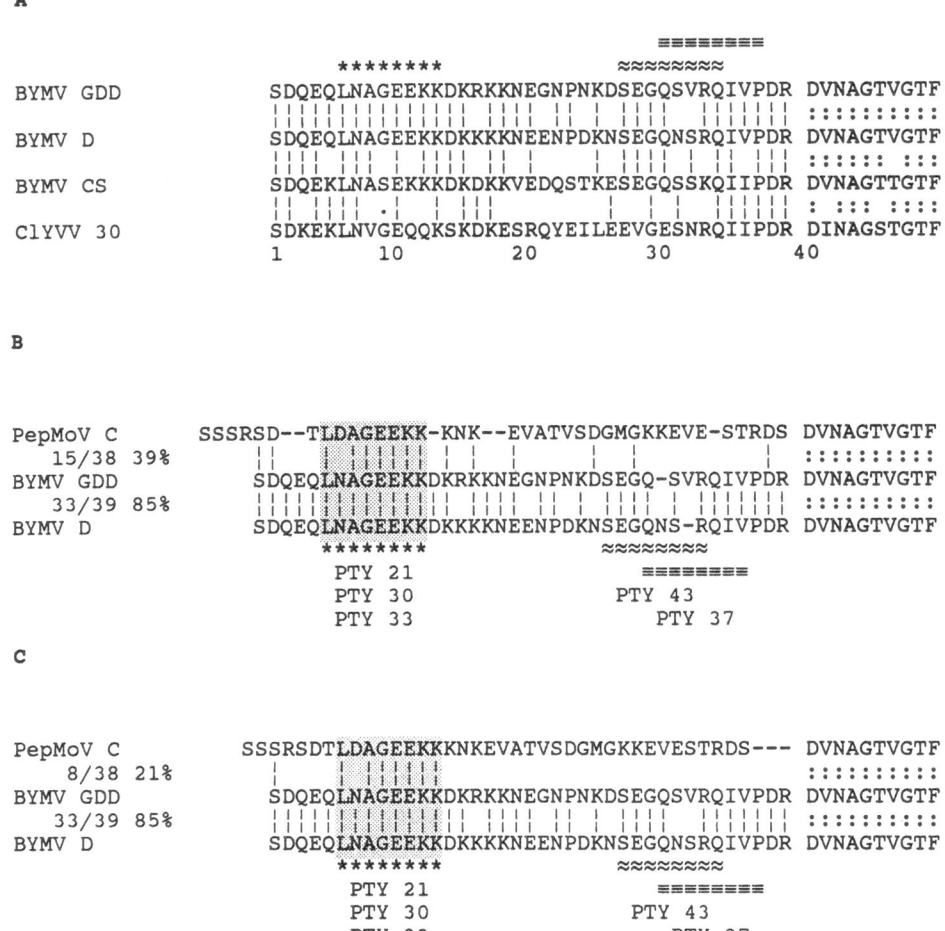

Fig. 2. Amino acid sequence comparisons of the amino termini of BYMV GDD [18], BYMV D [7], BYMV CS [47], ClYVV 30 [49], and PepMoV C [50], and delineation of predicted epitopes recognized by five PTY MAbs. The 38-39 N-terminal amino acids are separated (by a space) from the predicted starting amino acid of the trypsin-resistant core (TRC) protein, of which 10 amino acids are shown for reference. Identically aligned amino acids are indicated by a stroke in the N-termini, and colon in the TRC. The predicted amino acid residues within epitopes recognized by the MAbs are represented by wavy underlines (PTY 43), triple underlines (PTY 37), and asterisks (PTY 21, 30, and 33); sequence LN/ DAGEEKK is also highlighted. Alignments are of the four BYMV subgroup potyviruses (**A**); PepMoV C, BYMV GDD, and BYMV D with gaps (**B**); and without gaps (**C**)

and D, (and not found on BYMV CS and ClYVV 30), could not be identified using only N-terminal sequence alignments (Fig. 2A); especially in light of reports [14, 15] that only a few contact amino acid residues in an epitope are necessary for antibody binding. Amino acid sequence lengths of 7–11 to 15–21 amino acids, however, have generally been reported for antigenic determinants [15, 33].

As indicated above, MAbs PTY 21, 30, and 33 exhibit unique specificities in that they define N-terminal epitopes present on strains of BYMV (including

D, unpubl.) and on two isolates of PepMoV (NC 165, and C; these MAbs have not yet been tested against another PepMoV isolate [9]). Comparative alignment (with gaps) of the amino terminal 32 amino acids of this PepMoV [9] with BYMV GDD's amino terminal 39 amino acids showed only 16% identity (only five, dispersed, matches; data not shown) and only 28% identity (9 matches; data not shown) when this PepMoV [9] was aligned with PepMoV C [50]. However, when the N-termini of PepMoV C and BYMV GDD are aligned there is 38% identity (with gaps) (Fig. 2B). A potential epitope starting at amino acid 6 of GDD suggested from this alignment, is even more evident when the gaps are removed and the N-termini are realigned (Fig. 2C). The putative epitope for MAbs PTY 21, 30, and 33 encompasses amino acids 6 to 13 on GDD and contains the sequence LN/DAGEEKK (Fig. 2C). This sequence is not present on BYMV CS nor ClYVV 30, two sequenced BYMV subgroup viruses that these MAbs do not recognize. These three MAbs have differential reactivity with strains of BYMV, suggesting that specific contact residues for each MAb (and these specific residues are different for each MAb) are not present in all strains of BYMV.

This MAb-defined epitope (LN/DAGEEKK) also encompasses the putative aphid vector transmission recognition sequence (DAG or NAG) present in the majority of sequenced aphid-transmissible potyvirus coat proteins [39, and references therein]. MAbs PTY 21, 30, and 33 could be useful in experiments designed to inhibit aphid transmission of BYMV GDD and D, PepMoV NC165 and C. Unfortunately, these MAbs do not react with other DAG-containing potyviruses [30], presumably because the amino acids adjoining the DAG are inapproriate for antibody binding (i.e., not *LN/*DAG*EEKK*). Four to five amino acids of this epitope (i.e., LNAG and LNAGE) are also present in N-terminal peptides of BYMV S and ClYVV (isolate not identified), respectively, that were sequenced by Shukla et al. [42]. However, these isolates have not yet been tested with these MAbs.

Aphid-transmitted potyviruses: reference monoclonal antibodies

Shukla and Ward [44, 45, 52] have suggested that since the core region of potyvirus coat proteins show considerable sequence identity, antibodies targeted at this region should be capable of serving as broad-spectrum probes for the detection of most, if not all, potyviruses. As an example, polyclonal antisera raised against the denatured, truncated coat protein (devoid of N- and C-termini) of trypsin-treated particles of JGMV-JG reacted with 15 different aphid-transmitted potyviruses [42], and with mite- and whitefly-transmitted members of the *Potyviridae* [38]. A panel of well-characterized antisera produced to the denatured TRC proteins of unique and diverse potyviruses would be an additional suggestion.

Another recommendation would be to use a panel of well-characterized MAbs reactive to specific epitopes within the TRC that are highly conserved among the potyviruses. A proposal from this report is to recommend the use of MAbs PTY 1, 2, 3, 4, 10 and TBV 7 as the panel of reference MAbs for the identification of aphid-transmitted potyviruses. These six MAbs recognize different epitopes present, differentially, on distinct and diverse potyviruses. A panel of 30 PTY MAbs has already been shown to be useful in the examination of the intra-virus, inter-virus, and intra-group serological relationships among and between diverse potyviruses [20, 22, 30].

In the initial study [30], PTY MAbs 1–10 were able to recognize 15–33 of 33 diverse aphid-transmitted potyviruses (43 isolates) tested. None of the PTY MAbs (nor any other MAb reported in the literature), including PTY 1, have reacted with three non-aphid-transmitted potyviruses, namely the fungus-transmitted wheat spindle streak mosaic virus, the mite-transmitted wheat streak mosaic virus [30], and the whitefly-transmitted sweet potato mild mottle virus [20, 30]. PTY 1 has since been shown to react with over 135 isolates, representing 48 distinct potyviruses (Table 1).

Table 1. Aphid-transmitted potyviruses tested and recognized by monoclonal antibody
PTY 1

Alstroemeria mosaic	Pea seed-borne mosaic
Asparagus virus 1 (2)[a]	Peanut mottle (2)[c]
Bean common mosaic (22)	Peanut stripe (2)
Bean yellow mosaic (11)	Pepper mottle (2)
Blackeye cowpea mosaic	Plum pox
Carnation vein mottle	Pokeweed mosaic
Celery mosaic	Potato virus A
Clover yellow vein	Potato virus Y (5)
Cowpea aphid-borne mosaic	Soybean mosaic (2)
Dasheen mosaic	Statice virus Y (2)
Freesia mosaic	Sugarcane mosaic (4)
Garlic mosaic	Sweet potato feathery mottle (3)
Hyacinth mosaic	Sweet potato latent
Iris mild mosaic (2)	Tobacco etch (4)
Iris severe mosaic (2)	Tobacco vein mottling (2)
Lettuce mosaic	Tulip breaking (2)
Maize dwarf mosaic (2)	Tulip chlorotic blotch
Malva vein clearing	Turnip mosaic (7)
Narcissus yellow stripe	Vallota mosaic
Onion yellow dwarf	Watermelon mosaic (2)
Ornithogalum mosaic	White lupin mosaic
Papaya ringspot-P (4)	Yam mosaic
Papaya ringspot-W (14)[b]	Zucchini yellow mosaic (3)
Passionfruit woodiness	Other uncharacterized (7)
Pea mosaic (2)	

[a] In parenthesis, number of isolates tested, other than one
[b] 14 of 17 isolates recognized; see text
[c] 2 of 4 isolates recognized; see text

PTY 1 is not a 'universal potyvirus' MAb however, in that it has failed to detect up to eight different aphid-transmitted potyvirus isolates (i.e., PTY 1 has detected 135 of 143 isolates, tested to date). Apparently, the epitope recognized by PTY 1, although highly conserved, is altered in these isolates; most likely one or more of the contact amino acid residues is different in these viruses. Those isolates not detected include: nerine virus Y, nerine yellow stripe, and a partially characterized gloriosa virus isolate (from *Gloriosa rothschildiana*) [32] (although PTY 1 was able to detect two uncharacterized potyviruses in *Nerine* in New Zealand; G. Balasingham, pers. comm.); PRSV-W Florida isolates 2030, 2052, and 2040 [2] (PTY 1 'missed' these 3 isolates, but did detect the remaining 12 isolates tested by Baker et al. [2] and has detected two other isolates not tested by Baker et al. [30]); peanut mottle virus (PeMoV), neither isolates PeMoV-VS nor PeMoV-M were detected in tests done by Li et al. [34] (but PTY 1 did detect another isolate obtained from O. W. Barnett; unpubl.). In other tests (unpubl.), the isolates of PRSV and PeMoV not detected by PTY 1 were readily detected by PTY 4 and one or more of MAbs PTY 2, 3, and 10. Also, in some tests with some viruses, PTY 4, 10 or TBV 7 gave higher ELISA values that PTY 1 [20; Jordan, unpubl.].

All of the recommended MAbs are, or soon will be, available from Agdia, Inc (Elkhart, IN) (PTY 1), ATCC (Rockville, MD) (PTY 2-10, TBV 7), or IGEN (Rockville, MD) (TBV 7), or from the respective developers of each MAb. Further collaborative preliminary evaluation of this panel could be done with this author, as we have an APHIS permit for receipt of antigen-coated ELISA plates.

Concluding remarks

In this report I have presented a summary of the development of potyvirus virus-specific and group cross-reacting monoclonal antibodies. Information on the determination of the specific epitopes recognized by these MAbs as well as methods and results on the elucidation of an amino terminal epitope shared among strains of BYMV and PepMoV was also presented.

With the myriad of biological and molecular parameters available for use as determinants in potyvirus taxonomy, it is still surprising that the coat-protein gene plays so meaningful a role. Two important requirements in potyvirus taxonomy, group-specific characteristics and virus/strain discriminators [54], can in fact be met using coat-protein serology in conjunction with coat protein or coat-protein gene sequence information. Well characterized monoclonal antibodies that define virus-specific and group-common epitopes made to diverse and distinct potyviruses should be extremely useful reagents for the discrimination and identification of potyviruses. Some of the currently available MAbs have been very valuable and effective in examining specific intra-virus, inter-virus, and intra-group serological relationships among and between diverse potyviruses [20, 22, 30, 50].

Well characterized MAbs directed to conserved epitopes in the coat-protein core should be useful in determining potyvirus 'group status' of viruses of undetermined classification. A recommendation to use a specific panel of MAbs for defining and delineating new 'potyviruses' was presented in this report.

Acknowledgements

I wish to acknowledge and gratefully thank Mary Ann Guaragna for excellent technical assistance; John Hammond for collaboration, assistance, helpful discussions and critical reading of the manuscript; C. A. Baker, A. F. L. M. Derks, G. I. Mink, and C. Sutula for generously providing unpublished results of PTY 1 analyses; and O. W. Barnett, P. Berger, K. Boye, C. Niblett, I. Uyeda, K. Zagula, and all those listed in [30], for generously providing virus isolates.

The use of trade, firm or corporation names in this article does not imply the endorsement or approval by USDA-ARS of any product to the exclusion of others that may be suitable.

References

1. Allison RF, Dougherty WG, Parks TD, Willis L, Johnston RE, Kelly M, Armstrong FB (1985) Biochemical analysis of the capsid protein gene and capsid protein of tobacco etch virus: N-terminal amino acids are located on the virion's surface. Virology 147: 309–316
2. Baker CA, Lecoq H, Purcifull DE (1991) Serological and biological variability among papaya ringspot virus type-W isolates in Florida. Phytopathology 81: 722–728
3. Baker CA, Purcifull DE (1988) Reactivity of a monoclonal antibody to the amorphous inclusion protein of papaya ringspot virus type-W (PRSV-W). Phytopathology 78: 1537
4. Baker CA, Purcifull DE (1990) Reactivity of two monoclonal antibodies to the cylindrical inclusion protein of papaya ringspot virus type-W. Phytopathology 80: 1033
5. Barnett OW, Randles JW, Burrows PM (1987) Relationships among Australian and North American isolates of the bean yellow mosaic potyvirus subgroup. Phytopathology 77: 791–799
6. Boonekamp PM, Pomp H (1986) Problems concerning the production of monoclonal antibodies for plant diagnostic purposes. Acta Hortic 177: 103–109
7. Boye K, Jensen PE, Stummann BM, Henningsen KW (1990) Nucleotide sequence of cDNA encoding the BYMV coat-protein gene. Nucleic Acids Res 18:4926
8. Culver JN, Sherwood JL (1988) Detection of peanut stripe virus in peanut seed by an indirect enzyme-linked assay using a monoclonal antibody. Phytopathology 72: 676–679
9. Dougherty WG, Allison RF, Parks TD, Johnston RE, Feild MJ, Armstrong FB (1985) Nucleotide sequence at the 3'-terminus of the pepper mottle virus genomic RNA: evidence for an alternative mode of capsid protein gene organization. Virology 146: 282–291
10. Dougherty WG, Parks TD (1991) Post-translational processing of the tobacco etch virus 49 small nuclear inclusion polyprotein: identification of an internal cleavage site and delimitation of VPg and proteinase domains. Virology 183: 449–456
11. Dougherty WG, Willis L, Johnston RE (1985) Topographic analysis of tobacco etch virus capsid protein epitopes. Virology 144: 66–72

12. Fernandez-Northcote EN, Gugerli P (1985) Reaction of a broad spectrum of PVY isolates to monoclonal antibodies in ELISA. Phytopathology 75: 1353
13. Fortass M, Bos L, Goldbach RW (1991) Identification of potyvirus isolates from faba bean (*Vicia faba* L.), and the relationships between bean yellow mosaic virus and clover yellow vein virus. Arch Virol 118: 87–100
14. Geysen HM, Mason TJ, Rodda SJ (1988) Cognitive features of continuous antigenic determinants. J Mol Recogn 1: 23–41
15. Geysen HM, Rodda SJ, Mason TJ, Tribbick G, Schoofs PG (1987) Strategies for epitope analysis using peptide synthesis. J Immunol Methods 102: 259–274
16. Gugerli P, Fries P (1983) Characterization of monoclonal antibodies to potato virus Y and their use for virus detection. J Gen Virol 64: 2471–2477
17. Hammond J (1992) Potyvirus serology, sequences and biology. In: Barnett OW (ed) Potyvirus taxonomy. Springer, Wien New York, pp 123–138 (Arch Virol [Suppl] 5)
18. Hammond J, Hammond RW (1989) Molecular cloning, sequencing and expression in *Escherichia coli* of the bean yellow mosaic virus coat-protein gene. J Gen Virol 70: 1961–1974
19. Hammond J, Jordan RL, Kamo KK (1990) Use of chimeric coat-protein constructs and deletion mutants to examine potyvirus structure and coat-protein-mediated resistance. Phytopathology 80: 1018
20. Hammond J, Jordan RL, Larsen RC, Moyer JW (1992) Serological relationships between three filamentous viruses of sweet potato examined using polyclonal antisera and monoclonal antibodies. Phytopathology 82: 713–717
21. Hammond J, Lawson RH, Hsu HT (1985) Use of a monoclonal antibody reactive with several potyviruses for detection and identification in combination with virus-specific antisera. Phytopathology 75: 1353
22. Hampton RO, Shukla, DD, Jordan RL (1992) White lupin mosaic virus: comparative host range, serology and coat-protein peptide profiles. Phytopathology 82: 566–571
23. Hill EK, Hill JH, Durand DP (1984) Production of monoclonal antibodies to viruses in the potyvirus group: use in radioimmunoassay. J Gen Virol 65: 525–532
24. Himmler G, Brix U, Steinkellner H, Laimer M, Mattanovich D, Katinger HWD (1988) Early screening for anti-plum pox monoclonal antibodies with different specificities by means of gold-labelled immunosorbent electron microscopy. J Virol Methods 22: 351–358
25. Hsu HT, Franssen JM, Van der Hulst CTC, Derks AFLM, Lawson RH (1988) Factors affecting selection of epitope specificity of monoclonal antibodies to tulip breaking potyvirus. Phytopathology 78: 1337–1340
26. Jones FE, Hill JH, Durand DP (1988) Detection and differentiation of maize dwarf mosaic virus, strains of A and B, by use of different class immunoglobulins in a double-antibody sandwich enzyme-linked immunosorbent assay. Phytopathology 78: 1118–1124
27. Jordan RL (1989) Mapping of potyvirus-specific and group-common antigenic determinants with monoclonal antibodies by Western-blot analysis and coat-protein amino acid sequence comparisons. Phytopathology 79: 1157
28. Jordan RL (1992) Mapping of bean yellow mosaic virus-specific and potyvirus group-common coat-protein antigenic determinants defined by monoclonal antibodies. Phytopathology (submitted)
29. Jordan RL, Hammond, J (1986) Analysis of antigenic specificity of monoclonal antibodies to several potyviruses. Phytopathology 76: 1091
30. Jordan RL, Hammond J (1991) Comparison and differentiation of potyvirus isolates and identification of strain-, virus-, subgroup-specific and potyvirus group-common epitopes using monoclonal antibodies. J Gen Virol 72: 25–36
31. Jordan RL, Thornbury DW, Pirone TP (1990) Production and initial characterization of monoclonal antibodies to helper components of potato virus Y and tobacco vein mottling virus. Phytopathology 80: 1018

32. Langeveld SA, Dore JM, Memlelink J, Derks AFLM, van der Vlugt CIM, Asjes CJ, Bol JF (1991) Identification of potyviruses using the polymerase chain reaction with degenerate primers. J Gen Virol 72: 1531–1541

33. Lenstra JA, Kusters JG, Van der Zeijst BAM (1990) Mapping of viral epitopes with prokaryotic expression products. Arch Virol 110: 1–24

34. Li RH, Zettler FW, Elliot MS, Petersen MA, Still PE, Baker CA, Mink GI (1991) A strain of peanut mottle virus seedborne in bambarra groundnut. Plant Dis 74: 130–133

35. Miller AW, Surgeoner R, Mills PR (1988) Production and characterization of monoclonal antibodies to bean common mosaic virus. Phytopathology 78: 1585

36. Scott SW, McLaughlin MR, Ainsworth AJ (1989) Monoclonal antibodies produced to bean yellow mosaic virus, clover yellow vein virus, and pea mosaic virus which cross-react among the three viruses. Arch Virol 108: 161–167

37. Sherwood JL, Sanborn MR, Keyser GC (1987) Production of monoclonal antibodies to peanut mottle virus and their use in enzyme-linked immunosorbent assay and dot-blot immunobinding assay. Phytopathology 77: 1158–1161

38. Shukla DD, Ford RE, Tosic M, Jilka J, Ward CW (1989) Possible members of the potyvirus group transmitted by mites or whiteflies share epitopes with aphid-transmitted definitive members of the group. Arch Virol 105: 143–151

39. Shukla DD, Frenkel MJ, Ward CW (1991) Structure and function of the potyvirus genome with special reference to the coat-protein coding region. Can J Plant Pathol 13: 178–191

40. Shukla DD, Jikla J, Tosic M, Ford RE (1989) A novel approach to the serology of potyviruses involving affinity-purified polyclonal antibodies directed towards virus-specific N-termini of coat-proteins. J Gen Virol 70: 13–22

41. Shukla DD, McKern NM, Gough KH, Tracy SL, Letho SG (1988) Differentiation of potyviruses and their strains by high performance liquid chromatographic peptide profiling of coat-proteins. J Gen Virol 69: 493–502

42. Shukla DD, Strike PM, Tracy SL, Gough KH, Ward CW (1988) The N and C termini of the coat-proteins of potyviruses are surface-located and the N-terminus contains the major virus-specific epitopes. J Gen Virol 69: 1497–1508

43. Shukla DD, Tribbick G, Mason TJ, Hewish DR, Geysen HM, Ward CW (1989) Localization of virus-specific epitopes of plant potyviruses by systematic immunochemical analysis of overlapping peptide fragments. Proc Natl Acad Sci USA 86: 8192–8196

44. Shukla DD, Ward CW (1989) Structure of potyvirus coat-proteins and applications in the taxonomy of the potyvirus group. Adv Virus Res 36: 273–314

45. Shukla DD, Ward CW (1989) Identification and classification of potyviruses on the basis of coat-protein sequence data and serology. Arch Virol 106: 171–200

46. Slade DE, Johnston RE, Dougherty WG (1989) Generation and characterization of monoclonal antibodies reactive with the 49 proteinase of tobacco etch virus. Virology 173: 499–508

47. Takahashi T, Uyeda I, Ohshima K, Shikata E (1990) Nucleotide sequence of the capsid protein gene of bean yellow mosaic virus chlorotic spot strain. J Fac Agric Hokkaido Univ 64: 152–163

48. Turpen T (1989) Molecular cloning of a potato virus Y genome: nucleotide sequence homology in non-coding regions of potyviruses. J Gen Virol 70: 1951–1960

49. Uyeda I, Takahashi T, Shikata E (1991) Relatedness of nucleotide sequence of the 3′ terminal region of clover yellow vein potyvirus to bean yellow mosaic potyvirus RNA. Intervirology 32: 234–245

50. Vance VB, Jordan RL, Edwardson JR, Christie RG, Purcifull DE, Turpen T, Falk B (1992) Evidence that pepper mottle virus and potato virus Y are distinct viruses: analysis of the coat-protein and 3′ untranslated sequences of a California isolate of pepper mottle virus. In: Barnett OW (ed) Potyvirus taxonomy. Springer, Wien New York, pp 337–345 (Arch Virol [Suppl] 5)

51. Wang W, Mink GI, Silbernagel MJ (1985) A broad spectrum monoclonal antibody prepared against bean common mosaic virus. Phytopathology 75: 1352
52. Ward CW, Shukla DD (1991) Taxonomy of potyviruses: current problems and some solutions. Intervirology 32: 269–296
53. Wisler GC, Baker CA, Purcifull DE, Hiebert E (1989) Partial characterization of monoclonal antibodies to zucchini yellow mosaic virus (ZYMV) and watermelon mosaic virus (WMV-2). Phytopathology 79: 1213
54. Yao SK, Cai SH, Jia SR, Hsu HT (1985) Specificity of monoclonal antibodies to strains of potato virus Y. Phytopathology 75: 1356

Author's address: R. Jordan, USDA-ARS, BARC-West, FNCL, Building 004, Room 208, 10300 Baltimore Ave., Beltsville, MD 20705-2350, U.S.A.

Arch Virol (1992) [Suppl 5]: 97–122

Serological relationships involving potyviral nonstructural proteins

D. E. Purcifull and **E. Hiebert**

Department of Plant Pathology, Institute of Food and Agricultural Sciences,
University of Florida, Gainesville, Florida, U.S.A.

Summary. This report represents a compilation of many of the publications on antigenic properties of potyviral-specified nonstructural proteins. Polyclonal antisera have been prepared for use in characterization of six nonstructural proteins. These include antisera to the cylindrical inclusion proteins of at least 28 potyviruses, to small nuclear inclusion protein (protease) of four potyviruses, to large nuclear inclusion protein (putative replicase) of three viruses, helper component-protease or amorphous inclusion protein of at least four viruses, to the P1 protein (located at the N-terminus of the polyprotein) of one virus, and to the P3 protein (located between helper component protease and cylindrical inclusion protein) of one virus. Monoclonal antibodies also have been prepared to several of these nonstructural proteins. The evidence thus far indicates that cylindrical inclusions of different potyviruses have both conserved and unique epitopes. Nuclear inclusion proteins and amorphous inclusion proteins also may have conserved and unique epitopes. Antigenic relationships of potyviral nonstructural proteins have potential for the identification and classification of potyviruses.

Introduction

It has been known for many years that potyviruses induce inclusion bodies in their hosts [53]. In the past 25 years, considerable evidence has been obtained that virus-specified nonstructural proteins constitute major components of certain potyviral-induced cytoplasmic and nuclear inclusions [26, 27, 29, 30, 45–48, 54, 81]. This report deals primarily with the potyviral nonstructural proteins whose antigenic relationships have been studied to date: cylindrical inclusion protein, helper component-protease, and the small and large nuclear inclusion proteins. Most of the literature is summarized by presenting results of tests on antigenic relationships in tabular form.

Antigenic relationships of potyviral nonstructural proteins

Cylindrical inclusion proteins

All potyviruses induce cytoplasmic, cylindrical (pinwheel) inclusions in their hosts [25, 32–34, 61, 67]. The inclusion protein subunits have molecular weights ranging from 65–75 kDa [30, 46] and are encoded by the viral genome [29, 30]. There is significant sequence homology not only between the cylindrical inclusions of different potyviruses, but also between cylindrical inclusions and the nonstructural proteins induced by a range of other positive-stranded RNA viruses [28, 36, 39, 57]. The cylindrical inclusion proteins are helicase-like [57], and ATPase activity has recently been associated with plum pox cylindrical inclusion protein [58].

Cylindrical inclusions have been purified for immunological and other studies by various methods [15, 41, 48, 75]. Usually the inclusions are partially purified by chemical treatment of sap followed by centrifugation. The inclusions may then be dissociated into protein subunits by treatment with sodium dodecyl sulfate and the subunits purified further by preparative gel electrophoresis. This facilitates separation of the inclusion proteins from viral capsid proteins, other nonstructural viral proteins, and host proteins. In spite of these precautions, antisera produced to cylindrical inclusion proteins may contain antibodies to contaminating proteins. For example, antisera to wheat streak mosaic cylindrical inclusion protein also contained some antibodies reactive with a 66 kDa cell wall protein and to the large subunit of ribulose bisphosphate carboxylase [15]. Preparations of proteins from noninoculated plants have been used to remove antibodies to the host proteins by absorption [15]. Soybean mosaic virus capsid protein apparently can form dimers which co-migrate with cylindrical inclusion protein in polyacrylamide gel electrophoresis (E. Hiebert and D. Purcifull, unpubl.).

The antigenic nature and relationships of extracted cylindrical inclusion proteins were first studied in a series of papers in the early 1970's. All of these studies involved the use of double radial immunodiffusion tests in the presence of sodium dodecyl sulfate, which dissociated the inclusions into diffusible components which retained antigenic reactivity [6, 47, 69, 79–82]. Hiebert et al. [47] were the first to purify cylindrical inclusions for immunological studies. They reported that the purified inclusions were antigenically distinct from capsid protein and from host protein, and that the cylindrical inclusions induced by two different potyviruses (potato Y and tobacco etch) were antigenically distinct. In a subsequent study, some antigenic cross-reactivities were detected among the cylindrical inclusions of five potyviruses (bidens mottle, pepper mottle, potato Y, tobacco etch, and turnip mosaic), but the cylindrical inclusions of each virus had unique antigenic determinants and some antisera were specific for the cylindrical inclusion protein used for immunization [81] (Table 1). Reactions of partial identity (spur formation) were observed between related but antigenically different inclusions. Cross-reactive antibodies were removable by intragel cross-absorption. The propagative host did not influence antigenic specificity of the cylindrical inclusions. McDonald and Hiebert [69] obtained evidence that the cylindrical inclusion proteins of three turnip mosaic virus strains were antigenically more conserved than their capsid proteins. A strain which showed reactions of partial identity with respect to capsid-protein antisera showed reactions of identity with cylindrical inclusion antisera. Furthermore, tobacco etch virus cylindrical inclusions were found to be antigenically unrelated to virus-induced nuclear inclusions, another type of virus-specified nonstructural protein [54].

Table 1. Polyclonal antisera prepared to potyviral cylindrical inclusion protein (CIP) and examples of their cross reactivities with cylindrical inclusions induced by other potyviruses

Antiserum to CIP of	Cross reactivity with CIP of[a]		Serological test[b]	Ref.[c]
Agropyron mosaic	+	HoMV	gold label in situ	59
	–	MDMV-B	gold label in situ	59
	–	WSMV	gold label in situ	59
Bean yellow mosaic	–	ArjMV	IP	44
	(+)	AV1	WB	H
	–	ClYVV	ID	74
	–	ClYVV	ID	19
	+	ClYVV	indirect ELISA	19
	+	ClYVV	WB	H
	+	IMMV	WB	H
	+	ISMV	WB	H
	(+)	MDMV-A	WB	H
	+	MDMV-O	WB	H
	+	PSbMV	WB	H
	+	PepMoV	WB	H
	(+)	PkMV	WB	H
	+	PVA	WB	H
	+	PVY	WB	H
	+	SPFMV	WB	H
	(+)	SCMV-H	WB	H
	+	TEV	WB	H
	+	TVMV	WB	H
	+	TCBV	WB	H
	+	TuMV	WB	H
Bidens mottle	–	PRSV-W	ID	108
	+	PepMoV	ID	81, 108
	–	PVY	ID	108
	+	PVY	ID	81

Table 1 (continued)

Antiserum to CIP of	Cross reactivity with CIP of[a]		Serological test[b]	Ref.[c]
Bidens mottle	−	TEV	ID	81, 108
	−	TuMV	ID	81
	−	WMV-2	ID	108
Blackeye cowpea mosaic		NT	ID	62, 63
Clover yellow vein	(+)	AV1	WB	H
	+	BYMV	ID	19
	+	BYMV	indirect ELISA	19
	+	BYMV	WB	H
	+	IMMV	WB	H
	+	ISMV	WB	H
	(+)	MDMV-A	WB	H
	(+)	MDMV-O	WB	H
	+	PSbMV	WB	H
	+	PepMoV	WB	H
	(+)	PkMV	WB	H
	+	PVA	WB	H
	+	PVY	WB	H
	+	SPFMV	WB	H
	(+)	SCMV-H	WB	H
	+	TEV	WB	H
	+	TVMV	WB	H
	+	TCBV	WB	H
	+	TuMV	WB	H
Dasheen mosaic	+	ArjMV	IP	44
		NT	ID	107
Hordeum mosaic	−	WSMV	gold label in situ	59
Maize dwarf mosaic A	(+)	AV1	WB	H
	(+)	BYMV	WB	H
	(+)	ClYVV	WB	H
	(+)	IMMV	WB	H
	(+)	ISMV	WB	H
	+	MDMV-B	WB	50
	−	MDMV-O	WB	50
	(+)	MDMV-O	WB	H
	(+)	PSbMV	WB	H
	(+)	PepMoV	WB	H
	(+)	PkMV	WB	H
	(+)	PVA	WB	H
	(+)	PVY	WB	H
	(+)	SPFMV	WB	H
	+	SCMV-H	WB	H
	(+)	TEV	WB	H

Table 1 (continued)

Antiserum to CIP of	Cross reactivity with CIP of[a]		Serological test[b]	Ref.[c]
	(+)	TVMV	WB	H
	(+)	TCBV	WB	H
	(+)	TuMV	WB	H
Maize dwarf mosaic B	+	MDMV-A	WB	50
	−	MDMV-O	WB	50
	−	WSMV	gold label in situ	59
Narcissus yellow stripe	−	A1MV	dot-blot ELISA	70
	+	BYMV	dot-blot ELISA	70
	+	IMMV	dot-blot ELISA	70
	−	PVV	dot-blot ELISA	70
	−	PVY	dot-blot ELISA	70
	+	TEV	ELISA	BDH
	−	TCBV	dot-blot ELISA	70
	−	TuMV	dot-blot ELISA	70
Papaya ringspot P	+	PRSV-W	ELISA	103
	+	PRSV-W	ID	103
	+	PRSV-T	ID	83
	−	BCMV	ID	103
	−	BlCMV	ID	103
	−	BYMV	ID	103
	−	CABMV	ID	103
	−	LMV	ID	103
	−	PVY	ID	103
	−	TEV	ID	103
	−	TuMV	ID	103
	−	WMV-2	ID	103
	−	WMV-2	ELISA	103
Papaya ringspot W	(+)	ArjMV	IP	44
	−	WMV-M	ID	8
	+	WMV-M	ID	2
	−	WMV-2	ID	2, 8, 103
	−	WMV-2	ELISA	103
	+	WMV-2	ELISA	2
	+	WMV-2	electroblot ELISA	93
	+	ZYFV	IF	3
	+	ZYFV	ELISA	2
	+	ZYFV	ID	3
	−	ZYMV	ID	2
	+	ZYMV	ELISA	2
	+	ZYMV	electroblot ELISA	93
Papaya ringspot T	+	PRSV-P	ID	83, 84
	+	PRSV-W	ID	83, 84

Table 1 (continued)

Antiserum to CIP of	Cross reactivity with CIP of[a]		Serological test[b]	Ref.[c]
Papaya ringspot T	+	WMV-M	ID	83, 84
	–	WMV-2	ID	84
	–	ZYFV	ID	84
	–	ZYMV	ID	84
Passionfruit woodiness		NT	gold label in situ	64
Peanut mottle	–	BYMV	ID	99
	–	NT	IF	100
	–	PStV	ID	99
	–	TEV	ID	99
Pea seed-borne mosaic		NT	direct ELISA	1
		NT	indirect ELISA	1
Pepper mottle	(+)	ArjMV	IP	44
	+	BiMoV	ID	81
	+	PVY	ID	81
	–	PVY	ID	6
	+	TEV	ID	81
	–	TuMV	ID	81
Plum pox		NT	WB	68
Potato Y	+	BiMoV	ID	81
	+	PepMoV	ID	81
	+	TEV	ID	81
	–	TEV	ID	47
	+	TuMV	ID	81
Soybean mosaic	+	ArjMV	IP	44
Sugarcane mosaic (see Maize dwarf mosaic)				
Sweet potato feathery mottle	+	AV1	WB	H
	+	BYMV	WB	H
	+	ClYVV	WB	H
	(+)	IMMV	WB	H
	+	ISMV	WB	H
	(+)	MDMV-A	WB	H
	(+)	MDMV-O	WB	H
	+	PSbMV	WB	H
	+	PepMoV	WB	H
	+	PkMV	WB	H
	+	PVA	WB	H
	+	PVY	WB	H

Table 1 (continued)

Antiserum to CIP of	Cross reactivity with CIP of[a]		Serological test[b]	Ref.[c]
	(+)	SCMV-H	WB	H
	+	TEV	WB	H
	+	TVMV	WB	H
	+	TCBV	WB	H
	+	TuMV	WB	H
Tobacco etch	–	ArjMV	IP	44
	–	BiMoV	ID	6, 81
	–	PepMoV	ID	81
	+	PepMoV	ID	6
	+	PepMoV	IP	29
	+	PPV	WB	68
	–	PVY	ID	47, 81
	+	PVY	ID	6
	–	TuMV	ID	81
Tobacco vein mottling		NT	IP	38, 42
		NT	immunostaining	66
		NT	gold label in situ	71
Turnip mosaic	–	ArjMV	IP	44
	+	AV1	WB	H
	+	BYMV	WB	H
	–	BiMoV	ID	81
	+	ClYVV	WB	H
	+	IMMV	WB	H
	+	ISMV	WB	H
	(+)	MDMV-A	WB	H
	+	MDMV-O	WB	H
	+	PSbMV	WB	H
	–	PepMoV	ID	81
	+	PepMoV	WB	H
	+	PkMV	WB	H
	+	PVA	WB	H
	–	PVY	ID	81
	+	PVY	WB	H
	+	SCMV-H	WB	H
	+	SPFMV	WB	H
	–	TEV	ID	81
	+	TEV	WB	H
	+	TVMV	WB	H
	+	TCBV	WB	H
Watermelon mosaic 1 (see Papaya ringspot-W)				
Watermelon mosaic 2	(+)	ArjMV	IP	44
	–	PRSV-W	ID	8

Table 1 (continued)

Antiserum to CIP of	Cross reactivity with CIP of[a]		Serological test[b]	Ref.[c]
Watermelon mosaic 2	+	PRSV-W	electroblot ELISA	93
	–	WMV-M	ID	8
	+	ZYMV	electroblot ELISA	93
Watermelon mosaic M	–	BCMV	ID	7
(Moroccan isolate)	+	DsMV	ID	7
	–	LMV	ID	7
	+	PRSV-W	ID	8
	+	PRSV-P	ID	7
	–	SbMV	ID	7
	+	WMV-2	ID	8
Wheat streak mosaic	–	AgMV	gold label in situ	59
	–	HoMV	gold label in situ	59
	–	MDMV-B	gold label in situ	15, 59
	+	TEV	ELISA	BDH
	–	WSSMV	gold label in situ	15
Zucchini yellow mosaic	+	PRSV-W	electroblot ELISA	93
	–	PRSV-W	ID	78
	–	WMV-2	ID	78
	+	WMV-2	electroblot ELISA	93

[a] + Positive cross-reactivity; – no cross reactivity in test indicated; (+) very weak cross-reactivity in Western blots or immunoprecipitation; *NT* positive reaction with cylindrical inclusion protein of homologous virus, but not tested against cylindrical inclusions of other viruses

[b] *ID* Immunodiffusion; *IF* immunofluorescence; *IP* immunoprecipitation; *WB* Western blot

[c] *H* J. Hammond, pers. comm.; *BDH* D. Baunoch et al., pers. comm.

Virus acronyms: ArjMV araujia mosaic virus; AlMV alstroemeria mosaic virus; AgMV agropyron mosaic virus; AV1 asparagus virus 1; BCMV bean common mosaic virus; BiMoV bidens mottle virus; BlCMV blackeye cowpea mosaic virus; BYMV bean yellow mosaic virus; CABMV cowpea aphid-borne mosaic virus; ClYVV clover yellow vein virus; DsMV dasheen mosaic virus; HoMV hordeum mosaic virus; IMMV iris mild mosaic virus; ISMV iris severe mosaic virus; LMV lettuce mosaic virus; MDMV-A maize dwarf mosaic virus A; MDMV-B maize dwarf mosaic virus B; MDMV-O maize dwarf mosaic virus O (johnsongrass mosaic virus); PepMoV pepper mottle virus; PkMV pokeweed mosaic virus; PPV plum pox virus; PRSV-P papaya ringspot virus, papaya; PRSV-W papaya ringspot virus, watermelon mosaic virus 1; PRSV-T papaya ringspot virus, Tigre; PSbMV pea seed-borne mosaic virus; PStV peanut stripe virus; PVA potato virus A; PVV potato virus V; PVY potato virus Y; SbMV soybean mosaic virus; SPFMV sweet potato feathery mottle virus; SCMV-H sugarcane mosaic virus strain H (sorghum mosaic virus); TCBV tulip chlorotic blotch virus; TEV tobacco etch virus; TuMV turnip mosaic virus; TVMV tobacco vein mottling virus; WSMV wheat streak mosaic virus; WSSMV wheat spindle streak mosaic virus; WMV-M watermelon mosaic virus, Morocco; WMV-2 watermelon mosaic virus 2; ZYFV zucchini yellow fleck virus; ZYMV zucchini yellow mosaic virus

Since these initial reports on the antigenic properties and relationships of cylindrical inclusions, there have been reports with many additional viruses and with a variety of serological techniques. Examples are listed in Table 1 and in the following brief discussions of selected comparisons.

Yeh and Gonsalves [103, 105] found that papaya ringspot virus P and W gave reactions of identity in immunodiffusion tests with either capsid or cylindrical inclusion antisera. They introduced the use of enzyme linked immunosorbent assays (ELISA) for detecting cylindrical inclusion protein, and found that cylindrical inclusions of papaya ringspot virus P and W were serologically indistinguishable by indirect ELISA. Direct ELISA (double antibody sandwich) was problematic due to nonspecific background reactions. Neither the papaya ringspot P nor W cylindrical inclusion antiserum reacted with extracts from plants infected with watermelon mosaic virus 2 in immunodiffusion or indirect ELISA. Using different antisera and electroblot ELISA methods, Suzuki et al. [93] found that cylindrical inclusion proteins of papaya ringspot virus W, watermelon mosaic virus 2, and zucchini yellow mosaic virus were antigenically related but different, based on serological differentiation indices. They found that cylindrical inclusions were more antigenically conserved than capsid proteins. Based on reactions of partial identity in immunodiffusion tests, Quiot et al. detected antigenic differences between the cylindrical inclusions of papaya ringspot virus W and T, whose capsid proteins also show reactions of partial identity [83]. Western blot procedures have recently been used to test cross-reactivity of cylindrical inclusion proteins. Jensen and Staudinger [50] reported that groupings of sugarcane mosaic and maize dwarf mosaic virus strains based on cylindrical inclusion serology were correlated with groupings based on capsid-protein serology. Potyviral cylindrical inclusions of five viruses cross reacted with a range of potyviral cylindrical inclusions, based on Western blots [40; Hammond, pers. comm.] (Table 1).

An extensive antigenic characterization of cylindrical inclusion protein has been conducted by Baunoch et al. [12; and pers. comm.]. They prepared synthetic peptides to the cylindrical inclusion protein sequence of tobacco etch virus and tested the peptides against tobacco etch virus cylindrical inclusion protein antiserum in ELISA tests. At least 166 epitopes were detected, with the majority in the N-terminal portion of the protein. Antiserum to cylindrical inclusion proteins of either wheat streak mosaic virus or narcissus yellow stripe virus also reacted with the synthetic peptides. Western blot tests also showed that tobacco etch virus cylindrical inclusion protein has epitopes in common with the 126 kDa protein of tobacco mosaic virus and with the NS3 protein of flaviviruses. The finding that other viruses encode proteins that are antigenically related to potyviral cylindrical inclusion proteins is a factor to consider in assessing the taxonomic value of the antigenic relationships among potyviral nonstructural proteins.

Immunocytochemical methods have been applied for determining relationships of virus-specified antigens and for intracellular localization of several viral nonstructural proteins [9, 10, 11, 19, 22, 26, 48, 55, 64, 66, 71, 85, 87, 88]. Such techniques have also been useful as one means of assessing cross-reactivity of cylindrical inclusions induced by different potyviruses [3, 59] (Table 1).

Monoclonal antibodies have been employed for distinguishing potyviral capsid proteins [51, 56]. However, there are few reports on the preparation of monoclonal antibodies to cylindrical inclusion proteins. A monoclonal antibody prepared to papaya ringspot virus W cylindrical inclusion protein reacted with fewer isolates of this virus than polyclonal antiserum in Western blots [2, 5]. Although no cross-reactivity tests were reported, a monoclonal antibody also has been prepared to watermelon mosaic virus 2 cylindrical inclusion protein [92].

Nuclear inclusion proteins

Nuclear inclusions were reported by Kassanis [53] in cells infected with tobacco etch virus, and these inclusions have been the subject of many subsequent cytological studies [25, 32, 34]. The isolation and characterization of these inclusions has helped in understanding the roles their components play in virus replication. Knuhtsen et al. [54] isolated the nuclear inclusions and prepared antisera to them. The inclusions were antigenically distinct from capsid protein, cylindrical inclusion protein, and host protein. The inclusions contained subunits with molecular weights of about 49 kDa and 54 kDa in equimolar amounts. The subunits have now been determined to be 49 kDa and 58 kDa by sequence analysis [30]. In addition, minor components with molecular weights of approximately 95,600 and 101,400 are reported [54]. Dougherty and Hiebert [29] prepared antisera to the 49 kDa and 58 kDa proteins. The antisera were used in immunoprecipitation studies of the in vitro translation products of tobacco etch virus RNA to confirm that the proteins were virus encoded and serologically distinct from each other. Subsequent studies have implicated the small (49 kDa) nuclear inclusion protein (NIa) as a protease involved in proteolytic processing of the viral polyprotein [16, 18, 24, 31, 35, 43, 76]. Murphy et al. [72] suggested that either the VPg of tobacco etch virus is the 49 kDa NIa protein or the 24 kDa VPg represents the N-terminal portion of NIa. The C-terminal portion of the NIa protein represents the protease domain [31]. The large (58 kDa) subunit (NIb) is a putative replicase [30].

Polyclonal antisera prepared either to NIa or NIb of tobacco etch virus cross reacted with several other potyviruses (Table 2). Immunoprecipitation of in vitro translation products accounts for a significant portion of the evidence that tobacco etch virus nuclear inclusion protein antisera cross react with corresponding proteins induced by other potyviruses. In a comparison of antisera to NIa and NIb proteins of bean yellow mosaic virus, clover yellow vein virus, and tobacco etch virus by immunodiffusion, all antisera cross reacted with the corresponding proteins of heterologous viruses with the exception of tobacco etch virus NIa antiserum [23] (Table 2). Chang [20], in a unique study, compared the antigenic specificities of four proteins (capsid, cylindrical inclusion, and nuclear inclusion proteins NIa and NIb) for distinction of several isolates of bean yellow mosaic virus and clover yellow vein virus by sodium dodecyl sulfate immunodiffusion tests. The antisera to the large nuclear inclusion proteins (NIb) were the most suitable for identifying and distinguishing the two viruses. All 20 isolates of bean yellow mosaic virus reacted identically and were distinguished from the five clover yellow vein virus isolates when tested against antiserum to the bean yellow mosaic virus NIb protein. Antiserum to clover yellow vein virus NIb protein reacted identically with the five clover yellow vein virus isolates, and all isolates of bean yellow mosaic virus were distinguishable. Antisera to capsid protein, cylindrical inclusion protein, and small nuclear inclusion protein were more isolate specific.

Table 2. Reactivities of antisera prepared to potyviral nuclear inclusion proteins

Virus and immunogen used for antiserum preparation	Reactivity with antigens of [a]		Serological test [b]	Ref. [c]
Bean yellow mosaic NIa	+	BYMV	IVTP	IP	24
(49 kDa protein)	+	BYMV	NIa	ID	23
	−	BYMV	NIb	ID	23
	−	BYMV	CIP	ID	23
	−	ClYVV	CP	ID	23
	+	BYMV	sap	ID	20, 23
	+	BYMV	ppNI	WB	23
	+	BYMV	tissue	IF	21, 22
	+	ClYVV	sap	ID	23
	−	TEV	sap	ID	23
Bean yellow mosaic NIb	+	BYMV	IVTP	IP	23, 24
(54 kDa protein)	−	BYMV	NIa	ID	23
	+	BYMV	NIb	ID	23
	−	BYMV	CIP	ID	23
	−	BYMV	CP	ID	23
	+	BYMV	sap	ID	20, 23
	+	BYMV	ppNI	WB	23
	+	BYMV	tissue	IF	21, 22
	+	BYMV	sap	ELISA	23
	+	ClYVV	sap	ELISA	23
	+	ClYVV	sap	ID	23
	+	CABMV	sap	ELISA	23
	+	PepMoV	IVTP	IP	23
	+	PeMoV	sap	ELISA	23
	+	TEV	sap	ID	23

Table 2 (continued)

Virus and immunogen used for antiserum preparation	Reactivity with antigens of[a]			Serological test[b]	Ref.[c]
Bean yellow mosaic NIb (54 kDa protein)	+	TEV	sap	ELISA	23
	+	WMV-2	sap	ELISA	23
Clover yellow vein NIa (49 kDa protein)	+	BYMV	sap	ID	23
	+	ClYVV	NIa	ID	23
	−	ClYVV	NIb	ID	23
	−	ClYVV	CIP	ID	23
	+	ClYVV	ppNI	WB	23
	+	ClYVV	tissue	IF	23
	−	TEV	sap	ID	23
Clover yellow vein NIb (60 kDa protein)	+	BYMV	sap	ID	23
	+	BYMV	IVTP	IP	23
	-	ClYVV	NIa	ID	23
	+	ClYVV	NIb	ID	23
	−	ClYVV	CIP	ID	23
	−	ClYVV	CP	ID	23
	+	ClYVV	tissue	IF	23
	+	ClYVV	ppNI	WB	23
	+	TEV	sap	ID	23
	+	PepMoV	IVTP	IP	23
Plum pox NIa (49 kDa protein)	+	PPV	NIa	immunoblot	68
Tobacco etch NIa (49 kDa protein)	+	ArjMV	IVTP	IP	44
	−	BYMV	sap	ID	23
	+	BYMV	IVTP	IP	74
	+	ClYVV	IVTP	IP	74
	+	MDMV-A	IVTP	IP	14
	+	MDMV-B	IVTP	IP	14
	+	PPV	IVTP	IP	68
	+	PeMoV	IVTP	IP	101
	+	PepMoV	IVTP	IP	29
	+	PVY	IVTP	IP	49
	+	PRSV-T	IVTP	IP	83
	+	TEV	IVTP	IP	29
	+	TEV	NIa	WB	91
	−	TEV	NIb	WB	91
	+	TEV	sap	ELISA	11
	+	TEV	tissue	gold label	9, 11
	+	TVMV	IVTP	IP	42
	+	WMV-2	IVTP	IP	EH
	−	WSMV	IVTP	IP	EH
	+	ZYMV	IVTP	IP	60

Table 2 (continued)

Virus and immunogen used for antiserum preparation	Reactivity with antigens of[a]			Serological test[b]	Ref.[c]
Tobacco etch NIb (58 kDa protein)	+	ArjMV	IVTP	IP	44
	+	BYMV	IVTP	IP	23, 74
	+	BYMV	sap	ID	23
	+	ClYVV	IVTP	IP	74
	+	MDMV-A	IVTP	IP	14
	+	MDMV-B	IVTP	IP	14
	+	PepMoV	IVTP	IP	23
	+	PeMoV	IVTP	IP	101
	+	PPV	IVTP	IP	68
	+	PVY	IVTP	IP	49
	+	PRSV-W	FP	IP	73
	+	PRSV-W	FP	WB	73
	+	PRSV-T	IVTP	IP	83
	+	TEV	IVTP	IP	29
	−	TEV	NIa	WB	91
	+	TEV	NIb	WB	91
	+	TEV	sap	ELISA	11
	+	TEV	tissue	gold label	9
	+	TVMV	IVTP	IP	42
	+	WMV-2	IVTP	IP	EH
	+	WSMV	IVTP	IP	EH
	+	ZYMV	IVTP	IP	60
Tobacco etch (NIa+NIb)	+	PeMoV	sap	ID	99
	+	PeMoV	tissue	IF	99
	+	TEV	sap	ID	6, 54
	+	TEV	ppNI	ID	54
	−	TEV	CP	ID	6, 54
	−	TEV	CI	ID	6

[a] + Positive reaction; − no reaction in the serological test indicated. Antigen preparation: *CIP* cylindrical inclusion protein; *CP* capsid protein; *FP* fusion products; *IVTP* in vitro translation products; *NIa* small (49kDa) nuclear inclusion subunit preparation; *NIb* large (54–60kDa) nuclear inclusion subunit preparation; *ppNI* partially purified preparation containing nuclear inclusions; *sap* plant extracts; *tissue* plant tissue used for cytological examination

[b] *ID* Immunodiffusion; *IF* immunofluorescence; *IP* immunoprecipitation; *WB* Western blot

[c] *EH* E. Hiebert, unpubl. data

Virus acronyms: ArjMV araujia mosaic virus; BYMV bean yellow mosaic virus; CABMV cowpea aphid-borne mosaic virus; ClYVV clover yellow vein virus; MDMV-A maize dwarf mosaic virus A; MDMV-B maize dwarf mosaic virus B; PeMoV peanut mottle virus; PPV plum pox virus; PepMoV pepper mottle virus; PVY potato virus Y; PRSV-W papaya ringspot virus, watermelon mosaic virus 1; PRSV-T papaya ringspot virus, Tigre; TEV tobacco etch virus; TVMV tobacco vein mottling virus; WMV-2 watermelon mosaic virus 2; WSMV wheat streak mosaic virus; ZYMV zucchini yellow mosaic virus

Monoclonal antibodies have been prepared to both the NIa (49kDa) and NIb (58kDa) proteins of tobacco etch virus [91]. Western blots confirmed the antigenic distinction of NIa and NIb. One of the monoclonal antibodies to the NIa protease inhibited self processing of this protein from a larger protein precursor [91].

Helper component-protease and amorphous inclusion protein

Aphid-borne potyviruses encode a protein (helper component) which is required for aphid transmission. Some potyviral infections induce the formation of cytoplasmic amorphous or irregular inclusion bodies (aggregates of the helper component protein) in their hosts [10,25,26,32,34,67]. The subunits of helper component or amorphous inclusion proteins are about 51–58 kDa. Immuno-precipitational analysis of in vitro translation products, Western blot tests, and cytological studies indicate that the helper component and amorphous inclusion proteins are antigenically closely related, if not identical [10,26,27]. Based on the work of Carrington et al. [17] with tobacco etch virus, the carboxyl terminal portion of helper component protein also functions as a protease. The precise biological relationship between the helper component-protease protein and amorphous inclusion protein is not clear [30]. Nevertheless, these two proteins are considered together for the purpose of this report, because they are encoded by the same potyviral gene and because of their close serological relationship.

The antigenic properties of helper component have been used in determining the nature of the protein and in distinguishing it from other potyviral-encoded proteins. Govier et al. [37] used antiserum to the helper component to show that it was antigenically unrelated to capsid or cylindrical inclusion protein. Hellmann et al. [42] provided evidence that tobacco vein mottling virus helper component was encoded by the viral genome, based on the analysis of immunoprecipitation results obtained with in vitro translation products. Potato virus Y and tobacco vein mottling viruses were found to induce antigenically distinct helper components. Antisera to 53 kDa TVMV-HC and to 58 kDa PVY-HC inactivate their homologous helper components [77, 94, 95], but treatment with heterologous antiserum had little effect on either helper component. Hiebert et al. [49] studied the immunoprecipitation of in vitro translation products of 17 potyviruses with antiserum to tobacco vein mottling virus and potato Y virus helper components. Antiserum to tobacco vein mottling virus helper component immunoprecipitated the products of all viruses tested, including non-aphid transmissible viruses, but the efficiency of immunoprecipitation clearly varied, depending on the virus. Antiserum to potato Y helper component protein was much more specific than the antiserum to tobacco vein mottling virus. The potato Y helper component antiserum reacted strongly with the translation products of potato virus Y, but cross reacted very weakly with the products of three viruses and no detectable immunoprecipitation was observed with products of the other 13 viruses.

The amorphous inclusion proteins induced by pepper mottle virus and papaya ringspot virus W were isolated and used for antiserum preparation [26]. The amorphous inclusion proteins from the two potyviruses were antigenically different in immunodiffusion and Western blot tests. However, the amorphous inclusion protein antiserum of papaya ringspot virus W was cross-reactive with tobacco vein mottling virus helper component in Western blots, and pepper mottle virus amorphous inclusion protein antiserum was cross-reactive with helper component of potato virus Y. Amorphous inclusion proteins have been localized immunochemically in cells by immunofluorescence and immuno-gold labelling techniques [3, 10, 26] (Table 3).

Table 3. Reactivities of antisera prepared to helper component-protease or amorphous inclusion protein[a]

Virus and immunogen used for antiserum preparation	Reactivity with antigens of[b]		Serological test[c]	Ref.
Papaya ringspot P (amorphous inclusion protein)	+ PRSV-P	IVTP	IP	102, 104
Papaya ringspot W (amorphous inclusion protein)	+ BiMoV	extract	ELISA	2
	+ CABMV	extract	ELISA	2
	+ ClYVV	extract	ELISA	2
	+ PeMoV	extract	ELISA	2
	− PepMoV	AIP	WB	27
	− PepMoV	extract	ID	26
	+ PRSV-P	extract	ELISA	2
	+ PRSV-P	extract	ID	2
	+ PRSV-T	extract	ID	2, 83
	− PRSV-T	extract	ID	84
	+ PRSV-W	extract	ID	26
	+ PRSV-W	extract	ID	26, 84
	+ PRSV-W	tissue	IF	3, 26
	+ PRSV-W	extract	ELISA	2
	+ PRSV-W	IVTP	IP	27, 104
	+ PRSV-W	AIP	WB	26, 27
	− PRSV-W	CIP	WB	26
	− PRSV-W	CP	WB	26
	+ PStV	extract	ELISA	2
	− PVY	HC	WB	27
	+ PVY	extract	WB	2
	+ PVY	extract	ELISA	2
	+ PVY	extract	ID	2
	+ SbMV	extract	ID	2
	+ TEV	extract	ID	2
	+ TVMV	extract	ID	2
	+ TVMV	HC	WB	27

Table 3 (continued)

Virus and immunogen used for antiserum preparation	Reactivity with antigens of[b]			Serological test[c]	Ref.
Papaya ringspot W (amorphous inclusion protein)	−	WMV-2	extract	ID	84
	+	WMV-2	extract	ID	2
	+	WMV-2	extract	ELISA	2
	+	WMV-2	AIP	electroblot ELISA	93
	+	WMV-M	extract	ID	2
	−	WMV-M	extract	ID	83, 84
	+	WMV-M	extract	ELISA	2
	−	WSMV	extract	ELISA	2
	−	ZYFV	extract	ID	84
	+	ZYFV	extract	ID	2, 3
	+	ZYFV	extract	ELISA	2
	+	ZYFV	tissue	IF	3
	+	ZYMV	extract	ID	2
	−	ZYMV	extract	ID	84
	+	ZYMV	extract	ELISA	2
	+	ZYMV	AIP	electroblot ELISA	93
Papaya ringspot W (monoclonal antibody F10E-A5 to amorphous inclusion protein)	+	PRSV-P	extract	ELISA	2, 4
	+	PRSV-T	extract	ELISA	2, 4
	+	PRSV-W	extract	ELISA	2, 4
	+	WMV-2	extract	ELISA	2
	+	WMV-M	extract	ELISA	2
	+	ZYFV	extract	ELISA	2
	+	ZYMV	extract	ELISA	2
Papaya ringspot W (monoclonal antibody F19K-G3 to amorphous inclusion protein)	+	PRSV-P	extract	ELISA	2
	+	PRSV-T	extract	ELISA	2
	+	PRSV-W	extract	ELISA	2
	−	WMV-2	extract	ELISA	2
	−	WMV-M	extract	ELISA	2
	+	ZYFV	extract	ELISA	2
	−	ZYMV	extract	ELISA	2
Pepper mottle (amorphous inclusion protein)	+	PepMoV	extract	ID	26
	+	PepMoV	tissue	IF	26
	+	PepMoV	IVTP	IP	27
	+	PepMoV	AIP	WB	27
	−	PRSV-W	AIP	WB	27
	−	PRSV-W	extract	ID	26
	+	PVY	HC	WB	27
	−	TVMV	HC	WB	27
Potato Y (helper)	−	ArjMV	IVTP	IP	49
	−	BYMV	IVTP	IP	49
	−	CABMV	IVTP	IP	49
	−	ClYVV	IVTP	IP	49
	(+)	DsMV	IVTP	IP	49
	−	PRSV-P	IVTP	IP	49

Table 3 (continued)

Virus and immunogen used for antiserum preparation	Reactivity with antigens of[b]			Serological test[c]	Ref.
	−	PRSV-W	IVTP	IP	49
	−	PepMoV	IVTP	IP	49
	+	PVY	IVTP	IP	49
	−	PVY	HC	ID	37
	+	PVY	HC	inhibition	37, 94
	+	PVY	HC	absorption chromatography	94
	+	PVY	extract	WB	10
	+	PVY	tissue	gold label	10
	−	SbMV	IVTP	IP	49
	(+)	TEV	IVTP	IP	49
	−	TuMV	IVTP	IP	49
	(+)	TVMV	IVTP	IP	49
	−	TVMV	HC	absorption chromatography	94
	+	WMV-2	IVTP	IP	49
	+	WMV-M	IVTP	IP	49
	+	WSMV	IVTP	IP	49
	+	ZYMV	IVTP	IP	49
Potato Y (polypeptide purified by SDS-PAGE)	−	MDMV-A	IVTP	IP	14
	−	MDMV-B	IVTP	IP	14
	+	PVC	extract	WB	96
	+	PVY	extract	WB	96
	+	PVY	HC	absorption chromatography	95
	+	PVY	IVTP	IP	95
	−	TVMV	IVTP	IP	95
	−	TVMV	HC	absorption chromatography	95
Tobacco etch (helper component protease fragment from *E. coli*)	+	TEV	extract	immunoblot	18
Tobacco vein mottling (helper)	+	ArjMV	IVTP	IP	49
	+	BYMV	IVTP	IP	49
	+	CABMV	IVTP	IP	49
	+	ClYVV	IVTP	IP	49
	+	DsMV	IVTP	IP	49
	+	PRSV-P	IVTP	IP	49
	+	PRSV-W	IVTP	IP	49
	+	PepMoV	IVTP	IP	49
	+	PeMoV	IVTP	IP	101
	+	PVY	IVTP	IP	49
	−	PVY	HC	absorption chromatography	94
	+	SbMV	IVTP	IP	49

Table 3 (continued)

Virus and immunogen used for antiserum preparation	Reactivity with antigens of[b]		Serological test[c]	Ref.
Tobacco vein mottling (helper)	+	TEV IVTP	IP	49
	+	TuMV IVTP	IP	49
	+	TVMV IVTP	IP	49
	+	TVMV HC	absorption chromatography	94
	+	WMV-2 IVTP	IP	49
	+	WMV-M IVTP	IP	49
	+	WSMV IVTP	IP	49
	+	ZYMV IVTP	IP	49
Tobacco vein mottling (polypeptide purified by SDS-PAGE)	–	MDMV-A IVTP	IP	14
	–	MDMV-B IVTP	IP	14
	–	PVY IVTP	IP	95
	–	PVY HC	absorption chromatography	95
	+	TVMV HC	absorption chromatography	95
	+	TVMV IVTP	IP	42, 95
	+	TVMV extract	WB	13
	+	TVMV PPE	ELISA	65
	+	TVMV PP	immunostaining	66
Watermelon mosaic 2 (monoclonal antibody to amorphous inclusion protein)	+	WMV-2 extract	electroblot ELISA	92

[a] Antisera are polyclonal unless indicated otherwise

[b] + Positive reaction; – no reaction in the serological test indicated; (+) very weak reaction. Antigen preparation: *AIP* amorphous inclusion protein; *CIP* cylindrical inclusion protein; *CP* capsid protein; *extract* sap or other extract from infected plant tissue; *HC* helper component preparation; *IVTP* in vitro translation product; *PP* protoplasts; *PPE* protoplast extract; *tissue* plant tissue used for cytological examinations

[c] *ID* Immunodiffusion; *IF* immunofluorescence; *IP* immunoprecipitation; *WB* Western blot

Virus acronyms: ArjMV araujia mosaic virus; BiMoV bidens mottle virus; BYMV bean yellow mosaic virus; CABMV cowpea aphid-borne mosaic virus; ClYVV clover yellow vein virus; DsMV dasheen mosaic virus; MDMV-A maize dwarf mosaic virus strain A; MDMV-B maize dwarf mosaic virus strain B; PepMoV pepper mottle virus; PeMoV peanut mottle virus; PRSV-P papaya ringspot virus, papaya; PRSV-W papaya ringspot virus, watermelon mosaic virus 1; PRSV-T papaya ringspot virus, Tigre; PStV peanut stripe virus; PVC potato virus C; PVY potato virus Y; SbMV soybean mosaic virus; TEV tobacco etch virus; TuMV turnip mosaic virus; TVMV tobacco vein mottling virus; WMV-2 watermelon mosaic virus-2; WMV-M watermelon mosaic virus, Morocco; WSMV wheat streak mosaic virus; ZYFV zucchini yellow fleck virus; ZYMV zucchini yellow mosaic virus

Monoclonal antibodies have been prepared to amorphous inclusion protein of watermelon mosaic virus 2 [92] and papaya ringspot virus W [2, 4]. Differences were reported in the cross-reactivities of the two monoclonal antibodies to amorphous inclusion protein of papaya ringspot virus W, and neither monoclonal antibody was as cross-reactive as polyclonal antiserum (Table 3).

Some additional reports on the antigenic relationships of helper component protease or amorphous inclusion proteins are listed in Table 3. Helper component proteins have been detected serologically in transformed plants and bacteria [13, 52].

Other nonstructural proteins

Other potential nonstructural proteins for serological studies include the product encoded by the gene at the 5' end of potyviral RNA. The protein encoded by this portion of the genome is the P1 protein [90] and the estimated size of this protein ranges from 31 to 65 kDa [45]. The C-terminal portion of the P1 protein apparently represents a proteinase [97]. Yeh et al. [106] recently isolated a 112 kDa precursor protein from the amorphous inclusion fraction of plants infected with papaya ringspot virus. Monoclonal antibodies were prepared to the 112 kDa protein. One reacted with the 112 kDa protein and with the 51 kDa amorphous inclusion protein. The other monoclonal antibody reacted with 112 kDa protein, and with proteins of 70 kDa and 64 kDa, but not with the 51 kDa amorphous inclusion protein (second potyviral gene product). Rodriguez-Cerezo and Shaw [86] cloned the gene corresponding to the P1 protein of tobacco vein mottling virus, expressed it in *E. coli*, isolated a polypeptide of 34 kDa, and prepared polyclonal antiserum to this protein. The antiserum was used to detect a 31 kDa protein by immunoblots of extracts (presumably a membrane-rich fraction) from infected plants. It also reacted weakly with extracts from virus-infected protoplasts. No tests with potyviruses other than tobacco vein mottling virus were reported.

The third gene of potyviruses encodes a protein, P3 [90], of about 40-50 kDa. Rodriguez-Cerezo and Shaw [86] also prepared an antiserum to a 42 kDa protein isolated from *E. coli* transformed with cloned DNA corresponding to the P3 gene of tobacco vein mottling virus. This antiserum reacted with 42 kDa and 37 kDa proteins in certain fractions from virus-infected tobacco plants and with lysates from infected tobacco protoplasts.

Discussion and conclusions

Considerable emphasis has been placed on the properties of capsid proteins in the classification of potyviruses [89, 98]. Potyviral nonstructural proteins also merit evaluation in this regard. The potyviral nonstructural proteins studied so far (cylindrical inclusion, helper component protease/amorphous inclusion, nuclear inclusion NIa, and nuclear inclusion NIb) are antigenically distinct from each other and from capsid protein. However, a given nonstructural protein (e.g., cylindrical inclusion protein) of one virus may cross react with the same type of nonstructural protein of other potyviruses. Broad cross-reactivity of cylindrical inclusion proteins has been demonstrated, although there is considerable evidence that cylindrical inclusions of a given virus have both unique and conserved antigenic determinants. Different nonstructural proteins may show different degrees of antigenic conservation. The large nuclear inclusion protein (NIb) may be antigenically more conserved than the small nuclear inclusion protein (NIa). Antiserum to tobacco etch virus NIb protein gave strong reactions with the products from all potyviruses translated in vitro to date. By contrast, antiserum to tobacco etch virus NIa protein has given weak reactions with the in vitro translation products of most potyviruses, with some exceptions such as peanut mottle virus [100] and watermelon mosaic virus 2 (E. Hiebert, unpubl. data). Antiserum to helper component protein of tobacco vein mottling virus has shown broad spectrum cross-reactivity with the in vitro translation products of other potyviruses, but the intensity varied with the virus [49]. Antiserum to potato virus Y helper component has shown limited cross-reactivity [49].

In order for a target antigen to be suitable for extensive use in taxonomic studies, it must be feasible to purify immunogen from infected plants or transformed bacteria in order to prepare antisera, and the antigen must be detectable in a wide range of virus infections, in various hosts, and by various techniques. The cylindrical inclusion protein fulfills these requirements since it has been detected in all potyviral infections. The limited information obtained thus far indicates that serological comparisons of potyviral cylindrical inclusion proteins give similar, but not necessarily identical classifications of potyviruses in comparison to classifications based on antigenic relatedness of capsid proteins. The potyvirus cylindrical inclusions are somewhat more antigenically conserved than capsid proteins. Although different viruses may induce morphologically similar cylindrical inclusions, the antigenic properties of the inclusions may differ.

The other nonstructural proteins may not accumulate or aggregate as readily as the cylindrical inclusion proteins. Thus, cytoplasmic amorphous inclusions and nuclear inclusions are not detected in all potyviral infections and there have been fewer studies on their antigenic relationships. Many of the studies on cross-reactivity of helper component-protease/amorphous inclusion protein and the two nuclear inclusion proteins are based on immuno-precipitation of in vitro translation products of the viral RNAs.

The presence of both conserved and unique epitopes in the four potyviral nonstructural proteins studied to date indicates the potential value of the antigenic properties of these proteins for classification. Studies on the cross-reactivities of the nonstructural P1 and P3 potyviral proteins need to be conducted. Shukla et al. [90] have indicated that the P1 and P3 proteins are less conserved than other potyviral nonstructural proteins. Consequently, the P1 and P3 proteins are of considerable interest in terms of their potential in distinguishing and classifying potyviruses. If it is possible to detect all the potyviral nonstructural proteins in tissues by one or more techniques, then their antigenic relationships can be more thoroughly evaluated as a basis for potyviral classification.

Addendum. Narcissus latent virus has recently been reported [70a] to induce cylindrical inclusions which are structurally and antigenically related to those of potyviruses. However, the virus has a combination of properties (particle length of 650 nm, capsid protein of 45 kDa, antigenicity, and in vitro translation products) that prompted the authors to raise the possibility that the virus is different from both potyviruses and carlaviruses. It also was suggested that narcissus latent virus may be synonymous with narcissus yellow stripe virus. In electro-immunoblots, antiserum to narcissus yellow stripe virus reacted with a protein band of about 76 kDa in extracts from plants infected with either narcissus latent virus or maclura mosaic virus.

Acknowledgements

We thank the following for providing unpublished data or manuscripts in press: D. Baunoch, J. Carrington, P. Das, J. Hammond, V. Hari, E. Koonin, E. Rodriguez-Cerezo, J. Shaw, and J. Verchot.

References

1. Alconero R, Provvidenti R, Gonsalves D (1986) Three pea seedborne mosaic virus pathotypes from pea and lentil germ plasm. Plant Dis 70: 783–786
2. Baker CA (1989) Production and characterization of polyclonal and monoclonal antibodies to three virus-induced proteins of papaya ringspot virus type W. PhD dissertation, University of Florida, Gainesville
3. Baker CA, Purcifull DE (1987) Serological relationships of three proteins of papaya ringspot virus type W (PRSV-W) to antigens of zucchini yellow fleck virus (ZYFV). Phytopathology 77: 1722
4. Baker CA, Purcifull DE (1988) Reactivity of a monoclonal antibody to the amorphous inclusion protein of papaya ringspot virus-type W (PRSV-W). Phytopathology 78: 1537
5. Baker CA, Purcifull DE (1990) Reactivity of two monoclonal antibodies to the cylindrical inclusion protein of papaya ringspot virus type W. Phytopathology 80: 1033
6. Batchelor DL (1974) Immunogenicity of sodium dodecyl sulfate-denatured plant viruses and plant viral inclusions. PhD dissertation, University of Florida, Gainesville
7. Baum RH (1980) Purification, partial characterization, and serology of the capsid and cylindrical inclusion proteins of four isolates of watermelon mosaic virus. PhD dissertation, University of Florida, Gainesville
8. Baum RH, Purcifull DE (1981) Serology of cylindrical inclusions of several watermelon mosaic virus (WMV) isolates. Phytopathology 71: 202
9. Baunoch DA, Das P, Hari V (1988) Intracellular localization of TEV capsid and inclusion proteins by immunogold labelling. J Ultrastruct Mol Struct Res 99: 203–212
10. Baunoch, DA, Das P, Hari V (1990) Potato virus Y helper component protein is associated with amorphous inclusions. J Gen Virol 71: 2479–2482
11. Baunoch DA, Das P, Browning ME, Hari V (1991) A temporal study of the expression of the capsid, cytoplasmic inclusion and nuclear inclusion proteins of tobacco etch potyvirus in infected plants. J Gen Virol 72: 487–492

12. Baunoch, DA, Das P, Hari V (1991) Epitope analysis of the CIP of tobacco etch virus. Phytopathology 81: 1147

13. Berger PH, Hunt AG, Domier LL, Hellmann GM, Stram Y, Thornbury DW, Pirone TP (1989) Expression in transgenic plants of a viral gene product that mediates insect transmission of potyviruses. Proc Natl Acad Sci USA 86: 8402–8406

14. Berger PH, Luciano CS, Thornbury DW, Benner HI, Hill JH, Zeyen RJ (1989) Properties and in vitro translation of maize dwarf mosaic virus RNA. J Gen Virol 70: 1845–1851

15. Brakke MK, Ball EM, Hsu YH, Langenberg WG (1987) Wheat streak mosaic virus cylindrical inclusion body protein. J Gen Virol 68: 281–287

16. Carrington JC, Dougherty WG (1987) Small nuclear inclusion protein encoded by a plant potyvirus genome is a protease. J Virol 61: 2540–2548

17. Carrington JC, Cary SM, Parks DT, Dougherty WG (1989) A second proteinase encoded by a plant potyvirus genome. EMBO J 8: 365–370

18. Carrington JC, Freed DD, Oh C-S (1990) Expression of potyviral polyproteins in transgenic plants reveals three proteolytic activities required for complete processing. EMBO J 9: 1347–1353

19. Chang C-A (1986) Bean yellow mosaic and clover yellow vein viruses: purification, characterization, detection and antigenic relationships of their nuclear inclusion proteins. PhD dissertation, University of Florida, Gainesville

20. Chang C-A (1990) A one step identification of bean yellow mosaic and clover yellow vein potyviruses by detection of virus specific antigen in infected tissue. In: Proceedings of the VIII International Congress for Virology, Berlin, p 446

21. Chang C-A, Purcifull DE, Hiebert E (1985) Purification and partial characterization of nuclear inclusions induced by a pea mosaic isolate of bean yellow mosaic virus. Phytopathology 75: 499

22. Chang C-A, Purcifull DE, Hiebert E, Edwardson JR (1986) Immunofluorescence evidence for the origin of nuclear inclusion and cytoplasmic crystals induced by bean yellow mosaic virus. Phytopathology 76: 1061

23. Chang C-A, Hiebert E, Purcifull DE (1988) Purification, characterization, and immunological analysis of nuclear inclusions induced by bean yellow mosaic and clover yellow vein potyviruses. Phytopathology 78: 1266–1275

24. Chang C-A, Hiebert E, Purcifull DE (1988) Analysis of in vitro translation of bean yellow mosaic virus RNA: inhibition of proteolytic processing by antiserum to the 49K nuclear inclusion protein. J Gen Virol 69: 1117–1122

25. Christie RG, Edwardson JR (1977) Light and electron microscopy of plant virus inclusions. Fla Agric Exp Stat Monogr, no 9

26. deMejia MVG, Hiebert E, Purcifull DE (1985) Isolation and partial characterization of the amorphous cytoplasmic inclusions associated with infections caused by two potyviruses. Virology 142: 24–33

27. deMejia MVG, Hiebert E, Purcifull D, Thornbury DW, Pirone TP (1985) Identification of potyviral amorphous inclusion protein as a nonstructural, virus-specific protein related to helper component. Virology 142: 34–43

28. Domier LL, Shaw JG, Rhoads RE (1987) Potyviral proteins share amino acid sequence homology with picorna-, como-, and caulimovirus proteins. Virology 158: 20–27

29. Dougherty WG, Hiebert E (1980) Translation of potyvirus RNA in a rabbit reticulocyte lysate: identification of nuclear inclusion proteins as products of tobacco etch virus RNA translation and cylindrical inclusion protein as a product of the potyvirus genome. Virology 104: 174–182

30. Dougherty WG, Carrington, JC (1988) Expression and function of potyviral gene products. Annu Rev Phytopathol 26: 123–43

31. Dougherty WA, Parks TD (1991) Post-translational processing of the tobacco etch virus 49-kDa small nuclear inclusion polyprotein, identification of an internal cleavage site and delineation of VPg and proteinase domains. Virology 183: 449–456

32. Edwardson JR (1974) Some properties of the potato virus-Y group. Fla Agric Exp Stat Monogr, no 4

33. Edwardson JR, Christie RG (1978) Use of virus-induced inclusions in classification and diagnosis. Annu Rev Phytopathol 16: 31–55

34. Edwardson JR, Christie RG (1991) The potyvirus group. Fla Agric Exp Stat Monogr, no 16, vols I–IV

35. Garcia JA, Riechmann JL, Lain S (1989) Proteolytic activity of the plum pox potyvirus NIa-like protein in *Escherichia coli*. Virology 170: 362–369

36. Gorbalenya AE, Blinov VM, Donchenkoae, Koonin EV (1989) An NTP-binding motif is the most conserved sequence in a highly diverged monophyletic group of proteins involved in positive strand RNA viral replication. J Mol Biol 28: 256–268

37. Govier DA, Kassanis B, Pirone TP (1977) Partial purification and characterization of the potato Y helper component. Virology 78: 306–314

38. Graybosch R, Hellmann GM, Shaw JG, Rhoads RE, Hunt AG (1989) Expression of a potyvirus non-structural protein in transgenic tobacco. Biochem Biophys Res Comm 160: 425–432

39. Habili N, Symons RH (1989) Evolutionary relationship between luteoviruses and other RNA plant viruses based on sequence motifs in their putative RNA polymerases and nucleic acid helicases. Nucleic Acids Res 17: 9543–9555

40. Hammond J (1989) Antisera to cytoplasmic inclusion proteins of potyviruses contain cross-reactive antibodies. Phytopathology 79: 1174

41. Hammond J, Lawson RH (1988) An improved purification procedure for preparing potyviruses and cytoplasmic inclusions from the same tissue. J Virol Methods 20: 203–217

42. Hellmann GM, Thornbury DW, Hiebert E, Shaw JG, Pirone TP, Rhoads RE (1983) Cell-free translation of tobacco vein mottling virus RNA. Virology 124: 434–444

43. Hellmann GM, Shaw JG, Rhoads RE (1988) In vitro analysis of tobacco vein mottling virus NIa cistron: evidence for a virus-encoded protease. Virology 163: 554–562

44. Hiebert E, Charudattan R (1984) Characterization of araujia mosaic virus by in vitro translation analyses. Phytopathology 74: 642–646

45. Hiebert E, Dougherty WG (1988) Organization and expression of the viral genomes. In: Milne RG (ed) The plant viruses, vol 4, the filamentous viruses. Plenum, New York, pp 159–178

46. Hiebert E, McDonald JG (1973) Characterization of some proteins associated with viruses in the potato Y group. Virology 56: 349–361

47. Hiebert E, Purcifull DE, Christie RG, Christie SR (1971) Partial purification of inclusions induced by tobacco etch virus and potato virus Y. Virology 43: 638–646

48. Hiebert E, Purcifull DE, Christie RG (1984) Purification and immunological analysis of plant viral inclusion bodies. In: Maramorosh K, Koprowski H (eds) Methods in virology, vol 8. Academic Press, San Diego, pp 225–280

49. Hiebert E, Thornbury DW, Pirone TP (1984) Immunoprecipitation analysis of potyviral in vitro translation products using antisera to helper component of tobacco vein mottling virus and potato virus Y. Virology 135: 1–9

50. Jensen SG, Staudinger JL (1989) Serological grouping of the cytoplasmic inclusions of 6 strains of sugarcane mosaic virus. Phytopathology 79: 1215

51. Jordan R, Hammond J (1991) Comparison and differentiation of potyvirus isolates and identification of strain-, virus-, subgroup-specific and potyvirus group-common epitopes using monoclonal antibodies. J Gen Virol 72: 25–36

52. Karchi H (1989) Expression of the full-length genome of potato virus Y (PVY) in *Escherichia coli* cells: polyprotein processing and viral proteins activities. MS thesis, Hebrew University of Jerusalem, Jerusalem

53. Kassanis B (1939) Intranuclear inclusions in virus-infected plants. Ann Appl Biol 26: 705–709

54. Knuhtsen H, Hiebert E, Purcifull DE (1974) Partial purification and some properties of tobacco etch virus induced intranuclear inclusions. Virology 61: 200–209

55. Ko N-J (1987) LR White embedding for immunogold labelling of virus-infected plant tissues. Proc Nat Sci Council (Taiwan) [B]: 206–210

56. Koenig R (1988) Serology and immunochemistry. In: Milne RG (ed) The plant viruses, vol 4, the filamentous viruses. Plenum, New York, pp 111–158

57. Lain S, Riechmann JL, Martin MT, Garcia JA (1989) Homologous potyvirus and flavivirus proteins belonging to a superfamily of helicase-like proteins. Gene 82: 357–362

58. Lain S, Martin MT, Riechmann JL, Garcia JA (1991) Novel catalytic activity associated with positive-strand RNA virus infection: nucleic acid-stimulated ATPase activity of the plum pox potyvirus helicase-like protein. J Virol 65: 1–6

59. Langenberg WG (1991) Cylindrical inclusion bodies of wheat streak mosaic virus and three other potyviruses only self-assemble in mixed infections. J Gen Virol 72: 493–497

60. Lecoq H, Bourdin D, Raccah B, Hiebert E, Purcifull D E (1991) Characterization of a zucchini yellow mosaic virus isolate with a deficient helper component. Phytopathology 81: 1087–1091

61. Lesemann DE (1988) Cytopathology. In: Milne RG (ed) The plant viruses, vol 4, the filamentous viruses. Plenum, New York, pp 179–235

62. Lima JAA, Purcifull DE (1980) Immunochemical and microscopical techniques for detecting blackeye cowpea mosaic and soybean mosaic viruses in hypocotyls of germinated seeds. Phytopathology 70: 142–147

63. Lima JAA, Purcifull DE, Hiebert E (1979) Purification, partial characterization, and serology of blackeye cowpea mosaic virus. Phytopathology 69: 1252–1258

64. Lin N-S, Wang N, Hsu Y-H (1988) Sequential appearance of capsid protein and cylindrical inclusion protein in root-tip cells following infection with passion fruit woodiness virus. J Ultrastruct Mol Struct Res 100: 201–211

65. Luciano CS, Rhoads RE, Shaw JG (1987) Synthesis of potyviral RNA and proteins in tobacco mesophyll protoplasts inoculated by electroporation. Plant Sci 51: 295–303

66. Luciano CS, Gibb KS, Berger PH (1989) A general method for the detection of potyviral gene products in plant protoplasts and tissue. J Virol Methods 24: 347–356

67. Martelli GP, Russo M (1977) Plant virus inclusion bodies. Adv Virus Res 21: 175–266

68. Martin MT, Otin CL, Lain S, Garcia JA (1990) Determination of polyprotein processing sites by amino terminal sequencing of nonstructural proteins encoded by plum pox potyvirus. Virus Res 15: 97–106

69. McDonald JG, Hiebert E (1975) Characterization of the capsid and cylindrical inclusion proteins of three strains of turnip mosaic virus. Virology 63: 295–303

70. Mowat WP, Dawson S, Duncan GH (1989) Production of antiserum to a non-structural potyviral protein and its use to detect narcissus yellow stripe and other potyviruses. J Virol Methods 25: 199–210

70a. Mowat WP, Dawson S, Duncan GH, Robinson DJ (1991) Narcissus latent, a virus with filamentous particles and a novel combination of properties. Ann Appl Biol 119: 31–46

71. Murphy JF, Jarlfors U, Shaw JG (1991) Development of cylindrical inclusions in potyvirus-infected protoplasts. Phytopathology 81: 371–374

72. Murphy JF, Rhoads RE, Hunt AG, Shaw JG (1990) The VPg of tobacco etch virus RNA is the 49-kDa protease or the N-terminal 24-kDa part of the proteinase. Virology 178: 285–288

73. Nagel J, Hiebert E (1985) Complementary DNA cloning and expression of the papaya ringspot potyvirus sequence encoding capsid protein and a nuclear inclusion-like protein in *Escherichia coli*. Virology 143: 435–441

74. Nagel J, Zettler FW, Hiebert E (1983) Strains of bean yellow mosaic virus compared to clover yellow vein virus in relation to gladiolus production in Florida. Phytopathology 73: 449–454

75. Noda C, Maeda T, Inouye N (1988) Isopycnic separation of potyviral cylindrical inclusions by sucrose potassium tartrate density gradient centrifugation. Ann Phytopathol Soc Jpn 54: 319–322

76. Parks TD, Dougherty WG (1991) Substrate recognition by the NIa proteinase of two potyviruses involves multiple domains: characterization using genetically engineered hybrid proteinase molecules. Virology 182: 17–27

77. Pirone TP, Thornbury DW (1884) The involvement of a helper component in nonpersistent transmission of plant viruses by aphids. Microbiol Sci 1: 191–193

78. Provvidenti R, Gonsalves D, Humaydan HS (1984) Occurrence of zucchini yellow mosaic virus in cucurbits from Connecticut, New York, Florida, and California. Plant Dis 68: 443–446

79. Purcifull DE (1990) Ouchterlony double-diffusion tests in the presence of sodium dodecyl sulfate for detection of virion proteins and virus-induced inclusion body proteins. In: Hampton R, Ball E, DeBoer S (eds) Serological methods for detection and identification of viral and bacterial plant pathogens. American Phytopathological Society Press, St. Paul, Minn, pp 121–127

80. Purcifull DE, Batchelor DL (1977) Immunodiffusion tests with sodium dodecyl sulfate (SDS)-treated plant viruses and plant viral inclusions. Fla Agric Exp Stat Tech Bull 788

81. Purcifull DE, Hiebert E, McDonald JG (1973) Immunochemical specificity of cytoplasmic inclusions induced by viruses in the potato Y group. Virology 55: 275–279

82. Purcifull DE, Zitter TA, Hiebert E (1975) Morphology, host range, and serological relationships of pepper mottle virus. Phytopathology 65: 559–562

83. Quiot-Douine L, Purcifull DE, Hiebert E, deMejia MVG (1986) Serological relationships and in vitro translation of an antigenically distinct strain of papaya ringspot virus. Phytopathology 76: 346–351

84. Quiot-Douine L, Lecoq H, Quiot JB, Pitrat M, Labonne G (1990) Serological and biological variability of virus isolates related to strains of papaya ringspot virus. Phytopathology 80: 256–263

85. Restrepo MA, Freed DD, Carrington JC (1990) Nuclear transport of plant potyviral proteins. Plant Cell 2: 987–998

86. Rodriguez-Cerezo E, Shaw JG (1991) Two newly detected nonstructural viral proteins in potyvirus-infected cells. Virology 185: 572–579

87. Shepard JF, Shalla TA (1969) Tobacco etch virus cylindrical inclusions: antigenically unrelated to the causal virus. Virology 38: 185–188

88. Shepard JF, Gaard G, Purcifull DE (1974) A study of tobacco etch virus-induced inclusions using indirect immunoferritin procedures. Phytopathology 64: 418–425

89. Shukla DD, Ward CW (1989) Identification and classification of potyviruses on the basis of coat protein sequence data and serology. Arch Virol 106: 171–200

90. Shukla DD, Frenkel MJ, Ward CW (1991) Structure and function of the potyvirus genome with special reference to the coat protein coding region. Can J Plant Pathol 13: 178–191

91. Slade DE, Johnston RE, Dougherty WG (1989) Generation and characterization of monoclonal antibodies reactive with the 49-kDa proteinase of tobacco etch virus. Virology 173: 499–508

92. Suzuki N, Kudo T, Shirako Y, Ehara Y, Tachibana T (1989) Distribution of cylindrical inclusion, amorphous inclusion and capsid proteins of watermelon mosaic virus 2 in systemically infected pumpkin leaves. J Gen Virol 70: 1085–1091

93. Suzuki N, Shirako Y, Ehara Y (1990) Isolation and serological comparison of virus-coded proteins of three potyviruses infecting cucurbitaceous plants. Intervirology 31: 43–49

94. Thornbury DW, Pirone TP (1983) Helper components of two potyviruses are serologically distinct. Virology 125: 487–490

95. Thornbury DW, Hellman GM, Rhoads RE, Pirone TP (1985) Purification and characterization of potyvirus helper component. Virology 144: 260–267

96. Thornbury DW, Patterson CA, Dessens JT, Pirone TP (1990) Comparative sequence of the helper component (HC) region of potato virus Y and a HC-defective strain, potato virus C. Virology 178: 573–578

97. Verchot J, Koonin EV, Carrington JC (1991) The 35 kDa protein from the N-terminus of the potyviral polyprotein functions as a third virus-encoded protease. Virology 185: 527–535

98. Ward CW, Shukla DD (1991) Taxonomy of potyviruses – current problems and some solutions. Intervirology 32: 269–296

99. Xiong Z (1985) Purification and partial characterization of peanut mottle virus and detection of peanut stripe virus in peanut seeds. MS thesis, University of Florida, Gainesville

100. Xiong Z, Purcifull DE, Hiebert E (1985) Purification, serology, and cytology of peanut mottle virus. Phytopathology 75: 1334

101. Xiong Z, Hiebert E, Purcifull DE (1988) Characterization of the peanut mottle virus genome by in vitro translation. Phytopathology 78: 1128–1134

102. Yeh S-D, Bih F-Y (1989) Comparative studies on in vitro translation of a severe strain and a mild strain of papaya ringspot virus. Plant Protect Bull 31: 276–289

103. Yeh S-D, Gonsalves D (1984) Purification and immunological analyses of cylindrical-inclusion protein induced by papaya ringspot virus and watermelon mosaic virus 1. Phytopathology 74: 1273–1278

104. Yeh S-D, Gonsalves D (1985) Translation of papaya ringspot virus RNA in vitro: detection of a possible polyprotein that is processed for capsid protein, cylindrical inclusion protein, and amorphous-inclusion protein. Virology 143: 260–271

105. Yeh S-D, Gonsalves D, Provvidenti R (1984) Comparative studies on host range and serology of papaya ringspot virus and watermelon mosaic virus 1. Phytopathology 74: 1081–1085

106. Yeh S-D, Wang C-H, Chen M-J (1990) Purification of a 112 K protein of papaya ringspot virus produced in vivo. In: Proceedings of the VIII International Congress for Virology, Berlin, p 471

107. Zettler FW, Abo El-Nil MM, Hartman RD (1978) Dasheen mosaic virus. CMI/AAB Descriptions of Plant Viruses, no 191

108. Zurawski, DB (1979) Some biological and serological properties of bidens mottle virus isolated from Fittonia. MS thesis, University of Florida, Gainesville

Authors' address: D.E. Purcifull, Department of Plant Pathology, Institute of Food and Agricultural Sciences, University of Florida, Gainesville, FL 32611, U.S.A.

Arch Virol (1992) [Suppl 5]: 123–138

Potyvirus serology, sequences and biology

J. Hammond

United States Department of Agriculture, Agricultural Research Service,
Florist and Nursery Crops Laboratory, Beltsville Agricultural Research Center,
Beltsville, Maryland, U.S.A.

Summary. Amino acid sequences of the cytoplasmic cylindrical inclusion protein (CIP), large nuclear inclusion protein (NIb), and coat protein (CP) of potyviruses were re-examined in light of reported serological relationships, and correlated with known and deduced biological functions. No obvious correlations were observed between either amino acid sequences or epitopes recognized by monoclonal antibodies and the natural host ranges of the potyviruses examined. Whereas the identified sequence motifs of the RNA helicase (CIP) and replicase (NIb) are predicted to be antigenic, most of the conserved sequences and epitopes in the CIP, NIb and CP were presumed to be maintained for structural rather than functional reasons. Three possible potyvirus clusters are proposed on the basis of the length and composition of the virion surface-exposed amino terminal extension of the CP; these clusters do not correlate with overall CP sequence homology, host range, or vectors, but are of potential evolutionary significance and hence of possible taxonomic value.

Introduction

For many years the host range and serological relationships of the virions (coat protein, CP) were the best ways of classifying potyviruses, yet there were many conflicting reports, largely due to differences among antisera used and the specific virus isolates examined. More recently antisera have been produced to non-structural gene products as well as to CPs, and these have also been used in taxonomy. With the advent of monoclonal antibodies (MAbs) and the cloning and sequencing of several complete potyvirus genomes, it should now be possible to distribute standard reference materials to which new isolates can be compared. One question to be considered is which comparisons are meaningful. Those that relate to biological differences should be given more weight than those unrelated to biology.

Most previous comparisons of potyviruses have been made on the basis of a particular character, or on a correlation of a number of either physical or biological characteristics, to group or differentiate virus isolates. This communication is an attempt to synthesize results obtained in many different laboratories and by different methods, to combine features with functions and sequences with significance (i.e., consequent structural or biological effect). Thus, many of the observations are speculative and not supported by experimental data at this time. As the available sequence data for most of the viruses examined has been limited to the 3′ end of the genome, and most of the antisera produced are to gene products from this part of the genome, the discussion will also be largely limited to these genes, gene products, and antisera.

Sequence data and analysis

The sequences used for the analyses presented in this paper were from the following sources: the complete sequences of tobacco etch virus (TEV) [1]; tobacco vein mottling virus (TVMV) [8]; potato virus Y, N-strain (PVY-NF [42]; and plum pox virus strain NAT (PPV-NAT) [36] and coat-protein sequences of bean yellow mosaic virus strains GDD (BYMV-GDD) [18], PVC (BYMV-PVC) [5], and CS (BYMV-CS) [52]; clover yellow vein virus strain 30 (ClYVV-30) [57]; pea seed-borne mosaic virus (PSbMV) [54]; soybean mosaic virus (SbMV) isolates G1 (SbMV-G1) [46], N (SbMV-N) [11], G2 and G7 (SbMV-G2 and SbMV-G7) [27]; watermelon mosaic virus 2 (WMV) Australian isolate (WMV-2-Aus) [59], and American isolate (WMV-2-US) [39]; papaya ringspot virus strains P and W (PRSV-P, and PRSV-W) [38]; zucchini yellow mosaic virus (ZYMV) isolates from France (ZYMV-F) [39], Israel (ZYMV-I) [12], and Connecticut (ZYMV-CT) [15]; passionfruit woodiness virus (PWV) strains TB, S, and M (PWV-TB, PWV-S and PWV-M) [48]; turnip mosaic virus (TuMV) [55]; TEV-NAT [2]; pepper mottle virus (PepMoV) [9]; PVY strains D (PVY-D) [47], O (PVY-O) [6] and an isolate from Israel (PVY-I) [43]; ornithogalum mosaic virus (OrMV) [7]; sugarcane mosaic virus (SCMV) [13]; maize dwarf mosaic virus B (MDMV-B) [29]; and PPV strains AT (PPV-AT) [36], D (PPV-D) [41] and Rankovik (PPV-R) [34].

Sequence comparisons were made using programs in the PC-Gene package (IntelliGenetics Inc., Mountain View, CA). Multiple alignments of sequences were made by eye, with gaps introduced by eye to maximize the fit of PPV isolates. Prediction of antigenicity (program ANTIGEN) and flexibility (FLEXPRO) were also from the PC-Gene package.

Additional virus abbreviations used are: CABMV (cowpea aphid-borne mosaic virus), ISMV (iris severe mosaic virus), JGMV (johnsongrass mosaic virus), LMV (lettuce mosaic virus), PMosV (pea mosaic virus), PVA (potato virus A), SrMV (sorghum mosaic virus), SPFMV (sweet potato feathery mottle virus), TBV (tulip breaking virus), and TCBV (tulip chlorotic blotch virus).

Results of analysis

Cytoplasmic inclusion protein

Antisera to the cytoplasmic cylindrical inclusion proteins (CIP) of BYMV, ClYVV, TuMV, SPFMV, and ISMV are highly cross-reactive; reactions to the CIPs of up to 18 distinct potyviruses occurred on Western blots [17]. In contrast, Jensen and Staudinger [28] found that antisera to the CIPs of MDMV-A, MDMV-B (= SCMV) [50], SCMV-H (= SrMV) [50] and MDMV-O (= JGMV) [50] could be used to differentiate these viruses from each other in a similar assay; these antisera were collected after relatively short immunizations (S. G. Jensen, pers. comm.). Additional tests using Jensen's antiserum to the MDMV-A CIP showed cross-reactivity with SrMV CIP, and revealed very minor levels of cross-reactivity to the CIPs of 16 other potyviruses (J. Hammond, unpubl.).

Examination of the CIP amino acid sequences of TEV, TVMV, PVY, and PPV (which cross-react among CIPs) reveals much conserved sequence, but not in the regions predicted to be most antigenic (using programs ANTIGEN and FLEXPRO of the PC-Gene package). The cross-reactions observed may therefore be due either to minor epitopes within highly conserved sequences, or to weaker reactions with key contact amino acids. One function of the CIP that has been demonstrated recently is helicase activity [35], which is associated with the nucleotide binding motif (A/G)XXXXGK(S/T). The motif and additional residues are conserved among the four sequences (*GAVGSGKS*TGLP). Although the region is not predicted to be among the most antigenic portions of the CIP by the method of Hopp and Woods [24], it is predicted by the method of Karplus and Schulz [33] to be one of the most flexible domains, an attribute also correlated with antigenicity. Several other regions of high sequence homology also coincide with predicted flexibility, but the regions predicted to be most antigenic by the method of Hopp and Woods [24] have limited sequence homology and/or do not occur at equivalent positions in the four sequences. These differences may explain the virus specificity observed in Jensen and Staudinger's [28] tests with sera collected 5–6 weeks after immunization, which probably contain primarily antibodies to the most antigenic site, and the cross-reactivity observed with other sera collected after longer immunizations [17]. Longer immunizations may also result in production of more antibodies recognizing cryptotopes; conserved sequences in cryptotopes may be more important in maintaining CIP structure than in forming part of an active site.

Nuclear inclusion protein b

Polyclonal antiserum to the NIb (large nuclear inclusion protein) of TEV cross reacts with the equivalent translation product of at least 25 distinct potyviruses [37]. As NIb is presumed to be the viral replicase and contains the GDD motif, there is ample reason to expect elements of conserved structure for nucleotide binding and to a lesser extent for template recognition. There is in fact an average of over 60% amino acid sequence homology among the published sequences. Turpen [56] noted two blocks of conserved nucleotides in the 5' leader sequences of TEV, TVMV, and PVY that may have significance for virulence or translation functions, or could be (part of?) a replicase binding site; however, no complementary sequence or extended homologies have been observed in the 3' untranslated regions.

Examination of hydrophilicity of the NIb sequences by the method of Hopp and Woods [24] does not result in the prediction of an antigenic site common to all four complete sequences; one predicted site is shared by TEV and PVY, but differs by a single residue in PPV and TVMV where it is not predicted to be antigenic. Another predicted site varies in sequence between PVY, PPV and TVMV, and a third between PPV and TEV.

Prediction of flexibility [33] indicates several domains that may account for shared epitopes. The most flexible segment in each NIb is the fully conserved octapeptide (italics) in a larger conserved sequence (*GNNSGQPS*TVVDN), about 40 residues upstream from the GDD replicase motif. Further upstream is another perfectly conserved region YCD*ADGSQFD*SSL containing a heptapeptide (italics) predicted to be highly flexible (and hence antigenic). A number of other regions shared by at least two NIb sequences are predicted to be antigenic on the basis of flexibility. Several other groups of five or more residues are fully conserved among the available sequences, while many others have only conservative substitutions; these may form minor epitopes and thus also contribute to antigenic cross-reactions. They include, from N- to C-terminus, QLVTKH; WNGSLKAEL; FTAAP; KVCVDDFNN; VGMTKF; TPDGTI; N*GDD*L (including the replicase motif); and LWFMSH. Conservation of these groups of residues is likely to be important for structural and/or functional reasons.

Coat protein

Polyclonal antisera to coat proteins (CPs) of many potyviruses show significant cross-reactivity in antigen-coated plate forms of indirect ELISA; this is presumably due to the presence of antibodies to the highly conserved amino acid sequences in the trypsin-resistant core (TRC). Such cross-reactive antibodies are more common in sera collected after long immunizations [49, 51, 58] and in antisera to the more labile viruses such as PSbMV (J. Hammond, unpubl.). The N-terminus contains the major virus-specific epitopes [10, 49], and antiserum prepared to dissociated TRC protein reacts with many different potyviruses [49], presumably due to the induction of antibodies to conserved internal sequences. These internal conserved and conservatively substituted sequences are presumably maintained for structural and/or functional reasons [18]; a model suggested for the folding of PVY CP [49] presents the N- and C-termini as surface extensions as previously suggested by their sensitivity to trypsin [3,49] and results in positioning of several arginine residues implicated in RNA binding in or close to the two major loops predicted to be at the virion interior [49]. Comparison of a similar prediction of the folding of the BYMV CP to other potyvirus sequences (J. Hammond, unpubl.) shows that the first (more N-terminal) interior loop is quite variable with only two conserved residues, whereas the second interior loop is more highly conserved, with only a tyrosine-arginine dipeptide reversed in some isolates. Four arginine residues in or near the second loop are conserved, with the exception of the reversal of the tyrosine-arginine to an arginine-tyrosine dipeptide in the loop of some CPs. Conservation at many of these positions may be necessary to maintain subunit folding, and subunit–subunit or subunit–RNA interactions.

The ability of monoclonal antibodies (MAbs) to discriminate between antigens initially led many to expect that all MAbs would be highly virus-specific, if not strain-specific. It has since been shown that many MAbs cross-react with related antigens. Among MAbs produced to potyviruses several have been reported that cross-react with viruses not previously thought to be closely related [10,19,23,25,26,31,32]. However only MAb PTY 1 has been reported to have reactivity with an epitope conserved in a very broad range of potyviruses [32] despite the extent of homologies in the TRC. Other MAbs show varying cross-reactivity, but without any obvious correlation between CP epitopes and natural host range (Table 1), or vector specificity, or even with any previously determined serological relationships, except with virus-specific, strain cross-reactive MAbs [10,16] or within the BYMV subgroup [32,45]. Too few antigens were tested to determine the significance of three MAbs cross-reacting with four viruses infecting solanaceous hosts (TEV, TVMV, PepMoV and PVY) [10]. Jordan and Hammond [32] found no epitopes to be uniquely conserved among six legume potyviruses (BYMV, ClYVV, PMosV, SbMV, CABMV, and PSbMV), nor any common only to five viruses (PepMoV, PVA, PVY, TEV, and TVMV) naturally infecting solanaceous plants, although it is possible that epitopes might be correlated with particular host species; our current virus isolate and host range data do not permit generalizations beyond natural host plant family.

Although PepMoV was originally classified as an isolate of PVY on the basis of serology [60], and more recently on the basis of sequence similarity [47], fewer epitopes were shared by PepMoV and PVY (three) than differentiated them (eleven) when a number of cross-reactive MAbs were tested [32]. This is surprising as most of the epitopes differentiating PepMoV and PVY are crypto-topes presumed to be within the TRC; three metatopes are uniquely shared by PepMoV and BYMV (and not ClYVV or PMosV) and map to the N-terminus [20,32,30; R.L. Jordan and J. Hammond, unpubl.]. Epitopes recognized by several cross-reactive MAbs have been mapped to at least four distinct domains within the TRC and/or C-terminus of the BYMV CP by reaction with bacterially expressed deletion mutants [20; J. Hammond and R.L. Jordan, unpubl.].

Table 1. Reactivity of monoclonal antibodies with selected potyviruses, to show lack of correlation of epitopes with natural host range of the viruses

Host/Virus	Monoclonal antibody											
	TBV 4	TBV 7	PTY 1	PTY 2	PTY 4	PTY 5	PTY 10	PTY 21	TEV H-2	TEV H-7	PVY C-9	LMV L-5
Tulip												
TBV	+++	+++	+++	+++	++	−	+	−	NT	−	NT	NT
TCBV	+++	+++	+++	+++	−	++	++	−	NT	NT	NT	NT
TuMV	−	+++	+++	+++	−	++	++	−	NT	NT	NT	NT
Legumes												
BYMV	+++	(++)	+++	+++	++	+++	(+)	+++	NT	+++	NT	NT
ClYVV	−	+++	+++	+++	++	++	+	−	NT	+++	NT	NT
SbMV	++	−	+++	−	(+)	−	+	−	NT	+	NT	+
Solanaceae												
PepMoV	++	−	+++	−	−	−	+++	+++	+++	+++	NT	NT
PVY	−	−	+++	+++	+++	++	++	−	+++	+++	+++	NT
TEV	+	+++	+++	+++	−	++	−	−	+++	+++	NT	NT
TVMV	−	−	+++	−	−	−	+++	−	+++	+++	NT	NT
Cucurbits												
WMV-2	−	−	+++	−	+	−	+	−	−	+	NT	·NT
ZYMV	+	+++	+++	−	+++	−	(+)	−	NT	NT	NT	NT
Lettuce												
LMV	−	−	+++	−	−	−	−	−	NT	NT	NT	NT
Gramineae												
MDMV-A	NT	NT	+++	−	+	−	−	−	−	+	NT	+
MDMV-B	NT	NT	+++	+	++	+	−	(+)	−	+	NT	+

Antibodies TBV 4 and 7 [26; J. Hammond and R. L Jordan, unpubl.]; PTY 1, 2, 5, 10 and 21 [32]; TEV H-2 and H-7 [10; R. L. Jordan, unpubl.]; PVY C-9, [16]; LMV L-5 [23].
NT Not tested; (+) weak reaction; (++) with some isolates only

Coat-protein amino acid sequence

The CP N- and C-termini are not necessary for infection [49] (except via aphid transmission) and are therefore presumably not important in host range determination, at least at the level of establishment of infection. Results obtained by Salomon [44] with SPFMV suggest that the termini may have a role in virus replication or systemic movement. A proteolytic activity was induced in SPFMV-infected sweet potato and morning glory, even in leaves without detectable virus, but not in healthy plants, and this activity may contribute to the attenuation of symptoms and ultimate recovery from virus upon further growth [44]. However, the proteolytic activity in sweet potato may affect viral gene products and processes other than CP, so this is not clear evidence for a role of the N-terminus in movement apart from its important role in aphid transmission [3, 4] which is discussed elsewhere in this volume.

Further examination of the CP amino acid sequence (J. Hammond, unpubl.) revealed no obvious correlation between even single residues at equivalent positions in the TRC and host range; no residues are uniquely present at the same position in all of four viruses naturally infecting legumes (BYMV, ClYVV, SbMV, and PWV), or those infecting cucurbits (WMV-2, PRSV-W, PRSV-P, and ZYMV), or solanaceous plants (TEV, PepMoV, PVY, and TVMV). At some positions all of the viruses and isolates infecting one host family share a common residue, but other viruses not naturally infecting the same host family also share the same residue; at other positions no isolate *not* infecting that host type has the same residue, but not all isolates that are grouped by host type share a common residue (Fig. 1). In view of the apparent absence of sequence motifs or uniquely shared residues it may be assumed that CP amino acid sequence alone does not play a significant role in determination of host range, although it is possible that CP plays a coordinated role with other genes and gene products. This is supported by the results of Hellmann et al. [21] with chimeric infectious transcripts of TVMV; they found that the resistance-breaking phenotype of strain TVMV-S mapped to the 5' half of the genome, or at least that 5' terminal encoded gene products are essential for resistance-breaking in this system.

Dougherty et al. [10] and Shukla et al. [49] noted that the N and C termini are surface exposed and virus-specific, and Shukla et al. [51] showed that highly virus-specific antisera could be produced against the N terminal domain. Shukla and Ward [46] noted some homology in the N termini of SCMV-SC and TEV-HAT that might explain the reciprocal reactions observed with polyclonal antisera [51]; similar limited homologies may explain the unexpected serological reactions that have been observed between other potyviruses [46] including the reactions of three MAbs to both BYMV and PepMoV [32] that have been mapped to the N-terminal domain [30; R.L. Jordan and J. Hammond, unpubl.].

The N terminal domain of many potyvirus CPs is lysine-rich [22], and the length and sequence of the N-terminal domain show major differences among viruses [46]. Further examination of the N-terminal domains (J. Hammond, unpubl.) reveals additional differences and similarities; those viruses with significantly longer N termini (PPV, SCMV and MDMV-B) are not enriched for lysine in the trypsin-sensitive region, and are instead enriched for glycine and proline (Fig. 2). The viruses with lysine-rich extensions of 28-51 residues N terminal to the TRC all have a number of lysine residues separated typically by one and/or three other residues (Fig. 3); PRSV-W and PRSV-P have a multiple KEKEK motif, while SbMV isolates N, G2 and G7 have the sequence KEKE. Variations on this motif (EKEKKEREK in TuMV, KKDKE in ZYMV

```
                    100              110           120       125

Legumes

BYMV-GDD    E A W Y N G V K Q A Y E V E D S R M G I I L N G L M V W
BYMV-PVC    E A W Y S G V K Q A Y E V E D S Q M G I I L N G L M V W
BYMV-CS     K A W Y N G V K Q A Y E V E D S Q M S I I L N G L M V W
ClYVV-30    E A W H E G V K N A Y E V D D Q Q M E I I C N G L M V W
SbMV-G1     K A W H A A V M D A Y G I N E E D M K I V L N G F M V W
SbMV-N      E A W Y N A V K D E Y E L D D E Q M G V V M N G F M V W
            ---------------------------------------------------------

Cucurbits

WMV-2-AUS   E S W Y S A V K I E Y D L N D E Q M G V I M N G F M V W
WMV-2-US    E S W Y S A V K V E Y D L N D E Q M G V I M N G F M V W
PRSV-W      E K W Y E G V R N D Y G L N D N E M Q V M L N G L M V W
PRSV-P      E K W Y E G V R N D Y G L N D N E M Q V M L N G L M V W
ZYMV-F      A S W F N Q V K T E Y D L N D Q Q M G V V M N G F M V W
ZYMV-I      A S W F N Q V K T E Y D L N E Q Q M G V V M N G F M V W
ZYMV-CT     A S W F N Q V K T E Y D L N E Q Q M G V V M N G F M V W
            ---------------------------------------------------------

Passiflora (and legumes)

PWV-TB      A T W Y E G V K A E Y E L S D D Q M G V I M N P F M V W
PWV-S       A T W Y E G V K A E Y E L S D D Q M G V I M N P F M V W
PWV-M       A T W Y E G V K A E Y E L S D D Q M G V I M N P F M V W
            ---------------------------------------------------------

Solanaceous hosts

TEV-HAT     A A W H Q A V M T A Y G V N E E Q M K I L L N G F M V W
TEV-NAT     A A W H Q A V M T A Y G V N E E Q M K I L L N G F M V W
PepMoV      D T W Y E A V R V A Y D I G E T E M P T V M D G L M V W
PVY-NF      D T W Y E A V R M A Y D I G E T E M P T V M N G L M V W
PVY-D       D T W Y E A V R M A Y D I G E T E M P T V M D G L M V W
PVY-O       D T W Y E A V R M A Y D I G Q T E M P T V M N G L M V W
PVY-I       D T W Y E A V R M A Y D I G E T E M P T V M N G L M V W
TVMV        K A W H T N V M A E L E L N E E Q M K I V L N G F M I W
            ---------------------------------------------------------

Prunus

PPV-D       Q T W Y E G V K R D Y D V T D D E M S I I L N G L M V W
PPV-NAT     Q T W Y E G V K R D Y D V T D D E M S I I L N G L M V W
PPV-AT      Q T W Y E G V K R D Y D V T D D E M S I I L N G L M V W
PPV-R       Q T W Y E G V K R D Y D V T D D E M S I I L N G L M V W
            ---------------------------------------------------------

Graminaceous hosts

SCMV        Q F W Y N R V K K E Y D V D D E Q M R I L M N G L M V W
MDMV-B      D R W Y D A I K K E Y E I D D T Q M T V V M S G L M V W
```

Fig. 1. Coat protein amino acid sequences in part of the TRC; residues are numbered relative to the BYMV coat protein. There are no positions at which a unique residue is common to all viruses or isolates infecting a particular host type and not any viruses affecting another host type (e.g., A at residue 99 in the legume-infecting viruses). At other positions no isolate infecting other host types has the same residue (e.g., P at residue 116 in viruses with solanaceous hosts), but not all isolates infecting that host type have the same residue; the alternate residue at position 116 occurs also in SMV-G1 (K). Similar observations may be made in other domains

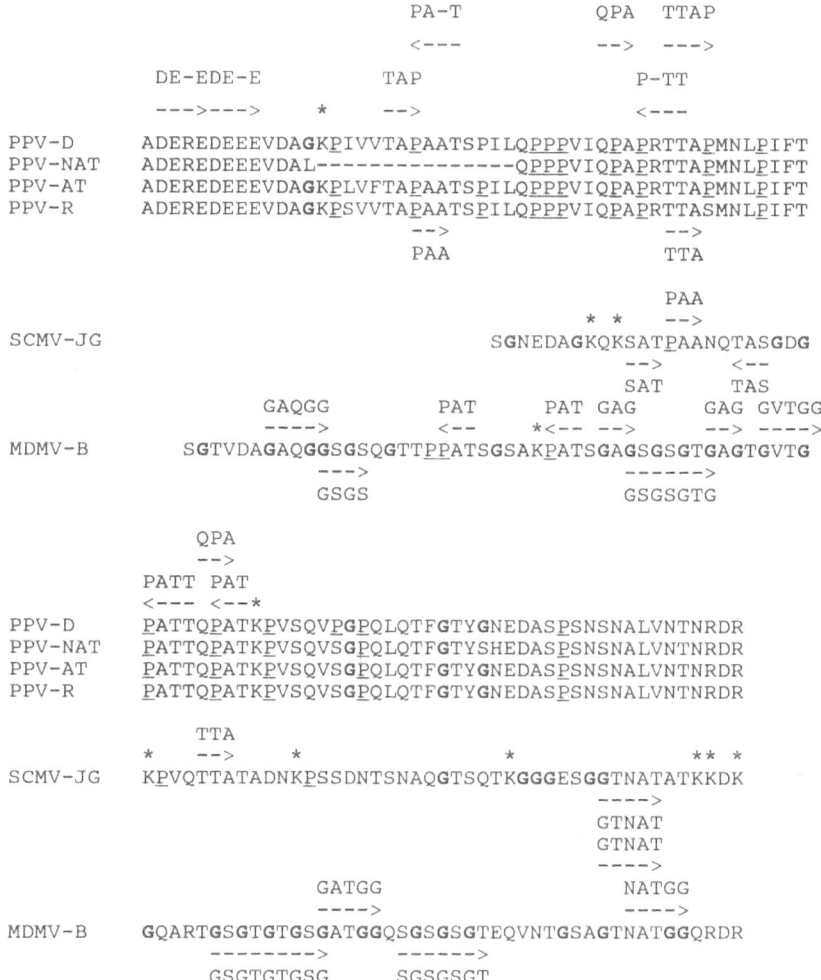

Fig. 2. The aminoterminal sequences of PPV, SCMV-JG and MDMV-B, with glycine (G) and proline (P) residues marked to emphasize their spacing (see text for discussion). Lysine (K) residues are marked with an asterisk. Direct and inverted repeats (or partial repeats) and short regions common to two of the three viruses are also indicated above and below each sequence. The final residue of each sequence shown is the presumed trypsin-sensitive site

isolates, KDK[R, K or D]K in BYMV, KEK and KSKDKE in ClYVV, KGKEKDK in PWV, KGKDK in most isolates of PVY; Fig. 3) suggest that it is derived from a common origin and may have some functional significance.

The only potyviral sequence with a shorter N-terminal extension that has been reported to date, ornithogalum mosaic virus (OrMV) [7] has only 22 residues N-terminal to the position equivalent to the trypsin-sensitive residue of other potyviruses. This residue is glutamine (Q) in OrMV, rather than lysine (K) or arginine (R), and is not trypsin-sensitive; the nearest R is four residues upstream, and there are no lysines at all in the N-terminal region. There are,

```
BYMV-GDD                            SDQEQLNAGEEKKDKRKKNEGNPNKDSEGQSVRQIVPDR
BYMV-PVC                            SDQEQLNAGEEKKDKKKKNEENPDKNSEGQNSRQIVPDR
BYMV-CS                             SDQEKLNASEKKKDKDKKVEDQSTKESEGQSSKQIIPDR
ClYVV-30                            SDKEKLNVGEQQKSKDKESRQYEILEEVGESNRQIIPDR
PSbMV                 AGDETKDDERRRKEEEDRKKREESIDASQFGSNRDKKNKNKESDTSNKLIVKSDR
SbMV-G1                             SNLQEVGDVKASAKKHQEYTNPALHPRRKDK
SbMV-N                              SGKEKEGDMDADKDPKKSTSSSKGAGTSSK
SbMV-G2                             SGKEKEGDMDAGKDPKKSTSSSKGAGTSSK
SbMV-G7                             SGKEKEGEMDAGKDPKKSTSSSKGAGTSSK
WMV-2-AUS                   SGKEAVENLDTGKDSKKDTSGKGDKPQNSQTGQGSKEQTKIGTVSK
WMV-2-US                    SGKETVENLDAGKESKKDASDKGNKPQNSQVGQGSKEPTKTGTVSK
PRSV-W             SKNEAVDTGLNEKFKEKEKQKEKEKEKQKEKEKDDASDGNDVSTSTKTGERDR
PRSV-P             SKNEAVDAGLNEKLKEKENQKEKEKEKQKEKEKDGASDGNDVSTSTKTGERDR
ZYMV-F                          SGTQPTVADARVTKKDKEDDKGENKDFTGSGSGEKTVVAAKKDK
ZYMV-I                          SGTQPTVADTGATKKDKEDDKGKNKDVTGSGSSEKTVAAVTKDK
ZYMV-CT                         SGTQPTVSDAGATKKDKEDDKGKNKDVTGSGSGEKTVAAVTKDK
PWV-TB                          --KDEIIDVGADGKKVVSKKDTQDAGEVNKGKEKDK
PWV-S                           --KDEIIDAEADAKKVVSKKDTQDAGEVNKGKEKDK
PWV-M                           --KDEIIDAGIDGKKGGGKKDTQDAGESNKGKEKDK
TuMV                      AGETLDADLTEEQKQAEKEKKEREKAEKERERQKQLAFKKGK
TEV-HAT                            SGTVDAGAAVGKKKDQKDDKVAEQASKDR
TEV-NAT                            GTTVDASADVGKKKDQKDDKVAEQASKDR
PepMoV                          ANDTIDTGGNSKKDVKPEQGSIQPSSNKGKEK
PVY-NF                          ANDTIDAGGSNKKDAKPEQGSIQPNPNKGKDK
PVY-D                           ANDTIDAGESSKKDARPEQGSIGVNPNKGKDK
PVY-OSA                         ANDTIDAGGNNKKDAKPEQSSIQSNLSKGKDK
PVY-I                           ANDTIDAGGSSKRDAKPEQGSIQPNPNGNKDK
TVMV                            SDTVDAGKDKARDQKLADKPTLAIDRTKDK
OrMV                            ADSMDAGGSSRPPAPLVRQQDQ

SCMV
SGNEDAGKQKSATPAANQTASGDGKPVQTTATADNKPSSDNTSNAQGTSQTKGGGESGGTNATATKKDK
```

Fig. 3. Alignment of N-terminal segments (by eye) to emphasize spacing of lysine residues (K). For each sequence shown the first residue is the predicted cleavage from the poly-protein (except for PWV [48]), and the final residue is that predicted to be sensitive to mild trypsin treatment (except for OrMV, where the residue in the equivalent position is shown; see text). Introduction of gaps into the sequences results in further alignment (data not shown)

however, two glycine and three proline residues among the N-terminal 22 amino acids.

The viruses that have long N-terminal extensions (69–95 residues before the trypsin-sensitive site; SCMV, MDMV-B and PPV) do not have a pattern of alternating lysines in this region; instead they are enriched for glycine and/or proline. In MDMV-B there is an extended pattern of alternating glycine residues, while PPV has mainly proline, typically at five to seven residue intervals. SCMV has more glycine than proline, at less regular intervals (Fig. 2). In addition to glycine and proline enrichment there are several repeats or imperfect direct and inverted repeats of three to seven amino acid residues, mainly including either glycine or proline; parts of these repeats occur in the N-termini of more than one of these three viruses (Fig. 2). This suggests that the N-terminal region of the potyvirus CP gene may be subject to replicase slippage, and possibly template jumping between positive and negative strands. Such occurrences could be a factor in the evolution of potyviruses with new CP properties.

Variation in gene product size and sequence of the 5' terminal gene and the gene between helper component (HC) and CIP (which vary more among the published sequences than do HC, CIP, NIb, and CP exclusive of the N-terminus) might also be due to replicase slippage and/or template jumping. Further evidence suggestive of replicase slippage or jumping is found in the deletions in PPV-NAT with respect to other PPV isolates, and in SbMV-N with respect to WMV-2.

The significance of lysine repeats versus glycine and proline repeats is not clear; however, it is possible that the positive charges of lysine at regular intervals contribute to the interaction between CP and HC. HC activity has been shown to require Mg^{++} ions under some buffer conditions [14, 40] and a cysteine-rich HC domain was identified as a possible metal-binding site [42]. It may be that the charge of the lysines affects CP–HC binding through this domain, in which Thornbury et al. [53] found one of two point mutations differentiating the HCs of PVY and PVC (which has a non-functional HC).

Although the only proven functions of CP are encapsidation of RNA and a role in aphid-transmissibility, sequence examination of the CP of different potyviruses suggests that two subgroups may be separated on the basis of the length and structure of the N-terminal extension – with one subgroup having patterns of alternating lysines within a probable alphahelical structure, while the second subgroup (MDMV-B, SCMV and PPV) has a longer extension rich in glycine and/or proline and a potentially more folded structure. Examination of further sequences may indicate that OrMV forms part of a third subgroup with shorter extensions lacking lysine. These subgroups are not consistent with phylogenetic trees constructed on the basis of CP amino acid sequence [7, 39], nor are there obvious correlations with serology, host range or vectors. Nonetheless there may prove to be an evolutionary significance to these differences, and hence value for taxonomy.

Lack of correlation of conserved epitopes with known biological functions suggests that the structure of the CP is more important for particle stability than for interactions involving host range or replication determinants, as most epitopes are thought to be surface exposed (flexibility being associated with both antigenicity and surface exposure) at least in terms of the subunit.

As more work is done with mutation of infectious cDNA clones, and more detail is gained from study of CP folding and structure, no doubt further understanding of the potential interactions between sequences, serology and biology will emerge.

Conclusions

Serological relationships between potyvirus virions, and other gene products, have long been recognized as very complex. The differences in apparent relationships reported by various laboratories can be largely explained by the use of different isolates, immunization protocols and types of assay. The increasing use of techniques that present antigens in denatured forms (e.g.,

Western blots and antigen-coated plate forms of indirect ELISA) is leading to greater recognition and detection of epitopes formed by conserved sequences that probably maintain the structure and/or function of potyviral gene products.

Our increased knowledge of gene functions, and in particular the developing ability to substitute genes and gene products of one virus into the genome of another, will lead to a better appreciation of the significance of conserved structures and of the differences that do exist. Insufficient work has yet been published to allow any clear picture of which genes and gene products contribute to host range determination and symptom expression of potyviruses, but work with chimeric infectious transcripts is likely to change that situation markedly over the next few years.

In vitro assays are being developed for other gene functions which will allow much greater understanding of gene product structure/function relationships, and hypotheses based on sequence comparisons and folding predictions will form the basis for much of this work. Topographical analysis of epitopes with monoclonal antibodies, and replacement set analysis using synthetic peptides or cassette mutagenesis will play a significant role in such studies.

Some of the observations on sequence similarities presented here have no current link to known aspects of potyvirus biology. Whether such similarities are indeed of biological significance remains to be seen, but the extensive homologies now known to exist between gene products of viruses previously thought to be only distantly serologically related suggests that conservation of form is associated with function.

Three subgroups of potyviruses are suggested here based on the length and composition of the amino-terminal extension of the capsid protein; these are not correlated with overall sequence homology, nor with groupings based on the cytology of the cytoplasmic inclusions. It is anticipated that a biological difference will be identified that correlates with this structural variation, possibly in vector specificity or retention, or perhaps in the infection process. Substitution of one type of N-terminus for another in infectious transcripts or CP-transgenic plants may afford one means of determining a biological effect.

The very diversity of the potyvirus group that has stymied taxonomic efforts over the years can now be used to determine biological effects through substitution in genetically engineered constructs. Through such experiments we may learn which features are of true taxonomic value, and which are merely curiosities; for characters without biological significance are of less value in a meaningful taxonomy. This is perhaps most true of viruses with RNA genomes such as the potyviruses, since the inherent infidelity of RNA replication may lead to diversity without correlation to biological activity.

Acknowledgements

I thank all of my colleagues with whom I have discussed aspects of taxonomy that form the topic of this paper, and in particular Ramon Jordan.

References

1. Allison R, Johnston RE, Dougherty WG (1986) The nucleotide sequence of the coding region of tobacco etch virus genomic RNA: evidence for the synthesis of a single polyprotein. Virology 154: 9–20
2. Allison RF, Sorenson JC, Kelly ME, Armstrong FB, Dougherty WG (1985) Sequence determination of the capsid protein gene and flanking regions of tobacco etch virus: evidence for synthesis and processing of a polyprotein in potyvirus gene expression. Proc Natl Acad Sci USA 82: 3969–3972
3. Allison RF, Dougherty WG, Parks TD, Willis L, Johnston RE, Kelly M, Armstrong FB (1985) Biochemical analysis of the capsid protein gene and capsid protein of tobacco etch virus: N-terminal amino acids are located on the virion's surface. Virology 147: 309–316
4. Atreya CD, Raccah B, Pirone TP (1990) A point mutation in the coat protein abolishes aphid transmissibility of a potyvirus. Virology 178: 161–165
5. Boye K, Jensen PE, Stummann BM, Henningsen KW (1990) Nucleotide sequence of cDNA encoding the BYMV coat protein gene. Nucleic Acids Res 18: 4926
6. Bravo-Almontacid F, Mentaberry AN (1989) Nucleotide cDNA sequence coding for the PVY$_O$ coat protein. Nucleic Acids Res 17: 4401
7. Burger JT, Brand RJ, Rybicki EP (1990) The molecular cloning and nucleotide sequencing of the 3' terminal region of Ornithogalum mosaic virus. J Gen Virol 71: 2527–2534
8. Domier LL, Franklin KM, Shahabuddin M, Hellmann GM, Overmeyer JH, Hiremath ST, Siaw MFE, Lomonossoff GP, Shaw JG, Rhoads RE (1986) The nucleotide sequence of tobacco vein mottling virus RNA. Nucleic Acids Res 14: 5417–5430
9. Dougherty WG, Allison RF, Parks TD, Johnston RE, Feild MJ, Armstrong FB (1985) Nucleotide sequence at the 3' terminus of pepper mottle virus genomic RNA: evidence for an alternate mode of potyvirus capsid protein gene organization. Virology 146: 282–291
10. Dougherty WG, Willis L, Johnston RE (1985) Topographic analysis of tobacco etch virus capsid protein epitopes. Virology 144: 66–72
11. Eggenberger AL, Stark DM, Beachy RN (1989) The nucleotide sequence of a soybean mosaic virus coat protein-coding region and its expression in *Escherichia coli*, *Agrobacterium tumefaciens* and tobacco callus. J Gen Virol 70: 1853–1860
12. Gal-On A, Antignus Y, Rosner A, Raccah B (1990) Nucleotide sequence of the zucchini yellow mosaic virus capsid-encoding gene and its expression in *Escherichia coli*. Gene 87: 273–277
13. Gough KH, Azad AA, Hanna D, Shukla DD (1987) Nucleotide sequence of the capsid and nuclear inclusion protein genes from the Johnsongrass strain of sugarcane mosaic virus RNA. J Gen Virol 68: 297–304
14. Govier DA, Kassanis B, Pirone TP (1977) Partial purification and characterization of the potato virus Y helper component. Virology 78: 306–314
15. Grumet R, Fang G (1990) cDNA cloning and sequence analysis of the 3' terminal region of zucchini yellow mosaic virus RNA. J Gen Virol 71: 1619–1622
16. Gugerli P, Fries P (1983) Characterization of monoclonal antibodies to potato virus Y and their use for virus detection. J Gen Virol 64: 2471–2477
17. Hammond J (1989) Antisera to cytoplasmic inclusion proteins of potyviruses contain cross-reactive antibodies. Phytopathology 79: 1174
18. Hammond J, Hammond RW (1989) Molecular cloning, sequencing and expression in *Escherichia coli* of the bean yellow mosaic virus coat protein gene. J Gen Virol 70: 1961–1974
19. Hammond J, Lawson RH, Hsu HT (1985) Use of a monoclonal antibody reactive with several potyviruses for detection and identification in combination with virus-specific antisera. Phytopathology 75: 1353

136   J. Hammond

20. Hammond J, Jordan RL, Kamo KK (1990) Use of chimeric coat protein constructs and deletion mutants to examine potyvirus structure and coat protein-mediated resistance. Phytopathology 80: 1018
21. Hellmann GM, Thornbury DW, Pirone TP (1990) Molecular analysis of tobacco vein mottling virus (TVMV) pathogenicity by infectious transcripts of chimeric potyviral cDNA genomes. Phytopathology 80: 1036
22. Hiebert E, Tremaine JH, Ronald WP (1984) The effect of limited proteolysis on the amino acid composition of five potyviruses and on the serological reaction and peptide map of the tobacco etch virus capsid protein. Phytopathology 74: 411–416
23. Hill EK, Hill JH, Durand DP (1984) Production of monoclonal antibodies to viruses in the potyvirus group: use in radioimmunoassay. J Gen Virol 65: 525–532
24. Hopp TP, Woods KR (1981) Prediction of protein antigenic determinants from amino acid sequences. Proc Natl Acad Sci USA 78: 3824–3828
25. Hsu HT, Franssen JM, Hammond J, Derks AFLM, Lawson RH (1986) Some properties of mouse monoclonal antibodies produced to tulip breaking virus. Phytopathology 76: 1132
26. Hsu HT, Franssen JM, Van Der Hulst CTC, Derks AFLM, Lawson RH (1988) Factors affecting selection of epitope specificity of monoclonal antibodies to tulip breaking potyvirus. Phytopathology 78: 1337–1340
27. Jayaram C, Hill JH, Miller WA (1991) Nucleotide sequences of the coat protein genes of two aphid-transmissible strains of soybean mosaic virus. J Gen Virol 72: 1001-1003
28. Jensen SG, Staudinger JL (1989) Serological grouping of the cytoplasmic inclusions of 6 strains of sugarcane mosaic virus. Phytopathology 79: 1215
29. Jilka JM (1990) Cloning and characterization of the 3′ terminal regions of RNA from select strains of maize dwarf mosaic virus and sugar cane mosaic virus. PhD dissertation, University of Illinois at Urbana-Champaign
30. Jordan RL (1989) Mapping of potyvirus-specific and group-common antigenic determinants with monoclonal antibodies by Western-blot analysis and coat protein amino acid sequence comparisons. Phytopathology 79: 1157
31. Jordan RL, Hammond J (1986) Analysis of antigenic specificity of monoclonal antibodies to several potyviruses. Phytopathology 76: 1091
32. Jordan RL, Hammond J (1991) Comparison and differentiation of potyvirus isolates and identification of strain-, virus-, subgroup-specific and potyvirus group-common epitopes using monoclonal antibodies. J Gen Virol 72: 25–36
33. Karplus PA, Schulz GE (1985) Prediction of chain flexibility in proteins. Naturwissenschaften 72: 212–213
34. Lain S, Riechmann JL, Mendez E, Garcia JA (1988) Nucleotide sequence of the 3′ terminal region of plum pox potyvirus RNA. Virus Res 10: 325–342
35. Lain S, Riechmann JL, Garcia JA (1990) RNA helicase: a novel activity associated with a protein encoded by a positive stranded RNA virus. Nucleic Acids Res 18: 7003–7006
36. Maiss E, Timpe U, Brisske A, Jelkmann W, Casper R, Himmler G, Mastanovich D, Katinger HWD (1989) The complete nucleotide sequence of plum pox virus RNA. J Gen Virol 70: 513–524
37. Nagel J, Hiebert E (1985) Complementary DNA cloning and expression of the papaya ringspot potyvirus sequences encoding capsid protein and a nuclear inclusion-like protein in *Escherichia coli*. Virology 143: 435–441
38. Quemada H, L'Hostis B, Gonsalves D, Reardon IM, Heinrikson R, Hiebert EL, Sieu LC, Slightom JL (1990) The nucleotide sequence of the 3′ terminal regions of papaya ringspot virus strains W and P. J Gen Virol 71: 203–210
39. Quemada H, Sieu LC, Siemeniak DR, Gonsalves D, Slightom JL (1990) Watermelon mosaic virus II and zucchini yellow mosaic virus: cloning of 3′ terminal regions, nucleotide sequences, and phylogenetic comparisons. J Gen Virol 71: 1451–1460.

40. Raccah B, Pirone TP (1984) Characteristics of and factors affecting helper component-mediated aphid transmission of a potyvirus. Phytopathology 74: 305–308
41. Ravelonandro M, Varveri C, Delbos R, Dunez J (1988) Nucleotide sequence of the capsid protein gene of plum pox potyvirus. J Gen Virol 69: 1509–1516
42. Robaglia C, Durand-Tardif M, Tronchet M, Boudazin G, Astier-Manifacier S, Casse-Delbart F (1989) Nucleotide sequence of potato virus Y (N strain) genomic RNA. J Gen Virol 70: 935–947
43. Rosner A, Raccah B (1988) Nucleotide sequence of the capsid protein gene of potato virus Y (PVY). Virus Genes 1: 255–260
44. Salomon R (1989) Partial cleavage of sweet potato feathery mottle virus coat protein subunit by an enzyme in extracts of infected symptomless leaves. J Gen Virol 70: 1943–1949
45. Scott SW, McLaughlin MR, Ainsworth AJ (1989) Monoclonal antibodies produced to bean yellow mosaic virus, clover yellow vein virus, and pea mosaic virus which cross-react among the three viruses. Arch Virol 108: 161–167
46. Shukla DD, Ward CW (1989) Identification and classification of potyviruses on the basis of coat protein sequence data and serology. Arch Virol 102: 171–200
47. Shukla DD, Inglis AS, McKern NM, Gough KH (1986) Coat protein of potyviruses. 2. Amino acid sequence of the coat protein of potato virus Y. Virology 152: 118–125
48. Shukla DD, McKern NM, Ward CW (1988) Coat protein of potyviruses. 5. Symptomatology, serology and coat protein sequences of three strains of passionfruit woodiness virus. Arch Virol 102: 221–232
49. Shukla DD, Strike PM, Tracy SL, Gough KH, Ward CW (1988) The N and C termini of the coat proteins of potyviruses are surface-located and the N-terminus contains the major virus-specific epitopes. J Gen Virol 69: 1497–1508
50. Shukla DD, Tosic M, Jilka J, Ford RE, Toler RW, Langham MAC (1989) Taxonomy of potyviruses infecting maize, sorghum and sugarcane in Australia and the United States as determined by reactivities of polyclonal antibodies directed towards virus-specific N-termini of coat proteins. Phytopathology 79: 223–229
51. Shukla DD, Jilka J, Tosic M, Ford RE (1989) A novel approach to the serology of potyviruses involving affinity-purified polyclonal antibodies directed towards virus-specific N-termini of coat proteins. J Gen Virol 70: 13–22
52. Takahashi T, Uyeda I, Ohshima K, Shikata E (1990) Nucleotide sequence of the capsid protein gene of bean yellow mosaic virus chlorotic spot strain. J Fac Agric Hokkaido Univ 64: 152–163
53. Thornbury DW, Patterson CA, Dessens JT, Pirone TP (1990) Comparative sequence of the helper component (HC) region of potato virus Y and a HC-defective strain, potato virus C. Virology 178: 573–578
54. Timmerman GM, Calder VL, Bolger LEA (1990) Nucleotide sequence of the coat protein gene of pea seedborne mosaic potyvirus. J Gen Virol 71: 1869–1872
55. Tremblay MF, Nicolas O, Sinha RC, Lazure C, Laliberte JF (1990) Sequence of the 3′ terminal region of turnip mosaic virus RNA and the capsid protein gene. J Gen Virol 71: 2769–2772
56. Turpen T (1989) Molecular cloning of a potato virus Y genome: nucleotide sequence homology in non-coding regions of potyviruses. J Gen Virol 70: 1951–1960
57. Uyeda I, Takahashi T, Shikata E (1991) Relatedness of nucleotide sequence of the 3′ terminal region of clover yellow vein potyvirus to bean yellow mosaic potyvirus RNA. Intervirology 32: 234–245
58. Van Regenmortel MHV, von Wechmar MB (1970) A re-examination of the serological relationship between tobacco mosaic virus and cucumber virus 4. Virology 41: 330–338
59. Yu M, Frenckel MJ, McKern NM, Shukla DD, Strike PM, Ward CW (1989) Coat protein of the potyviruses. 6. Amino acid sequences suggest watermelon mosaic virus 2 and soybean mosaic virus are strains of the same potyvirus. Arch Virol 105: 55–64

60. Zitter TA (1972) Naturally occurring pepper virus strains in Florida. Plant Dis Rep 56: 586–590

Author's address: J. Hammond, USDA-ARS, FNCL, Room 208, Building 004, BARC-West, 10300 Baltimore Avenue, Beltsville, MD 20705-2350, U.S.A.

Arch Virol (1992) [Suppl 5]: 139–170
© by Springer-Verlag 1992

Coat protein phylogeny and systematics of potyviruses

E. P. Rybicki[1] and **D. D. Shukla**[2]

[1] Department of Microbiology, University of Cape Town, Rondebosch, South Africa
[2] CSIRO Division of Biomolecular Engineering, Parkville Laboratory, Parkville, Victoria,
Australia

Summary. The feasibility of applying molecular phylogenetic methods of analysis to aligned coat-protein sequences and other molecular data derived from coat proteins or genomic sequences of members of the proposed taxonomic family *Potyviridae*, is discussed. We show that comparative sequence analysis of whole coat-protein sequences may be used reliably to differentiate between sequences of closely related strains, and to show groupings of more distantly related viruses; that coat proteins of putative *Potyviridae* cluster according to the proposed generic divisions, and, even if some are only very distantly related, the members of the family form a cluster distinct from coat proteins of other filamentous and rod-shaped viruses. Taxonomic revisions based on perceived evolutionary relationships, and the lack of feasibility of erecting higher taxa for these viruses, are discussed.

General introduction to virus taxonomy/systematics

Properties of viruses used in taxonomic constructions are particle morphology, genome type and homologies, serology, protein content, replication strategy, host range and effects on the host, and transmission mechanisms [46, 52]. The most important of these properties for purposes of initial identification are probably the morphology, genome type, preferred infection host(s), and mode of transmission. These properties can be used to "type" a virus down to and often below genus level (animal viruses) or group level (plant viruses). With plant viruses this sort of characterization would often be followed by serology and perhaps by more detailed characterization of virion components, which until recently has usually been sufficient for taxonomic placement of a virus. However, in this age of DNA and cDNA sequencing, "genome type" is often defined by the partial or entire genomic sequence. Comparative analysis of the currently available viral sequences has led recently to some profound changes in our understanding of virus evolution, and to a new science: molecular virus systematics [52, 57].

Molecular phylogenetics and virus classification

Carl Woese has written of ribosomal RNA comparisons between eukaryotes, archae- and eubacteria [80], that "From the extent (and nature) of the differences among a set of homologous sequences one can then reconstruct molecular genealogies, evolutionary trees of organisms." Indeed, it is increasingly taken for granted in animal virology that the deduced evolutionary history of a group of viruses, as arrived at from comparative sequence analysis, should be taken into account in their classification [6,52]. This is despite the presence in the "Guidelines for Delineation and Naming of Species" of the 1982 Fourth Report of the International Committee on Taxonomy of Viruses (ICTV) [46], of the statement "Virus taxonomy at its present stage has no evolutionary or phylogenetic implications." An important new development in animal virus classification is the recent proposal by Palmenberg [54] of a radical revision of the taxonomic family *Picornaviridae*, on the basis of phylogenetic reconstructions from coat and other protein sequence alignments. This proposal entails the establishment of up to 11 new subgenera based on coat-protein sequence homologies [68]. An equivalent phylogenetic study on plant viruses was done recently for geminiviruses by Howarth and Vandermark [31]. They produced phylogenies based on sequence similarities between replication-associated proteins and between coat proteins. These agree well with, and to some extent have probably helped to mold, current thinking on geminivirus classification (ICTV Plant Virus Subcommittee Proceedings, 1992), and should prove very useful in future taxonomic assignments within the group. Dolja and Koonin have shown [12] that a set of capsid proteins of viruses from several taxonomic groups of plant viruses (tombus-, diantho-, carmo-, sobemo-, and luteoviruses) "... forms a tight evolutionary cluster," presumably descended from a single common ancestor.

Taxonomy of potyviruses

New approaches to potyviral taxonomy

The potyvirus group is the largest and most rapidly growing taxonomic collection of plant viruses [46,78], and its taxonomy is in a chaotic state that has been only partially resolved by recent systematic application of sophisticated serological and physico-chemical techniques [35,39,59–62,78,79]. It has been suggested [35,60,64,78] that molecular biological approaches to potyvirus characterization may provide a more rational or perhaps systematic basis for identification and classification in this taxonomic family. These approaches include the use of monoclonal and affinity-purified monospecific antibodies, partial and complete genomic sequencing, and coat protein peptide profiling by high performance liquid chromatography (HPLC).

HPLC peptide profiling

The applications and/or limitations of serology and of genomic sequencing are discussed in companion papers in this volume [63, 79]. A technique that is not discussed in these papers that has proved very useful recently in potyvirus taxonomy, is HPLC profiling of peptides derived from coat proteins by trypsin (or other protease) digestion [64]. This has been used to good effect to show similarities and differences between a large number of strains and isolates of a number of potyviruses, including soybean mosaic virus strains and the sugarcane mosaic virus subgroup [33, 47–49, 64]. It has been shown elsewhere that the degree of similarity of coat-protein peptide profiles is a reliable indication of their sequence relatedness [47, 64]. Recently, it was shown that HPLC peak retention time data may be used for numerical analysis of virus coat-protein strain relationships, if not for distinct viruses [49]. Thus, any discussion of coat-protein and/or genomic sequence analysis as an aid to classification of potyviruses should also include mention of the actual and potential contribution of peptide profiling.

Sequence relationships between members of the *Potyviridae*

To date, genomic sequencing has confirmed that accepted members of the potyvirus group share a common genomic organization, and that the genetic distances between their coat proteins are generally in agreement with previously observed properties such as cross-protection and serological relationships [8, 59–62, 79]. An important observation from sequence comparisons is a bimodal distribution of coat-protein sequence homology, with known distinct potyviruses varying in sequence similarity between 38–71% and known strains of given viruses varying between 90–99% [59]. Subgroupings among potyviruses on the basis of coat-protein sequence relationships also have been proposed: examples are a "PWV/WMV/ZYMV" subgroup, a subgroup consisting of MDMV-A, SCMV-SC, and SrMV-SCH, and a BYMV/ClYVV subgroup [20, 74, 78] (Table 1).

A useful observation from sequence analysis that has found recent application is that the potyviral genomic 3' non-coding region (3' NCR) can be used as a probe for sensitive detection of related strains of a given potyvirus, and that sequence comparisons can quickly and reliably be used to show strain relationships, though 3' NCR sequences probably are not useful for showing relationships between distinct viruses [20, 21, 74]. Comparative analyses of the whole genome sequences of certain potyviruses show that all the genomes appear colinear, and each gene appears to have homologies in the other viruses [62, 78, 79]. Thus, sequence comparisons for parts of the genomes of totally sequenced viruses would give much the same picture of virus relationships no matter which part of the genome is compared [62; Rybicki, unpubl.]. Interestingly, much the same answer can be obtained from serological studies on potyviral non-capsid proteins: Suzuki et al. [71] showed that antisera to cylindrical in-clusion body proteins (CIP), amorphous inclusion proteins (AIP), and coat proteins of WMV2, ZYMV and PRSV-W all gave the same sort of relation-ships among the respective proteins in terms of serological differentiation indices (SDIs). This means that the viruses compared, and presumably most potyviruses, probably all diverged from a common ancestral virus without any addition or reorganization of genes or gene order. This is also apparently true for the mite-transmitted poty-like wheat streak mosaic virus (WSMV): this is clearly, though distantly, related in sequence and genome organization to potyviruses [53]. The situation for the fungus-transmitted poty-like barley yellow mosaic virus (BaYMV) is more complex. It has a bipartite genome, with the larger RNA 1 corresponding to and being colinear with the 3' two-thirds of the "normal" potyvirus genome, and RNA 2 corresponding roughly to the 5' one-third of the potyviral genome. Important differences in genetic organization between BaYMV and potyviruses are confined to the 5' end of RNA 2, which apparently has no analogue in other potyviruses [37, 38, 79]. Kashiwazaki et al. [36] also showed a convincing alignment between parts of the coat proteins of potato virus X (PVX, type member of the potexvirus group) and those of BaYMV and various potyviruses. This was repeated and extended by Dinant et al. [11], who compared and showed similarities between coat-protein sequences of several potyviruses, potexviruses, a carlavirus, and a tobamovirus. These reports raise the intriguing possibility that coat protein similarities among rod-like and filamentous viruses might even transcend current taxonomic group divisions, to the "supergroup" level. This possibility has been confirmed for spherical viruses [12]. Dolja et al. [13] were not able to get statistically significant alignment between CPs of rod-like and filamentous viruses, but suggested that the two groups descended separately from single ancestors.

"Supergroup" and superfamilial relationships

Sequence comparisons of non-capsid potyviral proteins has led certain work-ers to propose "supergroups" or "superfamilies" of single-strand positive-sense RNA viruses of plants and animals [25; for review, see 70]. Putative superfamilies of NTP-binding proteins or helicases from the RNA poty-, como-, nepo-, and flaviviruses, and the DNA gemini-, parvo-, and papova-viruses have also recently been proposed to be evolutionarily related to each other and to cellular proteins [26, 42]. Bruenn [7] has proposed a relationship scheme for positive- and double-strand RNA viruses based on their RNA-dependent RNA polymerases which is strongly at variance with the ones outlined above, and which defines luteovirus-like, flavi-, poty-, tobamo-, alpha-, picorna-, levi-, and dsRNA virus-like replicase superfamilies. Koonin [41] has examined relationships among polymerases of plus-strand ssRNA viruses, and has proposed three superfamilies, one of which incorporates poty-

and picornaviruses. Thus, potyviruses may be related on a number of different levels and in a number of different directions, depending on which protein is being compared, to a variety of other virus taxonomic groups, and even to cellular proteins. A "module linkage" diagram for indicating the homologies (and presumed lines of descent) of different parts of the genomes of viruses in different taxonomic families has been proposed elsewhere. In this scheme each taxonomic family would have a unique arrangement and collection of "genomic modules," most of which would be shared individually with at least one other family [57].

A taxonomic family Potyviridae?

The current problem in potyvirus taxonomy and classification is not only how to group and differentiate the accepted members (= aphid-transmitted viruses) of the potyvirus group, but also how to place these relative to other distinct groups of poty-like and obviously related viruses such as those related to WSMV, BaYMV, and the whitefly-transmitted sweet potato mild mottle virus (SPMMV), as well as to group all of these relative to other viruses with rod-like or filamentous particles. Analysis of the partial genomic sequence of WSMV has suggested that, although it was distantly related, it should be considered a part of the potyvirus group [53]. Kashiwazaki and co-workers [36–38] have suggested that BaYMV and related viruses be considered a new and distinct taxonomic group (the bymoviruses) on the strength of the rather distant genetic relationship to accepted potyviruses, the possession of a bipartite genome, and differences in genome organization. Interestingly, there is also serological evidence linking BaYMV-like, WSMV-like, and SPMMV-like viruses to each other and to potyviruses [65, 67]. An alternative to putting poty-like viruses into/out of the present potyvirus group taxon is to erect some higher order taxon(s) that will take into account perceived similarities and differences between these viruses. A growing body of opinion now holds that all of these viruses should be considered as a virus family, with different genera comprising the easily distinguishable groups. Thus, it has been proposed that the family *Potyviridae* would include the genera *Potyvirus* (potyviruses), *Bymovirus* (BaYMV-like), *Rymovirus* (WSMV-like), and a possible genus *Ipomovirus* comprising SPMMV and related viruses [3]. Objections to inclusion of the bymoviruses in such a family could be overcome by recourse to the precedent of the proposed family *Geminiviridae*, which includes viruses which differ not only in genome organization, but also in number of genomic components and genome size (ICTV Plant Virus Subcommittee Recommendations, 1992; [31]). In this case, as should perhaps also be true of the proposed *Potyviridae*, it is felt that similarity in overall genome structure, detectable similarities in sequences of replication-associated and coat proteins, and the perception that the viruses have a common evolutionary origin, are sufficient justification for grouping them together in a family.

Molecular phylogenetic approach to *Potyviridae* taxonomy: a proposal

Comparisons of coat and other proteins

Most previous studies on potyvirus genetic relationships have utilized comparisons of coat proteins [8,35,36,56,59–65,75,77–79]. This has been done by comparing peptide HPLC elution profiles, serological properties, and nucleotide and amino acid sequences; the amino acid sequences have been determined directly or derived from partial or complete nucleotide sequences. It has been argued that it is not a good idea to base a potyvirus or *Potyviridae* classification scheme mainly on the coat-protein gene, since in poty- and related viruses this comprises less than 10% of the genome [79,84]. This is a cogent point; however, the coat protein is a major determinant of classification because of its contribution to the filamentous particle morphology, to serological relationships, and to vector relations. All these properties are vital for taxonomic assignments of viruses for which more detailed properties may never be determined. In addition, the coat-protein sequence of poty- and poty-like viruses is almost uniquely easy to determine among plant viruses. This is partly because it may often be purified in relatively high yields for direct sequencing, and largely because the coat protein (CP) and the 3′ NCR are the sequences directly proximal to the genomic 3′ poly(A) tail, which is routinely used for oligo-dT priming for reverse transcription from viral RNA for subsequent cDNA cloning. As a result over 40 CP and flanking sequences have been determined for members of the *Potyviridae*, while relatively few genomes (<10) have been completely sequenced. The CP sequences of the proposed *Potyviridae* comprise probably the largest plant virus sequence database, and one that is easily comparable to the picornavirus sequence collection which has been used to such good effect in predicting relationships [54,68]. There are also HPLC coat-protein peptide profiles which embody a large amount of information for a large number of potyviruses [33,47–49,64]. Sequence data from the existing database has been used by two groups for the synthesis of degenerate oligonucleotide "consensus primers" for amplification by means of the polymerase chain reaction (PCR) of stretches of sequence from the NIb and CP coding regions of the genomes of generic potyviruses. Sequences may be amplified and determined relatively easily from the CP coding region of a wide variety of viruses, and less reliably from the NIb region [43; R. Brand and E. Rybicki, unpubl.]. Comparison of the partial CP region nucleotide sequences gives a very similar relationship picture to one obtained from the whole CP nucleotide or amino acid sequence, and certainly serves to unequivocally identify distinct viruses and strains of others. PCR technology therefore represents a reliable means of obtaining valuable partial CP sequence information from hard-to-purify viruses, or from viruses present in small leaf samples or archival and even perhaps non-infectious material.

Although others have pointed out that the putative viral protease (NIa) and polymerase (NIb) sequences are probably more conserved, and can be used to show relationships outside the potyvirus family [14, 25, 41] (see also above), there are simply too few sequences available to do as complete an analysis as is possible using CPs, even though the relationships that are shown by such analysis closely parallel those shown using CPs [62; E Rybicki, unpubl.]. Accordingly, analyses undertaken for this work were based on CP sequences, and to a limited extent on HPLC profiles (see below).

It is our intention in this article, given the above arguments, to show that detailed comparative analysis of all of the known coat-protein sequences of *Potyviridae* species provides an excellent basis for an accurate and detailed classification of the viruses from strain up to family level. We hope to demonstrate that a phylogeny of the viruses based on coat-protein sequences has great relevance to their classification and taxonomy, and should be taken into account in any future taxonomic assignments in, or revisions of, the proposed family.

Methods for sequence alignment and phylogeny reconstruction

Amino acid or nucleotide sequences?

The relative merits of using nucleotide or derived amino acid sequences for analysis of relationships have been exhaustively discussed elsewhere; it suffices for our purposes to point out that comparisons of protein sequences give a far better idea of distant relationships than do equivalent comparisons of nucleotide sequences. This is partly because polypeptides have 20 different "character states" as opposed to the four of nucleic acids, which results in far lower "noise levels" when searching for meaningful alignments, and partly because codon redundancy in nucleic acid coding sequences allows for variations that are not reflected in the encoded protein sequence [44]. It is also possible to use restriction enzyme cleavage profiles and maps of cDNA and data such as HPLC profiles for numerical analysis of phylogenetic relationships. Restriction endonuclease cleavage profiles will differ more than HPLC peptide profiles for strains of the same virus because restriction sites in cDNA will be less well conserved than endoproteinase cleavage sites in coat proteins, meaning that comparisons of peptide profiles will presumably give a better idea of distant relationships than comparison of equivalent restriction profiles. Other good reasons for comparing protein rather than nucleotide sequences are that some potyvirus CP sequences have been obtained directly from purified proteins, and that it is also easier to obtain purified protein and an HPLC profile than it is to generate sufficient cDNA for restriction analysis.

Multiple sequence alignment

Although it is possible to align sequences by eye using microcomputer (PC) text editing programs, less bias is introduced in alignments done using proven computer programs which are now readily available for both PCs and mainframes. One may use pairwise alignment routines such as GAP and BESTFIT in the GCG (Genetics Computer Group, WI, U.S.A. [9]) mainframe package, or subroutines in PC programs such as GENEPRO (Riverside Scientific Enterprises, WA, U.S.A.), to build up a multiple alignment [8]. However, given the large number of potyvirus CP sequences available, it is probably wiser to use one of the newer multiple sequence alignment packages for PC and mainframe [15a]. We recommend the use of a mainframe package because of its superior processing speeds and ability to handle greater numbers of longer sequences. We tested the same potyvirus CP data set on three mainframe packages on a VAX 8530. These were CLUSTAL [30], PAPA (Parsimony After Progressive Alignment) [15, 19], and TreeAlign [29]. Perhaps the most generally useful program in our hands has been CLUSTAL: this was easier to use and had a more permissive data input format than either of the others, could handle more and longer sequences than PAPA, and consistently gave better alignments (as judged by previously-published alignments) than TreeAlign. A newer and more versatile version (CLUSTALV) is now available from the EMBL fileserver; a similar program (called PileUp) is also included in version 7.0 of the GCG package.

Phylogeny reconstruction

Construction of phylogenetic trees or relationship dendrograms may be done fairly easily on a PC, using a number of software packages; however, one must beware of becoming embroiled in the often acrimonious debate raging between proponents of phenetic or "distance matrix" methods, and cladistic or "character state" maximum parsimony methods [44]. We make no recommendations in this regard, except to note that distance matrix methods are usually easier to apply, take far less computer time and processing power, give less ambiguous results, and are less subject to errors resulting from different rates of evolution in different branches [80]. Input is simply data matrices of all possible pairwise distance measurements: these can be values such as serological differentiation indices (SDIs), proportion of shared HPLC "peaks" between two profiles, the proportion of restriction fragments or map sites shared between cDNAs, or, more usually, sequence homologies. Popular programs tested by us are the KITSCH and FITCH method options in Felsenstein's Phylogeny Inference Package versions 3.1 and 3.3 (PHYLIP, [18]) and the program NJTREEE (Neighbor-Joining method, [58]). CLUSTAL produces a dendrogram calculated by the unweighted pair-group method using averages (UPGMA) distance matrix technique of Sneath and Sokal [66], as a by-product of the alignment process. CLUSTALV calculates a neighbor-joining tree from pairwise distance measurements made from aligned sequences. KITSCH was used by Burger et al. [8] for phylogenetic reconstruction from potyvirus CP sequences, while Shukla and Ward [60] used the UPGMA. If one wishes to more accurately reconstruct phylogenies, or to be aware of alternative tree topologies, then maximum parsimony (or "character state" analysis) methods may be more attractive. Input data for these methods is a matrix consisting of aligned sequences, or appropriately encoded restriction fragment or map profiles, or HPLC profiles. Useful programs include the protein parsimony program PROTPARS and the DNA parsimony options DNAPARS, DNAPENNY, and DNABOOT in PHYLIP (which are all extremely slow); HENNIG86 [17], which is extremely fast, but useful only for nucleic acid sequences; and the new versatile and fast Apple Macintosh version of PAUP (Phylogenetic Analysis Using Parsimony [72]). PAUP embodies all of the PHYLIP options while being far more user-friendly. Quemada et al. [56] used PAUP to calculate a tree for the potyvirus 3' CP coding region and NCR, and it appears to have become a standard in assessing the relatedness of HIV strains and isolates. PAPA and TreeAlign also both purportedly produce most parsimonious phylogenetic trees as by-products of the alignment algorithm. An attractive feature of these methods is that because they compare the entire sequence of each protein/nucleic acid (or whole restriction/HPLC profiles), they take into account single residues or short runs of sequence (or sites, or fragments) conserved between subgroups of viruses but not between all species being compared. This kind of information, which could be lost in the reduction of sequence similarity to a single number as happens with distance matrix methods, could potentially be valuable in subgrouping of potyviruses according to host plant species infected [28].

While not specifically underwriting any of the software mentioned, we have found it useful to do multiple sequence alignments with CLUSTAL on the mainframe and to draw a tree from the UPGMA or NJTREE dendrogram produced. We then use PAUP on the aligned sequences for maximum parsimony analysis, and also to calculate matrices of pairwise distances for input into distance programs such as NJTREE or FITCH/KITSCH.

Phylogeny and evolution

The relationship dendrograms produced by the various programs mentioned represent approaches to determining probable evolutionary relationships; however, only if one assumes that the sequences diverged according to a molecular clock can one call the dendrogram a phylogenetic tree, and indicate a root [44, 80, 81]. The mutability of RNA sequences relative to DNA has led some workers to decry any possibility of tracing RNA genome evolution. However, Yokoyama [81] has assumed that the evolution of the extremely mutable retroviral immunodeficiency agents may be traced by phylogenetic analysis, and has evidence from comparative rates of evolution of virus-captured and host-resident oncogenes that appears to back up his argument. Consequently, and in the absence of any evidence suggesting that this is not so, we have assumed coat- and other proteins of the *Potyviridae* also probably evolve with a constant rate of change, and that relationship dendrograms derived by comparative sequence analysis do in fact represent phylogenies, or lines of evolutionary descent.

Coat protein and other phylogenies for Potyviridae

Sequences and origins

Names and acronyms of viruses, and sources of sequences and/or coat-protein HPLC profiles were collected for 48 generic potyviruses, one rymo-, and one bymovirus, and seven other rod-shaped viruses from different taxonomic groups (Table 1).

Table 1. Names, acronyms, and taxonomic affiliations of viruses, and sources of sequences

Virus name and strain	Accepted acronym (synonym)	Sequence ref.[a]
Potyvirus		
Azuki bean mosaic	AzMV	48*
Blackeye cowpea mosaic Type	BlCMV-Ty	48*
Bean common mosaic NL3	BCMV-NL3	24
Bean common mosaic NL4	BCMV-NL4	77
Bean common mosaic NL8	BCMV-NL8	77
Bean yellow mosaic GDD	BYMV-GDD	in 8
Bean yellow mosaic S	BYMV-S	74
Clover yellow vein B	ClYVV-B	74
Clover yellow vein 30	ClYVV-30 (BYMV-30)	75
Johnsongrass mosaic JG	JGMV-JG (SCMV-JG)	in 60
Johnsongrass mosaic KS1	JGMV-KS1 (MDMV-KS1)	47
Johnsongrass mosaic MDO	JGMV-MDO (MDMV-O)	45
Lettuce mosaic O (common)	LMV	11
Maize dwarf mosaic A	MDMV-A	22
Ornithogalum mosaic	OrMV	8
Papaya ringspot P	PRSV-P (PRV-P)	55
Papaya ringspot W	PRSV-WM1 (PRV-W, WMV1)	5, 78
Passionfruit woodiness K	PWV-K	27
Passionfruit woodiness M	PWV-M	in 60
Passionfruit woodiness S	PWV-S	in 60
Passionfruit woodiness TB	PWV-TB	in 60
Passionfruit woodiness SA	PWV-SA	R. Brand
Pea seed-borne P1	PSbMV	78
Peanut stripe Stripe A	PStV-A	48*
Peanut stripe Stripe B	PStV-B	48*
Plum pox D	PPV-D	in 60
Plum pox NAT	PPV-NAT	in 60
Potato Y D	PVY-D	in 60
Potato Y 10	PVY-10	in 61
Potato Y 18	PVY-18	in 61
Potato Y 43	PVY-43	in 61
Potato Y I	PVY-I	in 61
Potato Y Nb	PVY-N	see 78
Pepper mottle[b]	PVY-PeM (PepMoV)	see 78

Table 1 (continued)

Virus name and strain	Accepted acronym (synonym)	Sequence ref.[a]
Sorghum mosaic SCH	SrMV-SCH (SCMV-H)	J. Jilka
Soybean 74	(Soy74)[c]	48*
Soybean PM	(SoyPM)[c]	48*
Soybean PN	(SoyPN)[c]	48*
Soybean mosaic N	SbMV-N (WMV-SN)	in 60, 78
Soybean mosaic VA	SbMV-VA (SbMV-V)	S. Tolin
Sugarcane mosaic MDB	SCMV-MDB (MDMV-B)	22
Sugarcane mosaic SC	SCMV-SC	22
Tamarillo mosaic	TamMV	16
Tobacco etch HAT	TEV-HAT	in 60
Tobacco etch NAT	TEV-NAT	in 60
Tobacco vein mottling AT	TVMV-AT	in 2
Tobacco vein mottling NAT	TVMV-NAT	in 2
Turnip mosaic	TuMV	40
Watermelon mosaic II Aus	WMV2-Aus (WMV2)	20
Watermelon mosaic II USA	WMV2-US (WMV2-FC)	56; see 78
Zucchini yellow mosaic F	ZYMV-F	56
Alfalfa mosaic virus group		
Alfalfa mosaic	AMV	14
Aphthovirus		
Foot and mouth disease	FMDV	14
Baymovirus		
Barley yellow mosaic	BaYMV	36
Bromovirus		
Brome mosaic	BMV	14
Cardiovirus		
Encephalomyocarditis	EMCV	14
Carlavirus		
Potato S	PVS	45
Closterovirus		
Apple chlorotic leafspot	ACLV	23
Comovirus		
Cowpea mosaic virus	CPMV	14
Cucumovirus		
Cucumber mosaic	CMV	14
Papillomavirus		
Human papillomavirus type 11	HPV-11	GenBank**
Human papillomavirus type 18	HPV-18	GenBank**
Enterovirus		
Polio virus	Polio	14
Potexvirus		
Potato X	PVX	32

Table 1 (continued)

Virus name and strain	Accepted acronym (synonym)	Sequence ref.[a]
Ryemovirus		
Wheat streak mosaic	WSMV	53
Tobamovirus		
Tobacco mosaic Vulgare	TMV	GenBank
Tobravirus		
Tobacco rattle CAM	TRV-CAM	GenBank
Togaviridae		
Alphavirus		
Sindbis	Sind	14

[a] Unannotated entries are amino acid sequence, whether derived from nucleotide sequence or obtained directly; * HPLC profile only; ** nucleotide sequence only

[b] Designated as a strain of PVY in [75]

[c] Virus name not yet assigned [46]

See [77, 78] for more information on grouping and strains

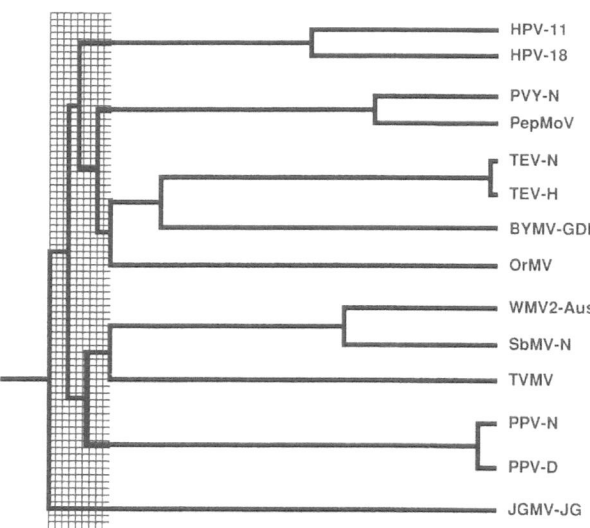

Fig. 1. A dendrogram produced using KITSCH from a distance data matrix (not shown) of pairwise percent sequence differences calculated using GAP for 3'-non-coding regions of 12 potyviruses and two sequences of length 341 bp from the L1 capsid protein gene of two human papillomaviruses, HPV-11 and HPV-18 [see 8]. Cross-hatched region indicates "region of uncertainty" where relationships are unreliable (=no significant sequence homology between species). Vertical distances are arbitrary; horizontal distances are proportional to percent sequence divergence and to evolutionary time (method assumes a molecular clock)

3′ terminal non-coding region relationships

Potyviral 3′ NCR sequences may be used reliably for assignments of a virus as a strain of another, or as a distinct virus, but among distinct viruses the 3′ NCRs probably have no significant sequence homology [20, 21]. A dendrogram drawn from KITSCH distance matrix analysis of pairwise distance data (Fig. 1) illustrates that homologies between 3′ NCRs of related strains or more closely related viruses are sufficiently high for these relationships to be distinguished from "background noise," whereas the homologies between NCRs of distinct potyviruses are sufficiently distant as to be non-significant when compared to each other and to partial CP sequences from the totally unrelated DNA human papillomaviruses [8]. Though the conclusions drawn are essentially the same, the value of the dendrogram in assessing relationships should be immediately apparent when one compares Fig. 1 with the table of homologies in Frenkel et al. [20].

Coat protein alignments

Alignments produced using CLUSTAL between the complete CPs of potyviruses, and parts of the coat proteins of a rymovirus (WSMV), a bymovirus (BaYMV), a potexvius (PVX), and a carlavirus (PVS), are shown in Fig. 2. The alignment is undoctored in that no adjustments have been made by eye. Two features that are immediately obvious, and that were observed in previous work [59, 60, 61, 62], are the apparently poor alignments in the region marked "N-terminal variable region," and the pronounced homologies in the region marked "constant region." It is obvious that the alignment of the variable N-terminal regions is at variance with those shown elsewhere [8, 60], and that at least one so-called motif in the variable region that has been aligned by other workers, is not necessarily aligned here. This is the "DAG" motif known to be important for aphid transmission [2, 34, 78, 79]. However, inspection of the aligned sequences in the region of the "DAG" box(es) shows other runs of residues whose alignments may be more important. Some viruses have more than one DAG box in the variable region (e.g., see TEV-HAT); however, the functionalities of these extra boxes have not been investigated. Alignments of all of the poty- and poty-like viruses in the constant region are very good because of strong sequence conservation. This conservation is presumably due to its being the region involved in forming the tertiary and quaternary structures necessary for virus assembly [11, 13, 60, 61, 79]. It is possible that the wide-spectrum cross-protection against potyvirus infection exhibited by transgenic plants expressing the SbMV-N coat-protein gene is related to this fact, as presumably surfaces related to assembly of coat protein with RNA would be highly conserved [11, 69]. Indeed, the similarly-structured PVX CP (and other potex- and carlavirus CPs) may be partially aligned with the putative potyviruses in the "constant region," as shown here and previously by

```
                                                                                                        ||
BYMV-GDD  -SDQEQLNAGEEKKD----------------------------------------KRKKNEGNPNKDSEGQSVRQIVP---------DRDVNAGTVG-TFSVPRLKKIAGK
BYMV-S    -SDQEKPNAGEKKKD----------------------------------------KDKKIEDNPSKDSDGQSSRRIVP-----------DRDINTGTVG-TFSIPRLKKIAGK
ClYVV-B   -SDKEKLNAGEQQKF----------------------------------------KDKEPRQRD-QEGENSNRQIIP----------DRDINAGTTG-TFSVPKLKKISGK
ClYVV-30  -SDKEKLNVGEQQKS----------------------------------------KDKESRQYEILEEVGESNRQIIP----------DRDINAGSTG-TFSVPKLKKISGK
PSbMV     AGDETKDDER--------------------------------------RRKEEEDRKKREESIDASQFGSNRDNKKNKNKESDTSNKLIVKSDRDVDAGSSG-TITVPRLEKISAK
PPV-NAT   ADEREDEEEVDA------------LQ---PPPVIQPAPRTTAPMLNPIFTPATTQPATKPVSQVSGPQLQTFGTYSHEDASPSNSNALVNTNRDRDVDAGSTG-TFTVPRLKAMTSK
PPV-D     ADEREDEEEVDAGKPIVVTAPAATSPILQ---PPPVIQPAPRTTAPMLNPIFTPATTQPATKPVSQVPGPQLQTFGTYGNEDASPSNSNALVNTNRDRDVDAGSTG-TFTVPRLKAMTSK
PRSV-P    SKN----EAVDAG------------------------------LNEKLKEKENQ--KEKEKEKQKEKEKDGASDGNDVSTSTKT----GERDRDVNVGTSG-TFTVPRIKSFTDK
PRSV-W    SKN----EAVDTG------------------------------LNEKFKEKEKQ--KEKEKEKQKEKEKDDASDGNDVSTSTKT----GERDRDVNVGTSG-TFTVPRIKSFTDK
TuMV      AGETLDAGLTEEQKQ-----------------AEKERKER-------------ERSEKERERQRQLALKK----------GKNAAQEEGERDNEVNAGTSE-LFSVPRLKSLTSK
PVY-D     -ANDTIDAGESSKKD----------------------------------------ARPEQGSIQVN---PNKG------KDKDVNAGTSG-THTVPRIKAITAK
PVY-N     -ANDTIDAGGSNKKD----------------------------------------AKPEQGSIQPN---PNKG------KDKDVNAGTSG-THTVPRIKAITSK
PepMoV    -ANDTIDTGGNSKKD----------------------------------------VKPEQGSIQPS---SNKG------KEKDVNAGTSG-THTVPRIKAITAK
LMV       -VDAKLDAGQGSKTD----------------------------------------DKQKNSADPKDNIITEKGSGSGQMKKDDDINAGLHG-KHTIPRTKAITQK
TEV-NAT   GGTVDASA-DVGKKK----------------------------------------DQKDDKVAE-----------QASKDRDVNAGTSG-TFSVPRINAMATK
TEV-HAT   SGTVDAGA-DAGKKK----------------------------------------DQKDDKVAE-----------QASKDRDVNAGTSG-TFSVPRINAMATK
TamMV     AGTLDAGEATAQKAE---------------------------------------GKKKEGEV--------SSGKAVVVKDKDVDLGTAG-THSVPRLKSMTSK
TVMV-AT   SDTVDAGKDKAR---------------------------------------DQKL------------ADKPTLAIDRTKDKDVNTGTSG-TFSIPRLKKAAMN
TVMV-NAT  SDTVDAEKDKAR---------------------------------------DQKL------------ADKPTLAIDRTKDKDVNTGTSG-TFSIPRLKKAAMN
OrMV      ASSMDAGG---------------------------------------SSRPPAPLVRQQDQDVNV----G-TFSVARVKALSDK
JGMV-JG   -SGNE------DAGKQKSATPAANQTASGDGKPVQTTATADNKPSSDN-----------------TSNAQGTSQTKGGGESGGTNATATKKDKDVDVGSTG-TFVIPKLKKVSPK
SCMV-MDB  SG--TVDAGAQGGSGSQGQGTTPPATGS----GAKPATSGAGSGSGTGAGTGVTGGQARTGSGTGTGSGATGGQSGSGSGTEQVNTSAGTNATGGQRDRDVDAGSTG-KISVPKLKAMSKK
SCMV-SC   AG--TVDAGAQGGGGGNAGTQPPATGAAAAQGGAQPPATGAAAAQPPTTQGSQLPQGGATGGGGAQTGAGGTGS------------------VTGGQRDKDVDAGTTG-KITVPKLKAMSKK
PWV-TB    -----------------------------KDE--IIDVGADGKKVVSKKDT-----QDAGEVNKGKE-----------KDKDVNAGSKG-SG-VPRLQKITKK
PWV-K     --------------------SGKDKDE--TLDAGADGKRSTGKKAA-----EGSSGGDSRKKS-------EEDTTQDKDVNAGSKG-NV-VPRLQKITKK
WMV2-AU   ----------------SGKEAVENLDTGKDSKKDTSGKGD----KPQNSQ----TGQGSKEQTKIGTVSKDVNVGSKG-KE-VPRLQKITKK
WMV2-US   ----------------SGKETVENLDAGKESKKDASDKGN----KPQNSQ----VGQGSKEPTKTGTVSKDVNVGSKG-KE-VPRLQKITKK
SbMV-N    --------------SGKEKEGDMDADKDPKKSTSSSKG---------AGTSSKDVNVGSKG-KV-VPRLQKITRK
PStV      --------------SGSTSTQSPVLNAGVDTAKDKKEKSN-----KGKGPESSEGSGNNSRGTENQSMRDKDVNAGSKG-KI-VPRLQKITKR
BCMV-NL4  AESALKTLYTNKKTKIEELARYLEVLDFDYEVGCGESVHLQSGSGHPPLPVVDAGVDTEKDKKDKSS-----RGKDPENKEETRNNSRGTENPTMRDKDVNAGSRG-KV-VPRLQRITKR
BCMV-NL8  AETALRKLYTDKDAKMEEMQEYLKQLEFGSDDEVCESVSTQSSKKEEEK---DAEA---DKREK-D-----KGKGPA--------DKDVGTGSKG-KV-VPRLQKITKK
ZYMV-F    SGT--------------------------QPTVADARVTKKDKE-----DDKGENKDFTGSGSGGEKTVVAAKKDKDVNAGSHG-KI-VPRLSKITKK
                                                                 .  . ...       .       ..    ..
WSMV                                                                            GRATDVQDQTPGLVFPAPKITTKAIY
BaYMV                                                                           TKQVNAGLTL--------------
```

```
BYMV-GDD  LNIPKIGGKIVFNLDHLLKYNPPQDDISNVIATQEQFEAWYNGVKQAYEVEDSR-MGIILNGLMVWCIENGTSGDLQ--GEWTMM----DGEEQVTYPLKPILDNAKPTFRQIMSHFSE
BYMV-S    LNIPKIGGKIVLNLDHLLDYNPPQDDISNTIATQAQFEAWYNGVKQAYEVDDSQ-MGIILNGLMVWCIENGTSGDLQ--GEWTMM----DGEEQVTYPLKPILDNAKPTFRQIMSHFSE
ClYVV-B   LSLPKIKGKGLLNLDHLLVYVPNQDDISNNIATQEQLEAWHEGVKNAYEVDDQQ-MEIICNGLMVWCIENGTSGDLQ--GEWTMM----DGEKQVTFPLKPILDFAKPTLRQIMAHFSQ
ClYVV-30  LSLPKIKGKGLLNLDHLLVYVPNQDDISNNIATQEQLEAWHEGVKNAYEVDDQQ-MEIICNGLMVWCIENGTSGDLQ--GEWTMM----DGEKQVTFPLKPILDFAKPTLRQIMAHFSQ
PSbMV     IRMPKHKGGVAISLQHLVDYNPAQVDISNTRATQSQFDNWWRRVSQEYGVGDNE-MQVLASGLMVWCIENGTSPNIN--GMWTMM----DGEEQVEYPLKPVMDNARPTFRQIMAHFSD
PPV-NAT   LSLPKVKGKAIMNLNHLAHYSPAQVDLSNTRAPQSCFQTWYEGVKRDYDVTDDE-MSIILNGLMVWCIENGTSPNIN--GMWVMM----DGETQVEYPIKPLLDHAKPTFRQIMAHFSN
PPV-D     LSLPKVKGKAIMNLNHLAHYSPAQVDLSNTRAPQSCFQTWYEGVKRDYDVTDDE-MSIILNGLMVWCIENGTSPNIN--GMWVMM----DGETQVEHPIKPLLDHAKPTFRRIVARFSD
PRSV-P    MVLPRIKGKTVLNLNMLLQVNPQQIDISNTRATHSQFEKWVEGVRNDYGLNDNE-MQVMLNGLMVWCIENGTSPDIS--GVVVMM----DGETQVDYPIKPIEHATPSFRQIMAHFSN
PRSV-W    MILPRIKGKSVLNLNMLLQVNPQQIDISNTRATHSQFEKWVEGVRNDYGLNDNE-MQVMLNGLMVWCIENGTSPDIS--GVVVMM----DGETQVDYPIKPIEHATPSFRQIMAHFSN
TuMV      MRVPKYEKRVALNLDHLILYTPEQTDLSNTRSTRKQFDTWFEGVMADYELTEDK-MQIILNGLMVWCIENGTSPNIN--GNWVMM----DGDDQVEFPLKPILDHAKPTFRQIMAHFSD
PVY-D     MRMPRSKGATVLHLEHLLEYAPQQIDISNTRATQSQFDTWYEAVRMAYDIGETE-MPTVMDGLMVWCIENGTSPNVN--GVVWMM----DGNEQVEYPLKPIVENAKPTLRQIMAHFSD
PVY-N     MRMPTSKGATVLNLEHLLEYAPQQIDISNTRATQSQFDTWYEAVRMAYDIGETE-MPTVMNGLMVWCIENGTSPNIN--GVVWMM----DGNEQVEYPLKPIVENAKPTLRQIMAHFSD
PepMoV    MRMPKSKGAAVLKLDHLLEYAPQQIDISNTRATQSQFDTWYEAVRVAYDIGETE-MPTVMDGLMVWCIENGTSPNIN--GVVWMM----DGSEQVEYPLKPIVENAKPTLRQIMAHFSD
LMV       MKLPMIRGKVALNLDHLLEYEPNQRDISNTRATQKQYESWYDGVKNDYDVDDSG-MQLILNGLMVWCIENGTSPNIN--GTWVMM----DGEEQVEYALKPIIEHAKPTFRQIMAHFSD
TEV-NAT   LQYPRRMKGEVVVNLNHLLGYKPQQIDLSNARATHEQFAAWHQAVMTAYGVNEEQ-MKILLNGFMVWCIENGTSPNLN--GTWVMM----DGEEQVSYPLKPMIENAKPTLRQIMTHFSD
TEV-HAT   LQYPRMRGEVVVNLNHLLGYKPQQIDLSNARATHEQFAAWHQAVMTAYGVNEEQ-MKILLNGFMVWCIENGTSPNLN--GTWVMM----DGEDQVSYPLKPMIENAKPTLRQIMTHFSD
TamMV     LTLPMLKGKSRCNLDHLLSYK-QTVDLSNARATHEQFQNWYDGVMASYELEESS-MEIILNGFMVWCIENGTSPDIN--GVVTMM----DDEEQISYPLKPMLDHAKPSLRQIMRHFSA
TVMV-AT   MKLPKVGGSSVVNLDHLLTYKPAQEFVVNTRATHSQFKAWHTNVMAELELNEEQ-MKIVLNGFMIWCIENGTSPNIS--GVVTMM----DGDEQVEYPIEPMVKHANPSLRQIMKHFSN
TVMV-NAT  MKLPKVGGSSVVNLDHLLTYKPAQEFVVNTRATHSQFKAWHTNVMAELELNEEQ-MKIVLNGFMIWCIENGTSPNIS--GVVTMM----DGDEQVEYPIEPMVKHANPSLRQIMKHFSN
OrMV      MMLPKVRGKTVLNLQHLVQYNPEQTEISNTRATRTQFNNWYDRVRDSYGVTDDQ-MAVILNGLMVWCIENGTSPNLN--GNWTMM----DGDEQIEYPLQPVLENAQPTFRQIMAHFSN
JGMV-JG   MRLPMVSNKAILNLDHLIQYKPDQRDISNARATHTQFWYNRVKKEYDVDDEQ-MRILMNGLMVWCIENGTSPDIN--GYWTMV----DGNNQSEFPLKPIVENAKPTLRQCMMHFSD
SCMV-MDB  MRLPKAKGKDVLHLDFLLTYKPQQQDISNTRATKEEFDRWYDAIKKEYIDDTQ-MTVVMSGLMVWCIENGCSPNIN--GNWTMM----DKDEQRVFPLKPVIENASPTFRQIMHHFSD
SCMV-SC   MRLPKAKGQDVLHLDFLLTYKPQQQDISNTRATREEFDRWYEAIKKEYELDDTQ-MTVVMNGLMVWCIENGCSPNIN--GSWTMM----DGDEQTVFPLKPVIENASPTFRQIMHHFSD
PWV-TB    MNLPKVGNMVLDLDHLIEYKPDQTKLFNTRATDAQFATWYEGVKAEYELSDDQ-MGVIMNPFMVWCIENGTSPDIN--GVVWMM----DGDEQVEYPLKPMVENAKPTLRQIMHHFSD
PWV-K     MNLPMVKGNMILNLEHLIEYKPEQTKLFNTRATDAQFSAWYDAVKEEYELTDDQ-MGVVMNGFMVWCIDNGTSPDVN--GVVWMM----DGDEQVEYPLKPMVENAKPTLRQIMHHFSD
WMV2-AU   MNLPTVGGKIILSLDHLLEYKPNQVDLFNTRATKTEFESWYSAVKIEYDLNDEQ-MGVIMNGFMVWCIDNGTSPDVN--GVWVMM----DGEEQVEYPLKPIVENAKPTLRQIMHHFSD
WMV2-US   MNLPTVGGKIILSLDHLLEYKPSQVDLFNTRATKTQFESWYSAVKVEYDLNDEQ-MGVIMNGFMVWCIDNGTSPDVN--GVWVMM----DGEEQVEYPLKPIVENAKPTLRQIMHHFSD
SbMV-N    MNLPMVEGKIILSLDHLLEYKPNQVDLFNTRATRTQFEAWYNAVKDEYELDDEQ-MGVVMNGFMVWCIDNGTSPDAN--GVWVMM----DGEEQIEYPLKPIVENAKPTLRQIMHHFSD
PStV      MNLPMVKGNVILNLDHLLDYKPEQTDLFNTRATKMQFEMWYNAVKGEYEIDDEQ-MSIVMNGFMVWCIDNGTSPDVN--GTWVMM----DGDEQVEYPLKPMVENAKPTLRQIMHHFSD
BCMV-NL4  MNLPMVKGNVILNLDHLLDYKPEQTDLFNTRATKMQFEMWYNAVKGEYEIDDDQ-MAIIMNGFMVWCIDNGTSPDVN--GVWVMM----DGDEQVEYPLKPMVENAKPTLRQIMHHFSD
BCMV-NL8  MNLPMVGGRMILNLDHLIEYKPQQTDLYNTRATKAQFERWYEAVKTEYELDDQQ-MGVVMNGFMVWCIDNGTSPDVN--GVEVMM----DGDEQIEYPLKPMVENAKPTLRQVMHHFSD
ZYMV-F    MSLPRVKGNVILDIDHLLEYKPDQIELYNTRASHQQFASWFNQVKTEYDLNDQQ-MGVVMNGFMVWCIENGTSPDIN--GVWFMM----DGNEQVEYPLKPIVENAKPTLRQIMHHFSD
          .  *                .*.          *          *        .  .*  *    .  .*  *     *   *.  . .  . * *.  .  *. ** .. **.
WSMV      MPKTVRDKIKPEMINNMIKYQPRTELIDNRYATTEQLNTWIKEASEGLDVTEDVFINTLLPGFVYHCIINTTSPENRALGTWRVVNNAGKDNEQQLEFKIEPMYKAAKPSLRAIMRHFGE
BaYMV     --K--------IPLNKLKSVPKSVMEHNNSVALESELKAWTDAVRTSLGITTDEAWIDALIPWIGWCCNNGTSDKHAENQV--MQIDSGKGAVTEMSLSPFIVHARMNGGLRRIMRNYSD
PVS       RRNPENPYSRFSIDELFKMEIRSVSNNMANTEQMAQ-ITADIAGLGVPTEHVAGVILK-VVIMCASVSSSVYLD------PAGTVEFPTGAVPLDSIIAIMKNRAGLRKVCRLYAP
PVX       SNAVATNEDLSK-IEAIWKDMKVPTDTMAQAAWD-LVRHCADVGSSAQTE------MIDTGPYSNG-ISRARLAAAIKEVCTLRQFCMKYAP
```

Fig. 2. Alignment produced using CLUSTAL of complete coat protein sequences of 32 generic potyviruses, and partial sequences of the rymovirus WSMV, the bymovirus BaYMV, the potexvirus PVX, and the carlavirus PVS. Asterisk indicates residue conserved between generic potyviruses only; dot indicates conservative substitution between generic potyviruses only. Arrows and legend indicate N-terminal variable region lost after tryptic digestion, and C-terminal "core" or constant region [59–61]. See Table 1 for explanation of virus acronyms

Kashiwazaki et al. [36] and Dinant et al. [11]. The most striking conservation between these sequences is in a region potentially involved in formation of a "salt bridge" [41].

The strong conservation of amino acid sequences in the CP gene constant region of generic potyviruses means that there is considerable similarity in genomic RNA sequences coding for this region. As discussed above, this allows the construction of degenerate oligonucleotide primers for use in the PCR which should anneal to most potyviral genomic RNAs/cDNAs. The core region motif that is especially obvious is a MVWCI$^D/_E$NG$^T/_C$S box conserved among all generic potyviruses, and partly conserved in BaYMV and WSMV as well. This conservation has been taken advantage of for construction of degenerate oligonucleotide primers for cDNA cloning and PCR [43; Brand and Rybicki, unpubl.].

There is no other region of comparable length in the aligned CP genes that is conserved to the same extent, meaning there is little chance of synthesizing other primers for the CP constant region that will be sufficiently specific to amplify only potyviruses. It is possible to detect both OrMV and PWV-SA, which are as distinct as any two other generic potyviruses, by PCR amplification from cDNA generated from total nucleic acid extracts from infected plants using a single "MVWCI box" primer, and an oligo-dT primer specific for the poly(A) tail (S. Pappu et al., manuscript in prep.). This means that a wide range of potyviruses may be detected by PCR [43], and the sequence subsequently obtained for most of the CP coding region and the whole 3′ NCR: this should allow rapid typing of potyviruses from even small amounts of infected tissue as sole virus source.

Coat protein relationships and phylogenies

A tree calculated for the same complete CP sequences by CLUSTAL using the UPGMA is shown in Fig. 3. The dendrogram in Fig. 3 represents presumed evolutionary relationships among all poty- and poty-like CPs sequenced to date: for the sake of convenience it and all subsequent dendrograms will be referred to as "trees," with the caveat that no statistical test of a molecular clock has been made for this data. This tree is a convenient graphical representation of the fact that aphid-transmitted (= generic) potyvirus CPs form a tight cluster distinct from all the other viruses. Although WSMV and BaYMV CPs are obviously distantly related to each other and to generic potyvirus CPs, the familial Potyviridae are still more closely related to each other than any of these viruses is to the "natural" outgroup PVX. If one assumes a molecular clock, the tree can be taken as indicating an ancient divergence of potex- and poty-like virus CPs, and subsequent divergence of the various genera within the Potyviridae. The branching allows a natural and evolutionary division of viruses into family Potyviridae (anything inside the PVX branch), separate genera of the family (each separate major branch), and "species" (each distinct branch in a genus). It is obvious that the generic potyvirus cluster can be subdivided into distinct potyvirus "subgenera" or subgroups. The largest and most distinct of these consists of viruses related to PWV; another deeply bifurcated subtree consisting solely of viruses of Gramineae; another obvious small subgroup contains two BYMVs and two ClYVVs. These subgroupings agree well with previous observations [59–62, 78, 79]. Comparison of this tree with the similarly constructed one for "core regions" of potyviral CPs [79] reveals how inclusion of the N-terminal variable region in the analysis allows better differentiation of the viruses. It is probably more informative to use total sequence information than partial sequences for tree construction when one wishes to make taxonomic assignments, especially when differences in the N-terminal region may be responsible for some of the biological differences that determine whether a virus isolate may be considered distinct or not.

Apart from its obvious value as a visual aid for rapid assessment of relationships, the UPGMA tree can also be used (Fig. 3) to make preliminary decisions on the taxonomic placement of a newly-sequenced virus. These may range from "specific" (e.g., is it a strain of BCMV?) to "generic" (is it a poty-

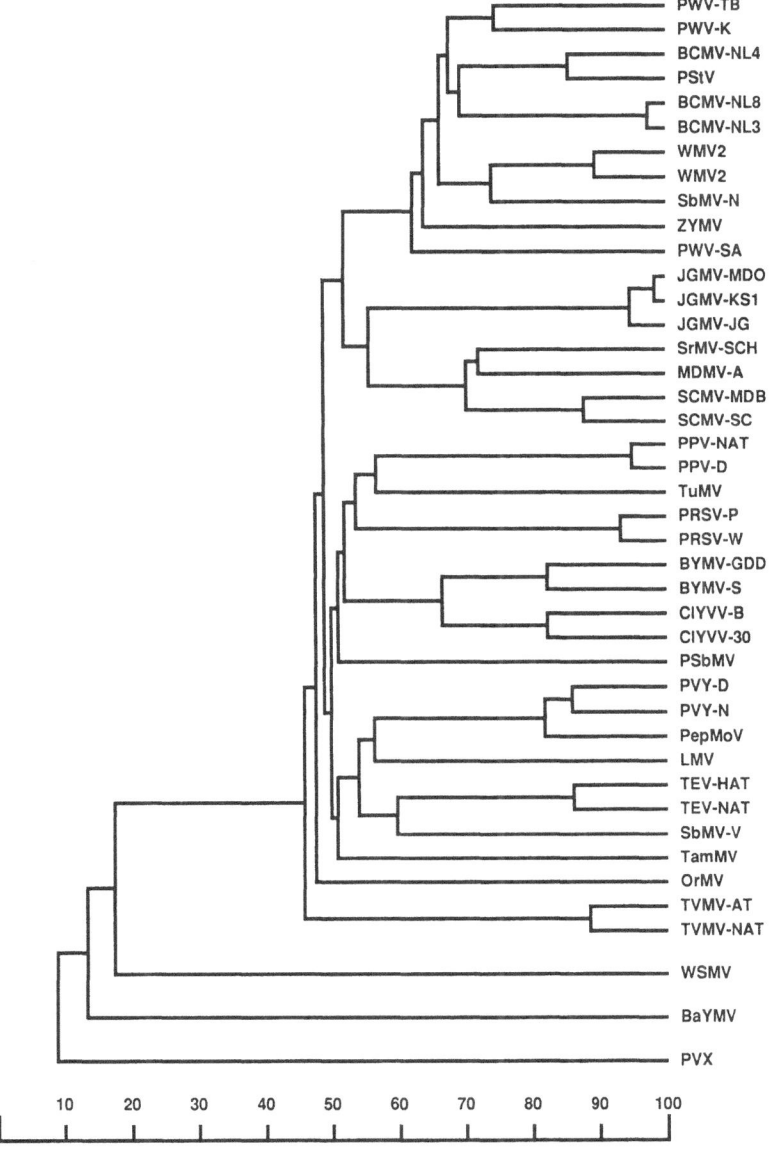

Fig. 3. Sequence relationship dendrogram (phylogenetic tree) produced using CLUSTAL from alignments of the total coat protein amino acid sequences of 39 generic potyviruses, WSMV (= genus *Rymovirus*), BaYMV (= genus *Bymovirus*) and PVX (a potexvirus). Vertical distances in the tree are arbitrary, horizontal distances are proportional to percent amino acid sequence divergence, shown in the scale below the figure

or bymovirus?) to "familial" (does it belong in the *Potyviridae*). In a practical example, it took 20 minutes to determine that the BCMV-NL3 CP sequence determined by Gilbertson et al. [24] was very closely related to BCMV-NL8, and in the "PWV subgroup" rather than related to the BYMV subgroup.

A parsimonious tree calculated from matrices of aligned sequences using PAUP is shown in Fig. 4a. The tree was calculated from consideration of aligned complete CP sequences of generic potyviruses, with the TVMV-AT sequence chosen as an outgroup because of its position in the UPGMA tree in Fig. 3. Comparison of Figs. 3 and 4a shows that PAUP analysis has the effect of accentuating internodal distances near the base of the tree. This clearly reveals subgroup clusters in Fig. 4a that are not immediately obvious in the UPGMA tree (Fig. 3), while confirming others such as the PWV, Gramineae virus, and BYMV subgroups. Distinct clustering was previously shown in a PAUP dendrogram calculated by Quemada et al. [56] for 3′ coding and NCR nucleotide sequences from several potyviruses. Interesting observations are the grouping of SbMV-VA and the TEVs, which is much less obvious from distance data, and the grouping of these with a cluster containing PVY. The juxtaposition of OrMV and the viruses of Gramineae is interesting as all infect monocotyledonous plants. There is an emergence of some clustering among the viruses that appeared distinct in Fig. 3, in particular of PSbMV and BYMV/ ClYVV group, and of the PPVs and the PRSVs. Some of these aspects will be explored in more detail elsewhere, particularly in relation to their biological relevance (M. Kyle et al., manuscript in prep.). Otherwise, the overall relationship picture was very similar to that shown by the UPGMA analysis, which indicates to us that the relationship picture is "robust."

The PAUP tree in Fig. 4b was drawn to explore the relevance of sequence alignments in the CP N-terminal or variable region alone. Comparison of Fig. 4b with Figs. 3 and 4a reveals that the alignments produced using CLUSTAL for this region produce a tree that agrees unexpectedly well in certain respects, given the degree of sequence divergence in this region, with those produced from alignments of the whole protein (Fig. 4a), or of the far more homologous constant or core region [79]. Comparison of Figs. 1 and 4b suggests that comparison of potyvirus N-terminal regions is a better measure of strain relatedness than equivalent analysis of 3′ NCR sequences. For example, all closely-related virus strains group closely in both the Fig. 1 and 4b trees; however, most higher-order or more distant homologies (such as of generic potyviruses vs. rymo- and bymoviruses) may still be discerned in the CP N-terminal tree in Fig. 4b, whereas 3′ NCR comparisons are not reliable for revealing even loosely clustered intra-generic relationships such as those in the PWV subgroup (Fig. 1 [Rybicki, unpubl.]).

The subgroupings revealed with the potyvirus genus by UPGMA and especially by PAUP analyses deserve further discussion, as they may yet be the subject(s) of further taxonomic proposals [78, 79]. Obvious subgroupings are of viruses related to PWV, viruses infecting graminaceous hosts, and a clear

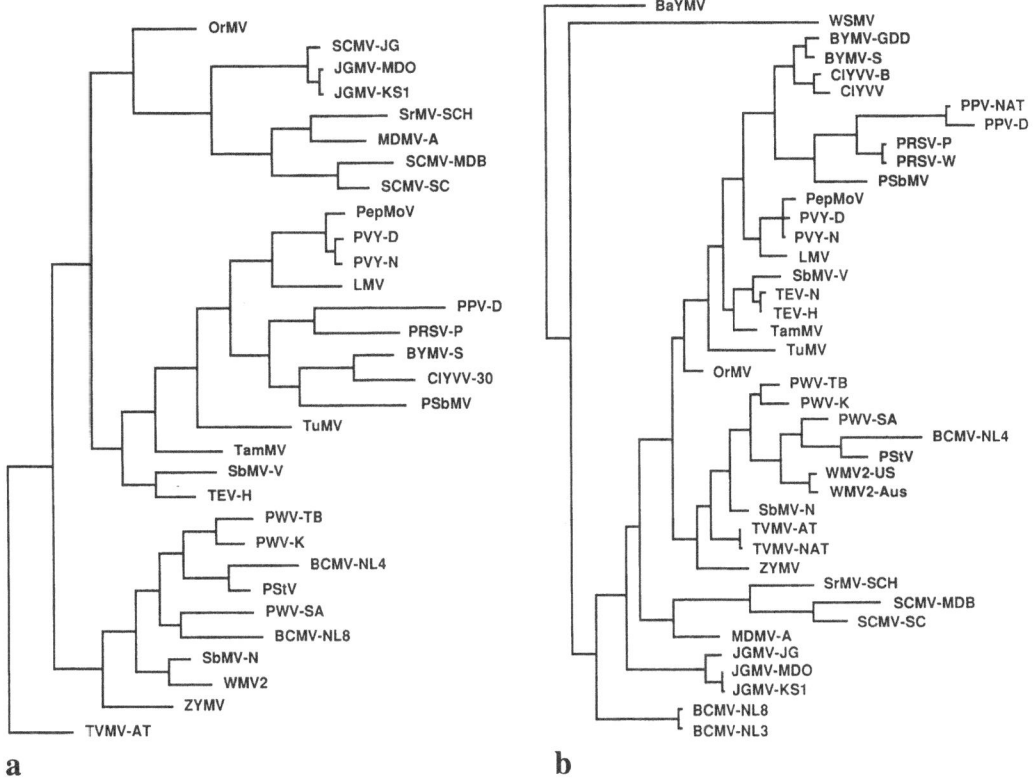

a   **b**

Fig. 4. Unrooted phylogenetic trees produced using PAUP from sequence alignments calculated using CLUSTAL for coat-protein sequences of *Potyviridae*. **a** One of 13 most parsimonious trees produced from matrix of aligned total coat proteins of 31 generic potyviruses: heuristic search option with branch-breaking and rearrangement. TVMV was chosen as an outgroup species from its position in the UPGMA tree in Fig. 3. This tree was selected from consideration of topological agreement with subtrees generated using more rigorous search routines (not shown). **b** Single most parsimonious tree produced from matrix of aligned N-terminal variable regions of generic potyviruses and WSMV and BaYMV, as shown in Fig. 2. Heuristic search option with nearest-neighbour branch breaking and rearrangement. BaYMV was chosen as outgroup. Vertical distances are arbitrary, horizontal distances are proportional to number of amino acid substitutions between ancestral (nodal) species and tip species

sub-subgrouping of BYMV and ClYVV strains, contained within a large and rather amorphous cluster including the PVYs, TEVs, PPVs, and other viruses (Fig. 4a).

The "PWV subgroup" is especially distinct, and contains some surprising relationships. For example, some of the various so-called PWVs are obviously only distantly related. Indeed, PWV-SA is as dissimilar to any other PWV as are any two other distinct viruses in the subgroup, indicating that it should perhaps be given a new name. Though PWV-K consistently clusters with other strains of PWV, these cluster far more closely with PWV-TB, indicating that PWV-K is a distantly related strain at best, and perhaps a distinct virus [27]. These relationships are shown in PAUP tree of the subgroup which includes two additional PWVs (Fig. 5a) [59,61]. PWV-SA is the only one from Africa in the comparison, as all the others come from Australasia; thus the major variation among viruses purported to be PWVs may be due to different viruses finding the same biological niche in different geographical areas.

The BCMVs also cluster in an interesting way, with the reportedly African NL3 and NL8 strains being related to, but quite distinct from, NL4 from the U.S.A. BCMV-NL4 is the closest relative of PStV; however, as BCMV-NL4 is from the U.S.A. and the sequenced isolate of PStV is from China [50], the two viruses obviously do not cluster according to geographical origin. It is worth noting that PWV-SA and BCMV-NL8 are more closely related in the PAUP tree in Fig. 5a than either is to the other PWVs or BCMVs, though the relationship is distant enough (compared to relationships such as that between WMV2 and SbMV-N) for the viruses to be considered distinct. The grouping of BCMV-

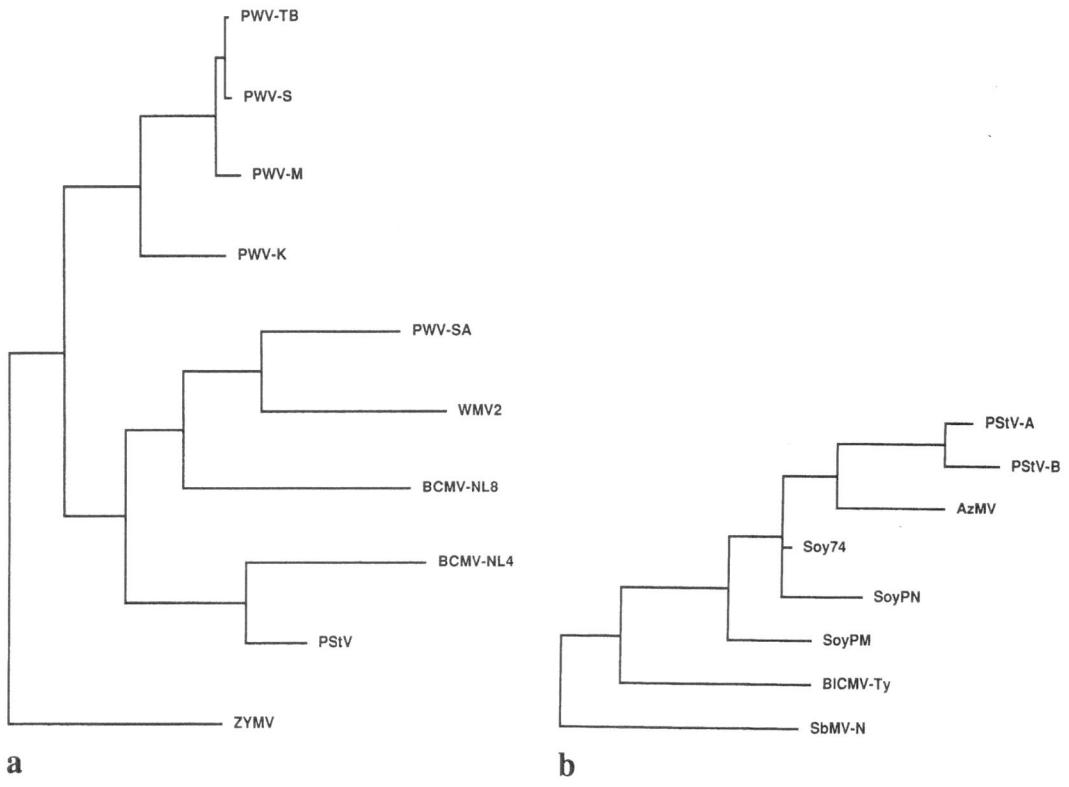

Fig. 5. Unrooted trees produced using PAUP from alignments of coat protein sequences or HPLC peptide elution profiles, for subsets of generic potyviruses in the PWV subgroup. **a** Most parsimonious tree produced from a matrix of total coat protein sequences aligned using CLUSTAL of distinct viruses in the PWV subgroup: branch-and-bound tree search option, furthest addition sequence. Outgroup was ZYMV. Extra PWV sequences obtained from [60]. Drawing format as in Fig. 4. **b** Most parsimonious tree produced from a matrix of aligned digitized (1 = present, 0 = absent) HPLC elution profile "peaks" from [48]. Exhaustive search option used. SMV-N was used as outgroup. Horizontal distance is proportional to number of peptides shared between profiles

NL4 and PStV is interesting in light of HPLC profile data for a number of unsequenced viruses in this subgroup, including two BlCMV stains, AzMV, four PStV strains, and some unnamed potyviruses infecting soybean. Profiles were subjected to numerical comparisons, which indicated that all of the viruses analyzed were strains of one virus [48]. We used digital transforms of the profiles, in which each elution peak is considered as a 1, and its lack a 0, to construct a PAUP consensus tree, shown in Fig. 5 b. This shows graphically that the SbMV-N profile, known to be distinct from PStV (see Figs. 3 and 4 a), is a good outgroup for a cluster which has the known PStV strains as closest relatives on one branch with AzMV and a cluster of soybean viruses, and the two BlCMVs as nearest relatives on another branch. Thus, although BlCMVs are closely related to the PStV strains, the relationship is less close than that of AzMV, for example. It has recently been proposed on the basis of similarities in host range, symptoms, and serology that cowpea aphid-borne mosaic virus (CABMV) be considered a strain of BlCMV [10]. Thus, BCMV-NL4 and closely-related BCMV strains form a complex of viruses/strains with PStV, BlCMV and CABMV, AzMV, and other unnamed soybean viruses, which is quite distinct from other purported BCMVs such as "strains" -NL3 and -NL8. Obviously, this particular sub-subgroup of viruses is ripe for a taxonomic overhaul. The utility of phylogenetic analysis of HPLC profile data for analysis of strain relationships is also evident, and will be explored in more detail elsewhere (Rybicki et al., in progress).

Still in the PWV subgroup, SbMV-N and the WMV2s are apparently as closely related in all of the trees shown as are known strains of other viruses. The relevance of this to the taxonomy of the viruses is discussed elsewhere [19, 53, 57–59, 73, 74, 78, 82]; it suffices to say that in cases such as this, assignment of a virus as a strain or as a distinct "species" must be done with the caveat that they are known to be closely related in terms of sequence. Of little doubt is the very clear difference and phylogenetic separation between SbMV-VA and SbMV-N. Recent evidence suggests SMV-VA is not representative of viruses causing soybean mosaic disease, and was probably a misidentified adventitiously-infecting potyvirus [33; S. Tolin, pers. comm.].

The "Gramineae virus subgroup" is seen in Fig. 3, and more clearly in Fig. 4 a. It consists of a closely-related group of three JGMVs, and a very distinct and less-related group consisting of an SrMV, two SCMVs, and MDMV-A. This grouping is in agreement with HPLC profiling and numerical analysis of ten virus isolates, and observations on serology of 17 viruses, with data suggesting that there are four distinct viruses comprising this group [47, 49, 62, 78, 79]. It appears safe to conclude that the JGMVs are sufficiently closely related to all be strains of the same virus, while being very different from other potyviruses infecting graminaceous hosts; that although MDMV-A, SrMV, and SCMV are together in a sub-subgrouping, they are all sufficiently distantly related (compared to accepted strains of, e.g., PVY, BYMV, ClYVV) to be distinct viruses.

The "large loose subgroup" visible in the PAUP tree in Fig. 4a, but not in Fig. 3, also contains some intriguing relationships and sub-subgroups. These are perhaps illustrated more clearly in the PAUP tree of the "subgroup" shown in Fig. 6. The close relationship between PepMoV and PVY strains has been the subject of some discussion recently [59–62, 78, 79]. Figure 7 shows an analysis by the distance matrix neighbor-joining method NJTREE in CLUSTAL V of the relationship between the whole CPs of six PVY strains, the previously-sequenced PepMoV, the newly-sequenced PepMoV-C [76], LMV, and TEV-NAT. In this TEV-NAT and LMV CPs appear as unrelated to each other as any is to any other virus, while PepMoV-C groups with – but distinct from – a more closely related group consisting of PepMoV, PVY-I, PVY-N, and the closely-related PVY-43, PVY-18, and PVYs -D and -10. It is not at all obvious, therefore, that the originally-sequenced PepMoV should be considered any more distinct an entity than PVY-I or PVY-N: however, the newly-sequenced PepMoV-C is definitely more distinct, even if related to the PVYs. If this virus is more representative of the biologically-characterized PepMoV strains, an explanation of the apparent lack of correlation of biological differences between PVY and PepMoV with the previous sequence data could be that the older PepMoV sequence was in fact obtained from a PVY strain incorrectly presumed to be a PepMoV [76]. It is interesting that LMV clusters with the PepMoV/PVY PAUP-generated sub-subgroup in Fig. 6, but not in Fig. 7: this

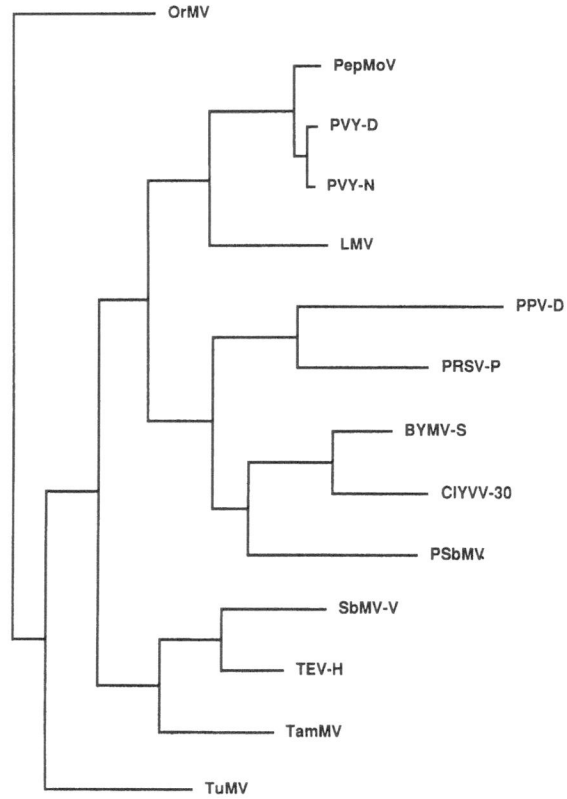

Fig. 6. Most parsimonious unrooted PAUP tree of relationships between total coat protein sequences of potyviruses in the "large loose" subgroup aligned using CLUSTAL (see text). OrMV was used as outgroup; search option was branch-and-bound with furthest addition sequence. Note that tree topology is not necessarily the same as in Fig. 4 a: the use of a different matrix (though the same CLUSTAL alignment as in Figs. 3–5) and a different and more rigorous search option may result in different trees

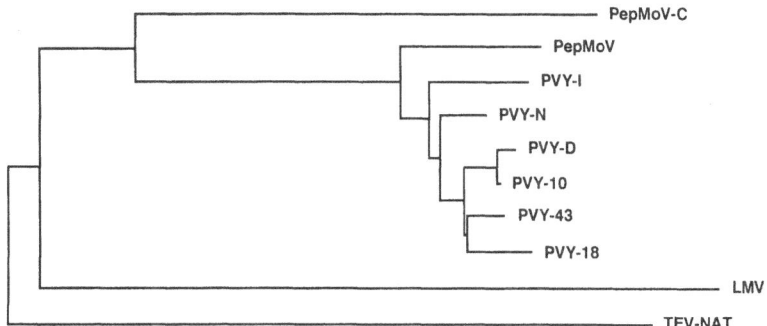

Fig. 7. Tree produced using NJTREE subroutine in CLUSTALV from percent sequence distance data matrix produced from an alignment calculated using CLUSTALV, for strains of PVY, PepMoV and PepMoV-C, LMV, and TEV-NAT. Sequences were taken from [59, 76]. Horizontal distances are proportional to percent sequence divergence of tip species from nodes, vertical distances are arbitrary. The tree was rooted with reference to Fig. 6

may be evidence of an as yet unrevealed biological relationship. The relatively close relationship between BYMV and ClYVV strains has also been noted previously [4, 74, 75]; the viruses form a clear sub-subgroup, which apparently is still not close enough for ClYVV and BYMV to be considered as strains of one virus on biological grounds [4, 74]. Evidence from 3′ NCR hybridization experiments suggests that pea mosaic virus (PMosV), a member of the BYMV subgroup of potyviruses [4, 74], is also a distinct virus. It is interesting that PSbMV clusters reliably with this sub-subgroup in different PAUP analysis runs (revealing a "robust" relationship): this will be explored elsewhere (Kyle et al., manuscript in prep.). The grouping together of SbMV-VA and the TEVs has been noted elsewhere [74]; in Fig. 6 it appears that TamMV may be related to these two. The loose subgroup also contains an interesting and reliable, though distinct, clustering of the PPVs with the PRSVs: it is not known if this has any biological relevance whatsoever.

Newly described and/or sequenced viruses such as OrMV, TamMV, PSbMV, TuMV, and LMV are all obviously distinct from each other and established members (Figs. 3, 4a, and 6), as their CPs are each separated by deep branches from any other virus in all trees, even when they are clustered with others. The taxonomic conclusion is that their designations as distinct viruses is correct in light of sequence relationships.

It is interesting that the recent CP sequence revision published for TVMV [2] does not change the previously published position of the virus CP relative to other accepted potyviruses [8, 61]. It is still the most distinct of these viruses, and still does not obviously cluster with other viruses infecting tobacco or even Solanaceae. This illustrates a potential error that can be made in trusting biological properties over sequence in classifying viruses: they should not be presumed to be close relatives merely because they have a particular host in common, or even similar host ranges. Examples in the *Potyviridae* are the

PWVs and SbMV-N and SbMV-VA; another apposite example is HIV-1 and HIV-2, which both cause immunodeficiency disease in humans, but are in fact only distantly related [80].

Higher-order relationships defined by CP sequences

Although we and others [11,37,41] have shown sequence similarities between CPs of other rod-shaped viruses and members of the *Potyviridae*, these are weak at best and presumably reflect extremely distant evolutionary divergences of helically-assembling proteins. In an exercise to determine whether the "familial clustering" of *Potyviridae* was in fact robust, and whether other relationships could be shown between viruses from different taxonomic groups but with similar morphology, we included the CP sequences of a potexvirus (PVX), a tobamovirus (TMV), a tobravirus (TRV), a carlavirus (PVS) and a closterovirus (ACLSV), together with "core sequences" of several of the *Potyviridae* in a CLUSTAL alignment and PAUP tree generation run. Figure 8 shows that the *Potyviridae* species can clearly be distinguished as a coherent family cluster distinct from all other viruses; that tobamoviruses, closteroviruses and tobraviruses are apparently more closely related to each other than each is to any other group; and that potex- and carlaviruses are apparently more closely related to each other than are (for example) WSMV and BaYMV. The PVX-PVS grouping could in fact be termed another "familial structure": this

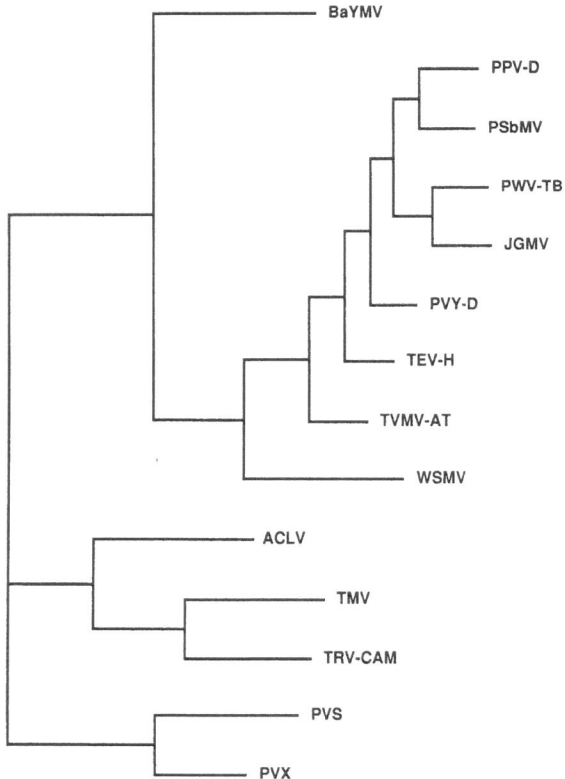

Fig. 8. Unrooted most parsimonious tree produced using PAUP from CLUSTAL-aligned matrix (not shown) of coat protein seuences of *Potyviridae* and plant viruses of similar morphology from other taxonomic groups. *Potyviridae* sequences were truncated by removal of the N-terminal variable region as it is shown in Fig. 2; all other sequences were complete. All non-potyviridae species were designated as outgroups; search routine was branch-and-bound with furthest addition sequence

has also been proposed elsewhere from comparisons of genome sequences and deduced genome organizations [13, 23, 32, 41, 45, 51, 83]. Similarly, the TRV-TMV grouping reflects similar genome structures [25]. Thus, the coat-protein sequences of filamentous and rod-shaped plant viruses may apparently be used in a quite reliable way to define apparent familial and even super-familial groupings of these viruses. Interestingly, this also appears to be true of the helically-assembling nucleocapsid proteins of the enveloped pneumo-, paramyxo-, rhabdo- and filoviruses, which were also aligned using CLUSTAL [5]. Supergroupings can also be demonstrated among CPs of plant viruses with spherical capsids, as discussed above [12]. It may therefore be possible to "type" a filamentous plant virus CP from super-familial to subgeneric level, simply by multiply aligning its protein in just such an exercise as was explored above. However, it is likely that such an exercise would only be used to confirm a taxonomic assignment, as it should be obvious from other properties to which group (= generic) taxon a new virus should be assigned.

Phylogenies based on other proteins

Presumed evolutionary relationships based on phylogenetic analysis of conserved "core sequence motifs" from alignments of putative replicase-related proteins, as originally described by Domier et al. [14], are illustrated in Fig. 9 a. This dendrogram shows super-familial relationships for six taxonomic groups of plant viruses, and two families and four genera of animal viruses. There is a clear division of the viruses into two "superfamilies"; namely, "picorna-like" and "alpha-like," based on similarities to replicases of either *Picornaviridae* or the *Alphavirus* genus of the *Togaviridae*. The potyviruses TEV and TVMV cluster with cowpea mosaic virus (CPMV) in the picorna-like superfamily. This is only part of a larger picture, as consideration of more polymerase sequences leads others to postulate more than two superfamilial groupings [7, 41]; however, the relevance of such analyses to potyvirus taxonomy is dubious (see below).

Higher-level taxa?

Before beginning to speculate on the possibilities for erecting even higher taxa than the familial for *Potyviridae* and other viruses, it is educational to overlay the dendrogram in Fig. 9a with another showing possible CP relationships between the same viruses (see Fig. 9b). It is readily seen that viruses in the same taxonomic families or groups have co-segregating "core polymerase motifs" and CPs, whereas viruses from different groups or families do not necessarily even have the same type of polymerase if they have similar types of CP. Thus, potyviruses and tobamoviruses share a similar coat protein, but have very different replicases and genomic organizations (Figs. 8 and 9; see also [57]). This and other sorts of graphic analysis can usefully be elaborated to include all

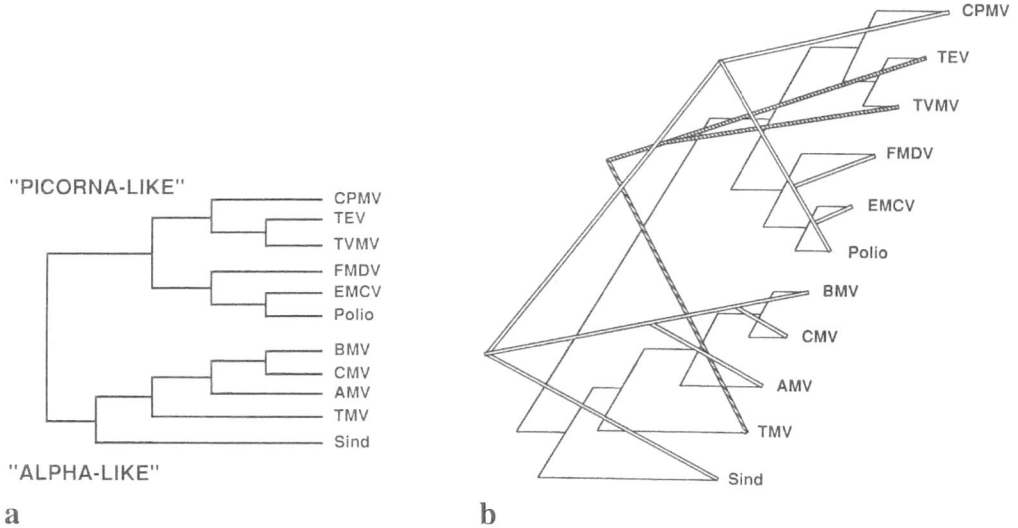

Fig. 9. Cladistic trees showing relationships between selected plant and animal viruses with single-stranded positive sense RNA genomes. **a** Unrooted tree calculated using PROPARS (PHYLIP 3.1) from "core conserved sequences" of presumed RNA-dependent RNA polymerases. Sequences are from plant poty- (TEV and TVMV), como- (CPMV), tobamo- (TMV), bromo- (BMV), cucumo- (CMV), and alfalfa mosaic (AMV) virus taxonomic groups, and animal *Picorna*- (polio, FMDV, EMCV) and *Togaviridae* (*Sind*, Sindbis virus, alphavirus) taxonomic families. The tree is arbitrarily bifurcated between the two largest clusters of species: these are designated "picorna-like" for those in a cluster with picorna-viruses, and "alpha-like" for those clustering with Sindbis virus. Both horizontal and vertical distances are arbitrary; the tree shows branching events but no evolutionary distances. The sequence data was obtained from [13], and alignments in the data matrix were optimized by eye before tree calculation. The tree was re-drawn and modified from one presented elsewhere [57]. **b** PROTPARS cladistic tree of polymerase sequences from **a** overlaid with trees showing hypothetical relationships among coat proteins of the different viruses (redrawn and modified from [57]). The latter two trees are drawn for illustrative purposes only, from evolutionary and functional speculations [25, 57, 60], and are not intended to represent true phylogenies. The overlaid trees indicate the different origins of parts of the genomes of viruses in different groups and families. Solid line: polymerase tree; unfilled line: coat proteins of spherical viruses; hatched filled line: coat proteins of rod-like and filamentous viruses

or most of the genes in a genome to show the respective affinities of different parts, in a sort of "module linkage" diagram that serves to define basic types or families of viruses as unique combinations of modules [54]. It appears obvious that it would be very difficult to group the viruses shown at any higher level than the family as it is now conceived.

The plant virus family: a definition

The conclusion to be drawn from these and other studies [25, 57] is that the erection of taxa at a higher level than the family for plus-strand RNA viruses is probably not feasible because of the "module shuffling" that apparently preceded the linear evolution of today's virus families [25]. This leads to a useful operational definition of a "family" taxon for plant or other viruses: this would be the highest grouping which could include all those viruses having a common morphology, sharing all or most of the genes in a "gene pool," and having a common genetic organization and mode of expression. The "gene pool" could be defined as a group of sequence cassettes or "modules" which have co-evolved or co-segregated during evolution, and could include some modules also present in other families or groups. One could then concisely define the virus taxonomic family as: *the largest unique group of viruses which could reasonably be assumed to have evolved from a common ancestor, without genome reorganization or addition of genes or modules from elsewhere.* This definition does not preclude the erection of higher-level taxa; however, the selection of criteria for their definition would be difficult and far more contentious.

Coat proteins and classification of *Potyviridae*: a summary

Sequence alignments and comparative analysis of CP sequences and HPLC profiles of *Potyviridae* species may be used to demonstrate groupings or clusters that correspond well with already accepted taxonomic groupings. We have clearly demonstrated that CPs of the generic potyviruses form a cluster distinct from CPs of the proposed *Bymo-* and *Rymovirus* genera, which in turn are more closely related to each other and to the *Potyvirus* genus than any of the *Potyviridae* are to familially distinct rod-shaped viruses such as potex-, carla-, clostero-, tobra- and tobamoviruses.

In several cases demonstrated here and elsewhere, the sequence/profile information should prompt taxonomic revisions, especially as concerns the SbMVs, the BCMV complex, the PWVs, and perhaps the PepMoV/PVY group. In all cases, however, it should be borne in mind *that analysis of sequence alignments and phylogenetic reconstruction of evolutionary relationships are merely adjuncts to the taxonomic process.* These methods of analysis of relationships are very valuable and provide perhaps the largest amount of comparative information of any of the taxonomic methods currently in use; however, where biological properties are felt to be important enough to justify describing a virus as a distinct "species," whatever its similarity with another virus, then obviously this must be given a fair weighting in the eventual decision. For this reason, it is not possible to do more than suggest that, according to the observed phylogenetic clustering, a given virus should be designated as a distinct virus or as a strain of another. It is a measure of the usefulness of the program PAUP that it allows for any number of extra characters apart from raw sequence, and differential weightings of them, to be included in the phylogenetic analysis. Thus, it may prove useful to include characters descriptive of host range and indicator typing, transmission mechanism, and relative virulence, together with sequence/HPLC profile data, in order to get a more comprehensive picture of relationships that is relevant to taxonomy.

With a growing movement to erect higher taxons for certain collections of plant viruses, the family *Potyviridae* provides an excellent test system for experimenting with different criteria for collecting viruses into a family or similar taxon. It is the premise of this paper, in agreement with Ward et al. [74], that analysis of coat-protein sequences alone so far appears to be sufficient to satisfactorily define the proposed plant virus taxonomic family *Potyviridae*.

Acknowledgements

EPR wishes to thank Cornell University Information Technology, J.Veronneau and M. Schrier for use of computer facilities and valuable assistance, and S.H. Howell and the Boyce Thompson Institute at Cornell University for providing sabbatical work facilities. We thank R. L. Gilbertson, S. Tolin, J. Jilka and R. Brand for unpublished sequences, and R. F. Doolittle, J. Hein, J. Felsenstein and E. Harley for computer software. EPR was supported by the Foundation for Research Development and the Mellon Foundation (RSA).

References

1. Agranovsky AA, Boyko VP, Karasev AV, Lunina NA, Koonin EV, Dolja VV (1991) Nucleotide sequence of the 3'-terminal half of beet yellows closterovirus RNA genome – unique arrangement of eight virus genes. J Gen Virol 72: 15–23
2. Atreya CD, Raccah B, Pirone TP (1990) A point mutation in the coat protein abolishes aphid transmissibility of a potyvirus. Virology 178: 161–165
3. Barnett OW (1991) *Potyviridae*, a proposed family of plant viruses. Arch Virol 118: 139–141
4. Barnett OW, Randles JW, Burrows PM (1987) Relationships among Australian and North American isolates of the bean yellow mosaic potyvirus subgroup. Phytopathology 77: 791–799
5. Barr J, Chambers P, Pringle CR, Easton AJ (1991) Sequence of the major nucleocapsid protein gene of pneumonia virus of mice: sequence comparisons suggest structural homology between nucleocapsid proteins of pneumoviruses, paramyxoviruses, rhabdoviruses and filoviruses. J Gen Virol 72: 677–685
6. Bishop DHL (1985) The genetic basis for describing viruses as species. Intervirology 24: 79–93
7. Bruenn JA (1991) Relationships among the positive strand and double-strand RNA viruses as viewed through their RNA-dependent RNA polymerases. Nucleic Acids Res 19: 217–226
8. Burger JT, Brand RJ, Rybicki EP (1990) The molecular cloning and nucleotide sequencing of the 3'-terminal region of ornithogalum mosaic virus. J Gen Virol 71: 2527–2534
9. Devereux J, Haeberli P, Smithies O (1984) A comprehensive set of sequence analysis programs for the VAX. Nucleic Acids Res 12: 387–395
10. Dijkstra J, Bos L, Bouwmeester HJ, Hadiastono T, Lohuis H (1987) Identification of blackeye cowpea mosaic virus from germplasm of yard-long bean and from soybean, and the relationships between blackeye cowpea mosaic virus and cowpea aphid-borne mosaic virus. Neth J Plant Pathol 93: 115–133

11. Dinant S, Lot H, Albouy J, Kuziak C, Meyer M, Astier-Manifacier S (1991) Nucleotide sequence of the 3' terminal region of lettuce mosaic potyvirus RNA shows a Gln/Val dipeptide at the cleavage site between the polymerase and the coat protein. Arch Virol 116: 235–252

12. Dolja VV, Koonin EV (1991) Phylogeny of capsid proteins of small icosahedral RNA plant viruses. J Gen Virol 72: 1481–1486

13. Dolja VV, Boyko VP, Agranovsky AA, Koonin EV (1991) Phylogeny of capsid proteins of rod-shaped and filamentous RNA plant viruses: two families with distinct patterns of sequence and probably structure conservation. Virology 184: 79–86

14. Domier LL, Shaw JG, Rhoads RE (1987) Potyviral proteins share amino acid sequence homology with picorna-, como-, and caulimoviral proteins. Virology 158: 20–27

15. Doolittle RF, Feng D-F (1990) Nearest neighbour procedure for relating progressively aligned amino acid sequences. In: Doolittle, RF (ed) Methods in enzymology, vol 183. Academic Press, San Diego, pp 659–669

15a. Doolittle RF (ed) (1990) Methods in enzymology, vol 183, molecular evolution: computer analysis of protein and nucleic acid sequences. Academic Press, San Diego

16. Eagles RM, Gardner RC, Forster RLS (1990) Nucleotide sequence of the tamarillo mosaic virus coat protein gene. Nucleic Acids Res 18: 7166

17. Farris JS (1988) HENNIG86, Version 1.5 Computer programme distributed by JS Farris, Port Jefferson Station, NY

18. Felsenstein, J (1990) PHYLIP (Phylogeny Inference Package) Version 3.3. Computer programme distributed by J Felsenstein, University of Washington, Seattle, WA

19. Feng D-F, Doolittle RF (1990) Progressive alignment and phylogenetic tree construction of protein sequences. In: Doolittle RF (ed) Methods in enzymology, vol 183. Academic Press, San Diego, pp 375–387

20. Frenkel MJ, Ward CW, Shukla DD (1989) The use of 3' non-coding nucleotide sequences in the taxonomy of potyviruses: application to watermelon mosaic virus 2 and soybean mosaic virus-N. J Gen Virol 70: 2775–2783

21. Frenkel MJ, Jilka JM, Shukla DD, Ward CW (1991) Differentiation of potyviruses and their strains by hybridisation with the 3' non-coding region of the viral genome. J Virol Methods 36: 51–62

22. Frenkel MJ, Jilka JM, McKern NM, Strike PM, Clark JM Jr, Shukla DD, Ward CW (1991) Unexpected diversity in the amino-terminal ends of the coat proteins of strains of sugarcane mosaic virus. J Gen Virol 72: 237–242

23. German S, Candresse T, Lanneau M, Huet JC, Pernollet JC, Dunez J (1990) Nucleotide sequence and genomic organization of apple chlorotic leaf spot closterovirus. Virology 179: 104–112

24. Gilbertson RL, Zambolim EM, Hidayat SH, Rybicki EP, Maxwell DP (1991) The molecular cloning and nucleotide sequence of the coat protein gene of the NL-3 strain of bean common mosaic potyvirus. Phytopathology 81: 1184

25. Goldbach RW (1987) Genome similarities between plant and animal RNA viruses. Microbiol Sci 4: 197–202

26. Gorbalenya AE, Koonin EV, Wolf YI (1990) A new superfamily of putative NTP-binding domains encoded by the genomes of small DNA and RNA viruses. FEBS Lett 262 : 145–148

27. Gough KH, Shukla DD (1992) Major sequence variations in the N-terminal region of the capsid protein of a severe strain of passionfruit woodiness virus. Arch Virol 124: 389–396

28. Hammond J (1992) Potyvirus serology, sequences and biology. In: Barnett OW (ed) Potyvirus taxonomy. Springer, Wien New York, pp 123–138 (Arch Virol [Suppl] 5)

29. Hein J (1990) Unified approach to alignment and phylogenies. In: Doolittle, RF (ed) Methods in enzymology, vol 183. Academic Press, San Diego, pp 626–669

30. Higgins DG, Sharp PM (1988) CLUSTAL: a package for performing multiple sequence alignment on a microcomputer. Gene 73: 237–244
31. Howarth AJ, Vandemark GJ (1989) Phylogeny of geminiviruses. J Gen Virol 70: 2717–2727
32. Huisman MJ, Linthorst HJM, Bol JF, Cornelissen BJC (1988) The complete nucleotide sequence of potato virus X and its homologies at the amino acid level with various plus-stranded RNA viruses. J Gen Virol 69: 1789–1798
33. Jain RK, McKern NM, Tolin SA, Hill JH, Barnett OW, Tosic M, Ford RE, Beachy RN, Yu MH, Ward CW, Shukla DD (1991) Similarity of coat protein peptide profiles of fourteen potyvirus isolates from soybean confirms that they are strains of the one virus. Phytopathology 81: 1167
34. Jayaram C, Hill JH, Miller WA (1991) Nucleotide sequences of the coat protein genes of two aphid-transmissible strains of soybean mosaic virus. J Gen Virol 72: 1001–1003
35. Jordan R, Hammond J (1991) Comparison and differentiation of potyvirus isolates and identification of strain-, virus-, subgroup-specific and potyvirus group-common epitopes using monoclonal antibodies. J Gen Virol 72: 25–36
36. Kashiwazaki S, Hayano Y, Minobe Y, Omura T, Hibino H, Tsuchizaki T (1989) Nucleotide sequence of the capsid protein gene of barley yellow mosaic virus. J Gen Virol 70: 3015–3023
37. Kashiwazaki S, Minobe Y, Hibino H (1991) Nucleotide sequence of barley yellow mosaic virus RNA 2. J Gen Virol 72: 995–999
38. Kashiwazaki S, Minobe Y, Omura T, Hibino H (1990) Nucleotide sequence of barley yellow mosaic virus RNA 1: a close evolutionary relationship with potyviruses. J Gen Virol 71: 2781–2790
39. Khan JA, Lohuis RW, Goldbach RW, Dijkstra J (1990) Distinction of strains of bean common mosaic virus and blackeye cowpea mosaic virus using antibodies to N- and C- or N-terminal peptide domains of coat proteins. Ann Appl Biol 117: 583–593
40. Kong L-J, Fang R-X, Chen Z-H, Mang K-Q (1990) Molecular cloning and nucleotide sequence of coat protein gene of turnip mosaic virus. Nucleic Acids Res 18: 5555
41. Koonin EV (1991) The phylogeny of RNA-dependent RNA polymerases of positive-strand RNA viruses. J Gen Virol 72: 2197–2206
42. Lain S, Riechmann JL, Martin MT, Garcia JA (1991) Homologous potyvirus and flavivirus proteins belonging to a superfamily of helicase-like proteins. Gene 82: 357–362
43. Langeveld SA, Dore J-M, Memelink J, Derks AFLM, van der Vlugt CIM, Asjes C, Bol JF (1991) Identification of potyviruses using the polymerase chain reaction with degenerate primers. J Gen Virol 72: 1531–1541
44. Li W-H, Graur D (1991) Fundamentals of molecular evolution. Sinauer, Sunderland,
45. Mackenzie DJ, Tremaine JH, Stace-Smith R (1989) Organization and interviral homologies of the 3′-terminal portion of potato virus S RNA. J Gen Virol 70: 1053–1063
46. Matthews REF (1982) Classification and nomenclature of viruses. Fourth report of the International Committee on Taxonomy of Viruses. Intervirology 17: 1–199
47. McKern NM, Whittaker LA, Strike PM, Ford RE, Jensen SG, Shukla DD (1990) Coat protein properties indicate that maize dwarf mosaic virus-KS1 is a strain of johnsongass mosaic virus. Phytopathology 80: 907–912
48. McKern NM, Shukla DD, Barnett OW, Vetten HJ, Dijkstra J, Whittaker LA, Ward CW (1991) Coat protein properties suggest that azuki bean mosaic virus, blackeye cowpea mosaic virus, peanut stripe virus and three isolates from soybean are all strains of the same potyvirus. Intervirology 33: 121–134
49. McKern NM, Shukla DD, Toler RW, Jensen SG, Tosic M, Ford RE, Leon O, Ward CW (1991) Peptide profiles of coat proteins confirm that the sugarcane mosaic subgroup consists of four distinct potyviruses. Phytopathology 81: 1025–1029

50. McKern NM, Edskes HK, Ward CW, Strike PM, Barnett OW, Shukla DD (1991) Coat protein of potyviruses. 7. Amino acid sequence of peanut strip virus. Arch Virol 119: 25–35

51. Meehan BM, Mills PR (1991) Nucleotide sequence of the 3'-terminal region of carnation latent virus. Intervirology 32: 262–267

52. Murphy FA, Kingsbury DW (1990) Virus taxonomy. In: Fields BN, Knipe DM, et al (eds) Virology, 2nd edn. Raven, New York, pp 9–25

53. Niblett CL, Zagula KR, Calvert LA, Kendall TL, Stark DM, Smith CE, Beachy RN, Lommel SA (1991) cDNA cloning and nucleotide sequence of the wheat streak mosaic virus capsid protein gene. J Gen Virol 72: 499–504

54. Palmenberg AC (1989) Sequence alignments of picornaviral capsid proteins. In: Semler, BL, Ehrenfelt, E (eds) Molecular aspects of picornavirus infection and detection. American Society for Microbiology, Washington, DC, pp 211–242

55. Quemada H, L'Hostis B, Gonsalves D, Reardon IM, Heinrikson R, Hiebert EL, Sieu LC, Slightom JL (1990) The nucleotide sequences of the 3'-terminal regions of papaya ringspot virus strains W and P. J Gen Virol 71: 203–210

56. Quemada H, Sieu LC, Siemieniak DR, Gonsalves D, Slightom JL (1990) Watermelon mosaic virus II and zucchini yellow mosaic virus: cloning of 3'-terminal regions, nucleotide sequences, and phylogenetic comparisons. J Gen Virol 71: 1451–1460

57. Rybicki E (1990) The classification of organisms at the edge of life or problems with virus systematics. S Afr J Sci 86: 182–186

58. Saitou N, Nei M (1987) The neighbour-joining method: a new method for reconstructing phylogenetic trees. Mol Biol Evol 4: 406–425

59. Shukla DD, Ward CW (1988) Amino acid sequence homology of coat proteins as a basis for identification and classification of the potyvirus group. J Gen Virol 69: 2703–2710

60. Shukla DD, Ward CW (1989) Structure of potyvirus coat proteins and its application in the taxonomy of the potyvirus group. Adv Virus Res 36: 273–314

61. Shukla DD, Ward CW (1989) Identification and classification of potyviruses on the basis of coat protein sequence data and serology. Arch Virol 106: 171–200

62. Shukla DD, Frenkel MJ, Ward CW (1991) Structure and function of the potyvirus genome with special reference to the coat protein coding region. Can J Plant Pathol 13: 189–191

63. Shukla DD, Lauricella R, Ward CW (1992) Serology of potyvirus: current problems and some solutions. In: Barnett OW (ed) Potyvirus taxonomy. Springer, Wien New York, pp 57–69 (Arch Virol [Suppl] 5)

64. Shukla DD, McKern NM, Gough KH, Tracy SL, Leth SG (1988) Differentiation of potyviruses and their strains by high-performance liquid chromatographic peptide profiling of coat proteins. J Gen Virol 69: 493–502

65. Shukla DD, Ford RE, Tosic M, Jilka J, Ward CW (1989) Possible members of the potyvirus group transmitted by mites or whiteflies share epitopes with aphid-transmitted definitive members of the group. Arch Virol 105: 143–151

66. Sneath PHA, Sokal RR (1973) Numerical taxonomy: the principles and practices of numerical classification. Freeman, San Francisco

67. Stanarius A, Proeseler G, Richter J (1989) Immunelektronenmikroskopische Untersuchungen zur serologischen Verwandtschaft des Gerstengelbmosaik-Virus (barley yellow mosaic virus) und des Milden Gerstenmosaik-Virus (barley mild mosaic virus) mit anderen gestreckten Viren. Arch Phytopathol Pflanzenschutz 25: 303–307

68. Stanway G (1990) Structure, function and evolution of picornaviruses. J Gen Virol 71: 2483–2501

69. Stark DM, Beachy RN (1989) Protection against potyvirus infection in transgenic plants: evidence for broad spectrum resistance. Bio/Technology 7: 1257–1262

70. Strauss EG, Strauss JH, Levine AJ (1991) Virus evolution. In: Fields BN, Knipe DM, et al (eds) Virology, 2nd edn. Raven, New York, pp 167–190

71. Suzuki N, Shirako Y, Ehara Y (1990) Isolation and serological comparison of virus-coded proteins of three potyviruses infecting cucurbitaceous plants. Intervirology 31: 43–49

72. Swofford, DL (1990) PAUP: Phylogenetic Analysis Using Parsimony, Version 3.3. Computer programme distributed by DL Swofford and the Illinois Natural History Survey, Champaign, Illinois

73. Timmerman GM, Calder VL, Bolger LEA (1990) Nucleotide sequence of the coat protein gene of pea seed-borne mosaic potyvirus. J Gen Virol 71: 1869–1872

74. Tracy SL, Frenkel MJ, Gough KH, Hanna PJ, Shukla DD (1992) Bean yellow mosaic, clover yellow vein and pea mosaic are distinct potyviruses: evidence from coat protein gene sequences and molecular hybridisation involving the 3′ non-coding regions. Arch Virol 122: 249–261

75. Uyeda I, Takahashi T, Shikata E (1991) Relatedness of the nucleotide sequence of the 3′-terminal region of clover yellow vein potyvirus RNA to bean yellow mosaic potyvirus RNA. Intervirology 32: 234–245

76. Vance VB, Jordan R, Edwardson JR, Christie R, Purcifull DE, Turpin T, Falk B (1991) Evidence that pepper mottle virus and potato virus Y are distinct viruses: analyses of the coat protein and 3′ untranslated sequences of a California isolate of pepper mottle virus. In: Barnett OW (ed) Potyvirus taxonomy. Springer, Wien New York, pp 337–345 (Arch Virol [Suppl] 5)

77. Vetten HJ, Lesemann D-E, Maiss E (1992) Serotype A and B strains of bean common mosaic virus are two distinct potyviruses. In: Barnett OW (ed) Potyvirus taxonomy. Springer, Wien New York, pp 415–431 (Arch Virol [Suppl] 5)

78. Ward CW, Shukla DD (1991) Taxonomy of potyviruses: current problems and some solutions. Intervirology 32: 269–296

79. Ward CW, McKern NM, Frenkel MJ, Shukla DD (1992) Sequence data as the major criterion for potyvirus classification. In: Barnett OW (ed) Potyvirus taxonomy. Springer, Wien New York, pp 283–297 (Arch Virol [Suppl] 5)

80. Woese CR (1991) The use of ribosomal RNA in reconstructing evolutionary relationships among bacteria. In: Selander RK, Clark AG, Whittam TS (eds) Evolution at the molecular level. Sinauer, Sunderland, pp 1–24

81. Yokoyama S (1991) Molecular evolution of human immunodeficiency viruses and related retroviruses. In: Selander RK, Clark AG, Whittam TS (eds) Evolution at the molecular level. Sinauer, Sunderland, pp 96–111

82. Yu MH, Frenkel MJ, Mckern NM, Shukla DD, Strike PM, Ward CW (1989) Coat protein of potyviruses. 6. Amino acid sequences suggest watermelon mosaic virus 2 and soybean mosaic virus-N are strains of the same potyvirus. Arch Virol 105: 55–64

83. Zavriev SK, Kanyuka KV, Levay KE (1991) The genome organization of potato virus M RNA. J Gen Virol 72: 9–14

84. Zettler FW (1992) Designation of potyvirus genera: a question of perspective and timing. In: Barnett OW (ed) Potyvirus taxonomy. Springer, Wien New York, pp 235–237 (Arch Virol [Suppl] 5)

Authors' address: E. P. Rybicki, Department of Microbiology, University of Cape Town, Private Bag, Rondebosch 7700, South Africa.

Virus biology and variation

Arch Virol (1992) [Suppl 5]: 173–176

Importance of host ranges and other biological properties for the taxonomy of plant viruses

Jeanne Dijkstra

Department of Virology, Wageningen Agricultural University,
Wageningen, The Netherlands

Summary. With reference to the confusion regarding the value of biological properties for the taxonomy of plant viruses, following proposal is made. Distinction between viruses should be based on the nucleotide sequences of the genome, whereas that between strains should take also biological characteristics into account.

*

With our present knowledge of the intrinsic properties of viruses, in general, and potyviruses, in particular, such as the complete nucleotide sequences of viral genomes (plum pox virus (PPV) [11], potato virus Y (PVY) [14], tobacco etch virus (TEV) [2], tobacco vein mottling virus (TVMV) [5]) we have acquired a good basis for a sound taxonomy of plant viruses. Nevertheless, there is still confusion about the classification of viruses. The cause of this confusion is that besides being macromolecules, viruses are also pathogens which means that they can infect plants in which they may induce symptoms and diseases. The results of infection are determined by the genome of the virus, but also by the genome of the host plant and the relationship between the two. Although basically these effects may be reduced to characteristics at the molecular level, many effects are only indirect; for instance, some metabolic changes in the host plant may be due to virus infection or environmental stresses or to the effects of the two together. Plants are complex organisms made of cells with many organelles (Fig. 1) and the cells are of different types and form different complex organs. Virus replication may be compartmentalized or general throughout the cell and organelles may be affected differentially. It is understandable, therefore, that the question arises: are host ranges important for the taxonomy of (poty) viruses? Further, are symptoms useful as criteria for classification?

There are many examples to show that minor changes at the molecular level may result in pronounced effects on host range and symptoms. A few amino acid substitutions in structural (capsid) protein or nonstructural

(130 kDa, 180 kDa, and 30 kDa) proteins of tobacco mosaic virus (TMV) may lead to different symptoms. A minor modification in the 30 kDa of TMV may stop cell to cell movement [10]. With tomato plants carrying the resistance genes *Tm*1 and *Tm*2 to tomato mosaic virus, there is a strong correlation between a few amino acid substitutions of the 130 kDa and 180 kDa proteins of the virus and its ability to overcome the *Tm*1 resistance [12]. An alteration of the 30 kDa (transport) protein affects the *Tm*2 resistance [13].

We also know that a single virus-coded protein may possess more than one function. Besides protecting the nucleic acid, coat protein may also play a role in the initiation of infection [17], transmission by vectors [7], and formation of symptoms [4]. A factor responsible for induction of the hypersensitive reaction in plants has been mapped in the coat-protein gene in TMV. The adaptation of brome mosaic virus and cowpea chlorotic mottle virus, both bromoviruses, to mainly monocotyledonous and dicotyledonous plants, respectively, is controlled by at least two genes [1].

In the potyviruses there are about eight genes (Fig. 1). The primary function of most of them is known [6], but it must be realized that the gene products are likely to have multiple roles. For example, the coat protein and non-structural proteins such as the cylindrical inclusion (CI), now known to have helicase activity [9], and nuclear inclusion (NI) proteins, with proteinase and polymerase functions, may be involved in symptom induction too; the

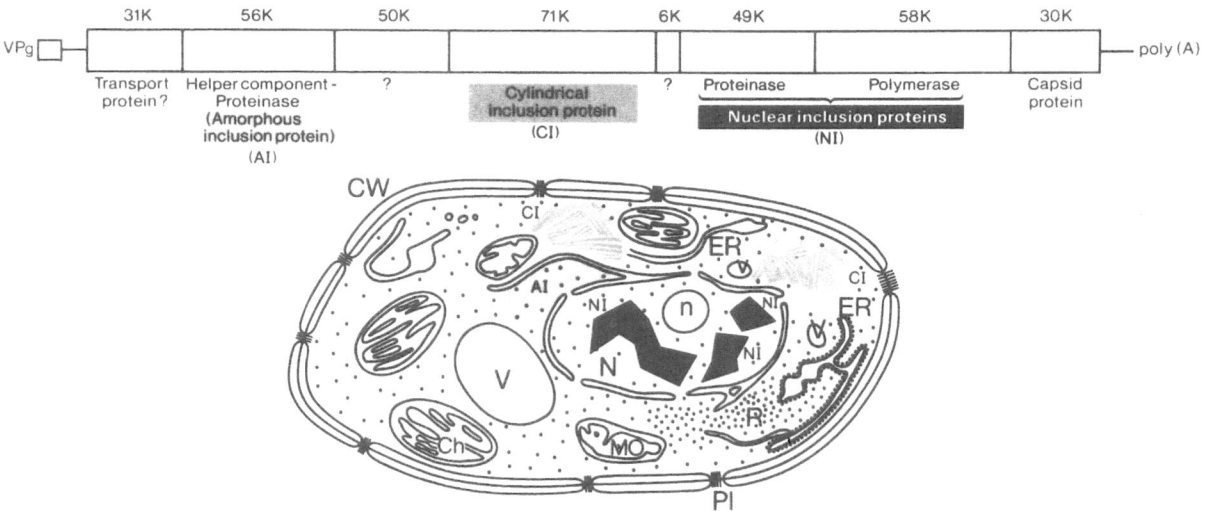

Fig. 1. Diagrammatic representations of the complexities of the genome of tobacco etch virus (TEV) and of a plant cell. Potyvirus genomes are translated into a polyprotein which is cleaved by virus-encoded proteases into at least seven proteins which may have multiple functions. Inside the host cells these genomes interact with numerous cellular metabolic pathways, cellular organelles, and plant organs as host plants are modified by virus infection

Arch Virol (1992) [Suppl 5]: 177–182

Clustering *Potyviridae* species on the basis of four major traits

K. M. Makkouk and **M. Singh**

International Center for Agricultural Research in the Dry Areas (ICARDA), Aleppo,
Syria

Summary. Cluster analysis was used to examine taxonomic relationships among 31 potyviruses, using four categorical variables; genome segmentation, vector, inclusion bodies produced and host range. Analysis showed that regardless of weight given to genome segmentation, the fungus-transmitted viruses clustered in one group and the rest of the viruses in another at 60% level of similarity. It has been concluded that the creation of one family to include both the bymoviruses and potyviruses seems to be a reasonable compromise at the present time.

Introduction

A classification system generally describes objects in some kind of relationship. With biological entities, the classification system needs to reflect phylogenetic relationships. When natural processes give rise to differentiated entities, overlapping classes with unclear boundaries are often created. This is true with plant viruses and particularly so with potyviruses, where virus strains seem to form a continuum in such a way that the borderlines separating species of potyviruses are difficult to define [1, 4, 10].

The early classifications of plant viruses failed because they were built around such superficial characteristics as symptomatology and host range. With the wealth of information available at present on viral structure, chemistry, serology, genome organization, and replication strategy in addition to characteristics such as vector transmission, symptomatology, and host range a more sound classification system can be established. Numerical analysis and computer facilities permit further exploration of taxonomic relationships [5].

The major problem with respect to potyviruses is to define a basis for differentiating a single virus, or species, in a consistent manner. This note is not to make an attempt in that direction, but rather to explore possibilities of grouping potyviruses using cluster analysis and to implement such analysis for 31 viruses selected as representative of the family *Potyviridae*.

Material and methods

The data matrix used for the analysis was a 31-virus by a 4-trait matrix. The traits included in the analysis were all categorical variables. The traits used were: genome organization, either monopartite or bipartite (G); vector, aphids, whiteflies, mites, or fungi (T); inclusion

Table 1. A list of viruses in the *Potyviridae* and their characteristics based on genome structure, transmission by types of vector, inclusion body type induced and hosts infected

Acronym	Virus name	Trait[a]			
		G	T	I	H
BaYMV	barley yellow mosaic virus	2	3	2	1
WYMV	wheat yellow mosaic virus	2	3	(2)	1
WSSMV	wheat spindle streak mosaic virus	2	3	4	1
OMV	oat mosaic virus	2	3	2	1
AgMV	agropyron mosaic virus	1	4	1	1
ONMV	oat necrotic mottle virus	1	4	1	1
SpMV	spartina mottle virus	1	4	(2)	1
WSMV[b]	wheat streak mosaic virus	1	4	4	1
HoMV	hordeum mosaic virus	1	4	1	1
PVY[c]	potato virus Y	1	1	4	2
PepMoV	pepper mottle virus	1	1	4	2
BCMV	bean common mosaic virus	1	1	1	2
BlCMV	blackeye cowpea mosaic virus	1	1	1	2
DsMV	dasheen mosaic virus	1	1	3	1
HMV	henbane mosaic virus	1	1	3	2
IMMV	iris mild mosaic virus	1	1	(2)	1
LYSV	leek yellow stripe virus	1	1	4	1
LMV	lettuce mosaic virus	1	1	2	2
PRSV	papaya ringspot virus	1	1	1	2
PeMoV[d]	peanut mottle virus	1	1	4	2
TEV	tobacco etch virus	1	1	2	2
GYSV	garlic yellow streak virus	1	1	(1)	1
ZYMV	zucchini yellow mosaic virus	1	1	1	2
ZYFV	zucchini yellow fleck virus	1	1	1	2
BYMV	bean yellow mosaic virus	1	1	2	2
BtMV	beet mosaic virus	1	1	2	2
ClYVV	clover yellow vein virus	1	1	2	2
PPV	plum pox virus	1	1	2	2
WMV2[d]	watermelon mosaic virus	1	1	4	2
IFMV	iris fulva mosaic virus	1	1	3	1
TVMV	tobacco vein mottling virus	1	1	1	2

[a] *G* Genome organization: monopartite (*1*), bipartite (*2*); *T* transmission by vectors: aphids (*1*), whiteflies (*2*), fungi (*3*), mites (*4*); *I* inclusion types based on Edwardson's classification into four divisions: 1, 2, 3, and 4; *H* hosts infected, either monocots (*1*) or dicots (*2*)
[b] Found more frequently in inclusion type IV, but also in III
[c] Found more frequently in inclusion type IV, but also in I
[d] Found more frequently in inclusion type IV, but also in II
In parentheses, tentative assignment to subdivision

bodies produced in infected cells based on the four subdivisions (I) established by Edwardson [3]; and whether the virus infects monocotyledonous or dicotyledonous plants (host range) (H). The 31 viruses and the traits used for their classification are presented in Table 1.

The viruses were classified using hierarchical clustering. The clustering required a matrix of similarities between all pairs of the viruses and a clustering method. The similarities were obtained from the formula based on test Type 2, appropriate for categorical variables, in the GENSTAD Statistical Package. Since various traits under consideration could vary in their importance, we introduced weights for traits in computing a modified similarity matrix, as follows. Let there be p traits (p = 4 in the present case). Let S_1 be the similarity matrix of order N (N is the number of viruses), obtained from ith trait (i = 1, 2, ..., p). Further, if to the ith trait a weight a_i is to be attached or a normalized weight $w_i = a_i/(a_1 + ... + a_p)$, then the combined similarity matrix S would be $S = W_1S_1 + W_2S_2 + ... + WpSp$. In the case of equal weight the formula becomes $S = p^{-1}(S_1 + ... + S_p)$. We used S and the single linkage method for cluster formation. The computation was done using GENSTAT 5 Release 1.2 on an IBM-PC.

Results

Cluster analyses of 31 viruses were performed using four traits receiving equal weights (Fig. 1) and two sets of variable weights reflecting possible changes in importance of the traits (Figs. 2 and 3).

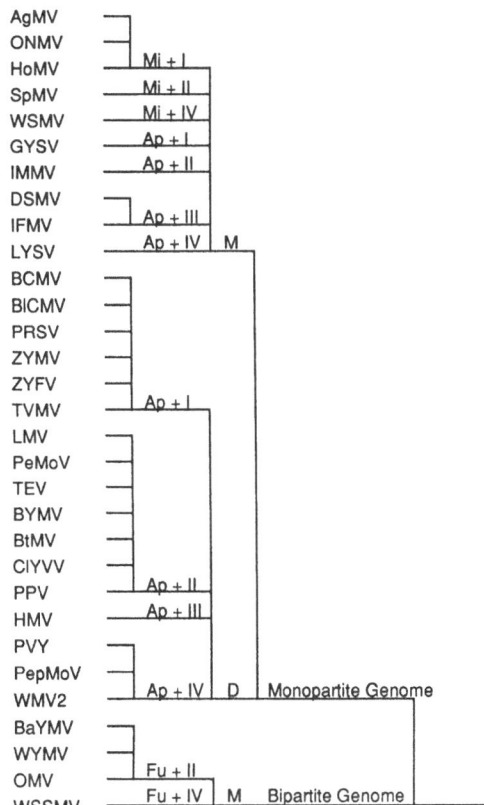

Fig. 1. Agglomerative hierarchical classification of 31 potyviruses based on genome segmentation, mode of transmission, type of inclusion bodies induced in host cells, and host range, with all four traits given equal weight (1:1:1:1). *Ap, Fu,* and *Mi* Aphid, fungal, and mite transmission; *I–IV* type of inclusions produced; *M* and *D* monocots and dicots, respectively

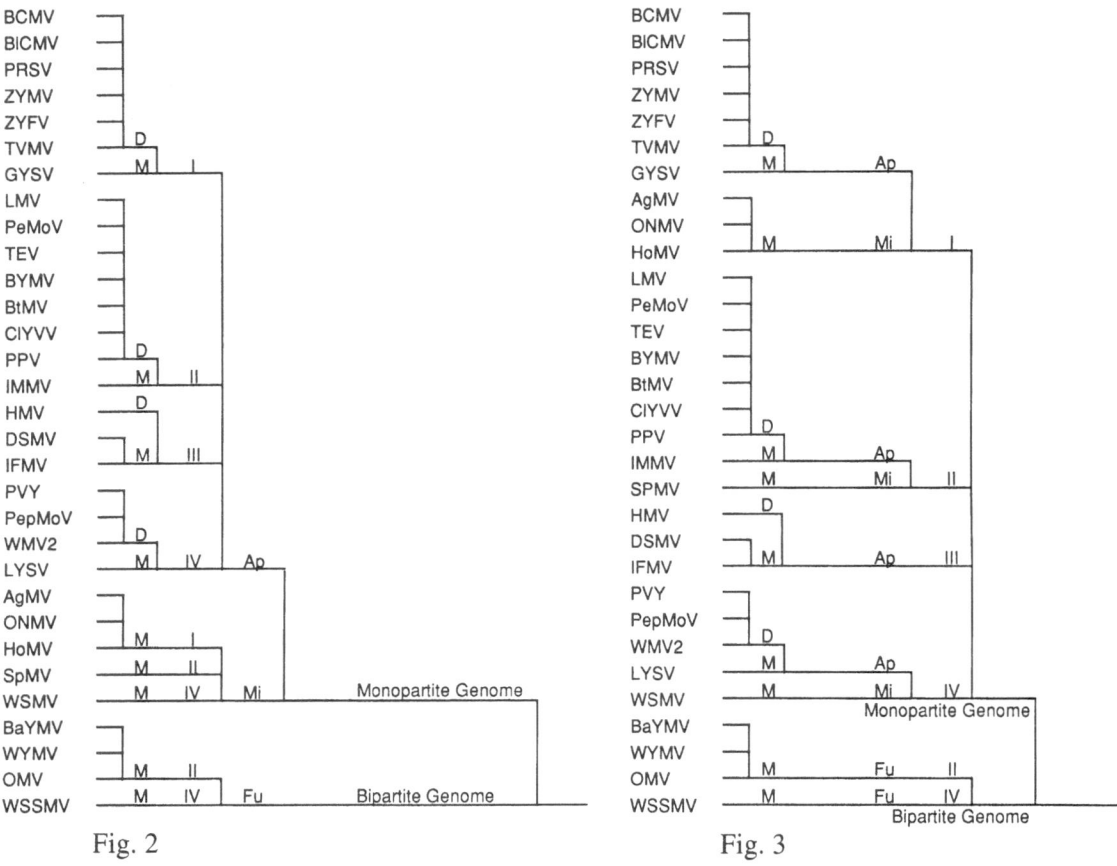

Fig. 2

Fig. 3

Fig. 2. Agglomerative hierarchical classification of 31 potyviruses based on genome segmentation, mode of transmission, type of inclusion bodies induced in host cells and host range, with all four traits given weights of 4, 3, 2, and 1, respectively

Fig. 3. Agglomerative hierarchical classification of 31 potyviruses based on genome segmentation, mode of transmission, type of inclusion bodies induced in host cells and host range, with all four traits given weights of 2, 3, 4, and 1, respectively

When equal weights were attached to genome segmentation (G), transmission by vectors (T), type of inclusions produced in infected cells (I) and host range (H), at any level of similarity higher than 75%, there were six clusters and seven viruses that remained ungrouped. For instance, barley yellow mosaic (BaYMV), oat mosaic (OMV), and wheat yellow mosaic (WYMV) viruses were in one cluster since they have the same properties with respect to the four traits. With a similarity level below 75% all viruses were grouped into two clusters; one consisting of the fungus-transmitted viruses (bymoviruses) BaYMV, OMV, WYMV, and WSSMV and the other of the remaining viruses (Fig. 1).

To see how the possible relationships among the viruses change with unequal weights, we attached weights of 4, 3, 2, and 1 to G, T, I, and H, respectively (in order of their possible importance). When the relationships were examined at a similarity level higher than 95%, the grouping was the same as in the case of equal weights, whereas the grouping pattern (merging of viruses into clusters) changed with a gradual decrease in the level of similarity down to 75% where the viruses split into three clusters based on type of vector transmission. At a similarity level of 70%, the same two groups of viruses were identified as with the use of equal weights (Fig. 2), the bymoviruses in one group and the rest of the viruses in the other.

Another exploration of relationships among viruses was conducted by interchanging the weights of I (production of inclusion bodies in host cells) and G (genome segmentation) and accordingly the weights given to G, T, I, and H were 2, 3, 4, and 1, respectively. At 95% similarity, the patterns of virus grouping remained the same as in the previous two cases. The same six clusters were formed down to 75% similarity level, and four viruses remained ungrouped. As in the second case, further merging occurred at lower similarity levels leaving again the same two groups, the bymoviruses in one and the rest of the viruses in the other, at 60% similarity.

Discussion

Some members of the fungus-transmitted viruses have a bipartite, ssRNA genome and no serological relationship seems to be present between this group of viruses and the insect- or mite-transmitted viruses which are serologically related [6, 8]. Murant and Harrison [7] suggested removing the fungus-transmitted viruses from the other viruses in the potyvirus group and establishing a group of bymoviruses. In this study, at the 60% level of similarity, using the 31 selected potyviruses the fungus-transmitted viruses cluster emerged separately from the other viruses regardless of the weight given to genome segmentation because two characters are different from all others compared.

Other workers argue that genome segmentation is a valid criterion for classification but its importance is overemphasized when used to create a separate virus group. Noteworthy homologies in amino acid sequences have been demonstrated between viruses with monopartite and bipartite genomes [11]. In addition all members of the family *Potyviridae*, including bymoviruses, produce similar inclusions in infected cells. The similarity between monopartite and bipartite potyviruses is also found in the genome structure [2]. In addition, the coat protein production of barley yellow mosaic virus is via polyprotein processing at a glutamine-alanine dipeptide in a 3' processing location, similar to coat-protein production by aphid-transmitted potyviruses with monopartite genomes. Thus, the similarities that exist between the fungus-transmitted and insect- and mite-transmitted potyviruses are an argument against their separation. The creation of one family to include both the bymoviruses and potyviruses seems to be a reasonable compromise at the present time.

In this study, analysis was conducted by employing four discontinuous variables. A more detailed analysis could be carried out to include data on continuous variables such as base composition, particle size, serological relatedness, molar percent and amino acid sequence of coat proteins, and total number of amino acid residues per subunit. Such analyses could prove powerful in differentiating between individual viruses and define those which make tight clusters. It could also prove to be useful in clarifying the situation of viral isolates which are presently considered as different entities, whereas data analysis might suggest that several should be considered as parts of one virus. Even with continuous variables, classes could be defined to differentiate between viruses. Shukla and Ward [9] when comparing coat-protein sequences of several potyviruses have shown the existence of a previously unrecognized discontinuity between different potyviruses. Distinct viruses exhibit sequence homologies ranging from 50 to 70% while strains of individual viruses exhibit sequence homologies of 90–99%. With continuous variables it should be possible to define boundaries that differentiate between species of the family *Potyviridae*.

References

1. Bos L (1970) The identification of three new viruses isolated from *Wisteria* and *Pisum* in The Netherlands, and the problems of variation within the potato virus Y group. Neth J Plant Pathol 76: 8–46
2. Brunt AA (1989) Viruses and virus-like pathogens transmitted by zoosporic fungi. EPPO Bull 19: 437–451
3. Edwardson JR (1974) Some properties of the potato virus Y group. Fla Agric Exp Stat Monogr Ser, no 4
4. Harrison BD (1985) Usefulness and limitations of the species concept for plant viruses. Intervirology 25: 71–78
5. Jellison, J (1987) Exploratory numerical taxonomy based on biochemical and biophysical characters of the tymoviruses. Intervirology 27: 61–68
6. Lesemann DE, Vetten HJ (1985) The occurrence of tobacco rattle and turnip mosaic viruses in *Orchis* spp., and of an unidentified potyvirus in *Cypripedium calceolus*. Acta Hortic 164: 45–54
7. Murant AF, Harrison BD (1988) Index to set. AAB Descriptions of Plant Viruses
8. Shukla DD, Ford RE, Tosic M, Jilka J, Ward CW (1989) Possible members of the potyvirus group transmitted by mites or whiteflies share epitopes with aphid-transmitted definitive members of the group. Arch Virol 105: 143–151
9. Shukla DD, Ward CW (1988) Amino acid sequence homology of coat proteins as a basis for identification and classification of the potyvirus group. J Gen Virol 69: 2703–2710
10. Van Regenmortel MHV (1989) Applying the species concept to plant viruses. Arch Virol 104: 1–17
11. Xiong Z, Lommel SA (1989) The complete nucleotide sequence and genome organization of red clover necrotic mosaic virus RNA-1. Virology 171: 543–554

Authors' address: K. M. Makkouk, International Center for Agricultural Research in the Dry Areas (ICARDA), P. O. Box 5466, Aleppo, Syria.

Arch Virol (1992) [Suppl 5]: 183–187

Specific infectivity and host resistance have predicated potyviral and pathotype nomenclature but relate less to taxonomy

R. O. Hampton[1] and **R. Provvidenti**[2]

[1] US Department of Agriculture, Agricultural Research Service, Department of Botany and Plant Pathology, Oregon State University, Corvallis, Oregon, U.S.A.
[2] New York State Agricultural Experiment Station, Department of Plant Pathology, Cornell University, Geneva, New York, U.S.A.

Summary. The names of potyviruses and viral-strains have represented the occurrence of predominant pathotypes on predominant crop genotypes. Thus virus nomenclature, but not viral taxonomy, has been decisively influenced by plant-genotype susceptibility and indirectly by host genetic resistance. Resistance to infection (i.e., host range) continues to serve a practical role in differentiating recognized viruses. Plant genes that confer disease tolerance or viral resistance remain a principal means of viral pathotype differentiation, as well as a principal control measure against major viral pathogens. Degrees of genetic diversity among isolates of recognized viruses should not be underestimated, and any system of viral taxonomy should be prepared for flexibility at the species level.

Introduction

Potyvirus pathotypes have been described as significant pathogens when they caused serious disease on specific crop hosts. (Pathotype is defined as a subspecies viral entity that is controlled by a host gene specific for that entity. Isolates of a given pathotype may vary in virulence, however, as illustrated by variants L (mild) and L1 (severe) of pea seed-borne mosaic potyvirus pathotype P-2. Although distinguishable in virulence on susceptible *Pisum* genotypes, they are both controlled by the same genes, either *sbm*-2 or *sbm*-3 [23].) The viral name typically designated symptoms induced in that crop host, and attempts were made to identify vectors and inoculum reservoirs.

Pathotype incidence implied that strategic crop genotypes contained alleles conferring susceptibility to systemic infection by the causal virus. Less damaging potyviruses or pathotypes were typically described later, probably because major crop genotypes contained alleles conferring resistance to a predominant viral pathotype or tolerance to the induced disease.

Virus pathotype–host relationship

Information development, an example

This pattern of information development is illustrated by pea seed-borne mosaic potyvirus (PSbMV), first found to occur naturally in *Pisum sativum* [18] and *Vicia faba* [13]. PSbMV, also readily aphid-transmissible [13], recurred increasingly as an exotic pathogen in breeding nurseries of private and public breeders in the U.S.A. from 1965 to 1975 [10]. PSbMV has a narrow host range and its preeminent inoculum reservoir is infected pea seed. Pathotype P2 (PSbMV-L) was discovered in 1980 in *Lens culinaris* germplasm from many origins [11] and was found not to infect most extant pea genotypes (i.e., those resistant to bean yellow mosaic virus).

Pathotype P2 also occurred extensively in commercial pea plantings on South Island, New Zealand (J. W. Ashby, pers. comm., 1981), and was later found in *Pisum* germplasm accessions [2] but was found in neither commercial lentil nor in pea plantings in northwestern U.S.A. Pathotype P4 [2] is known to occur in only a few *Pisum* germplasm accessions, and is capable of infecting even fewer *Pisum* genotypes than is P2. *Pisum* gene *sbm*-1, conferring resistance to pathotype Pl, was described in 1973 [7]. Genes *sbm*-2 and *sbm*-3 were found to confer resistance to pathotype P2 [23] and *sbm*-4 confer resistance to P4 [24].

Although PSbMV Pl is known from laboratory inoculations to infect other species of *Vicia* and *Lathyrus* and species of nine non-leguminous families (Apocyanaceae, Chenopodiaceae, Compositae, Cruciferae, Cucurbitaceae, Portulacaceae, Ranunculaceae, Solanaceae, and Urticaceae) [1], it has never been reported to occur naturally in any non-legume, i.e., it is uniquely a seed-borne pathogen of *Pisum* perennating and disseminating in infected seed. This pattern of pathotype-host-gene information development exists for other seed-borne, legume-infecting potyviruses including bean common mosaic virus in *Phaseolus* [5], soybean mosaic virus in *Glycine* [3, 15], and cowpea aphid-borne mosaic virus and blackeye cowpea mosaic virus in *Vigna* [19, 27].

Related examples

Pea mosaic virus (PMosV) was early considered a bean yellow mosaic virus (BYMV) variant with limited ability to infect snap/dry bean (*Phaseolus vulgaris*) cultivars [29]. It was subsequently learned that PMosV is able to infect only the *by*-1, *by*-1 genotype of *P. vulgaris* [26]. The inoculum reservoir from which aphids transmitted PMosV to peas (*Pisum sativum*) was typically one of several clover (*Trifolium*) species, and pea cultivars lacking gene *pmv*

[25] could be severely damaged by this virus. BYMV is commonly aphid-transmitted from red clover (*T. pratense*) to beans (*Phaseolus vulgaris*) [8], but rarely occurs in white clover [16, 17]. Snap bean genotypes lacking gene *By*-2 [4] and pea genotypes lacking gene *mo* [28] were sometimes severely damaged by BYMV. Clover yellow vein virus (ClYVV) commonly occurs in white clover (*T. repens*) over large agricultural areas [20] and is aphid transmissible from this reservoir to several food legume species [9]. Peas are protected against ClYVV by genes *cyv*-1 and *cyv*-2 [22], and beans are protected by gene *cyv* (*by*-3) [21].

Pathotype variants

The human desire to depend upon a predictable framework of information can lead to an erroneous mentality that viruses and viral pathotypes are fixed entities. Fortunately, type isolates of potyviruses have remained stable and provide dependable entities for characterization and comparison; otherwise there could be no orderly view of viral relationships. In any investigation of viral pathotypes, however, it becomes apparent that random isolates are single representatives of a heterogeneous population of "variants." For example, pathotypic comparisons among bean common mosaic virus isolates contributing to 1977 and 1981 BCM epidemics [12] demonstrated that the bean cultivars currently utilized to delineate BCMV pathotypes define only a portion of extant BCMV variants. It is less comforting, but more realistic and safer to consider viral isolates as points along a continuum of pathogenicity (ability to infect specific hosts) or virulence (severity of induced symptoms) variability.

Greater pathotypic and serotypic variation also exists among "native BYMV isolates" (e.g., a population of isolates occurring naturally in a selected bean cultivar) than may be generally recognized. Such variants are readily demonstrated by differences in infective capacity on selected bean cultivars and by numerous degrees of specific coupling with poly- or monoclonal immunoglobulin G to one or more selected BYMV isolates [14]. Although *Phaseolus* gene *By*-2 confers resistance to several defined BYMV isolates, the gene probably has not been tested against all variants of this virus.

Conclusions

As we are increasingly encouraged by more orderly virus-group relationships, let us remember that our taxonomic framework is fluid and is necessarily based on best-known type isolates. These best-known isolates generally have been artificially selected out of nature's reservoir of virus populations by "sieving" them through crop genotypes (i.e., virus selection by means of crop plant genes that permit or exclude infection when exposed to viruses). We should expect greater diversity than has yet been discovered and expect isolates from natural virus populations to express unique genetic profiles, with significant biological consequences and with probable sub-familial constraints on systems of viral taxonomy.

References

1. Aapola AA, Knesek JE, Mink GI (1974) The influence of inoculation procedure on the host range of pea seedborne mosaic virus. Phytopathology 64: 1003–1006
2. Alconero R, Provvidenti R, Gonsalves D (1986) Three pea seedborne mosaic virus pathotypes from pea and lentil germ plasm. Plant Dis 70: 783–786
3. Buzzell RI, Tu JC (1989) Inheritance of a soybean stem-tip necrosis reaction to soybean mosaic virus. J Hered 80: 400–401
4. Dickson MH, Natti JJ (1968) Inheritance of resistance of *Phaseolus vulgaris* to bean yellow mosaic virus. Phytopathology 58: 1450
5. Drijfhout E (1978) Genetic interaction between *Phaseolus vulgaris* and bean common mosaic virus with implications for strain identification and breeding for resistance (PhD Thesis). Agric Res Rep 872, IPO, Wageningen
6. Goodell JJ, Hampton RO (1984) Ecological characteristics of the lentil strain of pea seedborne mosaic virus. Plant Dis 68: 148–150
7. Hagedorn DJ, Gritton ET (1973) Inheritance of resistance to pea seed-borne mosaic virus. Phytopathology 63: 1130–1133
8. Hampton RO (1967) Natural spread of viruses infectious to bean. Phytopathology 57: 467–481
9. Hampton RO (1974) Natural spread of Oregon necrotic strains of bean yellow mosaic virus. Proc Am Phytopathol Soc 1: 37–38
10. Hampton RO, Mink GI, Hamilton RI, Kraft JM, Muehlbauer FJ (1976) Occurrence of pea seedborne mosaic virus in North American breeding lines and procedures for its elimination. Plant Dis 60: 455–459
11. Hampton RO (1982) Incidence of the lentil strain of pea seedborne mosaic virus as a contaminant of *Lens culinaris* germplasm. Phytopathology 72: 695–698
12. Hampton RO, Silbernagel MJ, Burke DW (1983) Bean common mosaic virus strains associated with bean mosaic epidemics in the northwestern United States. Plant Dis 67: 658–661
13. Inouye T (1967) A seed-borne mosaic virus of pea (in Japanese). Phytopathol Soc Jpn 33: 38–42
14. Jordan RL, Hammond J (1991) Comparison and differentiation of potyvirus isolates and identification of strain-, virus-, subgroup-specific, and potyvirus group-common epitopes using monoclonal antibodies. J Gen Virol 72: 25–36
15. Kiihl RAS, Hartwig EE (1979) Inheritance of reaction to soybean mosaic virus in soybean cultivars. Crop Sci 19: 372–375
16. McLaughlin MR, Boykin DL (1988) Virus diseases of seven species of forage legumes in the southeastern United States. Plant Dis 72: 539–542
17. McLaughlin MR, Ensign RD (1989) Viruses detected in forage legumes in Idaho. Plant Dis 73: 906–909
18. Musil M (1966) Über das Vorkommen des Virus des Blattrollens der Erbse in der Slowakei. Biologia Bratislava 21: 133–138
19. Patel PN, Mligo JK, Leyna HK, Kuwite C, Mmbaga ET (1982) Sources of resistance, inheritance, and breeding of cowpeas for resistance to a strain of cowpea aphid-borne mosaic virus from Tanzania. Indian J Genet 42: 221–229
20. Pratt MJ (1969) Clover yellow vein virus in North America. Plant Dis Rep 53: 210–212
21. Provvidenti R, Schroeder WT (1973) Resistance in *Phaseolus vulgaris* to the severe strain of bean yellow mosaic virus. Phytopathology 63: 196–197

22. Provvidenti R (1987) Inheritance of resistance to clover yellow vein virus in *Pisum sativum*. J Hered 78: 126–128

23. Provvidenti R, Alconero R (1988) Inheritance of resistance to a lentil strain of pea seed-borne mosaic virus in *Pisum sativum*. J Hered 79: 45–47

24. Provvidenti R, Alconero R (1988) Inheritance of resistance to a third pathotype of pea seed-borne mosaic virus in *Pisum sativum*. J Hered 79: 76–77

25. Provvidenti R (1990) Inheritance of resistance to pea mosaic virus in *Pisum sativum*. J Hered 81: 143–145

26. Schroeder WT, Provvidenti R (1968) Resistance of bean (*Phaseolus vulgaris*) to the PV2 strain of bean yellow mosaic virus conditioned by the single dominant gene *By*. Phytopathology 58: 1710

27. Taiwo MA, Gonsalves D, Provvidenti R, Thurston HD (1981) Partial characterization and grouping of isolates of blackeye cowpea mosaic and cowpea aphid-borne mosaic viruses. Phytopathology 72: 590–596

28. Yen DE, Fry PR (1956) The inheritance of immunity to pea mosaic virus. Aust J Agric Res 7: 272–280

29. Zaumeyer WJ (1957) Bean diseases and methods for their control. USDA Tech Bull 868

Authors' address: R. O. Hampton, USDA ARS, Department of Botany and Plant Pathology, Oregon State University, Corvallis, OR 97331, U.S.A.

Arch Virol (1992) [Suppl 5]: 189–211
© by Springer-Verlag 1992

Sources of resistance to viruses in the *Potyviridae*

R. Provvidenti[1] and R. O. Hampton[2]

[1] New York State Agricultural Experiment Station, Department of Plant Pathology,
Cornell University, Geneva, New York, U.S.A.
[2] US Department of Agriculture, Agricultural Research Service, Department of Botany
and Plant Pathology, Oregon State University, Corvallis, Oregon, U.S.A.

Summary. Resistance to 56 viruses in the family *Potyviridae* in 334 plant species was tabulated. Studies conducted in the last 60 years have elucidated the genetics and usefulness of 135 resistance genes, but no reports on the heritability of other sources of resistance are available. In most of the plant species, resistance to species of *Potyviridae* was simply inherited, either dominantly (60 genes) or recessively (39 genes). In some cases resistance was conferred by two or more genes. Symbols have been assigned to 86 genes, of which very few are duplicate entities. Resistance genes can be useful in determining relationships among these viruses, as well as for their identification. The role of conventional breeding and biotechnology in transferring genes from one species to another is discussed.

Introduction

Most of the viruses in the family *Potyviridae* are causal agents of serious diseases whose control is an important objective. Fungal and bacterial diseases can be controlled by the timely application of fungicides and bactericides, but the lack of viricides has hampered preventive and curative measures concerning viral infections. Historically, efforts have been directed toward elimination of vectors and eradication of, or isolation from, viral inoculum reservoirs. However, such provisional approaches are costly, only partially effective, and require continuous input. Conversely, use of resistance genes provides an efficient and economical solution for viral diseases. Time has demonstrated that these genetic factors are stable and durable, although they may be pathotype specific.

Resistance to these viruses has been found in: (*i*) existing cultivars, (*ii*) primitive cultivars or landraces, (*iii*) closely related species, wild and cultivated, and (*iv*) other genera of the same botanical family. The first two, classified as 'direct sources', can be quickly exploited in developing resistant cultivars. The other, 'indirect sources', have had very little appeal to plant breeders, because genetic incompatibility among different species can be an insurmountable barrier. Increasingly, however, we are relying on these last two sources for resistance genes. While traditional breeding methods are being augmented by useful techniques, such as early embryo rescue, tissue culture, protoplast regeneration, and cellular fusion, more benefits can be derived from genetic engineering. The expectations and prospectives for the role of recombinant DNA technology in redesigned plant genomes are increasing. Novel strategies for inducing viral resistance include the transferring into transgenic plants of complete viral genomes, viral coat-protein genes, antisense sequences, satellite sequences, ribozymes, putative viral replicase genes, defective interfering RNAs, antiviral antibodies, expression of pathogenesis-related proteins, and cloning of natural resistance genes [26, 65, 71].

Although sometimes sensationalized, most of these new methods have their limitations and in several instances, the resulting levels of resistance are not as effective as the action of naturally occurring resistance genes. More significant progress is probably achievable by transferring natural genes from one species to another. A number of approaches are being taken to clone them and one of the most promising consists of mapping resistance genes relative to polymorphic markers. Using restriction fragment length polymorphism (RFLP) for 'chromosome walking', genes are identified and cloned in vitro. The construction of polymorphic marker maps is being facilitated by the adoption of a novel technique based on the polymerase chain reaction (PCR). Natural resistance genes are the product of thousands or even millions of years of evolution, hence the cloning of resistance genes and transferring them from one species to another is likely to yield considerable benefit. This approach, however, would require a revised focus on host-plant genetics, rather than exclusive definition and exploitation of viral genetics. Linkage maps are available only for a few plant species and many more are needed for important crops. The experience and methodology that are deriving from the Human Genome Project should facilitate similar work for agricultural crops. For example, in inserting a foreign gene, a major problem is to specify where in the host's DNA this gene will become integrated. Placing it in a wrong location may interfere with the expression of native genes, or its action may be impaired. A very ingenious technique has been devised by human geneticists called 'homologous recombination', which tends to overcome these difficulties by allowing a specific gene to find an identical, or homologous sequence of DNA in the animal's genome and exchanging places with it. The same results have been obtained by plant breeders for many years. Using conventional breeding it has been possible to transfer a viral resistance gene into an homologous locus coding for susceptibility. However, the new technique eventually will facilitate the insertion of a valuable gene directly into a different species, thus bypassing the difficulty of intergeneric or interspecific breeding.

Resistance genes may be very useful to demonstrate anti-viral genetic mechanisms as well as relationships among viruses in the family *Potyviridae*, but only a few efforts have been directed toward this goal. Resistance to bean yellow mosaic virus and watermelon mosaic virus 2 is conferred by the same gene in *Pisum sativum* [176]. Resistance to papaya ringspot virus, strains P and W, is conferred by the same gene in *Cucumis metuliferus* [145]. In maize, the gene for resistance to maize dwarf mosaic virus and that for resistance to wheat streak mosaic virus appear to be either allelic or closely linked [102, 103]. The dominant gene for resistance to bean common mosaic virus [6] appears to be the same or closely linked to those conferring resistance to blackeye cowpea mosaic virus, cowpea aphid-borne mosaic virus, and passionfruit woodiness virus in *Phaseolus vulgaris* [148, 150]. Genes for resistance to potato virus Y in some *Solanum* species were reported to confer resistance also to potato virus A [36]. A similar situation occurs in peppers regarding resistance to potato virus Y and tobacco etch virus [45, 111, 213]. Thus, additional work is needed in this area to further understand evolutionary relationships among species of the *Potyviridae*.

Resistance genes also can be judiciously employed for virus identification, since host specificity is a species determinant. Such genes offer the only means by which pathotypes are identified. Genes conferring immunity can be utilized in separating individual viruses from naturally occurring mixtures.

In this paper we are reporting sources of resistance to 56 viruses of the *Potyviridae* in 334 plant species found in the last 60 years. Sources of resistance in different species of the same genus or family for the same *Potyviridae* species may be different or the same, or may be specific for different pathotypes of the same virus. In several instances, resistance to viruses in the *Potyviridae* has been demonstrated to be viral pathotype (strain) specific. Hence, a gene may confer resistance to one pathotype, but be totally ineffective against another pathotype of the same virus. Breeding for resistance is greatly facilitated if the heritability of each factor is known and its limitations are well understood. A large number of these studies have been undertaken and 136 resistance genes characterized. Of these, 60 were found to be single dominant and 39 were single recessive. In several hosts, resistance was found to be conferred by the complementary action of two dominant genes or two recessive genes, whereas in others, by three to five recessive genes with additive effect. Symbols were assigned to 86 genes (51 dominant and 35 recessive), and in a few rare instances, duplicate entities were identified [135, 140 (Table 1)]. A number of resistance genes already have been used in the development of new resistant cultivars, which in most of the cases have been able to eliminate or reduce damage caused by pertinent potyviruses. However, the term 'resistance' as used by different authors ranges from immunity to tolerance.

Table 1.

Virus and resistant species	Resistance gene(s)	Ref.
Azuki bean mosaic virus		
Vigna radiata		33
Agropyron mosaic		
Avena sativa		183
Hordeum jubatum		183
Araujia mosaic		
Asclepias humistrata		29
A. incarnata		29
A. syriaca		29
A. tuberosa		29
Dischidia sp.		29
Bean common mosaic		
Lablab purpureus		113
Phaseolus vulgaris	*I*	6
P. vulgaris	*bc*-u	56
P. vulgaris	*bc*-1, *bc*-1²	56
P. vulgaris	*bc*-2, *bc*-2²	56
P. vulgaris	*bc*-3	56
Pisum sativum	*bcm*	138
Macroptilium latyroides		142
Lupinus angustifolius		60
Bean yellow mosaic		
Glycine max		126
Lablab purpureus		113
Lupinus affinis		99
L. albococcineus		99
L. aridus		99
L. arboreus		46
L. barkeri		99
L. cumulicola		46
L. douglasii		99
L. elegans		99
L. hilarianus		99
L. hirsutissum		99
L. hartwegii		99
L. hybridus		99
L. lindleyanus		99
L. longiflorus		46
L. micranthus		99
L. mutabilis		99
L. perennis		46
L. polyphyllus		46
L. subcarnosus		99
L. succulentus		99

Table 1 (continued)

Virus and resistant species	Resistance gene(s)	Ref.
L. truncatus		99
L. villosus		46
Phaseolus coccineus		144, 169
P. coccineus	2–3 recessive	10
P. vulgaris	3 recessive	11, 190
P. vulgaris	*By-2*	54
Pisum sativum	*mo*	209
Trifolium ambiguum		14
T. pratense	*R*	53
T. repens		14
T. subterraneum		82
T. vesiculosum		119
Vicia faba	*bym-1*, *bym-2*	175, 176
Vigna aconitifolia		157
Beet mosaic		
Beta vulgaris	*Bm*	94
Bidens mottle		
Cichorium intybus		214
Lactuca sativa	*bi*	214
Blackeye cowpea mosaic		
Phaseolus vulgaris	*Bcm*	148
Vigna unguiculata	*bcm*	189
V. unguiculata	*blc*	200
V. unguiculata	1 dominant	156
Bryonia mottle		
Momordica charantia		97
Canna mosaic		
Canna generalis		22
Musa textilis		22
Carrot thin leaf virus		
Apium graveolens var. *dulce*		81
Celery mosaic		
Anthriscus sylvestris		199
Apium graveolens		199
A. graveolens var. *rapaceum*		20
Conium maculatum		188
Daucus carota		20
Pastinaca sativa		59
Chickpea bushy dwarf		
Arachis hypogaea		7
Glycine max		7

Table 1 (continued)

Virus and resistant species	Resistance gene(s)	Ref.
Chickpea bushy dwarf		
Phaseolus vulgaris		7
Pisum sativum		7
Vigna mungo		7
Vigna radiata		7
Clover yellow vein		
Phaseolus coccineus		144
P. vulgaris	*cyv*	133, 155
P. vulgaris	2 recessive	190
Pisum sativum	*cyv*	135
P. sativum	*cyv-2*	135
Cucurbita andreana		153
C. cordata		153
C. ficifolia		153
C. gracilior		153
C. pedatifolia		136
Trifolium ambiguum		14
T. pratense		96
T. repens		14
Vicia faba		63, 174
Cocksfoot streak		
Avena sativa		27
Festuca pratensis		27
Triticum aestivum		27
Colombian datura		
Capsicum annuum		85
Datura stramonium		85
Cowpea aphid-borne mosaic		
Phaseolus vulgaris	*Cam*	148
Vigna unguiculata		148
Eggplant severe mottle		
Nicotiana sylvestris		92
N. tabacum		92
Garlic yellow streak		
Allium fistulosum		106
Allium porrum		105
Guinea grass mosaic		
Agropyron junceiforme		194
Avena fatua		194
A. paniculata		194
A. strigosa		194
Bromus erectus		194

Table 1 (continued)

Virus and resistant species	Resistance gene(s)	Ref.
B. inermis		194
B. unioloides		194
Dactylis glomerata		194
Digitaria sanguinalis		194
Eleusine coracana		194
E. indica		194
E. tocussa		194
Hordeum murinum		194
Lolium multiflorum		194
L. rigidum		194
Miscanthus sinensis		194
Oryza sativa		194
Panicum capillare		194
Paspalum distichum		194
P. virgatum		194
Pennisetum japonicum		194
Poa pratensis		194
Sorghum vulgare		194
Trisetum flavescens		194
Triticum aestivum		194
T. durum		194
Leek yellow stripe		
Allium fistulosum		21
Lettuce mosaic		
Carduus arvensis		4
Lactuca sativa	*mo*	171
L. sativa	*g*	13, 117
Sonchus arvensis		4
S. oleraceous		4
Taraxacum officinale		4
Maize dwarf mosaic		
Avena fatua		166
A. sativa		166
Zea mays	*Rmd*[1]	159
Z. mays	2–3 dominant	55
Z. mays	*Mdm*1	102
Z. mays	5 additive	167
Z. mays	3 additive	115
Z. mays	1–2 dominant	84
Z. mays	1–2 dominant	97
Sorghum halepense		64
Oat mosaic		
Avena byzantina		25
A. sativa		28, 37
Lolium multiflorum		74

Table 1 (continued)

Virus and resistant species	Resistance gene(s)	Ref.
Onion yellow dwarf		
Allium cepa		21
A. fistulosum		21
Papaya ringspot-P		
Carica candamarcensis		3, 41
C. candicans		158
C. cauliflora		41, 100
C. papaya		42, 43, 95
C. pubescens		41, 77, 93
C. quercifolia		41, 77
C. stipulata		77
Cucumis metuliferus	*Wmv*	145
Papaya ringspot-W		
Benincasa hispida		127
Cucumis melo	*Wmv-1,Wmv-1²*	123, 206
C. sativus	*Wmv-1-1*	204
C. metuliferus	*Wmv*	151
Cucurbita ecuadorensis		153
C. ficifolia		153
C. foetidissima		153
Lagenaria siceraria		127
Parsnip mosaic		
Apium graveolens		109
Petroselinum crispum		109
Pimpinella anisum		109
Smyrnium olusatrum		109
Passionfruit ringspot		
Passiflora incarnata		52
P. mollissima		52
Passionfruit woodiness		
Passiflora edulis f. *flavicarpa*		197
Phaseolus vulgaris	1 dominant	150
Pisum sativum	1 recessive	151
Pea mosaic		
Phaseolus vulgaris	*By*	175
Pisum sativum		120
P. sativum	*pmv*	137
Pea seed-borne mosaic		
Cajanus cajan		5
Lens culinaris	*sbv*	72
Pisum sativum	*sbm*-1	73

Table 1 (continued)

Virus and resistant species	Resistance gene(s)	Ref.
P. sativum	*sbm*-2	140
P. sativum	*sbm*-3	140
P. sativum	*sbm*-4	141
Peanut mottle		
Arachis diogoi		105
Arachis spp.		105
Dolichos lablab		88
Glycine max	*Rpv, rpv*-2	159–162
G. max		49
Phaseolus vulgaris	*Pmv*	143
Vigna mungo		88
V. nilotica		88
V. radiata		88
V. unguiculata		18, 48
Peanut stripe		
Arachis diogoi		47
A. helodes		47
Glycine max		69, 205
G. max	*Pst*	32
Trifolium pratense		50
T. repens		50
Pepper mottle		
Capsicum annuum	*pmv*	213
C. chinense		213
Datura stramonium		213
Pepper veinal mottle		
Capsicum annuum		66
Plum pox		
Prunus besseyi		35
P. domestica		87
P. avium		34
P. cerasus		34
P. mahaleb		34
P. triloba		34
Potato A		
Solanum chacoense		208
S. commersonii		208
S. demissum		208
S. herrerae		208
S. kurtzianum		208
S. phureja		208

Table 1 (continued)

Virus and resistant species	Resistance gene(s)	Ref.
Potato A		
S. pinnatisectum		208
S. polyadenium		208
S. simplicifolium		208
S. stoloniferum	1 dominant	167
S. stoloniferum	Ry_{sto}, $Ry_{sto}{}^{n-1}$, $Ry_{sto}{}^{na}$	36
S. stoloniferum	$Ry_{stor}{}^{na}$, $Ry_{sto}{}^{n-2}$, Na_{sto}	36
S. sucrense		208
S. tuberosum		208
S. tuberosum	Na_{tbr}, Na_{sto}	36
Potato V		
Solanum cardiophyllum		61
S. stoloniferum		61
S. demissum		61
Potato Y		
Capsicum annuum	y^a	45
C. annuum		44, 110, 111
C. chinense	v^1, v^2	180
C. angulosus		182
C. barbatum		78
C. cardenasii		78
C. eximium		78
C. flexuosum		79
C. microcarpus		182
C. pubescens		78
Datura stramonium		78
Lycopersicon chilense		191
L. hirsutum	1 recessive	191, 192
L. peruvianum		191
L. pimpinellifolium		191
Nicotiana benavidesi	1 dominant	17
N. knightiana		179
N. leguiana		179
N. miersii		179
N. noctifera		179
N. otophora		179
N. raimondii		16
N. thyrsiflora		179
N. tabacum		57, 179
N. tomentosa		179
N. wigadioides		179
Solanum brevidens		67
S. cardiophyllum		207
S. chacoense		168, 207, 208
S. chacoense	Ny_{chc}	34
S. demissum		168, 205

Table 1 (continued)

Virus and resistant species	Resistance gene(s)	Ref.
S. hougasii	Ry_{hou}	36
S. microdontum	Ry_{hou}	36
S. phureja		207
S. stoloniferum	*Ry, Ryn*	168
S. stoloniferum	$Ry_{sto}, Ry_{sto}^{n-1}, Ry_{sto}^{na}$	36
S. stoloniferum	$Ry_{stor}^{na}, Ry_{sto}^{n-2}, Na_{sto}$	36
S. sambucinum		207
S. tuberosum		12, 35, 208
S. tuberosum	Na_{tbr}, Nc_{sto}	36
S. verrucosum		207
Ryegrass mosaic		
Agropyron repens		184
Lolium perenne	2 recessive	172
Hordeum vulgare		184
Oryza sativa		184
Secale cereale		184
Triticum aestivum		184
Soybean mosaic		
Glycine max		23, 31
G. max	Rsv, rsv^{t}	87
G. max	*Rsv-2, Rsv-3*	25
G. max	1 dominant	90
Phaseolus vulgaris	*Smv*	147
Sugarcane mosaic		
Saccharum officinarum		1, 112
S. sinensis		2
S. spontaneum		9
Sorghum bicolor	1 dominant	39
Telfairia mosaic		
Telfairia occidentalis		8
Tobacco etch		
Capsicum annuum	et^{a}	70
C. annuum		89, 110, 111
C. frutescens	et^{f}	70
C. frutescens	1 dominant	114
C. chinense		77
Lycopersicon esculentum	1 recessive	202
L. chilense		76
Nicotiana glauca		76
N. otophora		76
N. palmeri		76
N. raimondii		76
N. tabacum	2–3 recessive	170

Table 1 (continued)

Virus and resistant species	Resistance gene(s)	Ref.
Tobacco etch		
N. tomentosiformis		76
Solanum dulcamara		76
S. sanitwongsei		76
Tobacco vein mottling		
Capsicum annuum		122
C. frutescens		122
Tomato Peru		
Capsicum annuum		58
Datura stramonium		58
Lycopersicon esculentum	1 recessive	75
L. hirsutum	1 dominant	75
L. peruvianum		75
Nicotiana benavidesii		58
Solanum tuberosum		58
Tulip breaking		
Tulipa gesneriana		165
Tulipa spp.		165
Tulip chlorotic blotch		
Lilium formosanum		107
Turnip mosaic		
Brassica napus ssp *rapifera*	*Tum*	178
B. napus ssp. *oleifera*		185, 187, 196
B. napus ssp. *oleifera*	1 dominant	201
B. oleracea var. *gemmifera*	4 recessive	121
B. campestris ssp. *pekinensis*	2 dominant	116
B. campestris ssp. *pekinensis*		30, 68, 128
Cichorium intybus		154
Lactuca sativa	*Tu*	211
Impatiens balsamica		130
Mathiola incana	*rm*	83
Raphanus sativus		62
Watermelon mosaic		
Benincasa hispida		127
Citrullus colocynthis		132
Cucumis melo		186
Cucumis sativus	*Wmv*	38, 161
Cucurbita ecuadorensis		153
C. foetidissima		153
C. maxima		131
C. moschata		136
C. pedatifolia		136

Table 1 (continued)

Virus and resistant species	Resistance gene(s)	Ref.
Lagenaria siceraria		129
Phaseolus vulgaris	*Wmv*	91, 125
P. vulgaris	*Hsw*	91
Pisum sativum	*mo*	176
Watermelon mosaic Morocco		
Citrullus ecirrhosus		104
Coccinia sessifolia		104
Cucumis metuliferus		104
Luffa aegyptica		104
Wheat streak mosaic		
Agropyron amurense		101
A. cristatum		101
A. dasystachyum		101
A. divaricatus		101
A. elongatum		101
A. intermedium		101
A. lasianthum		101
A. repens		101
A. semicostatum		101
A. sibiricum		101
A. spicatum		101
A. trachycaulum		101
A. uganicum		101
Secale cereale		101
S. montanum		101
Triticum aestivum		177, 195
T. durum		195
Zea mays	*Wsm*1	103
White lupin mosaic		
Pisum sativum	*wlv*	149
Yam mosaic		
Dioscorea bulbifera		193
D. composita		193
D. floribunda		193
Zucchini yellow mosaic		
Citrullus colocynthis		132
C. lanatus	*zym*	139
C. melo	*Zym*	124
C. sativus	*zym*	134, 146
Cucurbita ecuadorensis	*Zym*	164
C. moschata	*Zym*	118
C. moschata	1 dominant	108
Lagenaria siceraria		146

202   R. Provvidenti and R. O. Hampton

References

1. Abbott EV (1958) Strains of sugarcane mosaic virus in Louisiana. Sugar Bull New Orleans 37: 49–51
2. Abbott EV (1961) A new strain of sugarcane mosaic virus. Phytopathology 51: 642
3. Adsuar J (1971) Resistance of *Carica candamarcensis* to the mosaic viruses affecting papayas (*Carica papaya*). J Agric Univ Puerto Rico 55: 265–266
4. Ainsworth GC, Ogilvie L (1939) Lettuce mosaic virus. Ann Appl Biol 26: 279-297
5. Alconero R, Provvidenti R, Gonsalves D (1986) Three pea seedborne mosaic virus pathotypes from pea and lentil germplasm. Plant Dis 70: 783–786
6. Ali MA (1950) Genetics of resistance to the common bean mosaic virus in bean (*Phaseolus vulgaris* L). Phytopathology 40: 69–79
7. Anjaiah V, Reddy DVR, Manohar SK, Naidu RA, Nene YL, Ratna AS (1989) Isolation and characterization of a potyvirus associated with bushy dwarf symptoms in chickpea, *Cicer arietinum*, in India. Plant Pathol 38: 520–526
8. Atiri GI, Varma A (1983) Development of improved lines of *Telfairia occidentalis* Hook resistant to mosaic disease. Trop Agric Trinidad 60: 95–96
9. Azab YE, Chilton SJP (1952) Transmission of mosaic resistance to progenies in crosses between certain sugarcane varieties. Phytopathology 42: 282
10. Baggett JR (1956) The inheritance of resistance to strains of bean yellow mosaic virus in the interspecific cross *Phaseolus vulgaris* × *P. coccineus*. Plant Dis Rep 40: 702–707
11. Baggett JR, Frazier WA (1957) The inheritance of resistance to bean yellow mosaic virus in *Phaseolus vulgaris*. Am Soc Hortic Sci Proc 70: 325–333
12. Bagnall RH, Bradley RHE (1958) Resistance to potato virus Y in the potato. Phytopathology 48: 121–125
13. Bannerot H, Buolidard L, Marrou J, Duteil M (1969) Etude de l'heredite de la tolerance au virus de la mosaique de la laitude chez la variete Gallega de Invierno. Ann Phytopathol 1: 219–226
14. Barnett OW, Gibson PB (1975) Identification and prevalence of white clover viruses and resistance of *Trifolium* species to these viruses. Crop Sci 15: 22–37
15. Barrios EP, Mosokar HI, Black LL (1971) Inheritance of resistance to tobacco etch and cucumber mosaic virus in *Capsicum frutescens*. Phytopathology 61: 1318
16. Berbec A (1988) Morphology, cytogenetics and resistance of amphidiploid *Nicotiana raimondii* MacBride × *N. tabacum* L. (F1 cv Zamojska 4 × cv LB-838) to potato virus Y. Genet Polon 29: 41–45
17. Berbec A, Glazewska Z (1988) Transfer of resistance to potato virus Y from *Nicotiana benevidesii* Goodspeed to *N. tabacum* L. Genet Polon 29: 323–333
18. Bijaisoedant M, Khun CW, Benner R (1988) Disease reaction, resistance, and viral antigen content in six legume species infected with eight isolates of peanut mottle virus. Plant Dis 72: 1042–1046
19. Bos L, Huijberts N, Huttinga H, Maat DZ (1978) Leek yellow stripe virus and its relationships to onion yellow dwarf virus: characteristics, ecology and possible control. Neth J Plant Pathol 84: 185–204
20. Bos L, Mandersloot HJ, Vader F, Steenbergen B (1989) An epidemic of celery mosaic potyvirus in celeriac (*Apium graveolens* var. *rapaceum*) in the Netherlands. Neth J Plant Pathol 95: 225–240
21. Brierley P, Smith FF (1946) Reaction of onion varieties to yellow dwarf virus and to three similar viruses isolated from shallot, garlic, and narcisus. Phytopathology 36: 292–296
22. Brierley P, Smith FF (1948) Canna mosaic in the United States. Phytopathology 38: 230–234

23. Buss GR, Chen PY, Roane CW, Tolin SA (1987) Genetics of reaction to soybean mosaic virus (SMV) in cultivars exhibiting differential reaction to SMV strains. Soybean Genet Newslett 14: 258:259

24. Buzzell RI, Tu JC (1989) Inheritance of soybean stem-tip necrosis reaction to soybean mosaic virus. J Hered 80: 400–401

25. Byrd BW, Graham D, Byrd WP (1971) Inheritance of tolerance to soil-borne oat mosaic virus in oats. Crop Sci 11: 875–877

26. Caplan A, Herrera-Estrella L, Inze D, Van Haute E, Van Montagu M, Schell J, Zambryski P (1983) Introduction of genetic material into plant cells. Science 222: 815–821

27. Catherall PL (1971) Cocksfoot streak virus. CMI/AAB Descriptions of Plant Viruses, no 59

28. Catherall PL, Valentine J (1987) Resistance to oat mosaic virus in autumn-sown oats. Ann Appl Biol 111: 483–487

29. Charudattan R, Zettler FW, Cordo H, Christie RG (1980) Partial characterization of a potyvirus infecting the milkweed vine *Morrenia odorata*. Phytopathology 70: 909–913

30. Chiu WF, Wang CK, Chang KP (1957) Factors influencing the development of the Chinese cabbage 'Kwuting'. Acta Phytopathol Sin 3: 45–53

31. Cho EK, Goodman RM (1979) Strains of soybean mosaic virus: classification based on virulence in resistant soybean cultivars. Phytopathology 69: 467–470

32. Choi SH, Green SK, Lee DR (1989) Linkage relationship between two genes conferring resistance to peanut stripe virus and soybean mosaic. Euphytica 44: 163–166

33. Choi, YM, Lee SH (1989) Identification of adzuki bean mosaic virus on *Phaseolus angularis*. Korean J Plant Pathol 5: 49–53

34. Christoff A (1958) The virus diseases of the fruit trees in Bulgaria. Phytopathol Z 31: 381–436

35. Cociu V, Minoiu N, Roman R, Cheorghiiu E, Isac M, Popescu I (1984) Investigations aimed at obtaining new plum varieties resistant to the plum pox virus. Probl Genet Theor Appl 16: 59–67

36. Cockerham G (1970) Genetical studies on resistance to potato virus X and Y. Heredity 25: 309–348

37. Coffman FA, Hebert TT, Gore UR, Byrd WP (1963) Sources of heritability of tolerance to soil-borne mosaic in winter oats. Plant Dis Rep 47: 54–57

38. Cohen S, Gertman E, Kedar N (1971) Inheritance of resistance to melon mosaic virus in cucumber. Phytopathology 61: 253–255

39. Conde BD, Moore RP, Fletcher DS, Teakle DS (1976) Inheritance of the resistance of Krish sorghum to sugarcane mosaic virus. Aust J Agric Res 27: 45–52

40. Conover R (1964) Distortion ringspot, a severe virus disease of papaya in Florida. Proc Fla State Hortic Soc 77: 440–444

41. Conover R (1976) A program for development of papayas tolerant to the distortion ringspot virus. Proc Fla State Hortic Soc 89: 229–231

42. Conover R, Litz RE (1978) Progress in breeding papayas with tolerance to papaya ringspot virus. Proc Fla State Hortic Soc 91: 182–184

43. Conover R, Litz RE (1981) Tolerance to papaya ringspot virus in papaya. Phytopathology 71: 868

44. Cook AA (1963) Genetics of response in pepper to three strains of potato virus Y. Phytopathology 53: 720–722

45. Cook AA, Anderson CW (1960) Inheritance of resistance to potato virus Y derived from two strains of *Capsicum annuum*. Phytopathology 50: 73–75

46. Corbett MK (1958) A virus disease of lupins caused by bean yellow mosaic virus. Phytopathology 48: 86–91

47. Culver JN, Sherwood JL, Melouk HA (1987) Resistance to peanut stripe virus in *Arachis* germplasm. Plant Dis 71: 1880–1082

48. Demski JW, Alexander AT, Stefani MA (1983) Natural infection, disease reactions and epidemiological implications of peanut mottle virus in cowpea. Plant Dis 67: 267–269

49. Demski JW, Kuhn CW (1975) Resistance and susceptible reactions of soybean to peanut mottle virus. Phytopathology 65: 95–99

50. Demski JW, Reddy DRV, Sowell G, Bays D (1984) Peanut stripe, a new seed-borne potyvirus from China infecting groundnut (*Arachis hypogaea*). Ann Appl Biol 105: 495–501

51. Demski JW, Sowell G (1981) Resistance to peanut mottle virus in *Arachis* spp. Peanut Sci 8: 43–44

52. De Wijs JJ (1974) A virus causing ringspot of *Passiflora edulis* in the Ivory Coast. Ann Appl Biol 77: 33–40

53. Diachun S, Henson L (1959) Inheritance of necrotic, mottle, and resistant reaction to bean yellow mosaic virus in clones of red clover. Phytopathology 49: 537

54. Dickson MH, Natti JJ (1968) Inheritance of resistance of *Phaseolus vulgaris* to bean yellow mosaic virus. Phytopathology 58: 1450

55. Dollinger EJ, Findley WR, Williams LE (1970) Inheritance of resistance to maize dwarf virus in maize (*Zea mays* L.). Crop Sci 11: 664–667

56. Drijfhout E (1978) Genetic interaction between *Phaseolus vulgaris* L. and bean common mosaic virus and its strains. Pudoc Wageningen, Agric Res Rep 872

57. Endemann W (1955) Resistance to strains of potato virus Y in tobacco. Ber Inst Tabak 2: 57–76

58. Fernandez-Northcote EN, Fulton RW (1980) Detection and characterization of Peru tomato virus strains infecting pepper and tomato in Peru. Phytopathology 70: 315–320

59. Freitag JH, Severin HHP (1945) Insect transmission, host range, properties of the crinkle-leaf strain of western celery mosaic virus. Hilgardia 16: 361–371

60. Frencel I, Pospieszny H (1979) Viruses in natural infections of yellow lupin (*Lupinus luteus* L) in Poland. IV. Bean common mosaic virus (BCMV). Acta Phytopathol Acad Sci Hung 14: 279–284

61. Fribourg CE, Nakashima J (1984) Characterization of a new potyvirus from potato. Phytopathology 74: 1363–1369

62. Fujisawa I (1990) Turnip mosaic virus strain in cruciferous crops in Japan. Jpn Agric Res Q 23: 289–293

63. Gadh IPS, Bernier CC (1984) Resistance in faba bean (*Vicia faba*) to bean yellow mosaic virus. Plant Dis 68: 109–111

64. Garrido MJ, Trujillo GE (1988) Identification of a new strain of maize dwarf mosaic virus (MDMV) in Venezuela. Fitopatol Venez 1: 73–81

65. Gasser CS, Fraley RT (1989) Genetically engineered plants for crop improvement. Science 244: 1293–1299

66. Gebre-Selassie K, Pochard E, Marchous G, Thouvenel JC (1986) New sources of resistance to pepper veinal mottle virus in pepper breeding lines. In: Eucarpia VI Meeting on genetics and breeding on *Capsicum* and eggplant. Zaragoza, Spain, pp 189–192

67. Gibson RW, Pehu E, Woods RD, Jones MGK (1990) Resistance to potato virus Y and potato virus X in *Solanum brevidens*. Ann Appl Biol 116: 151–156

68. Green SK, Deng TC (1985) Turnip mosaic virus strains in cruciferous hosts in Taiwan. Plant Dis 69: 28–31

69. Green SK, Lee DR (1989) Occurrence of peanut stripe virus on soybean in Taiwan, effect on yield and screening for resistance. Trop Pest Manag 35: 123–126

70. Greenleaf WH (1956) Inheritance of resistance to tobacco-etch virus in *Capsicum frutescens* and in *Capsicum annuum*. Phytopathology 46: 371–375

71. Grumet R (1990) Genetically engineered plant virus resistance. HortScience 25: 508–513

72. Haddad NI, Muehlbauer FJ, Hampton RO (1978) Inheritance of resistance to pea seed-borne mosaic virus in lentils. Crop Sci 18: 613–615

73. Hagedorn DJ, Gritton ET (1973) Inheritance of resistance to pea seed-borne mosaic virus. Phytopathology 63: 1130–1133

74. Hebert TT, Panizo CH (1975) Oat mosaic virus. CMI/AAB Descriptions of Plant Viruses, no 145

75. Hikida HR, Raymer WB (1972) Sources and inheritance of Peru tomato virus tolerance in tomato. Phytopathology 62: 764

76. Holmes FO (1946) A comparison of the experimental host range of tobacco etch and tobacco mosaic viruses. Phytopathology 36: 643–659

77. Horovitz S, Jimenez H (1967) Cruzamientos interspecificos e intergenericos en Caricaceas y sus implicaciones fitotecnicas. Agron Trop Maracay 17: 323–243

78. Horvarth J (1984) Virus resistance of species and varieties of pepper (*Capsicum* L.): new incompatible (hypersensitive) host-virus relations. Kertgazdasag 16: 93–95

79. Horvarth J (1986) Compatible and incompatible relations between *Capsicum* species and viruses. I. Review. Acta Phytopathol Entomol Hung 21: 35–47

80. Horvarth J (1986) Compatible and incompatible relations between *Capsicum* species and viruses. III. New incompatible host-virus relations (resistant and immune plants). Acta Phytopathol Entomol Hung 21: 59–62

81. Howell WE, Mink GI (1976) Host range, purification, and properties of a flexuous rod-shaped virus isolated from carrot. Phytopathology 66: 949–953

82. Hutton EM, Peak JW (1954) Varietal reactions of *Trifolium subterraneum* L to *Phaseolus* virus 2 Pierce. Aust J Agric Res 5: 598–607

83. Johnson BL, Barnhart D (1956) Resistance in *Mathiola incana* to turnip mosaic virus. Proc Am Soc Hortic Sci 67: 522–533

84. Josephson LM, Naidu B (1971) Reaction of diallele crosses of corn inbred (*Zea mays* L.) to maize dwarf mosaic virus. Crop Sci 11: 664–667

85. Kahn RP, Bartels R (1986) The Colombian datura virus, a new virus in the potato virus Y group. Phytopathology 58: 587–592

86. Kiihl RAS, Hartwig EE (1979) Inheritance of reaction to soybean mosaic virus in soybeans. Crop Sci 19: 372–375

87. Kleger H, Bauer E, Gruntzing M, Fuchs E, Verderevskaja TD, Bivol TF (1885) Detection of different types of resistance to plum pox virus in plum trees. Arch Phytopathol Pflanzenschutz 21: 339–346

88. Kuhn CW (1965) Symptomatology, host range, and effect on yield of a seed-transmitted peanut virus. Phytopathology 55: 880–884

89. Kuhn CW, Nutter FW, Padgett GB (1989) Multilevels of resistance to tobacco etch virus in pepper. Phytopathology 79: 814–818

90. Kwon SH, Oh JH (1980) Resistance to a necrotic strain of soybean mosaic virus in soybean. Crop Sci 20: 403–404

91. Kyle MM, Provvidenti R (1987) Inheritance of resistance to potato Y virus in *Phaseolus vulgaris* L. I. Two independent genes for resistance to watermelon mosaic virus 2. Theor Appl Genet 74: 595–600

92. Lapido JL, Lesemann DE, Koenig R (1988) Host range, serology, and cytopathology of eggplant and tomato strains of eggplant severe mottle virus, a new potyvirus found in Nigeria. J Phytopathol 122: 359–371

93. Larter L (1938) Plant pathology. Annu Rep Dept Agric Jamaica 1938:88-89

94. Lewellen RT (1973) Inheritance of beet mosaic virus resistance in sugarbeet. Phytopathology 63: 877–881

95. Lin CC, Su HJ, Wang DN (1989) The control of papaya ringspot virus in Taiwan ROC. Food Fert Tech Center Techn Bull 114

96. Lisa V, Dellavalle G (1983) Clover yellow vein virus in climbing bean (*Phaseolus vulgaris* L.). Phytopathol Mediter 22: 49–52

97. Lockhart BEL, Fisher HU (1979) Host range and some properties of *Bryonia* mottle virus, a new member of the potyvirus group. Phytopathol Z 96: 244–250

98. Loesch PJ, Zuber MS (1976) An inheritance study of resistance to maize dwarf mosaic virus in corn (*Zea mays* L.). Agron J 59: 423–426

99. Luchina NN, Sapun VM, Sergeenko MI (1981) Study of the resistance of lupin species of American group to *Fusarium* wilt and virus diseases. Sbornik Nau Trud Bel Nauch Issl Inst Zemledeliya 25: 149–155

100. Magdalita PM, Villegas VN, Pimentel RB, Bayot RG (1988) Reaction of papaya (*Carica papaya* L.) and related *Carica* species to ringspot virus. Phil J Crop Sci 13: 129–132

101. McKinney HH, Sand WJ (1951) Susceptibility and resistance to the wheat streak mosaic virus in the genera *Triticum, Agropyron, Secale* and certain hybrids. Plant Dis Rep 35: 476–479

102. McMullen MD, Louie R (1989) The linkage of molecular marker to a gene controlling the symptom response in maize to maize dwarf mosaic virus. Mol Plant Microbe Interact 2: 309–314

103. McMullen MD, Louie R (1991) Identification of a gene for resistance to wheat streak mosaic virus in maize. Phytopathology 81: 624–627

104. Meer FW van der, Garnett HM (1987) Purification and identification of a South African isolate of watermelon mosaic virus Morocco. J Phytopathol 120: 255–70

105. Melouk HA, Sanborn MR, Banks DJ (1984) Sources of resistance to peanut mottle virus in *Arachis* germplasm. Plant Dis 68: 563–564

106. Mohamed NA, Young BR (1981) Garlic yellow streak virus, a potyvirus infecting garlic in New Zealand. Ann Appl Biol 97: 65–74

107. Mowat WP (1985) Tulip chlorotic blotch virus, a second potyvirus causing flower break. Ann Appl Biol 106: 65–73

108. Munger HM, Provvidenti R (1987) Inheritance of resistance to zucchini yellow mosaic virus in *Cucurbita moschata*. Cucurb Genet Coop Ann Rep 10: 80

109. Murant AF, Munthe T, Gold RA (1970) Parsnip mosaic virus, a new member of the potato Y group. Ann Appl Biol 65: 127–135

110. Nagai H (1971) Novas variedades de pimentao resistentes ao mosaico causados por virus Y. Bragantia 30: 91–100

111. Nagai H, Smith P (1968) Reaction of pepper varieties to naturally occurring viruses in California. Plant Dis Rep 52: 928–930

112. Nagatomi S (1987) Studies on selection methods for sugarcane breeding. XVIII. Evaluation of varietal resistance to mosaic disease infection. Jpn J Trop Agric 31: 83–91

113. Narayan R, Poonan D (1987) Bean mosaic and bean yellow mosaic diseases of hyacinth bean (*Lablab purpureus* (L.) Sweet) and screening of available genotypes to find sources of resistance. India J Plant Pathol 5: 63–68

114. Nelson MR, Wheeler ER (1978) Biological and serological characterization and separation of potyviruses that infect peppers. Phytopathology 68: 979–984

115. Nikel MA, D'Arcy CJ, Rhodes AM, Ford RE (1984) Genetics of resistance of two dent corn inbreds to maize dwarf mosaic virus and transfer of resistance into sweet corn. Phytopathology 74: 467–473

116. Niu XK, Leung H, Williams PH (1983) Sources and nature of resistance to downy mildew and turnip mosaic virus in Chinese cabbage. J Am Soc Hortic Sci 108: 775–778

117. Pahlen der van A, Crnko J (1965) El virus del mosaico de la lechuga (*Marmor lactucae* (Holmes)) en Mendoza y Buenos Aires. Rev Invest Agropec 11: 25–31

118. Paris HS, Cohen S, Burger Y, Yodrph R (1988) Single gene resistance to zucchini yellow mosaic virus in *Cucurbita moschata*. Euphytica 37: 27–29

119. Pemperton J, Smith GR, McLaughlin MR (1989) Evaluation of arrowleaf clover for tolerance to bean yellow mosaic virus. Phytopathology 79: 230–234

120. Pierce WH (1935) The identification of certain viruses affecting leguminous plants. J Agric Res 51: 1017–1939

121. Pink DAC, Southerland RA, Walkey DGA (1986) Genetic analysis of resistance in Brussels sprout to cauliflower and turnip mosaic viruses. Ann Appl Biol 109: 199–208

122. Pirone TP (1989) Comparison of tobacco vein mottling and pepper veinal mottle viruses. Plant Dis 73: 336–339

123. Pitrat M, Lecoq H (1983) Two alleles for watermelon mosaic virus 1 resistance in melon. Cucurb Genet Coop Ann Rep 6: 52–52

124. Pitrat M, Lecoq H (1984) Inheritance of zucchini yellow mosaic virus resistance in *Cucumis melo* L. Euphytica 33: 57–61

125. Provvidenti R (1974) Inheritance of resistance to watermelon mosaic virus 2 in *Phaseolus vulgaris*. Phytopathology 64: 1448–1450

126. Provvidenti R (1975) Resistance in *Glycine max* to isolates of bean yellow mosaic virus in New York State. Plant Dis Rep 59: 166–168

127. Provvidenti R (1977) Evaluation of vegetable introductions from the People's Republic of China for resistance to viral diseases. Plant Dis Rep 61: 851–855

128. Provvidenti R (1980) Evaluation of Chinese cabbage cultivars from Japan and the People's Republic of China for resistance to turnip mosaic virus and cauliflower mosaic virus. J Am Soc Hort Sci 105: 571–573

129. Provvidenti R (1981) Sources of resistance to viruses in *Lagenaria siceraria*. Cucurb Genet Coop Ann Rep 4: 38–40

130. Provvidenti R (1982) A destructive disease of garden balsam caused by a strain of turnip mosaic virus. Plant Dis 66: 1076–1077

131. Provvidenti R (1982) Sources of resistance to viruses in accessions of *Cucurbita maxima*. Cucurb Genet Coop Ann Rep 5: 46–47

132. Provvidenti R (1986) Reaction of accessions of *Citrullus colocynthis* to zucchini yellow mosaic virus and other viruses. Cucurb Genet Coop Ann Rep 9: 82–83

133. Provvidenti R (1987) List of genes in *Phaseolus vulgaris* for resistance to viruses. Bean Improv Coop Annu Rep 30: 1–4

134. Provvidenti R (1987) Inheritance of resistance to a strain of zucchini yellow mosaic virus in cucumber. HortScience 22: 102–103

135. Provvidenti R (1987) Inheritance of resistance to clover yellow vein virus in *Pisum sativum*. J Hered 78: 126–128

136. Provvidenti R (1990) Viral diseases and genetic sources of resistance in *Cucurbita* species. In: Bates DM, Robinson RW, Jeffrey C (eds) Biology and utilization of Cucurbitaceae. Cornell University Press, Ithaca, pp 427–435

137. Provvidenti R (1990) Inheritance of resistance to pea mosaic virus in *Pisum sativum*. J Hered 81: 143–145

138. Provvidenti R (1991) Inheritance of resistance to the NL-8 strain of bean common mosaic virus in *Pisum sativum*. J Hered 82: 353–355

139. Provvidenti R (1991) Inheritance of resistance to the Florida strain of zucchini yellow mosaic virus in watermelon. HortScience 26: 407–408

140. Provvidenti R, Alconero R (1988) Inheritance of resistance to a lentil strain of pea seedborne mosaic virus in *Pisum sativum*. J Hered 79: 45–47

141. Provvidenti R, Alconero R (1988) Inheritance of resistance to a third pathotype of pea seedborne mosaic virus in *Pisum sativum*. J Hered 79: 76–77

142. Provvidenti R, Braverman SW (1976) Seed transmission of bean common mosaic virus in phasemy bean. Phytopathology 66: 1274–1275

143. Provvidenti R, Chirco EM (1987) Inheritance of resistance to peanut mottle virus in *Phaseolus vulgaris*. J Hered 78: 402–403

144. Provvidenti R, Dickson MH (1981) Kelvedon Marvel: a multiresistant cultivar of *Phaseolus coccineus* L. Bean Improv Coop Annu Rep 24: 124–125

145. Provvidenti R, Gonsalves D (1962) Resistance to papaya ringspot virus in *Cucumis metuliferus* and its relationship to resistance to watermelon mosaic virus 1. J Hered 73: 239–240

146. Provvidenti R, Gonsalves D, Humaydan HS (1984) Occurrence of zucchini yellow mosaic virus in cucurbits from Connecticut, New York, Florida and California. Plant Dis 68: 443–446

147. Provvidenti R, Gonsalves D, Ranalli P (1982) Inheritance of resistance to soybean mosaic virus in *Phaseolus vulgaris*. J Hered 73: 302–303

148. Provvidenti R, Gonsalves D, Taiwo MA (1983) Inheritance of resistance to blackeye cowpea mosaic virus and cowpea aphid-borne mosaic virus in *Phaseolus vulgaris*. J Hered 74: 60–61

149. Provvidenti R, Hampton RO (unpubl) Inheritance of resistance to white lupin mosaic virus in *Pisum sativum*

150. Provvidenti R, Niblett CL (unpubl) Inheritance of resistance to strains of passionfruit woodiness virus in *Phaseolus vulgaris*

151. Provvidenti R, Niblett CL (unpubl) Inheritance of resistance to a strain of passionfruit woodiness virus in *Pisum sativum*

152. Provvidenti R, Robinson RW (1977) Inheritance of resistance to watermelon mosaic virus 1 in *Cucumis metuliferus* (Naud.) Mey. J Hered 68: 56–57

153. Provvidenti R, Robinson RW, Munger HM (1978) Resistance in feral species to six viruses infecting *Cucurbita*. Plant Dis Rep 62: 326–329

154. Provvidenti R, Robinson RW, Shail JW (1979) Chicory: a valuable source of resistance to turnip mosaic virus for endive and escarole. J Am Soc Hortic Sci 104: 726–728

155. Provvidenti R, Schroeder WT (1973) Resistance in *Phaseolus vulgaris* to the severe strain of bean yellow mosaic virus. Phytopathology 63: 196–197

156. Quattara S, Chambliss OL (1991) Inheritance of resistance to blackeye cowpea mosaic virus in 'White Acre-BVR' cowpea. HortScience 26: 194–196

157. Rathore GS, Sharma RC, Agnihotri JP, Gour HN, Sharma RK (1986) Evaluation of moth bean germplasm for resistance to yellow mosaic virus. Indian J Virol 2: 81–86

158. Riccelli M (1963) Resistencia al virus del mosaic y adaptibilidad de tres especies se Caricaceae. Agric Trop Maracay 13: 89–94

159. Roane CW, Tolin SA, Aycock HS (1989) Genetics of reaction to maize dwarf mosaic virus strain A in several maize inbred lines. Phytopathology 79: 1364–1368

160. Roane CW, Tolin SA, Buss GR (1983) Genetics of reaction of five soybean cultivars to peanut mottle virus. Phytopathology 73: 968

161. Roane CW, Tolin SA, Buss GR (1986) Genetics of reaction to soybean mosaic virus (SMV) in the cultivars Kwang-gyo, Marshall and PI 96983. Soybean Genet Newslett 13: 134–135

162. Roane CW, Tolin SA, Buss GR (1986) Application of gene-for-gene hypothesis to soybean mosaic virus interactions. Soybean Genet Newslett 13: 136–137

163. Robinson RW, Munger HM, Whitaker TW, Bohn GW (1976) Genes of the Cucurbitaceae. HortScience 11: 554–568

164. Robinson RW, Weeden NF, Provvidenti R (1988) Inheritance of resistance to zucchini yellow mosaic virus in the interspecific cross *Cucurbita maxima* × *C. ecuadorensis*. Cucurb Genet Coop Annu Rep 11: 74–75

165. Romanov LR, Van Eijk JP, Eikrlboom W, Van Schdewijk AR, Peters D (1991) Determining level of resistance to tulip breaking virus (TBV) in tulip (*Tulipa* L.) cultivars. Euphytica 51: 273–280

166. Rosenkranz E (1981) Host range of maize dwarf mosaic virus. In: Gordon DT, Knobe JK, Scott GE (eds) Virus and virus-like diseases of maize in the United States. Coop Ser Bull 247, pp 152–162

167. Rosenkranz E, Scott GE (1984) Determination of the number of genes for resistance to maize dwarf virus strain A in five corn inbred lines. Phytopathology 74: 71–76

168. Ross H (1961) Breeding for virus resistance in the potato. Eur Potato J 1: 1–19

169. Rudorf W (1955) The transferring of resistance to bean mosaic virus 1 (common bean mosaic) and 2 (yellow bean mosaic) in *Phaseolus coccineus* into fertile hybrid plants from the cross *P. vulgaris* × *P. coccineus*. Naturwissenschaften 42: 19–20

170. Rufty RC, Wernsman EA, Gooding GV (1988) Inheritance of tobacco etch virus resistance found in *Nicotiana tabacum* cultivar Havana 307. Plant Dis 72: 879–882

171. Ryder FJ (1970) Inheritance of resistance to common lettuce mosaic. J Am Soc Hortic Sci 95: 378–379

172. Salehuzzaman M, Wilkins PW (1984) Components of resistance to ryegrass mosaic virus in a clone of *Lolium perenne* and their strain-specificity. Euphytica 33: 411–417

173. Schmidt HE, Geissler K, Karl E, Schmidt H (1986) A line of field bean (*Vicia faba* L.) with combined resistance to bean yellow mosaic virus, clover yellow vein virus, and *Aphis fabae* Scop. Arch Phytopathol Pflanzenschutz 221: 87–99

174. Schmidt HE, Rolewitz W, Schimanski HH, Kegler H (1985) Detection of resistance genes against bean yellow mosaic virus in *Vicia faba* L. Arch Phytopathol Pflanzenschutz 21: 83–85

175. Schroeder WT, Provvidenti R (1968) Resistance of bean (*Phaseolus vulgaris*) to the PV2 strain of bean yellow mosaic virus conditioned by the single dominant gene *By*. Phytopathology 58: 1710

176. Schroeder WT, Provvidenti R (1971) Common gene for resistance to bean yellow mosaic virus and watermelon mosaic virus in *Pisum sativum*. Phytopathology 61: 846–848

177. Seifers DL, Martin TJ (1988) Correlation of low level wheat streak mosaic virus resistance in Triumph 64 wheat with low virus titer. Phytopathology 78: 703–707

178. Shattuck VI, Stobbs LW (1987) Evaluation of rutabaga cultivars to turnip mosaic virus resistance and inheritance of resistance. HortScience 22: 935–937

179. Sievert RC (1972) Sources of resistance to potato virus Y in the genus *Nicotiana*. Tobacco Intelligence 174: 106–108

180. Simmonds NW, Harrison E (1959) The genetics of reaction to pepper veinbanding virus. Genetics 44: 1281–1289

181. Singh S, Chenulu VV (1980) Studies on resistance to virus diseases in *Capsicum* species. I. Sources of resistance to potato X and Y. Indian Phytopathol 33: 574–577

182. Singh S, Chenulu VV (1985) Studies on resistance to virus diseases in *Capsicum* species. III. Inheritance of resistance to potato virus Y. Indian Phytopathol 38: 479–483

183. Slykhuis JT, Bell W (1966) Differentiation of *Agropyron* mosaic, wheat streak, and an hitherto unrecognized *Hordeum* mosaic virus in Canada. Can J Bot 44: 1191–1208

184. Slykhuis JT, Paliwal YC (1972) Ryegrass mosaic virus. CMI/AAB Descriptions of Plant Viruses, no 86

185. Souza-Machado V, Shupe J, Stobbs LW (1988) TuMV resistance and chloro-triazine herbicide resistance in *Brassica napus*. Can J Plant Sc 68: 573

186. Sowell G, Demski JW (1981) Resistance to watermelon mosaic virus in muskmelon. FAO Plant Protect Bull 29: 71–73

187. Stobbs LW, Hume D, Forrest (1989) Survey of canola germplasm for resistance to turnip mosaic virus. Phytoprotection 70: 1–6

188. Sutabutra T (1968) Western celery mosaic virus. PhD Thesis University of California at Davis

189. Taiwo MA, Provvidenti R, Gonsalves D (1981) Inheritance of resistance to blackeye cowpea mosaic virus in *Vigna unguiculata* (L.) Walp. J Hered 72: 433–424

190. Tatchell SP, Baggett JR, Hampton RO (1985) Relationship between resistance to severe and type strain of bean yellow mosaic virus. J Am Soc Hortic Sci 110: 96–99

191. Thomas JE (1981) Resistance to potato virus Y in *Lycopersicon* species. Aust Plant Pathol 10: 67–68

192. Thomas JE, Mc Grath DJ (1988) Inheritance of resistance to potato virus Y in tomato. Aust J Agric Res 39: 475–479

193. Thouvenel JC, Fauquet C (1979) Yam mosaic, a new potyvirus infecting *Dioscorea cayenensis* in the Ivory Coast. Ann Appl Bio 93: 279–283

194. Thouvenel JC, Givord L, Pfeiffer P (1976) Guineagrass mosaic virus, a new member of the potato virus Y group. Phytopathology 66: 954–957

195. Timian RG, McMullen M (1986) The response of 13 hard red spring and 5 durum wheat cultivars to wheat streak mosaic virus. N Dakota Farm Res 44: 3–4

196. Tomlinson JA, Ward CM (1982) Selection for immunity in swede (*Brassica napus*) to infection by turnip mosaic virus. Ann Appl Biol 101: 43–50

197. Taylor RH, Greber RS (1973) Passionfruit woodiness virus. CMI/AAB Descriptions of Plant Viruses, no 122

198. Walkey DGA, Ward CM (1984) The reaction of celery (*Apium graveolens* L) to infection by celery mosaic virus. J Agric Sci 103: 415–419

199. Walkey DGA, Tomlinson J, Frowd JA (1970) Occurrence of western celery mosaic virus in umbelliferous crops in Britain. Plant Dis Rep 54: 370–371

200. Walker CA, Chambliss OL (1981) Inheritance of resistance to blackeye cowpea mosaic virus in *Vigna unguiculata* (L) Walp. J Am Soc Hortic Sci 106: 410–412

201. Walsh JA (1989) Genetic control of immunity to turnip mosaic virus in winter oilseed rape (*Brassica napus* ssp *oleifera*) and the effect of foreign isolates of the virus. Ann Appl Bio 115: 89–99

202. Walter JM (1956) Combination of resistance to tobacco etch and tobacco mosaic viruses in tomato breeding stocks. Phytopathology 46: 517–519

203. Van Eijk JP, Eikelboom W, Hogenboom (1986) The importance of wild species and old cultivars for breeding of flower bulbs. Acta Hortic 177: 399–403

204. Wang YJ, Provvidenti R, Robinson RW (1984) Inheritance of resistance to watermelon mosaic virus 1 in cucumber. HortScience 19: 587–588

205. Warwick D, Demski JW (1988) Susceptibility and resistance of soybeans to peanut stripe virus. Plant Dis 72: 19–21

206. Webb RE (1979) Inheritance of resistance to watermelon mosaic virus 1 in *Cucumis melo*. HortScience 14: 265–266

207. Webb RE, Hougas RW (1959) Preliminary evaluation of *Solanum* species and species hybrids for resistance to disease. Plant Dis Rep 43: 144–151

208. Webb RE, Schultz ES (1961) Resistance of *Solanum* species to potato viruses A, X, Y. Am Potato J 38: 137–142

209. Yen DE, Fry PR (1956) The inheritance of immunity to pea mosaic virus. Aust J Agric Res 7: 272–281

210. Zaklukiewicz K (1989) Resistance of Polish potato varieties to virus A. Biul Inst Ziemniaka 39: 27–36

211. Zink FW, Duffus JE (1973) Inheritance and linkage of turnip mosaic virus and downy mildew (*Bremia lactucae*) reaction in *Lactuca serriola*. J Am Soc Hortic Sci 98: 49–51

212. Zitter TA (1972) Naturally occurring pepper virus strains in South Florida. Plant Dis Rep 56: 586–590

213. Zitter TA, Cook AA (1973) Inheritance of tolerance to a pepper virus in Florida. Phytopathology 63: 1211–1212

214. Zitter TA, Guzman VL (1977) Evaluation of cos lettuce crosses, endive cultivars and chicory introductions for resistance to bidens mottle virus. Plant Dis Rep 61: 767–770

Authors' address: R. Provvidenti, Department of Plant Pathology, Cornell University, New York State Agricultural Experiment Station, D.W. Barton Laboratory, Geneva, NY 14456-0462, U.S.A.

Arch Virol (1992) [Suppl 5]: 213–215

Potential for using transgenic plants as a tool for virus taxonomy

W. K. Kaniewski

Monsanto Company, St. Louis, Missouri, U.S.A.

Summary. Transgenic plants resistant to viruses don't behave uniformly enough to be used as a convenient tool for virus taxonomy.

*

Transgenic plants expressing viral coat-protein genes have been developed primarily to achieve resistance to viral diseases [2, 3]. All experiments using these transgenic plants have been conducted to demonstrate and characterize viral resistance, not for any taxonomic purposes. The following statements and speculations are based on the results from these experiments.

It has been found in many instances that the viral resistance of transgenic plants is observed not only for the homologous virus from which the coat protein is derived, but also for different strains of the virus and even for related viruses. Plants which express viral coat proteins exhibit varying degrees of resistance, from near immunity to total susceptibility. An extremely resistant plant could be used to confirm virus identity but not to distinguish between viral strains. Kaniewski reported that transgenic potato resistant to potato virus Y (PVY) was not infectible by any PVY strain or isolate tested [4]. If the degree of resistance would be proportional to the relatedness of the viruses or their strains, transgenic plants could be a convenient tool for virus taxonomy. However, the experimental evidence to date indicates that transgenic plants will be of limited value for evaluating relatedness of viruses.

Results from several different experiments indicate these limitations. Shaw et al. [9] developed tobacco plants expressing tobacco vein mottling virus (TVMV) coat protein which showed some resistance to tobacco etch virus (TEV) but not to the severe strain of the homologous virus TVMV. Nejidat [6] challenged tobacco plants which expressed tobacco mosaic virus (TMV) coat protein with different viruses and their strains from the tobamovirus group. These plants had similarly high levels of protection against most tobamoviruses, although there were some distantly related members of the tobamovirus group against which protection was less effective. Nelson [7] field tested transgenic tomato expressing coat protein of TMV. He demonstrated effective resistance against TMV and strains of tomato mosaic virus (ToMV), but to correlate it to relatedness of these viruses would be difficult. Sanders [8] also reported high resistance in tomato plants transgenic for TMV coat protein against two TMV strains and found lower but still effective protection against different strains of ToMV. Anderson [1] reported that tobacco expressing TMV or alfalfa mosaic virus (AlMV) coat protein was partially protected against completely unrelated viruses such as potato virus X (PVX), PVY, or cucumber mosaic virus (CMV). He also observed some resistance in TMV coat protein expressing plants to AlMV, but no protection in AlMV coat protein expressing plants against TMV.

There is some evidence that protection mediated by expression of coat protein genes can be uniformly effective against different strains of a virus or against related viruses. Stark [10] showed similar levels of resistance in tobacco expressing coat protein of soybean mosaic virus (SbMV) against TEV and PVY. Ling [5] found that tobacco expressing coat protein of papaya ringspot virus (PRSV) was similarly resistant to TEV and PVY but not to CMV.

It is difficult to compare results from the cross protection experiments discussed above because of the varying methods used by different investigators to evaluate resistance. Still other reports of viral resistance in transgenic plants exist, but the results could not be included because of inconsistencies in their evaluation of resistance. In addition, the mechanism of protection mediated by expression of virus coat protein genes in transgenic plants is not well understood and probably differs from different viruses. It is possible that a mutation leading to a change in the coat-protein amino acid sequence of a virus could enable it to overcome the plant's resistance. Tumer [11] showed that tobacco plants expressing a modified AlMV coat protein containing a single amino acid change were susceptible to this virus, while plants with the unchanged coat protein were highly resistant.

To determine the potential usefulness of transgenic plants for virus taxonomy, experiments should be designed involving different transgenic plants expressing coat proteins of viruses from different virus groups. The plant species selected for the experiments should be natural hosts for most of the viruses from the group and have uniform susceptibility to them. Transgenic plants with high levels of resistance but not immunity to the homologous virus should be used. Experimental conditions should be standardized in these tests, and large sample sizes are necessary to facilitate statistical analysis. Plants should be carefully evaluated for different types of resistance, such as reduced incidence of infection, delay or attenuation of symptoms, lower virus titer, and decreased spread. Results from such experiments should be considered in relation to other known virus properties to judge if viral infectivity in transgenic plants could be considered a valid criterion for taxonomic purposes. Again, the current experimental evidence suggests that it is unlikely that transgenic plants will become a common tool for virus taxonomy in the near future.

References

1. Anderson E, Stark D, Nelson R, Powell P, Tumer N, Beachy R (1989) Transgenic plants that express the coat protein genes of tobacco mosaic virus or alfalfa mosaic virus interfere with disease development of some nonrelated viruses. Phytopathology 79: 1284–1290

2. Beachy R, Loesch-Fries S, Tumer N (1990) Coat protein-mediated resistance against virus infection. Annu Rev Phytopathol 28: 451–474

3. Hemenway C, Haley L, Kaniewski W, Lawson C, O'Connell K, Sanders P, Thomas P, Tumer N (1990) Novel methods of disease control (DNA recombinant technology). In: Mandahar CL (ed) Plant viruses, vol II, pathology. CRC Press, Boca Raton, pp 347–363

4. Kaniewski W, Sammons B, Lidell M, Tumer N (1990) Analysis of the mechanism of protection in transgenic Russet Burbank potato resistant to PVX and PVY. In: Proceedings of the VIIIth International Congress of Virology, Berlin, 1990, pp 80–020

5. Ling K, Namba S, Gonsalves C, Slightom J, Gonsalves D (1991) Protection against detrimental effects of potyvirus infection in transgenic tobacco plants expressing the papaya ringspot virus coat protein gene. Biotechnology 9: 752–758

6. Nejidat A, Beachy R (1990) Transgenic tobacco plants expressing a coat protein gene of tobacco mosaic virus are resistant to some other tobamoviruses. Mol Plant Microbe Interact 3: 247–251

7. Nelson R, McCormick S, Delannay S, Dube P, Layton J, Anderson E, Kaniewska M, Prokosch R, Horsch R, Rogers S, Fraley R, Beachy R (1988) Virus tolerance, plant growth, and field performance of transgenic tomato plants expressing coat protein from tobacco mosaic virus. Biotechnology 6: 403–409

8. Sanders P, Sammons B, Kaniewski W, Haley L, Layton J, LaVallee B, Delannay X, Tumer N (1992) Field resistance of transgenic tomatoes expressing the tobacco mosaic or tomato mosaic virus coat protein genes. Phytopathology 82: 683–690

9. Shaw J, Hunt AG, Pirone TP, Rhoads RE (1990) The organization and expression of potyviral genes. In: Pirone TP, Shaw JC (eds) Viral genes and plant pathogenesis. Springer, Berlin Heidelberg New York Tokyo, pp 107–123

10. Stark D, Beachy R (1989) Protection against potyvirus infection in transgenic plants: evidence for broad spectrum resistance. Biotechnology 7: 1257–1262

11. Tumer N, Kaniewski W, Haley L, Gehrke L, Lodge J, Sanders P (1991) The second amino acid of alfalfa mosaic virus coat protein is critical for coat protein-mediated protection. Proc Natl Acad Sci USA 88: 2331–2335

Author's address: W. Kaniewski, Monsanto Company, 700 Chesterfield Village Parkway, St. Louis, MO 63198, U.S.A.

Arch Virol (1992) [Suppl 5]: 217–219

A potyvirus in nature: indistinct populations

S.T. Ohki

College of Agriculture, University of Osaka Prefecture, Sakai, Osaka, Japan

Summary. Potyviruses occur in nature as a variable population. A number of strains have been reported for many potyviruses. Two or more viruses have been separated from "one virus" isolate. Experimental isolation and/or transmission often results in atypical viral isolates. A virus may be considered as a fuzzy population. We should properly understand the range of variation in one virus. The range of variation as well as typical characteristics of a virus should be included in future descriptions.

Introduction

What is a potyvirus? And how do we distinguish one from another? It is not simple to identify a virus species in the family *Potyviridae*, because this family includes so many viruses which are similar to each other and each virus has been independently described by criteria developed by each researcher. Some species have been adequately characterized, but there are still many which need to be better compared with other viruses in this family. Obviously, there is a need to examine and reorder the viruses in this large family. Therefore essential information for each virus in the family must be collected in different laboratories in ways that allow comparisons among viruses. Another problem is that a virus in nature is not a pure specimen. The aim of this brief communication is to consider that a virus in the field is not composed of identical particles but a population of particles.

Variation in a potyvirus

From an abstract standpoint, the classification "virus" seems to be clearly distinguishable and easily separable from other viruses. However, researchers who are experienced in detecting and identifying viruses have noticed that a virus in the field contains considerable variation and often occurs as a mixed infection of more than one virus. Consequently, a virus isolate should be considered a mixture first and then as a variable population rather than a pure specimen.

Various strains have been reported for a number of potyvirus species. For instance, many strains have been reported for bean yellow mosaic virus and soybean mosaic virus in Japan [2,6]. It is not easy to determine the strain status, since isolates which show intermediate characteristics between the described strains are often found.

We know that two or more viruses have been separated from a single disease. Sometimes earlier researchers did not distinguish between different isolates of what, at first, was considered a single virus. The watermelon mosaic disease was at first thought to be caused by "one virus." Watermelon mosaic virus-1 (now W strain of papaya ringspot virus [5]) and watermelon mosaic virus-2 [4] were shown to be too different to be strains of a single virus. Recently (1981) zucchini yellow mosaic virus was shown to be another distinct virus causing a mosaic disease of cucurbits [3].

And still we are often confused by the relationships between similar viruses [1, 8].

One of the reasons for this confusion is derived from the technique at the first step in the experimental processes. Although isolation is necessary for doing extensive characterization of viruses, minor variants are often selected in the isolation process. Isolated viruses do not always represent the disease agent and often cause symptoms different from the original field symptoms.

The single lesion transfer may not produce pure virus. Even if satellite RNAs are carefully eliminated from cucumber mosaic virus isolates, these RNAs are often found after several passages in host plants. Mutation of a virus while maintaining it in a greenhouse or a laboratory may result in the selection among minor variants. A part of host passage effects [7] is probably caused by this selection.

A virus should be considered as an indistinct or variable population which contains a major population of closely related variants and a minor population of more distinct variants. In an epidemic, variants may be continuously selected as the major population. It could be deemed a Gaussian population. One major strain causes the disease. However, variants are carried along in the population. If the environmental conditions change, one of the minor variants can be selected as the major component.

We should properly understand the range of variation in a virus in order to understand what a virus is and how it can change in the field, perhaps even causing widely different diseases.

Conclusion

Since a virus in the field is considered to contain minor variants, not only the characteristics typical of one isolate but also the range of variation of several isolates should be carefully observed and described hereafter. Without understanding the range of variation of a virus, we cannot identify it correctly. A database or a book which compiles the range of variation for each potyvirus species would be highly useful for identification and classification of viruses. Although it may take some time to prepare the database, I believe it is worth trying.

Which characteristics of a virus should be listed in the database? Basically, most of the essential characteristics for classification should be included. If the final objective of the classification is to identify the disease agent, biological traits such as host reactions and vector transmissibility should not be disregarded. Modern technology such as MAb and cDNA can detect differences among virus strains but we need more inclusive information on the total variability of a population.

We have not reached a consensus on what a virus "species" or "strain" is in practical terms. If we collect enough information for the variance in a virus, it may be easier to more exactly define the limits of a virus as a distinct taxon.

References

1. Demski JW, Reddy DVR, Wognkaew S, Kameya-Iwaki M, Saleh N, Xu Z (1988) Naming of peanut stripe virus. Phytopathology 78: 631–632
2. Inouye T (1973) Characteristics of cytoplasmic inclusions induced by bean yellow mosaic virus. Nogaku Kenkyu 54: 155–171
3. Lisa V, Lecoq H (1984) Zucchini yellow mosaic virus. CMI/AAB Descriptions of Plant Viruses, no 282
4. Purcifull D, Hiebert E, Edwardson J (1984) Watermelon mosaic virus 2. CMI/AAB Descriptions of Plant Viruses, no 293
5. Purcifull D, Edwardson J, Hiebert E, Gonsalves D (1984) Papaya ringspot virus. CMI/AAB Descriptions of Plant Viruses, no 292
6. Takahashi K, Tanaka T, Iida W, Tsuda Y (1980) Studies on virus diseases and causal viruses of soybeans in Japan. Bull Tohoku Natl Agric Exp Stat 62: 1–130
7. Yarwood CE (1979) Host passage effects with plant viruses. Adv Virus Res 25: 169–190
8. Xiang BC, Ohki ST, Inouye T (1990) Clover yellow vein virus and a carlavirus isolated from *Impatiens sultani* in Japan. Ann Phytopathol Soc Jpn 56: 557–560

Author's address: S. T. Ohki, College of Agriculture, University of Osaka Prefecture, Sakai, Osaka 593, Japan.

Arch Virol (1992) [Suppl 5]: 221–222

Potyvirus taxonomy:
potyviruses that affect solanaceous crops

E. N. Fernandez-Northcote

International Potato Center (CIP), Lima, Peru

Summary: Serology has been the main, or at least an important, tool for differentiating potyviruses that affect solanaceous crops. At present, analysis of the genome by hybridization techniques has supported the differentiation of viruses demostrated by serology. Phylogenetic groupings, based on nucleic acid sequences, should be combined with serological detection to make the groupings more usable.

*

The taxonomy of potyviruses should serve pathologists' needs. It should consider phylogenetic relationships determined on the basis of characters that can be detected reliably by simple techniques usable in unsophisticated laboratories.

For potyviruses which infect such solanaceous crops as potato, tomato, pepper, and tobacco, drop-microprecipitin tests with purified virus have shown that isolates with serological differentiation index (SDI) units equal to or greater than five are, for all practical purposes, different viruses [2, 4]. This criterion correlates well with results in sodium dodecyl sulfate-agar immuno-diffusion [5] using antigen in crude sap where formation of a weak spur indicates an SDI equal to or greater than five. These results also correlate well with a more recent technique like the standard direct, double antibody sandwich ELISA (DAS-ELISA) using proper polyclonal antibodies and crude sap. Isolates which, on the basis of drop-microprecipitin or SDS-agar double diffusion are considered to be different viruses, do not react in DAS-ELISA with polyclonal antibodies in the heterologous combination. Modified indirect ELISA on nitrocellulose membranes (NCM-ELISA) with crude sap, is less specific than DAS-ELISA; here different potyviruses are distinguished by SDI as in microprecipitin tests. When monoclonal antibodies (MAs) are used, some common epitopes are found among potyviruses with weak serological relationships with polyclonal antisera (different viruses but related), but a majority of MAs detect epitopes which are distinct for different viruses [3].

Serology has been the main, or at least an important, tool for differentiating potyviruses because serological characteristics of viruses are relatively stable.

This technique has been called into question prompting studies to analyze differences in the rest of the virus genome (about 10% of the genome codes for the coat protein) [1, 6, 7]. At present, analysis of the genome by hybridization techniques has supported the differentiation of viruses demonstrated by serology [8]. Even if differences are found in 90% of the noncoat-protein portion of the genome, these would be of secondary importance since the coat protein plays a critical role in the processes of vector-transmission and infection. Transmission and infection may determine an evolutionary process of the virus related to host specialization. This is, in the end, what is of interest to the pathologist.

Serological differences among isolates of a virus can be used for differentiating serotypes. These differences are important for appropriate virus detection.

The aphid-transmitted viruses of the old potyvirus group are considered a "genus" and each of the viruses a "species". Pathogenic differences within a virus allow for grouping isolates (here differences in vector relationships could also be considered) into strains. Virulence to specific genes for resistance within hosts can be used to differentiate pathotypes.

Nucleic acid technology can give a phylogenetic arrangement based on sequence homology within and between potyvirus species. These phylogenetic groupings, based on nucleic acid sequences, should be combined with serological detection to make the groupings more usable [1, 7] because serological tests are easily used in most field laboratories.

References

1. Baulcombe DC, Fernandez-Northcote EN (1988) Detection of strains of potato virus X and of a broad spectrum of potato virus Y isolates by nucleic acid spot hybridization (NASH). Plant Dis 72: 307–309
2. Fernandez-Northcote EN (1978) Detection and characterization of Peruvian tomato virus strains affecting pepper and tomato in Peru. PhD Thesis. University of Wisconsin, Madison, Wisconsin, USA
3. Fernandez-Northcote EN (1987) Reaction of a broad spectrum of potato virus Y isolates to monoclonal antibodies in ELISA. Fitopatologia 22: 33–36
4. Fernandez-Northcote EN, Fulton RW (1980) Detection and characterization of Peru tomato virus strains affecting pepper and tomato in Peru. Phytopathology 70: 315–320
5. Purcifull DE, Batchellor DL (1977) Immunodiffusion tests with sodium dodecyl sulfate (SDS)-treated plant viruses and plant viral inclusions. Fla Agric Exp Stat Bull 788
6. Robaglia C, Durand-Tardif M, Tronchet M, Boudazin G, Astier-Manifacier S, Casse-Delbert F (1989) Nucleotide sequence of potato virus Y (N strain) genomic RNA. J Gen Virol 70: 935–947
7. Shukla DD, Ward CW (1988) Amino acid sequence homology of coat proteins as a basis for identification and classification of the potyvirus group. J Gen Virol 69: 2703–2710
8. van der Vlugt R, Allefs S, de Haan P, Goldbach R (1989) Nucleotide sequence of the 3'-terminal region of potato virus Y^N RNA. J Gen Virol 70: 299–233

Author's address: E. N. Fernandez-Northcote, International Potato Center (CIP), Apartado 5969, Lima, Peru.

Arch Virol (1992) [Suppl 5]: 223–227

Biological variants of tobacco etch virus that induce morphologically distinct nuclear inclusions

R. G. Christie and **J. R. Edwardson**

Plant Virus Laboratory, Department of Agronomy, University of Florida,
Gainesville, Florida

Summary. The presence of distinctive nuclear inclusions has been used for many years as a diagnostic character for tobacco etch virus (TEV). Cytological examinations of isolates of TEV in both weeds and solanaceous crops from areas widely separated geographically have demonstrated the presence of a variety of nuclear inclusions that vary considerably in form. Such inclusions can, in many cases, be related to differences in symptom expression. It is suggested that distinctive nuclear inclusions may be used to select biological variants of TEV that may be useful in the study and manipulation of closely related "strains" of this potyvirus.

Introduction

A number of potyviruses induce nuclear inclusions (NIs). Some of the more distinctive NIs are associated with potyviruses belonging to subdivision II, i.e., those viruses that have rigid laminated aggregates (plates) associated with the cylindrical inclusions [7]. The cytology of a number of these viruses has been reviewed in a monograph on plant virus cytology [2]. Tobacco etch virus (TEV) was one of those viruses. All isolates of TEV studied to date have been found to induce NIs.

Nuclear inclusions

The NIs induced by TEV consist of two proteins that are products of the viral genome [5]. Both proteins play a significant role in TEV replication. One, a protease, cleaves the coat, nuclear inclusion, and cylindrical inclusion proteins from a polyprotein generated during the replication process. The other is a putative polymerase believed to be involved in the replication of the viral RNA. These important proteins are capable of interlocking in such a way as to produce structures (inclusions) within the nucleus of the cell. These inclusions are readily visible in the light microscope when properly stained [3]. With this instrument one is easily able to determine the gross morphology of the inclusions, since through-focus studies allow for three dimensional interpretations. Variants are determined by observing the inclusions in both face view and side view as demonstrated in Fig. 1b. Cytological observations with the light microscope have revealed that a number of isolates of TEV induce NIs that differ in size, shape, or number.

The fact that TEV isolates induce NIs with different structure has been known for a long time. In 1939, Kassanis [9] reported that a severe isolate of tobacco etch virus (TEV S) induced numerous thin, rectangular inclusions within the nuclei of infected host cells that were visible in the light microscope. Figure 1a demonstrates an isolate in our possession that produces a thin plate with rectangular structure. Since the inclusions were not detected in healthy plants and were not induced by other viruses, Kassanis considered them to be diagnostic for tobacco etch infections.

A few years later, Bawden and Kassanis [1] compared the properties of TEV S with an isolate of etch provided by W.C. Price that induced much milder symptoms on tobacco. Based on differences in symptom expression, serology, and cross protection data, it was concluded that the mild isolate (TEV M) should be considered a closely related strain. They also noted that TEV M induced fewer and larger NIs and that many of them appeared as bi-pyramids rather than thin, rectangular plates. They did not specifically state that the distinctive cytological appearance of these two etch isolates should be used to differentiate them as strains, although this would appear to be implied in the presentation of their data.

For many years following this report, no mention was made concerning NI variation of TEV isolates. This situation may have arisen from the fact that the electron microscope had come into general use and cytological studies were carried out with thin sections where three dimensional structure is difficult to visualize. Therefore, variant shapes were probably overlooked.

The standard isolate of TEV used at our laboratory was obtained from R.W. Fulton about 1965 and has since become designated as PV-69 by the American Type Culture Collection. This isolate induces thin, square plates (Fig. 1b) rather than the rectangular ones described by Kassanis for his TEV S isolate and does not produce as many per nucleus. The shape of the PV-69 NIs are similar in structure in all infected host species. PV-69 induces moderately severe symptoms on tobacco, but is rather mild on most pepper cultivars. This was the isolate first used for the purification and characterization of both cytoplasmic and nuclear inclusions of TEV. Antisera against coat, cytoplasmic cylindrical, and NI proteins have been prepared from this isolate.

In 1974, a study was undertaken to determine the resistance of tobacco variety V 20 to TEV [4]. Seven TEV isolates were used to challenge this variety. PV-69 was used as the control to which the other isolates were compared. V 20 proved to be immune to six of the seven isolates including PV-69. However, V 20 was susceptible to isolate TEV 6. Nuclei infected with

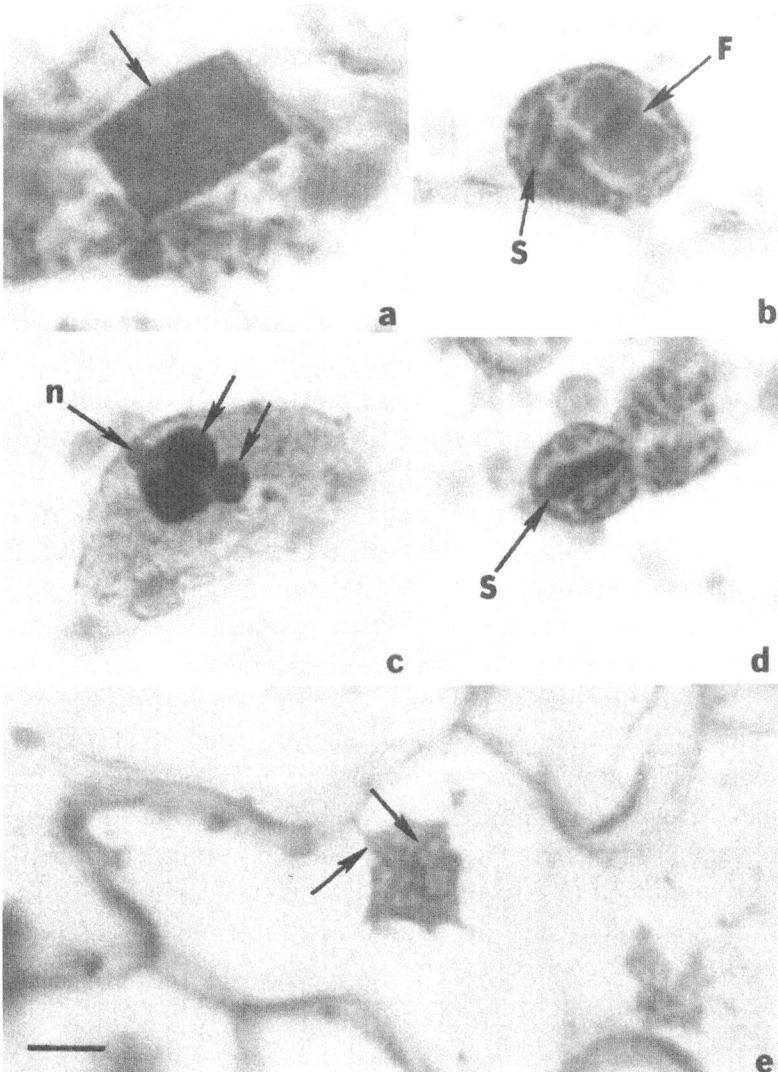

Fig. 1. Bar: 5 µm. **a** Rectangular nuclear inclusion (arrow) induced by a TEV variant causing severe distortion of a *Capsicum annum* variety possessing the VR-2 gene which is considered immune to PV-69. **b** *Nicotiana tabacum* var. Turkish NN cell nucleus containing four PV-69 induced thin, square plates. Two overlapping inclusions are shown in face view (*F*) and two closely aligned in side view (*S*). **c** Two pairs of base to base pyramids (arrows) induced by TEV-6 flanked by a nucleolus (*n*) in the nucleus of *N. tabacum* var. V 20 which is immune to PV-69. **d** *Capsicum chinense* nucleus infected with the TEV TP isolate from California which induces thick, square plates in side view (*S*). *C. chinense* is considered immune to PV-69. **e** Fusiform plates (arrows) in the nucleus of a *Capsicum frutescens* species infected with TEV Baja

this isolate contained pyramidal or dome-shaped inclusions that were often paired at the base (Fig. 1c). These inclusions were very similar to those described for TEV M of Bawden and Kassanis. As with PV-69, the inclusions maintained their characteristic appearance regardless of the host infected. The symptoms induced by TEV 6 on other tobacco cultivars were rather mild, although symptoms on pepper from which it was originally isolated were severe. TEV 6 was indistinguishable from PV-69 based on serological studies which included coat, cylindrical inclusion, and NI antisera.

In 1986, a virus-infected pepper sample was collected from a field near San Quintin in the Baja Peninsula of Mexico. The virus, identified serologically as TEV, caused very severe puckering of leaves and blistering of the fruit. The NIs induced by TEV Baja were fusiform plates that severely distorted the nucleus (Fig. 1e). Symptoms on the tobacco cultivars available for testing were almost imperceptible.

During the last few years, virus disease surveys have been carried out in pepper (*Capsicum annum* and/or *C. frutescens*) fields in California, Florida, and Mexico. Virus incidence was recorded in perennial weeds, as well as the pepper crops themselves, in order to determine potential sources of inoculum. Light microscopy was used as part of the survey procedure, since multiple viral infections could easily be determined through recognition of the characteristic inclusions induced by different viruses [6]. The presence of TEV and several other viruses was verified with appropriate antisera. TEV proved to be one of the predominate viruses encountered. NIs similar in structure to TEV PV-69 (thin, square plates), TEV 5 (bipyramidal), and TEV Baja (fusiform plates) were consistently detected in infected crops and solanaceous weeds that were widely separated geographically. In addition to these inclusions, a startling array of other forms were also detected. To date, over a dozen distinctive shapes have been recognized. Some of this variation is demonstrated in Fig. 1. It has been possible to maintain several of these TEV isolates containing nuclear variants, however, most of them still exist only in the wild. The TEV variants that have been characterized are serologically indistinguishable from PV-69 and each other. However, host range studies indicate that symptom expression induced by these isolates as they appear on different pepper and tobacco germplasm varies considerably. For instance, *Capsicum chinense* which is highly resistant to PV-69 (thin, square plate) is extremely susceptible to a TEV variant (TEV TP) from California inducing a thick (side view) square plate (Fig. 1d) and TEV Baja (fusiform plates). Several other TEV isolates containing NI variants also induce different symptom responses depending on the pepper or tobacco cultivar challenged.

Conclusions

Gooding and Rufty [8] in their study of burley tobacco in North Carolina concluded that TEV apparently consists of a population of isolates that vary from mild to severe based on symptoms they cause on commercial cultivars. These authors state that the number of definable strains probably would be limited only by the number of commercial entries and precision of severity measurements. Whether variations in NI structure are as extensive as the strain variations proposed by Gooding and Rufty [8] is unknown. However, as noted in the case of V 20 tobacco and *C. chinense* pepper, TEV isolates, inducing NI variants, can infect either a common species or cultivar and produce a radically different symptom response. Since these isolates are serologically indistinguishable, the presence of specific cytological markers may be helpful in distinguishing biological variants of TEV. The recognition of such variants could be very useful in determining specific pathotypes to be used in programs for breeding resistance to TEV. Such markers may also be helpful in differentiating among closely related "strains" of TEV which could subsequently be compared with more definitive nucleic acid and protein probes.

References

1. Bawden FC, Kassanis B (1941) Some properties of tobacco etch viruses. Ann Appl Biol 28:107–118
2. Christie RG, Edwardson JR (1977) Light and electron microscopy of plant virus inclusions. Fla Agric Exp Stat Monogr Ser, no 9
3. Christie RG, Edwardson JR (1986) Light microscopic techniques for detection of plant virus inclusions. Plant Dis 70: 273–279
4. Christie SR, Purcifull DE, Dean CE (1974) Resistance in V 20 tobacco to tobacco etch virus. Plant Dis 58: 658–659
5. Dougherty WG, Hiebert E (1980) Translation of potyvirus RNA in a rabbit reticulocyte lysate: identification of nuclear inclusion proteins as products of tobacco etch virus RNA translation and cylindrical inclusion protein as a product of the potyvirus genome. Virology 104: 174–182
6. Edwardson JR, Christie RG (1978) Use of virus-induced inclusions in classification and diagnosis. Annu Rev Phytopathol 16: 31–55
7. Edwardson JR, Christie RG, Ko NJ (1984) Potyvirus cylindrical inclusions – subdivision IV. Phytopathology 74: 1111–1114
8. Gooding GV, Rufty RC (1987) Distribution, incidence, and strains of viruses on burley tobacco. Plant Dis 71: 38–40
9. Kassanis B (1939) Intranuclear inclusions in virus infected plants. Ann Appl Biol 26: 705–709

Authors' address: R. G. Christie, Department of Agronomy, Plant Virus Laboratory, University of Florida, Gainesville, FL 32611, U.S.A.

Arch Virol (1992) [Suppl 5]: 229–234

Biological variability of potyviruses, an example: zucchini yellow mosaic virus

H. Lecoq[1] and **D. E. Purcifull**[2]

[1] INRA, Station de Pathologie Végétale, Montfavet, France
[2] Institute for Food and Agricultural Sciences, Department of Plant Pathology,
University of Florida, Gainesville, Florida, U.S.A.

Summary: Potyviruses present an important variability which may affect biological properties such as host range, symptomatology, virulence towards resistance genes, and transmissibility by vectors. A brief account of this potential is presented and illustrated by some aspects of the biological variability of zucchini yellow mosaic virus (ZYMV).

Introduction

Potyviruses commonly infect cultivated crops as well as wild plants which are in very different ecological environments. They cause significant economic losses and seem well-adapted to the intensive, modern agriculture of temperate regions, but also flourish in crops cultivated in more traditional ways in the tropics [7]. One of the most striking properties of these viruses is their high potential for biological variability. Whether this characteristic is a consequence or a cause of their ubiquity and adaptability to different hosts and environments remains to be elucidated. One can only speculate on the origin of the observed variability of potyviruses. The potyvirus polymerase, as other RNA virus polymerases, is probably subject to high error rates in the process of synthesis of viral RNA, leading to heterogeneous RNA populations within infected plants [15]. The efficiency of transmission of potyviruses by aphids is such that probably a single particle or just a few virus particles are enough to initiate infection in a new host because aphids need to acquire no more than 15–500 virus particles to be able to transmit potyviruses [13]. Under these conditions every vector transmission to a new host could be an opportunity to select a variant. Some nucleotide sequence variations will be silent and involve no phenotypic changes in the newly infected plants while others may lead to major evolutions.

Characters involved in strain differentiation

In the CMI/AAB Descriptions of Plant Viruses (1970–1990), biological variants (often referred to as strains) are mentioned for 75% of the potyviruses (38 out of 51). Among viruses for which no strains were reported, some have a very restricted geographical distribution and were not extensively studied in that respect, while updated information regarding others now reveals the occurrence of biological variants.

The most common biological variation observed in potyviruses (63% of the variants reported in the CMI/AAB descriptions) is in symptom expression on a given host: symptoms may be more severe, atypical (very often necrotic or wilting), or very mild (32%). This latter form of variation has been exploited for virus control, and mild variants of several potyviruses have been successfully used in cross-protection experiments to reduce the incidence of severe strains [5, 9].

The second character involved in strain differentiation is the host range or more precisely the ability of the isolate to infect a specific differential host (42%). Special attention has been paid to isolates on assessment of the durability of resistance genes. Virulent strains, i.e., strains able to infect hosts possessing a resistance gene, may indeed destroy the plant breeders' efforts to create resistant cultivars.

A third group of variants is defined by isolates presenting atypical relationships with their aphid vector (18%). Some are not transmitted by specific aphid species, while others are poorly transmitted or not transmitted by any aphid species. There are differences in transmission rates for potyviruses which are transmitted through dry seeds.

More variations probably would be revealed by comparing virus multiplication rates in a given host under different environmental conditions (temperature, photoperiod or mineral nutrition). Such differences could have major influences on the accessibility of the virus to vectors and on the adaptability of the virus to new ecological niches.

Biological variability in potyviruses may be revealed when comparing collections of field isolates (what may be referred to as 'virus gene banks') for specific characters. This was the approach used for studying the variability of papaya ringspot virus (PRSV) isolates, originating either from different parts of the world [14] or from a more limited geographical region, Florida, U.S.A. [2]. In these two studies important variations were observed between isolates in regard to both their biological and serological properties. However, such approaches will reflect divergences in virus populations which may have occurred on a relatively long time scale. Isolates differing in one character under study will probably also differ in many others, sometimes rendering correlations between properties, such as biological and serological, difficult to draw.

Biological variability may also be observed in the laboratory after limited numbers of subcultures either through single local lesions or successive mechanical inoculation transfers. Loss of aphid transmissibility or of the ability to infect a differential host often occurs after repeated mechanical inoculations [12]. Variations may also be induced by serial passages through specific hosts; this phenomenon observed in different groups of plant viruses is often referred to as the 'host passage effect' [16]. Pea mosaic virus (PMosV) and clover yellow vein virus (ClYVV) variants differing in aggressiveness, i.e., inducing more or less severe mosaic symptoms, were obtained after repeated subcultures on specific hosts [3]. Similarly, virulent isolates may be obtained after a single transfer on resistant plants or to plants possessing a resistance gene in the heterozygous form.

Variants can also be obtained through induced mutations; nitrous acid treatment was used to obtain mild variants of PRSV [5] and site-directed mutagenesis will be a very useful tool to understand the molecular basis of biological variability whenever infective cDNAs are available. This approach has already been used to show the importance of the amino acid triplet asp-ala-gly (located at the N'-terminal end of the coat protein) for the aphid transmissibility of tobacco vein mottling virus (TVMV) [1].

Zucchini yellow mosaic virus

Zucchini yellow mosaic virus (ZYMV), which was first recognized in 1973, is a potyvirus that naturally infects mostly cucurbit crops [11]. Although this virus has been associated with severe epidemics for only 10 years, a number of biological variants already have been differentiated when comparing collections of field isolates. These variants differ mostly in their symptomatology in susceptible hosts by inducing mild symptoms or atypical mosaic, necrosis, or wilting on some cultivars [11].

Variants grouped in the pathotype 2 of ZYMV, are able to overcome the *Zym* resistance gene found in melon line PI414723. They have been relatively easy to obtain from several isolates; back-inoculations from inoculated resistant plants showing systemic necrotic spots usually give variants able to induce vein-clearing, stunting, and leaf deformations when subsequently inoculated to resistant plants [10]. Interestingly, this virulent form of ZYMV does not seem to be frequent in nature: only one isolate from more than 100 field collections from different geographical origins was found to belong to pathotype 2. Apparently many ZYMV isolates possess the potential to evolve towards more virulent forms (pathotype 2), but in the absence of selective pressure (there are currently no commercial melon cultivars possessing the *Zym* resistance gene being cultivated), this evolution is unlikely to occur. Nevertheless, these observations cause some uncertainty about the durability of this resistance gene.

Poorly or non-aphid-transmissible isolates of ZYMV have also been obtained after several subcultures in the greenhouses. One isolate produces non-functional helper-component and the other possesses a capsid protein with an amino acid change in the asp-ala-gly triplet required for aphid transmissibility

[4, 8]. Interestingly, these two variants having the same phenotype (loss of aphid transmissibility) are the result of mutational events occurring in two different genes of the virus.

Variation may also occur in the laboratory as a succession of apparently independent events. ZYMV variants obtained successively from a single field isolate collected in 1979 during the first severe ZYMV outbreak reported in France, illustrate this phenomenon (Fig. 1). In a first series of cloning through single local lesions on *Chenopodium amaranticolor*, two phenotypically different subcultures were obtained: one (ZYMV-E15) induced severe symptoms similar to those of the original severe isolate, while another (ZYMV-E15I) produced only mild symptoms. Subsequently, a variant (ZYMV-D41) overcoming the *Zym* resistance gene, was selected from ZYMV-E15 through passage on a resistant plant [10]. The poorly-aphid-transmitted strain ZYMV-PAT, deficient in the helper component function [8] occurred spontaneously after a limited number of subcultures of ZYMV-E15 in melon. It was not possible to determine exactly when the modification from a transmissible to a non-transmissible form occurred, but it occurred after less than five successive mechanical transfers [8]. Finally, a variant (ZYMV-WK) that induced mild symptoms on leaves of melon and squash was obtained from ZYMV-PAT [9];

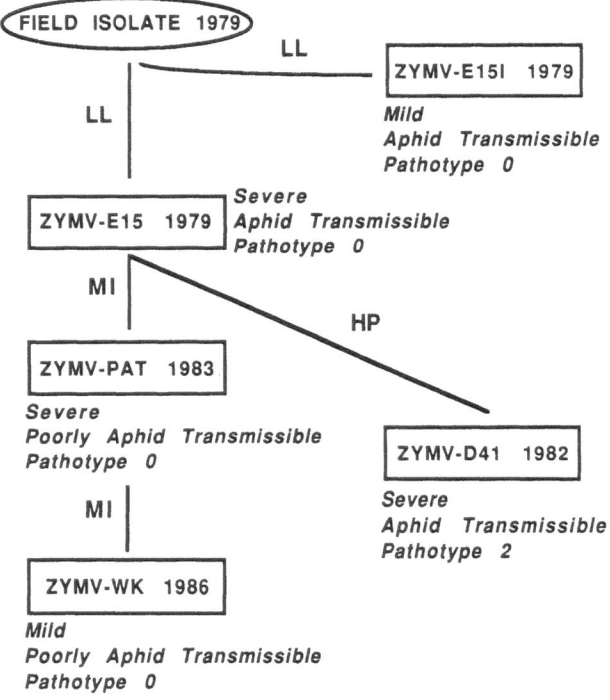

Fig. 1. ZYMV variants derived from a single field isolate. Years the variants occurred and biological properties are indicated. Variants occurred after single local lesion transfer on *Chenopodium amaranticolor* (*LL*) or successive mechanical transfers in susceptible melon (*MI*) or host passage selection in resistant melon (*HP*)

it was recovered from an axillary branch with attenuated symptoms on a melon plant infected with this isolate. Extracts from this branch were mechanically inoculated to a melon plant that subsequently exhibited mild symptoms. ZYMV-WK retained the poor aphid transmissibility of ZYMV-PAT from which it derives; this property makes it different from the mild isolate ZYMV-E15I obtained in 1979 from the original field isolate. This suggests that the two mild strains, ZYMV-E15I and ZYMV-WK, have evolved from severe forms of the virus during two different mutational events.

An interesting point is that many of the types of variability described among French isolates originating from a temperate climatic region, also occur in Florida, U.S.A., a semi-tropical climatic region. Indeed the characterization of six ZYMV isolates from different regions of Florida revealed that half of them were able to induce a rapid wilt and necrosis of melon cultivars possessing the *Fn* gene (such as Doublon or Honeydew). A similar proportion (47%) was found when characterizing 19 French isolates. Also a mild strain of ZYMV was isolated in Florida from a naturally infected plant, and a variant overcoming the *Zym* resistance was selected from a severe strain from this state. The potential for biological variation of ZYMV may therefore be expressed in completely different ecological conditions.

If ZYMV is to be considered as a single entity essentially based on coat-protein serological properties which is very convenient for the field virologist, this entity would include variants able to induce very different types of disease in the field.

Conclusions

Biological variability of potyviruses causes problems for taxonomists and raises a critical question: how should biological properties be taken into account by taxonomists in the classification of potyviruses, and what relative importance should be given to each of these properties? Attempts have been made to use biological properties, along with serological or molecular characters for grouping different PRSV related isolates through a factorial analysis of multiple correspondence [14].

Progress in the knowledge of the molecular background of biological variability will probably help to answer this question. Already we know that a single amino acid change in the coat protein may be enough to render a strain non-aphid transmissible and consequently to alter completely its biology [1,6]!

Whatever the criteria retained for differentiating potyvirus species, they should take into account the amazing potential for biological variability within each of these individual entities. Yet, assignment of a virus isolate to a given potyvirus must remain simple enough to be applied on large numbers of isolates or samples from the field and entail use of rather simple equipment. It is in the field or at a plant clinic that potyvirus identification must be done quickly and accurately in order to propose appropriate measures to the farmers.

References

1. Atreya CD, Raccah B, Pirone TP (1990) A point mutation in the coat protein abolishes aphid transmissibility of a potyvirus. Virology 178: 161–165
2. Baker CA, Lecoq H, Purcifull DP (1991) Biological and serological variability among papaya ringspot virus type-W isolates from Florida. Phytopathology 81: 728–734
3. Barnett OW, Burrows PM, McLaughlin MR, Scott SW, Baum RH (1985) Differentiation of potyviruses of the bean yellow mosaic subgroup. Acta Hortic 164: 209–216
4. Gal-On A, Antignus Y, Rosner A, Raccah B (1990) Nucleotide sequence comparison between zucchini yellow mosaic virus (ZYMV) strains differing in their multiplication rate and transmissibility by aphids. In: Abstracts of the VIIIth International Congress of Virology, Berlin, 1990, p 468
5. Gonsalves D, Garnsey SM (1989) Cross-protection techniques for control of plant viruses in the tropics. Plant Dis 73: 592–596
6. Harrison BD, Robinson DJ (1988) Molecular variations in vectorborne plant viruses: epidemiological significance. Philos Trans R Soc Lond [Biol] 321: 447–462
7. Hollings M, Brunt AA (1981) Potyviruses. In: Kurstak E (ed) Handbook of plant virus infection: comparative diagnosis. Elsevier/North Holland, Amsterdam, pp 731-807
8. Lecoq H, Bourdin D, Raccah B, Hiebert E, Purcifull D (1991) Characterization of a zucchini yellow mosaic with a deficient helper component. Phytopathology 81: 1087–1091
9. Lecoq L, Lemaire JM, Wipf-Scheibel C (1991) Control of zucchini yellow mosaic virus in zucchini squash by cross protection. Plant Dis 75: 208–211
10. Lecoq H, Pitrat M (1984) Strains of zucchini yellow mosaic virus in muskmelon (*Cucumis melo* L.). Phytopathol Z 111: 165–173
11. Lisa V, Lecoq H (1984) Zucchini yellow mosaic virus. CMI/AAB Descriptions of Plant Viruses, no 282
12. Murant AF, Raccah B, Pirone TP (1988) Transmission by vectors. In: Milne RG (ed) The plant viruses, vol 4, the filamentous plant viruses. Plenum, New York, pp 237–273
13. Pirone TP, Thornbury DW (1988) Quantity of virus required for aphid transmission of a potyvirus. Phytopathology 78: 104–107
14. Quiot-Douine L, Lecoq H, Quiot JB, Pitrat M, Labonne G (1990) Serological and biological variability of virus isolates related to strains of papaya ringspot virus. Phytopathology 80: 256–263
15. Steinhauer DA, Holland JJ (1987) Rapid evolution of RNA viruses. Annu Rev Microbiol 41: 409–433
16. Yarwood CE (1979) Host passage effects with plant viruses. Adv Virus Res 25: 169–190

Authors' address: H. Lecoq, INRA, Station de Pathologie Végétale, BP 94, F-84143 Montfavet Cedex, France.

Arch Virol (1992) [Suppl 5]: 235–237

Designation of potyvirus genera:
a question of perspective and timing

F. W. Zettler

Plant Pathology Department, University of Florida, Gainesville, Florida, U.S.A.

Summary. It may be premature to subdivide the *Potyviridae* into genera (*Bymovirus, Potyvirus, Rymovirus*) since definitive serological and vector interrelationships between them remain to be established. Monogenic proposals for classifying *Potyviridae* should also be viewed askance until their practical value in agricultural settings can be demonstrated.

*

Species of the *Potyviridae* constitute the largest group of plant viruses and collectively cause the greatest agricultural losses. The inherent diversity among viruses within this family has created considerable confusion with respect to establishing interrelationships. Although technological advances are giving us new opportunities for establishing criteria for classifying these viruses, the information derived from this research ultimately should be assessed from the standpoint of its applicability to real agricultural situations where *Potyviridae* species have an effect on commerce.

The increased resolving power of current technology for genome analysis can either reduce or add to the current confusion that exists in the taxonomy of the *Potyviridae*. Anything can be classified by a variety of systematic arrangements. For organisms, there is a consensus that the focus should be directed towards phylogenetic relationships. Potyvirus taxonomists can choose whether they want to devise a system of classification representative of the entire genome or parts thereof.

The capsid protein, which is encoded by less than 10% of the potyvirus genome, has received disproportionate attention in potyvirus taxonomy. Clearly, this reflects the ability to obtain large quantities of it through routine methods of purification and its proven diagnostic value in serological tests. However, this protein alone does not constitute an infallible representation of *Potyviridae* species as plant pathogens, as evidenced by the variability in capsid epitopes between and among different viruses. Papaya ringspot virus types P and W, for example, are indisputably different pathogens despite close similarities in their capsid and other proteins [2]. The vernacular names for these two viruses need not create confusion, however, as long as the type P and W designations remain intact when referring to them.

One should bear in mind that the capsid gene is only one of eight in the potyvirus genome. Although the precise functions of each of these genes is only partially understood, it is logical to assume that they all contribute in some way to the overall disease process. The corollary to this assumption is that any classification scheme that is based upon the characteristics of a single gene may be too narrow to be applicable in an agricultural setting. Accordingly, all monogenic proposals for classification should be viewed askance until their practical value can be ascertained.

At face value, the proposition to subdivide the potyvirus group into genera according to their natural modes of transmission is logical and probably more practical in agricultural settings than other criteria being considered. Obviously, the ability of a virus to be acquired and transmitted efficiently is vital to its survival, and this process involves more than one component of the viral genome. It is not clear, however, how exclusive vector/potyvirus interactions are. Can, for example, *Rymovirus* species be transmitted by aphids or *Polymyxa graminis*; can *Potyvirus* species be transmitted by eriophyid mites or *Polymyxa graminis*; or can *Bymovirus* species be transmitted by aphids or eriophyid mites? Reports of whitefly-borne *Potyviridae* species also need to be substantiated. Based on the current body of knowledge regarding potyvirus/ vector interactions, the ICTV may be premature should it act on this proposal now. The susceptibility of crops such as maize or wheat to species in different *Potyviridae* genera appear to present ideal opportunities for testing the validity of the aforementioned taxonomic proposals, heterologous encapsidation possibilities notwithstanding. Surprisingly little information about such vector relationships was provided in the original Proceedings of the Potyvirus Taxonomy Workshop, which was distributed in December, 1990.

Similarly, some of the serological comparisons cited in the Proceedings between proposed genera of *Potyviridae* are questionable, suggesting a currently limited knowledge base. Jordan [3] pointed out that the broad-spectrum monoclonal antibody (PTY1), which reacts to almost all aphid-transmitted potyviruses, did not react with any of the non-aphid-transmitted viruses tested, thereby lending support to those who would argue in favor of subdividing the *Potyviridae* according to vectors of the viruses. Contrasting reports of positive serological relationships between genera of the *Potyviridae* seem inconclusive and await publication and/or confirmation. Lesemann and Vetten [4], for example, reported "weak" reactions between ryegrass mosaic and turnip mosaic viruses in immuno electron microscopy tests, but could not confirm these results by indirect ELISA. In any event, this paper, cited twice in the Proceedings, was not intended as an in-depth serological study between genera of the *Potyviridae*. Considering the current availability of diverse mono- and polyclonal potyvirus antisera to both structural and nonstructural proteins, it is inevitable that our knowledge base will grow substantially in the near future, a factor the ICTV should take into account in its forthcoming deliberations.

Another issue the ICTV should consider is possible correlations between the respective phylogenies of *Potyviridae* species, their hosts, and their vectors. Whereas the suscepts of the *Potyviridae*, the angiosperms, are monophyletic (i.e., they are presumed to stem from a common ancestor), their chief natural vectors (aphids, eriophyid mites, *Polymyxa graminis*) clearly are polyphyletic. Although aphids and eriophyid mites are both arthropods, their respective classes, the Hexapoda and Arachnida, diverged long before angiosperms originated. It stands to reason that if subsequent virus research divulges indisputable interrelationships among species of *Potyvirus*, *Rymovirus*, and *Bymovirus*, a host-associated possibly monophyletic origin for potyviruses is plausible. If, however, such relationships cannot be shown, a vector-associated presumably polyphyletic origin would seem more likely.

It would be nice if all species of the *Potyviridae* could be conveniently subdivided into discrete genera according to their modes of transmission. Unfortunately, biological systems are not always that accommodating. I question whether we should be making sweeping taxonomic decisions based on what we currently know about these viruses. Earlier proposals for virus classification were discarded simply because they were not useful. Whatever decisions are ultimately reached by the ICTV, the importance of potyviruses as agents of disease should not be forgotten. In discussing the subdivision of potyviruses according to cylindrical inclusion morphologies, Edwardson [1] argued that his intention was to reduce the number of comparisons and tests required for identification, rather than to give this criterion taxonomic status. A criterion does not require taxonomic status to be useful to a pathologist.

References

1. Edwardson JR (1990) Inclusion bodies. In: Barnett OW (ed) Proceedings of the Potyvirus Taxonomy Workshop, Braunschweig, 1990, pp 16–17
2. Hiebert E, Purcifull DE (1990) Papaya ringspot type P (papaya ringspot isolate) and W (watermelon mosaic 1). In: Barnett OW (ed) Proceedings of the Potyvirus Taxonomy Workshop, Braunschweig, 1990, p 97
3. Jordan R (1992) Potyviruses, monoclonal antibodies and antigenic sites. In: Barnett OW (ed) Potyvirus taxonomy. Springer, Wien New York, pp 81–95 (Arch Virol [Suppl] 5)
4. Lesemann DE, Vetten HJ (1985) The occurrence of tobacco rattle and turnip mosaic viruses in *Orchis* spp., and of an unidentified potyvirus in *Cypripedium calceolus*. Acta Hortic 164: 45–54

Author's address: F. W. Zettler, Plant Pathology Department, University of Florida, Gainesville, FL 32611, U.S.A.

Arch Virol (1992) [Suppl 5]: 239–250
© by Springer-Verlag 1992

Fungal transmission of a potyvirus: uredospores of *Puccinia sorghi* transmit maize dwarf mosaic virus

M. B. v. Wechmar, R. Chauhan, and **E. Knox**

Department of Microbiology, University of Cape Town, Rondebosch, South Africa

Summary. Maize dwarf mosaic virus (MDMV) and maize rust, *Puccinia sorghi* Schw., occur as natural infections on cultivated maize in South Africa. *P. sorghi* often occurs as a secondary late infection on maize plants which have already been infected with MDMV earlier in the season, either seed or aphid transmitted. When MDMV isolates from maize plants naturally infected by both virus and fungus were propagated by sap inoculation in plant growth rooms, residual uredospores in the sap gave rise to the development of uredia under conditions of high humidity. When uredospores developing on MDMV-B-infected plants were germinated on virus free maize seedlings, these plants became infected with MDMV-B. Similarly, when uredospores, originating from maize plants infected with MDMV-A, were scattered onto virus free maize seedlings, these plants became infected with MDMV-A.

The presence of virus on uredospores in infected plant tissue was visualized by indirect immunofluorescence. Identification of virus infection was by DAS-ELISA and immunoelectro-blotting utilizing strain-specific antisera. Virus transmission occurred between closely situated plants which had no actual contact (unaided transmission). MDMV-B transmission by uredospores, to new maize seedlings, has been maintained for three successive years (1988–1991) in a plant growth room. The MDMV-B isolate remained sap and non-persistently aphid transmissible.

Introduction

In a program aimed at the characterization of maize dwarf mosaic virus (MDMV) and sugarcane mosaic virus (SCMV) strains in South Africa, numerous field samples were brought to the laboratory from various maize and sugarcane producing regions. Distinct MDMV-A and MDMV-B maize infecting strains [3], as well as SCMV-B and SCMV-D sugarcane infecting strains [18] were identified [17]. Strain identification of South African isolates was previously done by direct double antibody sandwich enzyme-linked immunosorbent assay (DAS-ELISA) and immunoelectroblotting (IEB) utilizing anti-sera from other laboratories in conjunction with our own [3].

In 1987 leaf samples with MDMV symptoms originating from maize breeder blocks in the Natal Midlands were gathered for strain identification. These leaves were co-infected with (maize rust) *Puccinia sorghi* Schw. During sap transmission to maize seedlings, uredospores of the maize rust were transferred with the sap. The spores germinated in the moist environment of the plant growth rooms and rust pustules developed, resulting in the infection of maize seedlings with both MDMV and maize rust. Virus-infected plants were usually separated by non-infected seedlings serving as controls. The maize control seedlings adjacent to the MDMV- and maize rust-infected plants developed MDMV symptoms, seemingly spontaneously, but few seedlings showed signs of rust infection. This happened several times in succession with new batches of control seedlings. When all possible avenues of contamination had been checked and the barrier plants continued to become infected with MDMV, the occurrence was investigated more closely. Uredospores from the sap-inoculated plants (MDMV and maize rust) dropped off, or were washed off during normal watering, onto the young seedlings in the barrier rows where spores germinated in waterdrops lodging in the cup formed by young developing leaves.

Previous work showed that uredospores of *P. graminis tritici* (stemrust of wheat) transmitted brome mosaic virus (BMV) when uredospores developed on plants simultaneously infected with stemrust and BMV [15, 5, 6].

This paper presents results of experiments conducted to examine whether maize rust uredospores could act as a vector of MDMV under certain conditions.

Materials and methods

Maize seedlings

Pregerminated seed of the cultivar Potchefstroom Pearl, which is susceptible to MDMV and maize rust, were planted in sterilized soil and grown in plant growth rooms under growlux fluorescent light (Sylvania F96 T12/GRO/VHO/WS) at 20 °C/24 °C, 10 h dark/14 h light. Seedlings were usually grown to the three-leaf stage before inoculation with virus and/or rust spores.

Virus strains

Maize seedlings were sap-inoculated with MDMV-B-Krug (MDMV-B) and MDMV-A-Jg (MDMV-A) [3] and MDMV-A-Bant [17]. Symptoms normally appeared five days after inoculation. The MDMV-A-Jg strain caused only a mild infection in maize and virus titers remained low [3; von Wechmar unpubl. results]. MDMV-A-Bant isolated from sweet corn was therefore included as a second MDMV-A strain.

Maize rust propagation

An isolate of maize rust from non-virus-infected maize leaves was used. The fungus was identified according to reported characteristics [14]. To start the infection cycle uredospores were suspended in distilled water containing 0.01% Tween-20 and sprayed with an atomizer onto young maize leaves [7]. Inoculated plants were kept in a plant growth room with high relative humidity (approx. 70%) maintained with an automatic humidifier. Once the rust fungus had established itself, new maize seedlings were added to the infected ones allowing the rust to spread naturally. This infection cycle was easily maintained.

Antisera

Immune sera utilized in this study were previously prepared and described [3]. The antisera were MDMV-A and MDMV-B specific [17, 18]. Immune sera consisted of late bleedings taken three months or later after initial immunizations. Host absorbed antisera and immuno-globulins (IgG) were prepared as described [13].

Immunological techniques

DAS-ELISA was done by the described method [4]. Host-absorbed IgG preparations were used throughout. Immunoelectroblotting (IEB) was performed as described [12]. An indirect fluorescent antibody staining technique [19] was adapted for use on leaf sections with uredospore pustules. Fluorescein isothiocyanate (FITC) labelled antibodies were used [11]. Labeling of antibodies was done as described [8]. Leaf sections with rust pustules were prepared as follows: 1 cm^2 sections were incubated in an enzyme cocktail consisting of 1% cellulase, 1% macerozyme, 1% driselase, 0.5 M mannitol, 10 mm calcium chloride, pH 5.5, for 2 h at 37 °C; sections were washed three times with saline containing 0.05% Tween 20 (saline-T20) before decolorizing in acetone for approximately 30 min. Older sections required more time. When they appeared transparent, sections were removed and washed carefully in saline. Two milliliters anti-MDMV-A or anti-MDMV-B IgG were added at a concentration of 0.01 mg/ml ($A_{280nm} = 0.014$). Sections were incubated at 37 °C for 2 h and washed with saline-T20. The sections were drained, and fluorescein isothiocyanate goat antirabbit IgG (GAR-FITC-IgG) added as a 1:50 dilution in 0.02 M sodium carbonate, pH 9.8, and incubated for 30 min in 37 °C in the dark. Sections were viewed with a Zeiss-inverted UV microscope at magnifications of 2.5×, 6.3×, 16× Plan, and 40× Nuofluar lenses and photographed with a Contax RTS Camera on 400 ASA color slide film. Controls consisted of prepared leaf sections treated with normal rabbit serum (NRS) and GAR-FITC-IgG; prepared leaf sections treated with GAR-FITC-IgG only; and leaf sections without antibody treatment to check for auto-fluorescence. Leaves infected with maize rust only (negative virus controls) were treated as above.

Virus transmission by uredospores

The following experiments were designed to determine the possible transmission of MDMV by uredospores: (i) Rust spores developing on MDMV-infected plants were allowed to drop/wash off naturally onto uninfected maize seedlings. (ii) Leaf sections in petri dishes were infected with rust spores from an MDMV-infected source. (iii) Rust spores shaken off MDMV-infected plants were suspended in water with 0.05% Tween-20 and pipetted into seedling leafwhorls. (iv) MDMV-infected maize plants without rust infection were grown in close proximity to uninfected maize seedlings to assess whether virus contamination could occur by leaf contact only. The above experiments were first performed with the MDMV-B strain and were repeated including the MDMV-A and MDMV-A-Bant strains.

Virus transmission by aphids

To determine whether repeated transmission of MDMV-B by uredospores influenced subsequent transmission by aphids, this was checked. *Rhopalosiphum padi* aphids, (from a laboratory maintained clone) starved overnight at 5 °C, were allowed a 5–10 min acquisition feeding on young leaf sections (with mosaic symptoms but without uredospores) taken from a uredospore-infected plant. After an inoculation feeding of 12 h the plants were sprayed with insecticide and returned to the plant growth room. Virus from the same source was also sap transmitted to maize seedlings.

Results and discussion

Uredospores dropping/washing off MDMV-infected maize onto healthy maize seedlings resulted in MDMV symptoms developing in many plants. Uredospore pustule development was not necessarily associated with virus infection and was moisture dependent. Seedlings with mosaic symptoms often had only one visible fully developed rust pustule on the tip of an older leaf indicating early infection with virus-contaminated rust spores. DAS-ELISA tests (Table 1) in conjunction with IEB (Fig. 1A and B) were used to identify

Table 1. DAS-ELISA results of uredospore transmission of MDMV-B

		No. positive/no. tested	
		MDMV-B	MDMV-A
1	Maize inoculated with MDMV-B and with uredospores	9/10[a]	0/10
2	Maize inoculated with MDMV-B contaminated with spores from 1	10/10[b]	0/10
3	Maize uninoculated, contaminated with spores from 1 and 2	10/10[c]	0/10
4	Maize uninoculated, contaminated with spores from MDMV-A infected plants	0/4	0/4[d]
5	Control maize no contamination	0/4	0/4
6	MDMV-A inoculated plant (control)	0/2	2/2
7	MDMV-B inoculated plant (control)	2/2	0/2

[a] Average A_{405nm}: 0.854
[b] Average A_{405nm}: 0.762
[c] Average A_{405nm}: 0.9376
[d] Concentration of MDMV-A in maize often very low and not detected by DAS-ELISA

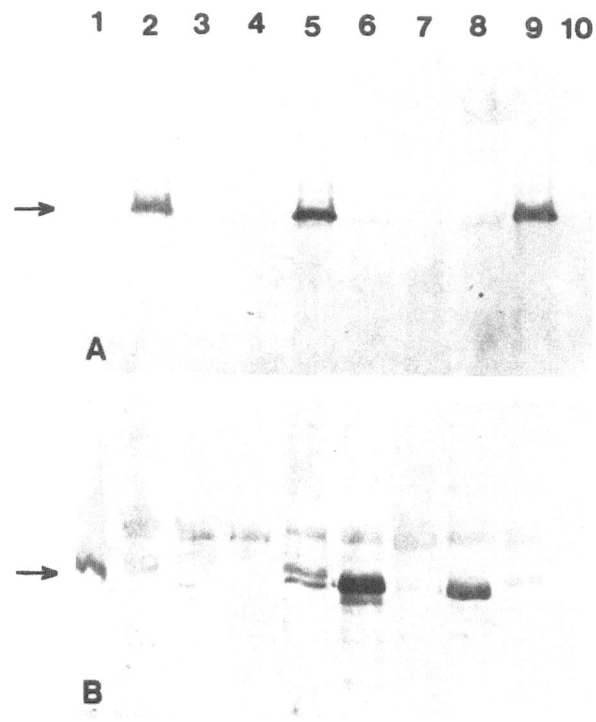

Fig. 1 A, B. Immunoelectroblot of MDMV-A- and MDMV-B- infected plants used for immunofluorescent staining. The blots were probed with **A** anti-MDMV-B and **B** anti-MDMV-A IgG. *1* MDMV-A control; *2* MDMV-B control; *3* leaf infected with old uredospores from MDMV-B-infected source (see Fig. 2 F); *4* healthy maize infected with control uredospores; *5* sap from maize leaf infected with MDMV-B and inoculated with control uredospores; *6* sap from maize leaf infected with MDMV-A and inoculated with control uredospores; *7* and *10* uninfected maize; *8* sap from maize leaf infected with MDMV-A-Bant; *9* maize sap infected with MDMV-B

the MDMV-strains and to support immunofluorescent assay. In all tests conducted, spores from the different virus-infected sources transmitted only the strain corresponding to that of its origin, i.e., spores from MDMV-B-infected plants transmitted MDMV-B and similarly spores from MDMV-A and MDMV-A-Bant-infected plants, transmitted MDMV-A and MDMV-A-Bant, respectively.

Suspensions of newly developed rust spores dropped into leaf cups showed lesions after 48 h. Examination of leaf sections corresponding to the areas in contact with the inoculum containing the spore suspension from MDMV-infected plants revealed very bright fluorescence around and inside stomata and inside adjacent cells with the strain-homologous antiserum (Fig. 2 B–D). Seedlings not used for assays developed mosaic symptoms five days after inoculation. The bright localized fluorescence identified areas of virus accumulation and indicated sites of virus replication. This was confirmed by IEB tests done on similar sections taken 48 h post inoculation. Lanes 7 and 8 in Fig. 3 A illustrate the high concentration of virus protein present at an early stage of infection. Considering the small quantity of leaf material available from the infection sites, rapid virus replication must have occurred to yield such strong bands on the IEB. Leaf sections prepared with heterologous

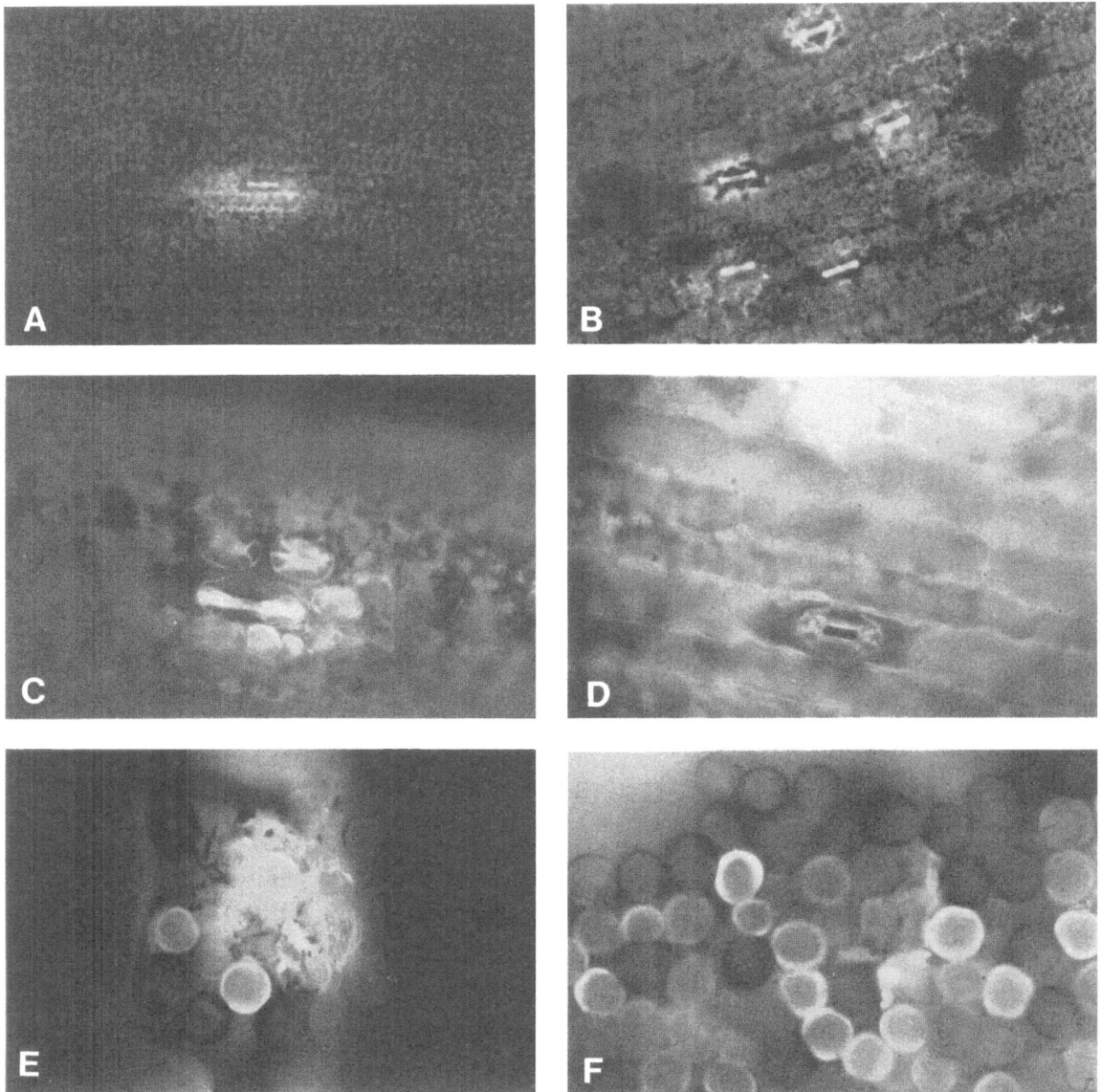

Fig. 2 A–F. Leaf sections prepared by indirect immunofluorescent staining with anti-MDMV-B IgG and GAR-FITC-IgG. **A** MDMV-B-infected leaf; **B** and **C** leaf infected with uredospores from MDMV-B-infected source after 48 h incubation; **D** as in **B** and **C** but after longer incubation; **E** newly developed uredospore pustule and **F** older uredospore pustule on MDMV-B- infected leaves. Photographs taken at the following magnifications: A, 8×; B, 20×; C and D, 50×; E and F, 128×. Black/white photos reproduced from color slides

antisera and other controls did not fluoresce and showed no bands on IEB (Fig. 3B). Control rust spores and uninfected maize reacted negatively (Fig. 3A, lane 9). Repeat experiments were conducted in separate plant growth rooms to avoid possible cross-contamination by drifting rust spores. Uredospores from virus-negative plants were transferred to MDMV-A (Fig. 1B, lane 6 and 8) and MDMV-B (Fig. 1A, lane 5) sap-infected maize plants. Transfer of these spores to uninfected plants resulted in MDMV-A and MDMV-B infection, respectively.

Early infection by sap inoculation of MDMV-B is illustrated in a leaf section in Fig. 2A. The presence of virus 'positive' uredospores on maize leaves infected with MDMV-B and maize rust is illustrated in Fig. 2E and F. Young rust pustules (Fig. 2E) produced stronger fluorescing spores compared to older more mature pustules (Fig. 2F). This finding supports the observation that uredospores from newly established rust infections on MDMV-B-infected plants produced the most efficient virus transmission. Considering the fact that

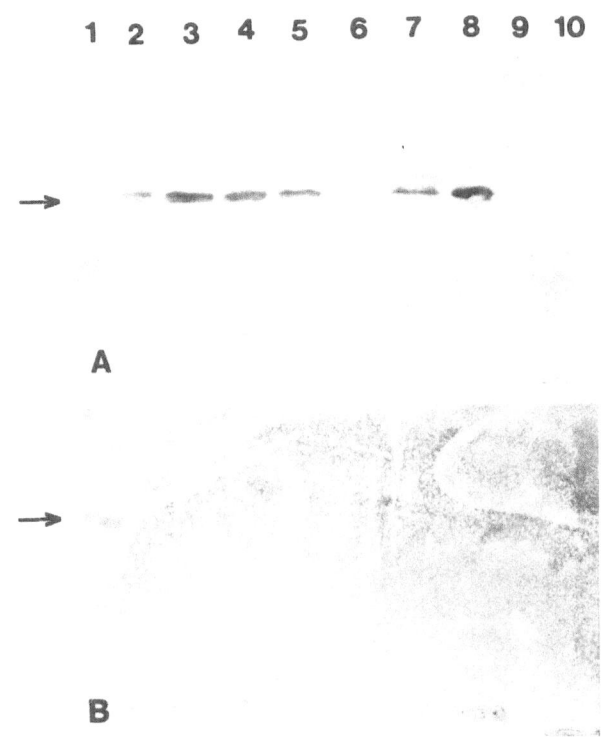

Fig. 3A, B. Immunoelectroblots of MDMV-B transmitted by uredospores. The blots were probed with **A** anti-MDMV-B and **B** anti-MDMV-A IgG. *1* MDMV-A control; *2* MDMV-B control; *3, 4,* and *6* MDMV-B sap-infected leaves inoculated with uredospores; *5* MDMV-B sap-infected leaf infected with rust fungus and harvested prior to uredospore formation; *7* uninfected maize leaf inoculated with uredospores from the sample in *6* (48h incubation); *9* control uredospores from uninfected maize; *10* uninfected maize control. The protein bands are at the 37kDa position

the preparation of leaf sections involves an enzyme digesting step prior to antibody treatment, and also includes several cycles of washing, suggests that the MDMV attachment to the uredospore is more stable than just a surface contamination.

To establish whether MDMV could spread by leaf contact only, rows of seedlings of Potchefstroom Pearl and of three commercial hybrids were inoculated with MDMV-A and MDMV-B, respectively. Rows of inoculated plants were alternated with control rows, allowing leaves to touch. The plants were watered from the top and harvested after two weeks. The juice was extracted and analyzed by DAS-ELISA (Table 2). Virus spread by leaf contact did not take place and low levels of "apparent" MDMV-B seed-borne infection (Ta-

Table 2. DAS-ELISA results to determine possible "spread" of MDMV-A and MDMV-B to adjacent control plants in the absence of rust spores

		No. positive/no. tested	
		MDMV-B	MDMV-A
I	Sap inoculated with MDMV-A	3/16	10/16
	Controls, adjacent, and leaves touching	1/8	0/8
	Hybrid 1	2/8	0/8
	Hybrid 2	2/8	0/8
	Hybrid 3	1/8	0/8
	Potchefstroom Pearl	0/7	0/7
	Controls, separate table		
	Hybrid 1	0/16	0/16
	Hybrid 2	1/16	0/16
	Hybrid 3	1/16	0/16
	Potchefstroom Pearl	4/16	0/16
II	Sap-inoculated with MDMV-B	11/16	0/16
	Controls, adjacent and leaves touching		
	Hybrid 1	1/8	2/8
	Hybrid 2	0/8	0/8
	Hybrid 3	0/8	0/8
	Potchefstroom Pearl	0/8	0/8
	Controls, separate table		
	Hybrid 1	0/16	0/16
	Hybrid 2	1/16	0/16
	Hybrid 3	1/16	0/16
	Potchefstroom Pearl	4/16	0/16

ble 2) [16, 17] did not differ significantly between experimental plants and healthy plants grown on a separate table.

To assess whether uredospore transmission of MDMV-A and MDMV-B could play a role in commercial hybrids, rust spores originating from virus-infected source plants were transmitted to seedlings of three different hybrids. Rust spores transmitted MDMV-B to Potchefstroom Pearl seedlings, but not to any of the hybrids (Fig. 4A) (results of Hybrid 3 not shown). The three hybrids were susceptible to MDMV when sap inoculated, and although they had an apparent tolerance to rust infection, no MDMV transmission occurred. Judging from the experimental evidence, it would appear that susceptibility to *P. sorghi* is essential for uredospores to carry virus into maize host cells, although development of rust pustules was not essential for the initiation of MDMV-B infection by uredospores (originating from a virus-infected source) in a maize rust and virus susceptible host such as Potchefstroom Pearl maize (Fig. 3A, lane 5). This observation has important epidemiological implications and should be examined in greater detail at the cellular level involving hybrids susceptible and resistant to *P. sorghi* [9].

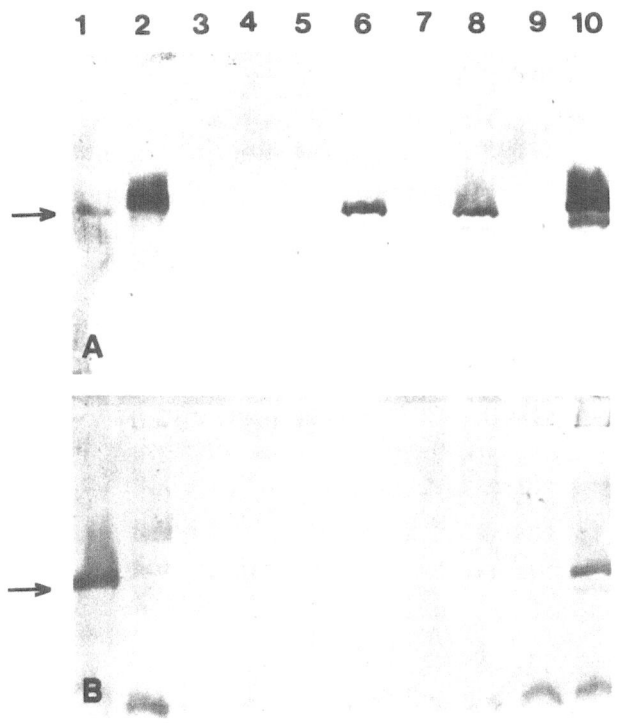

Fig. 4A, B. Immunoelectroblot of MDMV-B transmitted by uredospores. The blots were probed with **A** anti-MDMV-B and **B** anti-MDMV-A IgG. *1* MDMV-A control; *2* MDMV-B control; *3, 5* and *7* maize hybrid 1; *9* hybrid 2 inoculated with uredospores from a MDMV-B-infected source; *4, 6, 8,* and *10* Potchefstroom Pearl maize inoculated with uredospores from a MDMV-B-infected source

Occasionally confusing results were noted (Fig. 1B, lane 5; Fig. 4A and B, lane 10) where an apparent MDMV-A type contaminant (not homologous to the anti-MDMV-A serum used) appeared to be present in a plant inoculated with MDMV-B (Fig. 1A, lane 5). This phenomenon was noticed previously and considering the general confusion in strain identification by serological techniques, justifies a brief discussion. Seed-borne MDMV infections can cause symptomless infections occurring in maize [16, 3, 17]. The presence of such inapparent MDMV infections in seed sources used for growing plants for propagation of MDMV strains for antiserum production, resulted in antisera containing antibodies to more than one MDMV strain. This problem could be resolved by propagating MDMV strains in plants grown from carefully selected seed sources, and/or producing antisera from virus isolated from original source plants only, i.e., MDMV-A-Jg (S.A) from *Sorghum halepense* and SCMV-B (S.A) and SCMV-D (S.A) isolates from specific sugarcane clones of known origin. IEB results produced with a selection of our own antisera, and with antisera received from elsewhere, with South African MDMV and SCMV strains and isolates, were previously discussed [17, 18].

Repeated virus transmission by uredospores over a period of three years did not alter the aphid transmissibility of the MDMV-B isolate. Sap transmissibility of the same isolate remained unchanged. Symptoms appeared after five days for both aphid and sap transmission.

The results presented in this paper are limited to a laboratory study. In later field observations it was noted that *P. sorghi* infections often occur on the first two leaves of commercial hybrid maize seedlings and again on older leaves later in the season. Considering the very active and efficient transmission of MDMV by uredospores in the laboratory, it would appear very likely that this mechanism also operates in nature and may explain the curious distribution and transmission patterns of MDMV. Although aphids are generally considered to be the major vectors of MDMV, evidence to this effect was not always found [10], especially in the irregular occurrence of MDMV-A and MDMV-B strains in certain regions [2, 20]. In this context the observation of Arny et al. [1] is of interest. While assessing sweet corn Plant Introduction accessions and commercial hybrids for susceptibility to MDMV under field conditions they noted that many non-inoculated rows became infected with MDMV, and that in some entries common rust was more severe on MDMV-inoculated plants than on non-inoculated plants. They further recorded that the epidemic of *P. sorghi* developed during the course of their trials. One can only speculate that *P. sorghi* possibly played a role in causing entries in non-inoculated rows to become infected with MDMV.

In summary, this is the first report of *P. sorghi* uredospores transmitting MDMV. MDMV was previously only known to be sap, seed and aphid transmissible. It is also the first report of a new "vector" mechanism of virus transmission in the family *Potyviridae*.

Acknowledgements

We acknowledge the technical assistance of Mrs. S. Lindsey with immunological assays, Mr. M. A. Jaffer for photographic work and the financial assistance from the University of Cape Town and the Department of Agriculture Development.

References

1. Arny DC, Grau CR, Suleman PE (1980) Occurrence of maize dwarf mosaic in Wisconsin and reaction of sweet corn plant introduction accessions and commercial hybrids. Plant Dis 64: 85–87
2. Berger PH, Zeyen RJ, Groth JV (1987) Aphid retention of maize dwarf mosaic virus (potyvirus): epidemiological implications. Ann Appl Biol 111: 337–344
3. Chauhan R (1985) A study of filamentous viruses in maize and small grains. MSc Thesis, University of Cape Town, Cape Town
4. Clark MF, Adams AN (1977) Characteristics of the microplate method of enzyme-linked immunosorbent assay for the detection of plant viruses. J Gen Virol 34: 475–483
5. Erasmus DS (1982) The association of bromegrass mosaic virus (BMV) with *Puccinia graminis tritici*. MSc Thesis, University of Cape Town, Cape Town
6. Erasmus DS, von Wechmar MB (1983) The association of brome mosaic virus and wheat rusts: I Transmission of BMV by uredospores of wheat stem rust and leaf rust. J Phytopathol 108: 26–33
7. Erasmus DS, von Wechmar MB (1983) Brome mosaic virus (BMV) reduces susceptiblity of wheat to stem rust (*Puccinia graminis tritici*). Plant Dis 67: 1196–1198
8. Erasmus DS, Rybicki EP, von Wechmar MB (1983) The association of brome mosaic virus and wheat rust: II Detection of virus in/on uredospores. J Phytopathol 108: 34–40
9. Hughes, FL, Rijkenberg FHJ (1985) Scanning electron microscopy of early infection in the uredial stage of *Puccinia sorghi* in *Zea mays*. Plant Pathol 34: 61–68
10. Knoke JK, Louie R, Madden LV (1983) Spread of maize dwarf mosaic virus from Johnsongrass to corn. Plant Dis 67: 367–370
11. Otsuki Y, Takebe I (1969) Fluorescent antibody staining of tobacco mosaic virus antigen in tobacco mesophyll protoplasts. Virology 38: 497–499
12. Rybicki EP, von Wechmar MB (1982) Enzyme-assisted immune detection of plant virus proteins electroblotted onto nitrocellulose paper. J Virol Methods 5: 267–278
13. Rybicki EP, von Wechmar MB (1985) Serology and immunochemistry. In: Francki RIB (ed) The plant viruses, vol I. Plenum, New York, pp 207–244
14. Shurtleff MC (ed) (1980) Compendium of corn diseases, 2nd edn. American Phytopathological Society, St. Paul
15. von Wechmar MB (1980) Transmission of brome mosaic virus by *Puccinia graminis tritici*. Phytopathol Z 99: 289–293
16. von Wechmar MB, Chauhan R, Knox E (1984) Seed transmitted viruses in maize. In: Proceedings of the 6th South African Maize Breeding Symposium. Department of Agriculture, Technical Communication, pp 52–55
17. von Wechmar MB, Chauhan R, Knox E (1988) Application of immunoelectroblotting to differentiate between isolates of maize dwarf virus and sugarcane mosaic virus. In: Proceedings of the 8th South African Maize Breeding Symposium. Department of Agriculture, Technical Communication, pp 41–44
18. von Wechmar MB, Knox E (1988) Distribution of maize dwarf mosaic virus strains A and B in South African maize, sweet corn, sorghum, several grasses and sugarcane. In: Proceedings of the 8th South African Maize Breeding Symposium. Department of Agriculture, Technical Communication, pp 44–48

19. Walker FD, Batty I, Thomson RD (1971) The localization of bacterial antigens by the use of fluorescent and ferritin labelled antibody techniques. In: Norris JR, Ribbons DW (eds) Methods in microbiology, vol V. American Society of Microbiology, Washington, DC, pp 119–247
20. Zeyen RJ, Stromberg EL, Kuehnast EL (1987) Long range aphid transport hypothesis for maize dwarf mosaic virus: history and distribution in Minnesota, USA. Ann Appl Biol 111: 325–336

Authors' address: M. B. von Wechmar, Department of Microbiology, University of Cape Town, Private Bag, Rondebosch 7700, South Africa.

Arch Virol (1992) [Suppl 5]: 251–255

The usefulness of aphid transmission as a taxonomic criterion for potyviruses

P.H. Berger

Department of Plant, Soil, and Entomological Sciences, Division of Plant Pathology, University of Idaho, College of Agriculture, Moscow, Idaho, U.S.A.

Summary. In the past vector relationships have been one of the criteria used for delineating plant virus taxa. The proposed family, the *Potyviridae*, continues that practice. Aphid transmission of viruses within the genus *Potyvirus* is a useful characteristic in terms of identification, but is of only limited use in terms of taxonomy. This conclusion is based on a greater understanding of the molecular biology of potyviruses. The molecular basis of aphid transmission is not well understood at the present, but these data suggest that, beyond disease diagnosis, virus identification and characterization, and potential identification of genome microheterogeneity, aphid transmission should only be considered as a minor taxonomic criterion.

Introduction

A proposal to the International Committee on Taxonomy of Viruses recommends that a family of plant viruses, the *Potyviridae*, be established to include viruses in the former potyvirus group which meet the established criteria of particle morphology, genome organization, and induction of cytoplasmic inclusions [3]. The genera in this family are to be the genus *Potyvirus* containing the aphid-borne viruses, the genus *Bymovirus* composed of the fungal-transmitted viruses, the genus *Rymovirus* composed of the mite-transmitted viruses, and the possible genus *Ipomovirus* to include the whitefly-borne virus. While vector type seems to be correlated with other characters used to define the genera, lack of transmission by a vector does not preclude assignment to a genus because other molecular or serological properties are reliable for this task. The proposal continues a practice that has been followed for many years in that vector relationships will remain an important taxonomic factor.

The need for this reorganization comes about for several reasons, including our inability to classify viruses that were listed as possible members of the potyvirus group. An example is wheat streak mosaic virus (WSMV), which for virtually all other characteristics is a potyvirus with the significant exception that it is transmitted by the mite, *Eriophyes tulipae*. Further, WSMV appears to have a circulative relationship with its vector rather than a noncirculative relationship such as found with the aphid-transmitted potyviruses. Does this divergence of vector associations represent molecular differences which are phylogenetically important in taxonomy?

Vector associations will be examined for their usefulness as taxonomic tools. What role should this characteristic play and what weight should it be given? Indeed, there are numerous strains and isolates of common, normally aphid-transmitted potyviruses that have for one reason or another lost the ability to be transmitted by their natural vectors. These viruses have certainly not ceased to be potyviruses yet have lost an essential characteristic. If transmission has little value for taxonomy then does the type of vector or association with the vector have a value?

Aphid transmission of potyviruses

In the last twenty years, we have begun to understand how and why aphids transmit potyviruses. Aphids also are the only insects which transmit potyviruses efficiently. Transmission is dependent on the presence of a biologically active viral gene product, the helper component (HC) protein. In virus transmission, the HC appears to act by binding virus particles to sites in the aphid alimentary canal anterior to the esophageal valve from where it is subsequently eluted during ingestion-egestion while probing [4]. Thus, there must be an interaction between HC and virus coat protein (CP) which is essential for transmission. There must also be an interaction between one or both components of this complex and the putative receptor in the aphid's food canal. The details of this interaction are still obscure, although single amino acid substitutions in either HC [9] or CP [1] can reduce or eliminate aphid transmission.

The transmission process

In order for aphid transmission to occur, virus must be acquired from a suitable source, retained for a finite period of time, and inoculated into a suscept. If any one of these steps fails, transmission does not occur; numerous factors impinge upon these steps that go far beyond the relationship of just HC, virus, and vector. Potyviruses are intrinsically infectious, yet purified virus is not aphid transmissible after feeding aphids on solutions through a membrane that do not contain HC. Addition of a protein fraction which contains HC (a factor not present in healthy plants) will allow for aphid transmission of virus by membrane feeding on solutions of purified virus. Aphids must acquire HC either prior to or simultaneously with virus in order for transmission to occur [10].

The second cistron from the 5′ end of the potyvirus genome translates a bifunctional protein. This protein acts as an aphid transmission factor (HC) and as a proteinase in a manner analogous to the poliovirus 3C proteinase [5]. It catalyzes at least one and possibly both of the proteolytic cleavages involved in production of three mature viral proteins encoded by the 5′ third of the *Potyvirus* genome [5]. HC may also play a role in the virus infection process and may act upon sites within the aphid's stylets and foregut during the transmission process [10]. The exact mechanism of HC action with respect to transmission is still not fully understood. Nevertheless, in order for HC activity to be manifested in terms of virus transmission, it appears that it must interact with the viral CP. Studies localizing virus in aphids suggest that HC acts as a binding agent, perhaps in a manner analogous to a lectin [4]. At least one potyvirus defective in HC activity, potato virus C (PVC), has been sequenced. In the PVC nucleotide sequence, there are several amino acid changes that could account for inactivation of HC [11].

A strong correlation has been observed between the amino acid sequence of potyvirus coat proteins and aphid transmission. The sequence DAG appears near the amino terminus of coat proteins of aphid transmissible potyviruses while in non-transmissible isolates this triplet is usually changed, with the change most commonly occurring in the third residue [2]. In at least one case, working with mutated infectious in vitro transcripts, Atreya et al. [1] report that virus recovered from infected plants had reverted to wild type, in terms of the DAG triplet. Thus it would appear that there is relatively strong selection for this sequence.

Vector specificity

Aphid transmission of potyviruses is without doubt a useful tool for identifying these viruses. To be an effective tool or criterion for taxonomic purposes, a certain level of specificity is prerequisite. Unlike circulative viruses where there is a high degree of vector specificity, the specificity of *Potyvirus* transmission is low. Also unlike the noncirculative leafhopper-borne maize chlorotic dwarf virus (MCDV), there is no evidence for a phylogenetic relationship with aphid vectors [9]. There are numerous examples of potyviruses that are transmitted experimentally or naturally by many aphid species [7]. These lists are probably incomplete since only a limited number of aphid species are tested based on our perceptions of which species are important and because of limitations of time and materials. Nonetheless, the scientific literature contains numerous examples of virus isolates differing in either vector efficiency or transmissibility. The real significance of these differences is difficult to interpret; there is no set of standardized laboratory conditions for aphid transmission bioassays and even if that were possible, it is impossible to "standardize" aphid clones.

There are several possible explanations for these differences. Transmission-competency of different clones of a single aphid species can vary. This may be manifested in terms of insect behavior or biochemical and/or physiological makeup. Either of these characteristics may greatly affect transmission efficiency. Indeed, it is likely that many reports of non-transmission by certain virus-vector combinations may be a result of aphid behavior and not due to an intrinsic "incompatible" interaction. The virus strains or isolates tested can influence the outcome of these tests. Although the coat-protein amino acid triplet DAG may be directly correlated with transmissibility, the context in which it exists may also be a critical factor. Some other as yet unidentified amino acid substitutions could result in enhancement or inhibition of aphid transmission. Additionally, alterations of HC amino acid sequence can alter or inactivate its activity. Indeed, so little is known about the natural diversity of potyviruses that a definitive assessment of all the factors involved in aphid transmission is not possible.

Virus variability

The ability of a potyvirus to be aphid transmitted can be lost by serial transmission by mechanical inoculation [8]. This change is understandable since single amino acid changes in the HC or CP can alter aphid transmission of these viruses.

Then why should species of *Potyviridae* be grouped by genome structure characteristics which coincide with vector type? Goldbach [6] proposes that recombination events are important in genome formation during evolution. The 3' end of the genome contains a replication module which is present in all members of the *Potyviridae* which have been sequenced. The 5' end of the genome differs among genera but is similar for species within a genera. Bymoviruses, which have bipartite genomes instead of monopartite genomes contained by the other *Potyviridae* genera, lack a protein similar to HC. Perhaps a helper function is not required for fungal transmission and other functions in the bymoviruses.

Potyviruses have cistrons near the 3' and 5' end of the genome which are both required for aphid transmission. Aphid transmission is required for potyvirus spread even when seed transmission occurs. Could vector transmission be a means of stabilizing the viruses by limiting major recombination events?

Conclusions

The process by which aphids transmit potyviruses, although increasingly better understood, is still far from completely clear. As a tool for identification and diagnosis, it is still of great importance. Obviously, this interaction is a key component in terms of the epidemiology and economic importance of potyviruses. Vector relationships have had significant influences on the evolution of plant virus taxonomy (and vice versa). Yet we now have sufficient information on the molecular basis of the potyvirus–aphid interaction to see that it reflects but a small portion of the genome. Perhaps aphid transmission is based on c. 15–20 amino acids of the coat protein and 50–100 amino acids of HC and provides a reasonable taxonomic characteristic. Indeed, much is based on coat-protein serology which also represents only a fraction of the viral genome. Yet serological studies provide data that can be relative to each other while aphid transmission is more often an all-or-none situation. Cases where transmission efficiency is intermediate (e.g., TEV-PAT) are known, but quantitation is much less precise than immunologically based approaches.

Considering the lack of precision in aphid transmission bioassays, a relative lack of comparative data and standardized assays, and the fact that the aphid–virus interaction in terms of transmission is representative of only a fraction of the viral genome, the use of this characteristic as a taxonomic tool is limited. However, vector transmission might be an important means of stabilizing viruses at the family or genus level.

References

1. Atreya PL, Atreya CD, Pirone TP (1991) Amino acid substitutions in the coat protein that result in loss of insect transmissibility of a plant virus. Proc Natl Acad Sci USA 88: 787–791
2. Atreya CD, Raccah B, Pirone TP (1990) A point mutation in the coat protein abolishes aphid transmissibility of a potyvirus. Virology 178: 161–165
3. Barnett OW (1991) *Potyviridae*, a proposed family of plant viruses. Arch Virol 118: 139–141
4. Berger PH, Pirone TP (1986) The effect of helper component on uptake and localization of potyviruses in *Myzus persicae*. Virology 153: 256–261
5. Carrington JC, Cary SM, Parks TD, Dougherty WG (1989) A second proteinase encoded by a plant potyvirus genome. EMBO J 8: 365–370
6. Goldbach R (1992) The recombinative nature of potyviruses: implications for setting up a true phylogenetic taxonomy. In: Barnett OW (ed) Potyvirus taxonomy. Springer, Wien New York, pp 299–304 (Arch Virol [Suppl] 5)
7. Kennedy JS, Day MF, Eastop VF (1962) A conspectus of aphids as vectors of plant viruses. Commonwealth Institute of Entomology, London
8. Lecoq H, Purcifull DE (1992) Biological variability of potyviruses, an example: zucchini yellow mosaic virus. In: Barnett OW (ed) Potyvirus taxonomy. Springer, Wien New York, pp 229–234 (Arch Virol [Suppl] 5)
9. Nault LR, Madden LV (1988) Phylogenetic relatedness of maize chlorotic dwarf virus leafhopper vectors. Phytopathology 78: 1683–1687
10. Pirone TP, Thornbury DW (1984) The involvement of a helper component in nonpersistent transmission of plant viruses by aphids. Microbiol Sci 1: 191–193
11. Thornbury DW, Patterson CA, Dessens JT, Pirone TP (1990) Comparative sequence of the helper component (HC) region of potato virus Y and a HC-defective strain, potato virus C. Virology 178: 573–578

Author's address: P. H. Berger, Division of Plant Pathology, Department of Plant, Soil, and Entomological Sciences, Room 242 Ag Science, University of Idaho, College of Agriculture, Moscow, ID 83843, U.S.A.

Genome and sequence relationships

Arch Virol (1992) [Suppl 5]: 259–267

Viruses of the *Potyviridae* with non-aphid vectors

J. R. Edwardson

Plant Virus Laboratory, Department of Agronomy, University of Florida,
Gainesville, Florida, U.S.A.

Summary. The large majority of members of the family *Potyviridae* are aphid-transmitted. However, 17 viruses whose vectors are unknown have been classified as members of the genus *Potyvirus*. Loss of aphid transmissibility has been observed in some strains of several potyviruses. There are currently 11 members of the *Potyviridae* whose vectors are not aphids. These viruses with non-aphid vectors exhibit most of the characteristics of the family. Viruses of the *Potyviridae* induce cytoplasmic cylindrical inclusions in their hosts whether their vectors are aphids, non-aphids, or are unknown. The virus genome produces the inclusion protein and thus the viruses have related inclusion body gene sequences. Non-aphid-transmitted viruses of the *Potyviridae* also are serologically related to aphid-transmitted potyviruses.

Introduction

Prior to the Fourth Report of the International Committee on Taxonomy of Viruses (ICTV) [34] the principal concerns of potyvirus taxonomists seem to have been the extreme size of the potyvirus group and which of the viruses assigned to this group were actually strains or synonyms. We seem to have retained these problems. Nonpersistent transmission by aphids is considered by some workers to be a requirement for assigning a virus to the potyvirus group [18, 19]. However, no vectors have been reported for a considerable number of potyviruses, but the ICTV has assigned 16 of them as possible members and one has been assigned as a member of the potyviruses [12].

Significance of vectors

Since aphid transmissibility can be lost it does not appear to be a suitable requirement for assigning viruses to groups. Loss of aphid transmissibility has been reported for isolates of bean yellow mosaic [44], peanut mottle [35], potato virus C [28], sugarcane mosaic [30], tobacco etch [41], turnip mosaic [37] and zucchini yellow mosaic viruses [2]. Taxonomists classifying viruses in groups other than the potyviruses have recognized different vectors within nine groups (Table 1). The family *Potyviridae* has been proposed to encompass viruses in the potyvirus group.

There are several viruses whose characteristics are those of the family *Potyviridae* but whose vectors are not aphids. They are listed as possible members of the potyvirus group by the ICTV [34] and in different genera by a recent proposal, i.e., fungal-borne bymoviruses (barley yellow mosaic, oat

Table 1. Virus groups with more than one type of vector

Groups	Vectors and viruses	Refs.
Bromovirus	beetles (some members)	34
	fungus (brome mosaic)	48
	aphids (brome mosaic)	39
Carlavirus	aphids (some members)	34
	whiteflies (cowpea mild mottle)	23
	whiteflies (groundnut crinkle)	10
Carmovirus[a]	beetles (turnip crinkle)	33
	fungus (melon necrotic spot)	45
Closterovirus	aphids (some members)	34
	whiteflies (diodia vein chlorosis)	31
	whiteflies (lettuce infectious yellows)	11
Geminivirus	leafhoppers (some members)	34
	whiteflies (some members)	34
	treehoppers (pseudo curlytop)	8, 42
Luteovirus	aphids (some members)	34
	frit flies (barley yellow dwarf)	25
Plant Reovirus	leafhoppers (phytoreoviruses)	34
	planthoppers (fijiviruses)	34
Plant Rhabdovirus	aphids (lettuce necrotic yellows)	34
	lace bugs (beet leaf curl)	34
	leafhoppers (maize mosaic)	34
	mites (coffee ringspot)	34
	planthoppers (barley yellow striate mosaic)	34
Potyvirus family	aphids (most members)	34
	fungus (some members)	34
	mites (some members)	34
	whiteflies (sweet potato mild mottle)	34
Sobemovirus	aphids (blueberry shoestring)	38
	beetles (some members)	34
	mirids (velvet tobacco mottle)	13

[a] Virus group proposed by the Executive Committee of the ICTV [7]

mosaic, rice necrosis mosaic, wheat spindle streak mosaic, wheat yellow mosaic); mite-borne rymoviruses (agropyron mosaic, oat necrotic mottle, ryegrass mosaic, spartina mottle, wheat streak mosaic); and the whitefly-borne ipomoviruses (possible genus; sweet potato mild mottle). Measurements of properties of aphid-transmitted potyviruses are, with one exception, in agreement with those of viruses transmitted by other vectors (Table 2). The only property for which disagreements in measurements occur is in the molecular weights (M_rs) of RNAs of fungal-borne and aphid-transmitted viruses. The combined M_rs of RNA-1 and RNA-2 of the fungal-borne bymoviruses is approximately 4×10^6 while the upper limit for the range of M_rs of RNA in aphid-transmitted potyviruses is 3.65×10^6 (Table 2). The larger M_r of the fungal-borne *Bymovirus* RNA may be explained at least in part by results of recent work of Japanese researchers (D. D. Shukla, pers. comm.). Their studies have shown barley yellow mosaic virus (BaYMV) genes occurred in the same order as those of aphid-transmitted potyviruses and the divided genome contained duplications of the 3' and 5' ends.

Recently, Kashiwazaki and coworkers [27] proposed separating fungal-borne bymoviruses from the aphid-transmitted potyviruses. These workers stated that the BaYMV capsid gene is similar to that of the potyviruses in its 3' proximal location and in its manner of expression by processing from a polyprotein precursor. Futher, BaYMV shares with potyviruses the induction of cylindrical cytoplasmic inclusions [27]. These features suggested to Kashiwazaki and coworkers close evolutionary relations between fungal-borne bymoviruses and aphid-transmitted potyviruses [27]. However, they list the following characteristics where the bymoviruses differ from the potyviruses: (*i*) fungus transmission, (*ii*) bipartite genome, (*iii*) only small blocks of capsid proteins exhibiting homologies between BaYMV and aphid-transmitted potyviruses, (*iv*) absence of serological relationships between BaYMV and aphid-transmitted potyviruses.

The significance of vectors in grouping plant viruses has already been noted (Table 1). It is interesting to note that when an isolate of *Polymyxa graminis*-transmitted BaYMV was repeatedly transmitted mechanically it could no longer be acquired and/or transmitted by the fungus [1].

Differences in genome segmentation are certainly well established in the geminivirus group where viruses in subgroup A (leafhopper-transmitted) contain a monopartite genome of ssDNA and viruses in subgroup B (whitefly-transmitted) contain a bipartite ssDNA genome [7], and in the plant reoviruses where subgroup-1 viruses (leafhopper-transmitted) possess 12 genome segments and where subgroup-2 viruses (planthopper-transmitted) have 10 genome segments [34]. Also in higher plants where taxonomy rests on a firmer base than does viral taxonomy, differences in genome segmentation (chromosome numbers) are well documented within many families and many genera [9]. Extensive amino acid homologies between carnation mottle and turnip crinkle carmoviruses and barley yellow dwarf luteoviruses and red clover

J.R. Edwardson

Table 2. Property measurement ranges of viruses in the *Potyviridae* [12]

Virus	RNA (%)	Sedimentation coefficients RNA (S)	Sedimentation coefficients virus (S)	Molecular weights RNA (× 10⁶)	Molecular weights coat protein subunits (Da)	Optical density 260/280	Buoyant density (g/cm³)	Nucleotides (%)
Aphid-transmitted Potyviruses								
Potyvirus ranges	3–4 to 7	24.2–41.4	133–176	2.7–3.65	21,300–45.000–48,000	0.87–2.0	1.245–1.49	G=20.2–33.9 A=23.0–44.0 C=14.9–27.2 U=15.6–30.9
Mite-transmitted Rymoviruses								
Agropyron mosaic (oat necrotic mottle)		– –	165 [5] –	– –	– –	– –	– 1.33 [15] 1.40 [14]	–
Ryegrass mosaic	5.5 [36]	–	166 [36]	2.7 [36]	29,200 [36]	1.25 [36]	1.307 at 15°C 1.325 at 25°C [36]	G=31 A=23 C=24 U=22 [36]
Wheat streak mosaic		23.5 formaldehyde 40.0 undenatured [6]	159–173 [4] 165 [5]	2.8 [6]	–	1.37 [5]	1.49 [6]	–

Fungus-transmitted *Bymoviruses*

Virus							
Oat mosaic	—	—	RNA-1=2.8 RNA-2=1.8 [36]	26,500 & 30,000 [47]	—	—	—
Barley yellow mosaic	5 [46]	—	RNA-1=2.5 [26] RNA-1=2.6 RNA-2=1.4 [26] RNA-2=1.5 [47]	26,500 & 33,000 [47] 29,000 35,000– 36,000 [22] 31,600– 33,100 [26]	0.88 [46] 1.14 [24]	1.29 [24] 1.294 [46]	—
Rice necrosis mosaic	—	—	—	—	—	—	—
Wheat spindle streak mosaic	—	—	RNA-1=2.6 RNA-2=1.4 [47]	26,500 & 33,000 [47] 53,500 [29]	1.25 [16]	1.284 [29]	—
White yellow mosaic	5 [46]	—	RNA-1=2.6 RNA-2=1.5 [47]	33,000 + minor bands [47]	—	1.281 [46]	—
Barley mild mosaic	—	—	RNA-1= 2.7–2.8 RNA-2= 1.4–1.5 [22] RNA-1=2.6 RNA-2=1.3 [26]	29,000 & 35,000– 36,000 [22] 30,900 [26]	—	—	—

Whitefly-transmitted *Ipomovirus* (possible genus)

Virus							
Sweet potato mild mottle	—	155 [21]	—	37,700 [3]	1.33 [21] 1.73–1.80 [20]	—	—

necrotic mosaic dianthovirus have been reported [49]. Xiong and Lommel [49] noted that the homologies between the monopartite carmoviruses and luteoviruses and the bipartite dianthovirus decrease the importance of genome segmentation as a major criterion for virus classification.

Limited homologies between the capsid proteins of BaYMV and those of aphid-transmitted potyviruses exist [27], but their significance in the taxonomy of the *Potyviridae* is yet to be fully understood.

An absence of serological relationship between BaYMV and aphid-transmitted potyviruses does not constitute much support for this separation. There are at present eight viruses which have been assigned to the genus *Potyvirus* as members [34] which have not been reported to be related serologically to any potyvirus (carrot thin leaf, cocksfoot streak, commelina mosaic, iris mild mosaic, iris severe mosaic, narcissus degeneration, parsnip mosaic, tamarillo mosaic viruses) [12]. Failure to establish serological relationships between a virus isolate (exhibiting characteristics of the genus) and recognized potyviruses has for a long time been regarded as indicating that this isolate was a distinct potyvirus species.

Serological relationships between aphid-transmitted potyviruses and viruses of other genera whose vectors are not aphids have also been established. Stanarius and coworkers [43] reported that the coat protein of barley mild mosaic virus (formerly the M strain of BaYMV), which is fungus-transmitted and has a divided genome, is serologically related to the coat proteins of bean yellow mosaic and turnip mosaic viruses. Both of these viruses are aphid-transmitted and are long-standing members of the genus *Potyvirus*. The mite-transmitted ryegrass mosaic virus and the aphid-transmitted turnip mosaic virus coat proteins are reported to be related serologically [32] and the coat protein of the mite-transmitted wheat streak mosaic virus is serologically related to that of the aphid-transmitted johnsongrass mosaic virus [40].

Nuclear inclusion proteins of the aphid-transmitted tobacco etch virus are serologically related to proteins induced by the mite-transmitted wheat streak mosaic virus (E. Hiebert, pers. comm.).

Antiserum to the helper-component from the aphid-transmitted tobacco vein mottling virus precipitated a product with a M_r of 78,000 Da in cell-free translation of the mite-transmitted wheat streak mosaic virus. This translation product has the same M_r as that of the tobacco vein mottling virus helper component [17].

The coat proteins of the whitefly-transmitted sweet potato mild mottle and of the aphid-transmitted johnsongrass mosaic virus are reported to be related serologically [40].

Conclusion

A considerable number of viruses with different types of vectors, some viruses with different genome segmentations and several viruses without demonstrated serological relationships to definitive potyviruses have been separated from the former potyvirus group. Portions of the genomes of the fungus-, mite-, and whitefly-transmitted potyviruses induce cylindrical inclusions in the cytoplasms of their hosts as do portions of aphid-transmitted potyvirus genomes [12]. Serological relationships have been established between aphid-transmitted and non-aphid-transmitted viruses which induce cylindrical inclusions [17,32,40,43]. Thus these non-aphid-transmitted and aphid-transmitted viruses are related and should be placed together at some level. The family *Potyviridae* seems to be an appropriate level which retains all these viruses together due to their related characteristics and allows for their differences through establishment of genera.

References

1. Adams MJ, Swaby AG, Jones P (1988) Confirmation of the transmission of barley yellow mosaic virus (BaYMV) by the fungus *Polymyxa graminis*. Ann Appl Biol 112: 133–141

2. Antignus Y, Raccah B, Gal-On A, Cohen S (1989) Biological and serological characterization of zucchini yellow mosaic and watermelon mosaic virus-2 isolates in Israel. Phytoparasitica 17: 289–298

3. Barton RJ (1973) Analysis of virus capsid polypeptides. Glasshouse Crops Res Inst Annu Rep 1973: 123–124

4. Brakke MM (1958) Estimation of sedimentation constants of viruses by density-gradient centrifugation. Virology 6: 96–114

5. Brakke MK (1971) Wheat streak mosaic virus. CMI/AAB Descriptions of Plant Viruses, no 48

6. Brakke MK, van Pelt N (1970) Properties of infectious ribonucleic acid from wheat streak mosaic virus. Virology 42: 699–706

7. Brown F (1987) Minutes of the 17th Meeting of the Executive Committee of the International Committee on Taxonomy of Viruses, Edmonton

8. Christie RG, Ko NJ, Falk BW, Hiebert E, Lastra R, Bird J, Kim KS (1986) Light microscopy of geminivirus-induced nuclear inclusion bodies. Phytopathology 76: 124–126

9. Darlington CD, Wylie AP (1956) Chromosome atlas of flowering plants. Macmillan, New York

10. Dubern J, Dollet M (1981) Groundnut crinkle virus, a new member of the carlavirus group. Phytopathol Z 101: 337–340

11. Duffus JE, Larsen RC, Liu HY (1986) Lettuce infectious yellows virus – a new type of whitefly-transmitted virus. Phytopathology 76: 97–100

12. Edwardson, JR, Christie RG (1991) The potyviruses. Fla Agric Exp Stat Monogr Ser, no 16

13. Gibbs KS, Randles JW (1989) Non-propagative translocation of velvet tobacco mottle virus in the mirid, *Cyrtopeltis nicotianae*. Ann Appl Biol 115: 11–15

14. Gill CC (1971) Purification of oat necrotic mottle virus with silver nitrate as clarifying agent. J Gen Virol 12: 259–270

15. Gill CC (1976) Oat necrotic mottle virus. CMI/AAB Descriptions of Plant Viruses, no 169

16. Haulfer KZ, Fulbright DW (1985) Purification and partial characterization of wheat spindle streak mosaic virus (WSSMV). Phytopathology 75: 1349–1350

17. Hiebert E, Thornbury DW, Pirone TP (1984) Immunoprecipitation analysis of potyviral in vitro translation products using antisera to helper component of tobacco vein mottling virus and potato virus Y. Virology 135: 1–9

18. Hollings M, Brunt AA (1981) Potyviruses. In: Kurstak E (ed) Handbook of plant virus infections and comparative diagnosis. Elsevier/North Holland, New York, pp 731–807

19. Hollings M, Brunt AA (1981) Potyvirus group. CMI/AAB Descriptions of Plant Viruses, no 245

20. Hollings M, Stone OM, Bock KR (1971) Sweet potato virus T (SPV-T). Glasshouse Crops Res Inst Annu Rep 1970: 155–156

21. Hollings M, Stone OM, Bock KR (1976) Purification and properties of sweet potato mild mottle, a white-fly borne virus from sweet potato (*Ipomoea batatas*) in East Africa. Ann Appl Biol 82: 511–528

22. Huth W, Lesemann DE, Paul HL (1984) Barley yellow mosaic virus. Purification, electron microscopy, serology, and other properties of two types of the virus. Phytopathol Z 111: 37–54

23. Iizuka N, Rajeshwari R, Reddy DVR, Goto T, Muniyappa V, Bharathan N, Ghanekar AM (1984) Natural occurrence of a strain of cowpea mild mottle virus on groundnut (*Arachis hypogaea*) in India. Phytopathol Z 109: 245–253

24. Inouye T, Saito J (1975) Barley yellow mosaic virus. CMI/AAB Descriptions of Plant Viruses, no 143

25. Jess S, Mowat DJ (1986) Transmission of barley yellow dwarf virus by larvae of frit fly, *Oscinella frit* (L.), and the effects of sward-killing herbicides on transmission. Rec Agric Res 34: 5760

26. Kashiwazaki S, Ogawa K, Usugi T, Omura T, Tsuchizaki T (1989) Characterization of several strains of barley yellow mosaic virus. Ann Phytopathol Soc Jpn 55: 16–25

27. Kashiwazaki S, Hayano Y, Minobe Y, Omura T, Hibino H, Tsuchizaki T (1989) Nucleotide sequence of the capsid protein gene of barley yellow mosaic virus. J Gen Virol 70: 3015–3023

28. Kassanis B, Govier DA (1971) The role of the helper virus in aphid transmission of potato aucuba mosaic virus and potato virus C. J Gen Virol 13: 221–228

29. Kendall TL, Lommel SA (1985) Partial characterization of wheat spindle streak mosaic virus and its involvement in a disease of winter wheat in Kansas. Phytopathology 75: 964

30. Koike H (1979) Loss of aphid transmissibility in an isolate of sugarcane mosaic virus strain H. Plant Dis Rep 63: 373–375

31. Larsen RC, Kim KS, Scott HA (1991) Properties and cytopathology of Diodia vein chlorosis virus – a new whitefly-transmitted virus. Phytopathology 81: 227–232

32. Lesemann DE, Vetten HJ (1985) The occurrence of tobacco rattle and turnip mosaic viruses in *Orchis* spp., and of an unidentified potyvirus in *Cypripedium calceolus*. Acta Hortic 164: 45–54

33. Martini C (1958) The transmission of turnip viruses by biting insects and aphids. In: Proceedings of the 3rd Conference on Potato Virus Diseases, Lisse-Wageningen, 1957, pp 106–113

34. Matthews REF (1982) Classification and nomenclature of viruses. Fourth report of the International Committee on Taxonomy of Viruses. Intervirology 17: 1–199

35. Paguio OR, Kuhn CW (1976) Aphid transmission of peanut mottle virus. Phytopathology 66: 473–476

36. Paliwal YC, Tremaine JH (1976) Multiplication, purification and properties of ryegrass mosaic virus. Phytopathology 66: 406–414

37. Pound GS, Tochihara A, Shepherd RJ (1932) Relationship between turnip mosaic virus and the radish P virus in Japan. Phytopathology 52: 373

38. Ramsdell DC (1979) Blueberry shoestring virus. CMI/AAB Descriptions of Plant Viruses, no 204

39. Rybicki EP, von Wechmar MB (1982) Characterization of an aphid-transmitted virus disease of small grains. Phytopathol Z 109: 245–253

40. Shukla DD, Ford RE, Tosic M, Jilka J, Ward CW (1989) Possible members of the potyvirus group transmitted by mites or whiteflies share epitopes with aphid-transmitted definitive members of the group. Arch Virol 105: 143–151

41. Simons JN (1976) Aphid transmission of a nonaphid-transmissible strain of tobacco etch virus. Phytopathology 66: 652–654

42. Simons JN, Coe DM (1958) Transmission of pseudocurly top virus in Florida by a treehopper. Virology 6: 43–46

43. Stanarius A, Proeseler G, Richter J (1989) Immunelektronenmikroskopische Untersuchungen zur serologischen Verwandtschaft des Gerstengelbmosaik-Virus (barley yellow mosaic virus) und des Milden Gerstenmosaik-Virus (barley mild mosaic virus) mit anderen gestreckten Viren. Arch Phytopathol Pflanzenschutz 4: 303–307

44. Swenson KG (1957) Transmission of bean yellow mosaic virus by aphids. J Econ Entomol 50: 727–731

45. Tomlinson JA, Thomas BJ (1986) Studies on melon necrotic spot virus disease of cucumber and on the control of the fungus vector (*Olpidium radicale*). Ann Appl Biol 108: 71–80

46. Usugi T, Saito Y (1976) Purification and serological properties of barley yellow mosaic virus and wheat yellow mosaic virus. Ann Phytopathol Soc Jpn 42: 12–20

47. Usugi T, Kashiwazaki S, Omura T, Tsuchizaki T (1989) Some properties of nucleic acids and coat proteins of soil-borne filamentous viruses. Ann Phytopathol Soc Jpn 55: 26–31

48. von Wechmar MB (1980) Transmission of brome mosaic virus by *Puccinia graminis tritici*. Phytopathol Z 99: 289–293

49. Xiong Z, Lommel SA (1989) The complete nucleotide sequence and genome organization of red clover necrotic mosaic virus RNA-l. Virology 171: 543–554

Author's address: J. R. Edwardson, Department of Agronomy, Plant Virus Laboratory, University of Florida, Gainesville, FL 32611, U.S.A.

Arch Virol (1992) [Suppl 5]: 269–276

Potyviridae: genus *Rymovirus*

K. R. Zagula[1], **C. L. Niblett**[2], **N. L. Robertson**[3], **R. French**[3], and **S. A. Lommel**[1]

[1] Department of Plant Pathology, North Carolina State University, Raleigh, North Carolina
[2] Plant Pathology Department, University of Florida, Gainesville, Florida
[3] USDA-ARS, Department of Plant Pathology, University of Nebraska, Lincoln, Nebraska, U.S.A.

Summary. The genus *Rymovirus* of the family *Potyviridae* is comprised of seven rod-shaped viruses with the shared characteristic of being transmitted by mites. Aside from this distinguishing feature, rymoviruses are similar to aphid-transmitted potyviruses in that they share a similar particle morphology, some similar antigenic determinants, similar physico-chemical properties, the ability to induce the formation of cytoplasmic cylindrical inclusions, and the ability to infect only graminaceous hosts. In vitro translation studies with wheat streak mosaic virus (WSMV) suggest that this rymovirus uses a potyviral proteolytic processing strategy to express the 3′ terminal capsid protein. At the molecular level, limited nucleotide sequence data for WSMV show similarities with aphid-transmitted potyviruses in the potyviral capsid protein, large nuclear inclusion and cylindrical inclusion regions. Thus, given the similarities between the rymoviruses and the potyviruses, it is appropriate to include this genus within the family *Potyviridae*.

Introduction

Potyviruses comprise the largest and most economically important group of plant viruses, accounting for nearly one-quarter of all known plant virus diseases [21]. The taxonomy of the potyvirus group is complex and not well organized due to the growing numbers and diversity of possible new members. In 1982, the Fourth Report of the International Committee on Taxonomy of Viruses recognized 48 viruses as definite members and 67 viruses as possible members of this group [16]. Two years later, the editors of the CMI/AAB Descriptions of Plant Viruses subdivided the Potyvirus Group into four subgroups based on vectors. In September, 1990, participants at the Potyvirus Taxonomy Workshop agreed to establish the *Potyviridae* as a family of plant viruses subdivided into the following genera, based on nucleic acid and amino acid sequence relationships, which coincided with mode of transmission: *Potyvirus* (aphid-borne), *Bymovirus* (fungal-borne), *Rymovirus* (mite-borne), and possibly *Ipomovirus* (whitefly-borne). In this paper we define and discuss the physical, biological and chemical properties, and the molecular organization of the genus *Rymovirus* of the *Potyviridae*.

Genus *Rymovirus*

Members of the genus *Rymovirus* include ryegrass mosaic virus (RGMV), the type member, agropyron mosaic virus (AgMV), hordeum mosaic virus (HoMV), oat necrotic mottle virus (ONMV), and wheat streak mosaic virus (WSMV), as well as possible members brome streak mosaic and spartina mottle viruses. These viruses all have known or suspected mite vectors.

Physicochemical properties

Rymoviruses have flexuous, filamentous particles 11–15 nm wide × 680–725 nm in length, slightly shorter than most potyviruses. The genome consists of one linear positive-sense ssRNA molecule c. 8.5 kb [4] with 3′ terminal poly(A) of variable length.

Rymovirus virions are comprised of a single capsid protein species which ranges in size from 29 to 47 kDa. The capsid protein of RGMV is 29.2 kDa [21], similar in size to capsid proteins of the aphid-vectored potyviruses [10]. Several isolates of WSMV, to date the most extensively characterized rymovirus, possess a major capsid protein of 42–47 kDa [3, 20]. In addition, minor protein species of 36, 33, 32, and 31 kDa have been identified by SDS-PAGE from several WSMV isolates. These smaller proteins are considered to be subsets of the major capsid protein, as they react with antiserum to intact virions in Western blotting [3, 20].

Sedimentation coefficients for rymoviruses are within the range for aphid-transmitted potyviruses: 166 S for RGMV, 165 S for WSMV and 160 S for spartina mottle virus. A buoyant density in CsCl of 1.33 g/cm³ was reported for RGMV. In plant sap the thermal inactivation point for rymoviruses is 50–60 °C (10 min); the dilution endpoint is 10^{-3} [1, 7, 11, 18, 26, 28].

Most rymoviruses are moderately immunogenic. Although serological studies have revealed distant relationships among several rymoviruses, most members appear to be serologically unrelated. Weak serological reactions were reported between WSMV and AgMV, between AgMV and HoMV [27], and between WSMV and ONMV [8]. Antisera to AgMV and HoMV did not react with ONMV [8]; similarly, RGMV did not react with AgMV, HoMV, or WSMV antisera [28]. Spartina mottle virus reacted weakly with antiserum against AgMV but not at all with antisera against ONMV, HoMV, WSMV, or RGMV [11].

Some serological relationships between aphid-transmitted potyviruses and the rymoviruses have been reported. Turnip mosaic virus antiserum reacted weakly in immunosorbent electron microscopy with RGMV [15]. Shukla et al. [22] reported a serological reaction between WSMV capsid protein and a broad spectrum antiserum to johnsongrass mosaic virus capsid protein. Using antiserum to the helper component (HC) of tobacco vein mottling virus (TVMV), Hiebert et al. [9] immunoprecipitated two in vitro translation products of WSMV, one of which had the same molecular weight (78 kDa) as TVMV HC. Antiserum to tobacco etch virus (TEV) large nuclear inclusion protein (NIb) also reacted with WSMV translation products (E. Hiebert, unpubl.).

Biological properties

The main feature distinguishing rymoviruses from other potyviruses is that they are mite-transmitted. Eriophyid mites were first found on wheat plants showing varying degrees of leaf chlorosis, necrosis, and stunting in Canada and the Great Plains of the USA in the early 1950's. Slykhuis [25] first demonstrated that the wheat curl mite *Eriophyes (Aceria) tulipae* could transmit WSMV. Non-viruliferous colonies of *E. tulipae* reared from eggs hatched on healthy plants were transferred to manually inoculated, symptom-bearing source plants and allowed to feed for 10 days. Mites were then transferred to healthy test plants. Symptoms of WSMV developed on all plants colonized with mites that had fed on wheat manually inoculated with several isolates of WSMV, whereas plants colonized with mites that had not fed on diseased wheat remained healthy. Adult and all nymphal stages of the mite can transmit WSMV [25]. Another eriophyid mite, *Abacarus hystrix*, is the vector for AgMV and RGMV [26,28]. *E. tulipae* is unable to transmit AgMV [26]. No vectors have been reported for the other rymoviruses, although ONMV and spartina mottle virus were not transmissible by several aphid and leafhopper species [7,11]. Rymoviruses are readily sap transmissible but apparently not seed transmissible [1].

Rymoviruses are limited to but widespread within the Gramineae, causing mild to severe mosaics and stunting in infected plants. Most of the diseases caused by rymoviruses are not of major economic significance, except with severe epidemics of WSMV in the Great Plains states which resulted in significant wheat yield losses [24]. WSMV infects most varieties of wheat, oats, barley and rye, some varieties of maize and millets, and many species of wild grasses, but not *Elytrigia (Agropyron) repens* [1]. WSMV, AgMV, and HoMV have similar host ranges. However, AgMV infects *E. repens* but not oats, and only HoMV can systemically infect *Hordeum jubatum* [26]. Oat necrotic mottle virus infects species of oats and wild grasses, while RGMV infects ryegrass species and oats; neither infects barley, wheat, rye, or *E. repens* [7,28]. Brome streak mosaic virus was found naturally infecting *Bromus mollis* and *H. murinum*, and is mechanically transmissible to wheat, oats, and barley [18, 19]. Spartina mottle virus only affects species of the perennial saltmarsh grass *Spartina*; mechanical inoculation to a range of other graminaceous or dicotyledonous species was unsuccessful [11]. No local lesion or dicotyledonous hosts have been identified for any rymovirus, except that AgMV produced local lesions on *Chenopodium quinoa* [26].

All *Rymovirus* species induce characteristic cytoplasmic cylindrical (pinwheel) inclusions (CI) in infected cells, which are similar to those of aphidtransmitted potyviruses. Because structures associated with pinwheels show virus-specific variations independent of host species, Edwardson [5] separated the potyviruses into three subdivisions based on inclusion morphology. WSMV was assigned to Subdivision I, in which inclusions induce only scrolls (tubes in cross section) attached to the central portion of the pinwheel. In subsequent studies, both scrolls and laminated aggregates were detected in cells infected with WSMV [17] and brome streak mosaic virus [18], placing these viruses in Subdivision III.

The CI induced by WSMV is composed of a 66 kDa protein [2], similar in size to CI proteins of potyviruses [10]. Antiserum to the CI protein of WSMV did not immunolabel ultrathin sections containing CI proteins of AgMV or HoMV, nor did AgMV or HoMV antiserum react with CIs of WSMV [14]. However, in agreement with the serological data indicating relatedness between their capsid proteins, HoMV CIs were immunolabelled with antibodies to AgMV CI. Only a small percentage of the viruses within the *Potyviridae* produce discrete nuclear inclusions (NIs), most notably strains of TEV. None of the rymoviruses produce NIs.

Molecular characterization

Although little is known about the molecular organization of other rymoviruses, considerable information on the WSMV genome has been obtained [20, 29]. Approximately 70% of the WSMV genome has been cloned. T7 RNA transcripts from 3' proximal cDNA clones generated capsid-protein immunoprecipitable polypeptides, indicating that they contained capsid-protein sequence [20]. The translation product profiles obtained from WSMV RNA and the T7 transcripts were complex, suggesting that, like aphid-borne potyviruses, WSMV uses a translational strategy based on polyprotein processing.

Similar to other potyviruses, the WSMV genome has a 3' terminal poly(A) tract. Over 1.8 kb of 3' terminal sequence of WSMV RNA has been determined [20, 29] which includes the region encoding the capsid protein. With alignment maximization using gaps, the deduced amino acid sequence of the WSMV capsid protein showed limited but significant (20–25%) identity within the highly conserved central and C-terminal domains of the capsid-protein regions of at least two of the following four aphid-borne potyviruses: TEV, TVMV, potato virus Y (PVY) and plum pox virus (PPV) [20]. Virtually no amino acid sequence similarity was identified between the WSMV capsid protein and those predicted for BaYMV (a *Bymovirus* species) or potato virus X.

Amino acid sequence similarities 5′ to the capsid-protein region are more pronounced among WSMV and TEV, TVMV, PVY and PPV. The deduced 185 amino acid sequence of a primer extension clone generated from a 3′ terminal cDNA clone shared 49% identity (one gap introduced) with four aphid-transmitted potyviruses within the large nuclear inclusion (NIb) region [29]. The WSMV sequence contained the highly conserved "GDD" motif and represented a portion of the putative RNA-dependent RNA polymerase of potyviruses [12]. When conservative amino acid substitutions are considered, WSMV shares 67% similarity with the aphid-borne potyviruses in the NIb region. Further toward the 5′ terminus, a deduced 332 amino acid sequence shared 29% identity (six gaps introduced) with the four potyviruses within the N terminus of the potyviral CI region [29]. Alignment with the other potyviruses increases to 51% if conservative amino acid substitutions are allowed, which is close to the approximately 60% homology shared by the other potyviruses in this region. The most notable domain of alignment among the WSMV sequence and the potyviral CI sequences includes the consensus sequence GXXGXGKS, the putative nucleotide (phosphate) binding site [13]. A comparison of the putative genome organization and nucleotide sequence similarities between TEV, a typical aphid-transmitted *Potyvirus* species, and WSMV is presented in Fig. 1.

In an attempt to further investigate the relationships among the rymoviruses, Robertson and French (unpubl.) recently compared approximately 250 nucleotides of the 3′ noncoding regions of four WSMV isolates with HoMV and AgMV by direct RNA sequencing and the polymerase chain reaction. Potyviruses have been classified as either distinct viruses or strains of a single

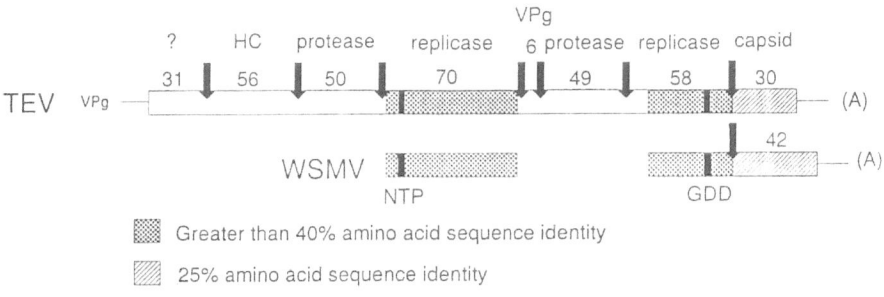

Fig. 1. Comparison of the genome organization and nucleotide sequence similarities between tobacco etch virus (TEV), an aphid-transmitted *Potyvirus* species, and wheat streak mosaic virus (WSMV), a mite-transmitted *Rymovirus* species. Hatched boxes indicate the approximate percentage of identity between the two viruses within several regions of the potyviral genome as described above. The locations of the "GDD" and NTP motifs are indicated; putative polymerase cleavage sites are indicated with arrows

virus by comparing nucleotide sequences in this region [6]. The 3' terminal sequences of the four WSMV isolates were greater than 90% identical to each other (Robertson and French, unpubl.) and to the 147 base 3' non-coding region of the WSMV H81 isolate which had been sequenced previously [20]. In contrast, the 3' noncoding regions of HoMV and AgMV showed little homology with WSMV isolates, further demonstrating that WSMV, HoMV, and AgMV are distinct viruses. The HoMV 3' terminal sequence is 73% identical to that of AgMV. Within the *Rymovirus* genus, based on 3' terminal sequence homology and some serological cross-reactivity between their capsid and CI proteins, it appears that HoMV and AgMV are the most closely related (yet distinct) rymoviruses.

Discussion

Except for WSMV, relatively little is known about individual species of the genus *Rymovirus*. All definitive members share a similar particle morphology, mode of transmission, and ability to produce pinwheel inclusions; all infect graminaceous hosts exclusively (with one possible exception), and some members show weak serological relatedness. At the molecular level, nucleotide sequence information exists for approximately 70% of the WSMV genome and 250 nucleotides of the 3' terminal regions of four other WSMV isolates, AgMV and HoMV. All viruses appear to be distinct species (not strains of WSMV as was previously thought for AgMV and HoMV), yet two (AgMV and HoMV) show a high degree of relatedness in their 3' terminal regions. In addition, WSMV shows significant sequence homology with potyviruses in the potyviral NIb and CI cistrons.

Some researchers have suggested that potyvirus-like viruses which are not aphid-transmitted should be excluded from any association with potyviruses. One may concur with this in the case of the rymoviruses given the following considerations: rymoviruses are mite-transmitted, their genome is smaller than the 10 kb of most potyviruses, the capsid protein of at least one rymovirus (WSMV) is much larger than most potyviral capsid proteins, and homology between capsid proteins of WSMV and potyviruses is only 20–25% (10% if no gaps are used) compared with sequence identities of 38–71% among potyviruses [23]. However, given the similarities between the rymoviruses and potyviruses, i.e., particle morphology, an apparent translation strategy based on polyprotein processing, the ability to induce CI proteins in infected cells, and the degree of sequence homology between WSMV and potyviral NIb and CI cistrons, it seems logical to include the genus *Rymovirus* within the family *Potyviridae*.

References

1. Brakke MK (1971) Wheat streak mosaic virus. CMI/AAB Descriptions of Plant Viruses, no 48

2. Brakke MK, Ball EM, Hsu YH, Langenberg WG (1987) Wheat streak mosaic virus cylindrical inclusion body protein. J Gen Virol 68: 281–287

3. Brakke MK, Skopp RN, Lane LC (1990) Degradation of wheat streak mosaic virus capsid protein during leaf senescence. Phytopathology 80: 1401–1405

4. Brakke MK, Van Pelt N (1970) Properties of infectious ribonucleic acid from wheat streak mosaic virus. Virology 42: 699–706

5. Edwardson JR (1974) Host ranges of viruses in the PVY-group. Fla Agric Exp Stat Monogr Ser, no 4

6. Frenkel MJ, Ward CW, Shukla D (1989) The use of 3'-noncoding nucleotide sequences in the taxonomy of potyviruses: application to watermelon mosaic virus 2 and soybean mosaic virus-N. J Gen Virol 70: 2775–2783

7. Gill CC (1976) Oat necrotic mottle virus. CMI/AAB Descriptions of Plant Viruses, no 169

8. Gill CC (1976) Serological properties of oat necrotic mottle virus. Phytopathology 66: 415–418

9. Heibert E, Thornbury DW, Pirone TP (1984) Immunoprecipitation analysis of potyviral in vitro translation products using antisera to helper component of tobacco vein mottling virus and potato virus Y. Virology 135: 1–9

10. Hollings M, Brunt AA (1981) Potyvirus group. CMI/AAB Descriptions of Plant Viruses, no 245

11. Jones P (1980) Leaf mottling of *Spartina* species caused by a newly recognised virus, spartina mottle virus. Ann Appl Biol 94: 77–81

12. Kamer G, Argos P (1984) Primary structural comparison of RNA-dependent polymerases from plant, animal and bacterial viruses. Nucleic Acids Res 12: 7269–7282

13. la Cour TFM, Nyborg J, Thirup S, Clark BFC (1985) Structural details of the binding of guanosine diphosphate to elongation factor Tu from *E. coli* as studied by X-ray crystallography. EMBO 4: 2385–2388

14. Langenberg WG (1991) Cylindrical inclusion bodies of wheat streak mosaic virus and three other potyviruses only self-assemble in mixed infections. J Gen Virol 72: 493–497

15. Lesemann D-E, Vetten HJ (1985) The occurrence of tobacco rattle and turnip mosaic viruses in *Orchis* ssp., and of an unidentified potyvirus in *Cypripedium calceolus*. Acta Hortic 164: 4554

16. Matthews REF (1982) Classification and Nomenclature of Viruses. Fourth report of the International Committee on Taxonomy of Viruses. Intervirology 17: 1–199

17. McMullen CR, Gardner WS (1980) Cytoplasmic inclusions induced by wheat streak mosaic virus. J Ultrastruct Res 72: 65–75

18. Milicic D, Kujundzic M, Wrischer M, Plavsic B (1980) A potyvirus isolated from *Bromus mollis*. Acta Bot Croat 39: 27–32

19. Milicic D, Mamula D, Plazibat M (1982) Some properties of brome streak mosaic virus. Acta Bot Croat 41: 7–12

20. Niblett CL, Zagula KR, Calvert LA, Kendall TL, Stark DM, Smith CE, Beachy RN, Lommel SA (1990) cDNA cloning and nucleotide sequence of the wheat streak mosaic virus capsid protein gene. J Gen Virol 72: 499–504

21. Paliwal YC, Tremaine JH (1976) Multiplication, purification and properties of ryegrass mosaic virus. Phytopathology 66: 406–414

22. Shukla DD, Ford RE, Tosic M, Jilka J, Ward CW (1989) Possible members of the potyvirus group transmitted by mites or whiteflies share epitopes with aphid-transmitted definitive members of the group. Arch Virol 105: 143–151

23. Shukla DD, Ward CW (1988) Amino acid sequence homology of coat proteins as a basis for identification and classification of the potyvirus group. J Gen Virol 69: 2703–2710

24. Sim T IV, Willis WG, Eversmeyer MG (1988) Kansas plant disease survey. Plant Dis 72: 832-836
25. Slykhuis JT (1955) *Aceria tulipae* Keifer (Acarina: Eriophyidae) in relation to the spread of wheat streak mosaic. Phytopathology 45: 116–128
26. Slykhuis JT (1973) Agropyron mosaic virus. CMI/AAB Descriptions of Plant Viruses, no 118
27. Slykhuis JT, Bell W (1966) Differentiation of Agropyron mosaic virus, wheat streak mosaic, and a hitherto unrecognized Hordeum mosaic virus in Canada. Can J Bot 44: 1191–1208
28. Slykhuis JT, Paliwal YC (1972) Ryegrass mosaic virus. CMI/AAB Descriptions of Plant Viruses, no 86
29. Zagula KR, Kendall TL, Lommel SA (1990) Wheat streak mosaic virus genomic RNA shares sequence homology with potyviral cylindrical inclusion cistrons. Phytopathology 80: 1036

Authors' address: K. R. Zagula, Department of Plant Pathology, North Carolina State University, Box 7616, Raleigh, NC 27695, U.S.A.

Arch Virol (1992) [Suppl 5]: 277–282

How important is genome division as a taxonomic criterion in plant virus classification?

M. A. Mayo

Scottish Crop Research Institute, Invergowrie, Dundee, Scotland

Summary. The number of nucleic acid components that constitute a virus genome has been used as an important discriminatory character in defining groups of plant viruses. However, with some virus groups, in particular potyviruses, recent results of nucleotide sequencing have reinforced previously deduced tentative relationships among viruses with different numbers of genome parts. A convenient solution is to classify these different types into groups or genera within a family (e.g. *Potyvirus* and *Bymovirus* in the family *Potyviridae* [1]).

Introduction

Since the first proposals for placing plant viruses in what are now accepted as taxonomically valid groups [13] the adoption of the number of nucleic acid components that make up a virus genome as a taxonomic criterion has proved to be very useful. For example, any attempt to classify a new virus with an apparently monopartite genome in the comovirus or tobravirus groups would probably be regarded with suspicion and technical reasons would be sought for the supposed loss of a second genome part. However, recent findings arising mainly from the determination of the nucleotide sequences of virus RNAs have started to erode this simple position and there are theoretical grounds for questioning the inflexibility of the idea. These findings and speculations are discussed largely in the light of the impending adoption of the family taxon into the classification of the majority of plant viruses.

The taxonomic relevance of genome division

Divided genomes are relatively common among plant viruses. Of the 30 groups or families of RNA-containing plant viruses described in the Fifth Report of the International Committee on Taxonomy of Viruses (ICTV) [6], 14 contain viruses with multipartite genomes. There are several possible evolutionary advantages to a virus in the possession of a divided genome [9, 21], but whatever the actual reasons for its success during natural selection, the character appears to have proved advantageous a number of times during virus evolution. Genome subdivision has also been used to categorize RNA-containing plant viruses. It has been one of a few key distinguishing features in virus taxonomy [4,19, 20] and the recent series entitled *The Plant Viruses* [5] has been organized into separate volumes on the basis of particle shape and genome subdivision. Similarly, although arguing persuasively for a polythetic definition of virus species, Van Regenmortel [26] clustered genera together largely on the basis of the number of nucleic acid species in the virus genome.

The separate RNA species in a virus genome can be thought of as analogous to the separate chromosomes that comprise the genome of a eukaryotic organism. This analogy is reinforced by the use of pseudo-recombination experiments to locate the determinants of phenotypic characters in the different nucleic acid species of the virus genome; typically, plants are inoculated with heterologous mixtures of RNA species and the distribution of characters between the different resulting pseudo-recombinant progeny is assessed [14]. This analogy between genome parts and chromosomes perhaps contributes to the feeling that genome division is a character of fundamental taxonomic importance. However, some well-established groups of plant viruses contain members with different numbers of genome parts. Reoviruses that infect plants have genomes of either 10 (genus *Fijivirus*) or 12 (genus *Phytoreovirus*) dsRNA species and geminiviruses have genomes of either one (e.g., maize streak virus) or two (e.g., African cassava mosaic virus) ssDNA species.

Recently a further example has been found of viruses that are strong candidates for being grouped together but which have either monopartite or bipartite genomes. Barley yellow mosaic virus (BaYMV) is transmitted by the fungus *Polymyxa graminis*; it forms filamentous particles and cytoplasmic, cylindrical (pinwheel) inclusions that resemble those of aphid-transmitted potyviruses but whereas they have monopartite genomes, BaYMV has a bipartite genome. Because of this Usugi et al. [25] suggested that BaYMV and similar viruses be placed in a separate virus group for which they suggested the name 'bymovirus'. However, nucleotide sequences in the genome RNA of BaYMV show marked similarities with sequences of potyvirus RNA [16, 17]. In the Fifth ICTV Report [6] BaYMV and similar viruses are described as 'possible members' of the potyvirus group.

The recent decision by the Potyvirus Study Group of the Plant Virus Subcommittee of ICTV to propose that BaYMV and other fungus-transmitted, potyvirus-like viruses and the definitive aphid-transmitted potyviruses each comprise a genus in the family *Potyviridae* [2] provides a solution to this taxonomic problem. Thus viruses in different genera of the family *Potyviridae* can have different numbers of genome parts. This arrangement is mirrored by both the family *Reoviridae*, in which the genera have 10 (*Orthoreovirus, Orbivirus, Cypovirus* and *Fijivirus*), 11 (*Rotavirus, Aquareovirus*) or 12 (*Coltivirus, Phytoreovirus*) genome segments, and the proposed family

Geminiviridae (G.P. Martelli, pers. comm.) in which the viruses with monopartite or bipartite genomes are placed in different genera. Of course families can comprise genera containing viruses all with the same number of genome parts. For example, the family grouping of viruses with tripartite genomes that resemble bromoviruses has long been thought of as a natural and logical arrangement [27]. However, genome division should perhaps be considered as essentially a generic character which may apply to a family, but which does not necessarily do so. There should be no compunction in classifying viruses with differently subdivided genomes into the same family.

The definition of a virus genome

If the subdivision of a genome is to be used as a taxonomic criterion then it is important to consider the question 'what constitutes the genome of a virus?'. The reasonable answer would seem to be that the genome is the nucleic acid comprising the genotype that specifies the phenotype of the virus. Thus the genome is the nucleic acid that encodes or otherwise specifies those functions needed to infect a plant, to induce the changes in the host characteristic of the virus disease, and to replicate the virus to produce progeny which can repeat the virus infection cycle. Genomes of the great majority of (+) sense RNA viruses are a linear array of genes (albeit often overlapping) in a single piece of nucleic acid. There is little problem in defining what constitutes such genomes. However, when the genome is divided the definition becomes more difficult. Thus, for example, the essential functions of comoviruses and nepoviruses are divided between a larger RNA component, which encodes functions such as helicase, protease, and polymerase activity and a smaller RNA component, which encodes the particle coat protein as well as a protein that mediates movement between host cells. The larger RNA on its own can infect and multiply in isolated protoplasts but it is not encapsidated and cannot spread from cell to cell in whole plants unless the smaller RNA is present [10,23]. Tobacco rattle virus differs somewhat in that its larger RNA-1 can, on its own, infect both protoplasts and whole plants because it can move from cell to cell presumably because it encodes a transport protein. However, like the larger RNA molecules of comoviruses and nepoviruses, it does not encode coat protein and the infection therefore does not result in the synthesis of virus particles [12] and the virus is not transmitted by its nematode vector. In contrast, the larger RNA of red clover necrotic mosaic dianthovirus encodes a coat protein but not a movement protein; it can therefore infect protoplasts to produce virus particles but these cannot spread from cell to cell to infect plants [22].

In these examples it is clear that both RNA species are genome parts because both are required for systemic infection of plants and for the production of virus particles. Thus the larger RNA species of the bipartite genomes are deficient in that one or more of the essential viral characteristics are lacking. Not so clear-cut are examples of viruses which have lost a transmission character. Some isolates of wound tumor virus which can multiply in plants lack two of the normal complement of 12 dsRNA genome segments and cannot infect cells of the leafhopper vector and are therefore not transmitted by it. Presumably the genome segments lost during repeated passage in plant hosts encode functions essential for the infection of the insect vector [11]. The transmission by the fungus *Polymyxa betae* of beet necrotic yellow vein virus is strongly, though not absolutely, dependent on the presence in the virus culture of a small RNA species that is not required for the virus to infect and complete a normal infection cycle in manually inoculated *Chenopodium quinoa* [24]. It is not clear whether this RNA should be regarded as a satellite [7] or a genome part. Thus what constitutes the virus genome will depend on the particular biological test used to assay the virus.

A possible example of the opposite trend is what appears to have been the acquisition of a genome part during the evolution of pea enation mosaic virus (PEMV). PEMV resembles luteoviruses in its particle morphology and in being transmitted by aphids in a persistent, circulative fashion. However, unlike luteoviruses, PEMV is mechanically transmissible, invades mesophyll cells in infected plants and has a bipartite genome [15]. Recently, the results of nucleotide sequencing [3] have shown that PEMV does indeed resemble a luteovirus because the arrangement of the genes in the larger RNA of PEMV, and parts of their sequences, closely resemble those of luteovirus genes. Thus it may well be that PEMV evolved from a luteovirus by the acquisition of a second genome nucleic acid that encodes transport into and between mesophyll cells (and thereby acquired the ability to infect a plant following manual inoculation).

It would seem that the most sensible classification to accommodate these relationships between luteoviruses and PEMV would be the same as that adopted for the potyviruses and bymoviruses; the viruses should be classified in distinct genera of the same family. However, classification of luteoviruses is further complicated by strong resemblances between luteovirus polymerase genes and those of either carmoviruses or sobemoviruses [18]. Nevertheless, the classification into families of bymoviruses with potyviruses and possibly of PEMV with luteoviruses shows that whereas the subdivision of a genome or the adoption of extra genome parts might obscure a relationship, consideration of the molecular organization of the virus genomes can give a useful basis for virus classification into families.

From these examples it can be seen that even when adopting a pragmatic definition of a virus genome it may not be clear how many genome parts a virus has or how much evolutionary distance exists between viruses with different numbers of genome parts. This uncertainty is further compounded by the probability that genes can be exchanged between different nucleic acid species in the genome. This is vividly illustrated by tobacco rattle virus in which the same genes are present in both genome RNA species of one strain of the virus [1].

Conclusions

Recent findings have shown that viruses with different numbers of genome parts can seem quite closely related and therefore should be classified together. At least for potyviruses and similar viruses this can be achieved by clustering genera into the family *Potyviridae*. Genome subdivision therefore seems to be a genus-level criterion. However, because RNA viruses seem to vary quite rapidly and evolve by modular recombination as well as by progressive mutation [8], it is quite likely that exceptions even to this simplification will be found. It should be emphasized that no phylogenetic significance is intended by applying this particular criterion at the genus level. That it gives useful results is its sole recommendation.

References

1. Angenent GC, Linthorst HJM, Van Belkum AF, Cornelissen BJC, Bol JF (1986) RNA 2 of tobacco rattle virus strain TCM encodes an unexpected gene. Nucleic Acids Res 14: 4673–4682
2. Barnett OW (1991) *Potyviridae*, a proposed family of plant viruses. Arch Virol 118: 139–141
3. Demler SA, De Zoeten GA (1991) The nucleotide sequence and luteovirus-like nature of RNA 1 of an aphid non-transmissible strain of pea enation mosaic virus. J Gen Virol 72: 1819–1834
4. Fenner F (1976) Classification and nomenclature of viruses. Second report of the International Committee on Taxonomy of Viruses. Intervirology 7: 1–115
5. Fraenkel-Corat H, Wagner RR (1985 on) (eds) The plant viruses, vols 1–4. Plenum, New York
6. Francki RIB, Fauquet CM, Knudson DL, Brown F (eds) (1991) Classification and nomenclature of viruses. Fifth report of the International Committee on Taxonomy of Viruses. Springer, Wien New York (Arch Virol [Suppl] 2)
7. Fritsch C, Mayo MA (1989) Satellites of plant viruses. In: Mandahar CL (ed) Plant viruses, vol I, structure and replication. CRC Press, Boca Raton, pp 289–321
8. Gibbs AJ (1987) Molecular evolution of viruses; 'trees', 'clocks' and ' modules'. J Cell Sci [Suppl] 7: 319–337
9. Goldbach RW (1986) Molecular evolution of plant RNA viruses. Annu Rev Phytopathol 24: 289–310
10. Goldbach R, Rezelman G, Van Kammen A (1980) Independent replication and expression of B-component RNA of cowpea mosaic virus. Nature 286: 297–300
11. Harrison BD, Murant AF (1984) Involvement of virus-coded proteins in transmission of plant viruses by vectors. In: Mayo MA, Harrap KA (eds) Vectors in virus biology. Academic Press, London, pp 1–36
12. Harrison BD, Robinson DJ (1986) Tobraviruses. In: Van Regenmortel MHV, Fraenkel-Conrat H (eds) The plant viruses, vol 2. Plenum, New York, pp 339–369
13. Harrison BD, Finch JT, Gibbs AJ, Hollings M, Shepherd RJ, Valenta V, Wetter C (1971) Sixteen groups of plant viruses. Virology 45: 356–363
14. Harrison BD, Murant AF, Mayo MA, Roberts IM (1974) Distribution of determinants for symptom production, host range and nematode transmissibility between the two RNA components of raspberry ringspot virus. J Gen Virol 22: 233–247
15. Hull R, Lane LC (1973) The unusual nature of the components of a strain of pea enation mosaic virus. Virology 55: 1–13

16. Kashiwazaki S, Minobe Y, Omura T, Hibino H (1990) Nucleotide sequence of barley yellow mosaic virus RNA 1: a close evolutionary relationship with potyviruses. J Gen Virol 71: 2781–2790
17. Kashiwazaki S, Minobe Y, Hibino H (1991) Nucleotide sequence of barley yellow mosaic virus RNA 2. J Gen Virol 72: 995–999
18. Martin RR, Keese PK, Young MJ, Waterhouse PM, Gerlach WL (1990) Evolution and molecular biology of luteoviruses. Annu Rev Phytopathol 28: 341–363
19. Matthews REF (1979) Classification and nomenclature of viruses. Third report of the International Committee on Taxonomy of Viruses. Intervirology 12: 133–296
20. Matthews REF (1981) Classification and nomenclature of viruses. Fourth report of the International Committee on Taxonomy of Viruses. Intervirology 17: 1–199
21. Mayo MA (1987) A comparison of the translation strategies used by bipartite genome RNA plant viruses. In: Rowlands DJ, Mayo MA, Mahy BWJ (eds) The molecular biology of the positive strand RNA viruses. Academic Press, London, pp 177–205
22. Osman TAM, Buck KW (1987) Replication of red clover necrotic mosaic virus RNA in cowpea protoplasts: RNA 1 replicates independently of RNA 2. J Gen Virol 68: 289–296
23. Robinson DJ, Barker H, Harrison BD, Mayo MA (1980) Replication of RNA-1 of tomato black ring virus independently of RNA-2. J Gen Virol 51:317–326
24. Tamada T, Abe H (1989) Evidence that beet necrotic yellow vein virus RNA-4 is essential for efficient transmission by the fungus *Polymyxa betae*. J Gen Virol 70: 3391–3398
25. Usugi T, Kashiwazaki S, Omura T, Tsuchizaki T (1989) Some properties of nucleic acids and coat proteins of soil-borne filamentous viruses. Ann Phytopathol Soc Jpn 55: 26–31
26. Van Regenmortel MHV (1989) Applying the species concept to plant viruses. Arch Virol 104: 1–17
27. Van Vloten-Doting L, Francki RIB, Fulton RW, Kaper JM, Lane LC (1981) *Tricornaviridae* – a proposed family of plant viruses with tripartite, single-stranded RNA genomes. Intervirology 15: 198–203

Author's address: M. A. Mayo, Scottish Crop Research Institute, Invergowrie, Dundee DD2 5DA, Scotland.

Arch Virol (1992) [Suppl 5]: 283–297

Sequence data as the major criterion for potyvirus classification

C.W. Ward, **N.M. McKern**, **M.J. Frenkel**, and **D.D. Shukla**

CSIRO, Division of Biomolecular Engineering, Parkville, Victoria, Australia

Summary. Recent knowledge of the structure of the potyvirus particle and its components appears to have resolved what was thought to be an intractable problem of plant virology. This review describes how coat-protein and gene sequence data can be used to provide an hierarchical classification of potyviruses. This classification puts the aphid and non-aphid-transmitted potyviruses into a single family, divides this family into four genera that correspond to the four modes of vector transmission, discriminates distinct potyvirus species from strains, and provides a basis for the formation of subgroups composed of closely related species within a genus.

Introduction

The development of effective control strategies against plant viruses is dependent on the availability of reliable methods of identification and detection. To date this has not seemed possible for the potyvirus group of plant viruses, because of its size, complexity, and immense variation. This single group of plant viruses is the most rapidly growing and by far the largest of the 34 different plant virus groups or families [34]. It has been pointed out repeatedly by taxonomists and reviewers that the taxonomy of this group is in a very unsatisfactory state and that successful resolution of potyvirus identification is a major challenge for plant virologists [13, 14, 19, 20, 41]. In a detailed review Francki et al. [14] pointed out the importance of defining those characters that are required for assigning viruses to the potyvirus group and those characters that distinguish distinct potyviruses from strains.

With regard to the first question, 'what defines a potyvirus?', the key characteristics considered in the past were particle morphology, cytopathology, and transmission mechanism. To these should be added molecular characteristics such as genome structure and organization, coat-protein sequence, and serology. Genotype should be the ultimate criterion on which to assign viruses to their groups. The nature of the viral genome (RNA or DNA), its complete nucleotide sequence (indicating the number, nature, and order of coding regions), and the mechanism of virus replication and assembly should collectively permit unambiguous assessment of plant virus group status. The next most valuable group-specific criteria are partial nucleotide sequences (particularly for the coat-protein coding region) or coat-protein sequences, as the coat protein has an amino acid composition that is characteristic of the group [12] and is the only virus product which shows little sequence identity with the corresponding protein of other virus groups [9].

The value of phenotypic characteristics (such as particle morphology, cytoplasmic inclusion morphology, transmission mechanism, or coat-protein properties) as group-specific parameters depends on how accurately they reflect the genotype and whether they are unique to the potyvirus group. The capacity to induce pinwheel cylindrical cytoplasmic inclusions [for a review, see 30] seems to be a unique phenotypic characteristic of potyviruses [59] and the appearance of these pinwheel cylindrical cytoplasmic inclusions is now recognized by the International Committee for the Taxonomy of Viruses [32] as a diagnostic feature of the potyvirus group. Filamentous particle morphology is also an important trait [6]. Most potyviruses are more flexuous than potex- and carlavirus particles but less flexuous than closteroviruses, and generally appear narrower in diameter than members of the other filamentous plant virus groups [for reviews, see 14, 20]. Aphid transmission is a third phenotypic characteristic that for a long time was considered an essential criterion for potyvirus group membership [20, 34]. However, recent molecular data has shown that the non-aphid-transmitted possible members have a number of characteristics which show they should be grouped with the aphid-transmitted potyviruses. Shukla et al. [59], using a broadly cross-reactive antiserum targeted to the conserved core region of the potyvirus coat protein, showed that wheat streak mosaic virus (mite vector) and sweet potato mild mottle virus (whitefly vector) have coat proteins that share epitopes with those of definitive potyviruses. This was the first evidence for sequence similarities between the coat proteins of these possible potyviruses and definitive members of the group. Subsequently Stanarius et al. [65] demonstrated serological cross-reactions between the aphid-transmitted potyviruses bean yellow mosaic virus (BYMV) and turnip mosaic virus and the fungal-transmitted barley mild mosaic virus. More recently, nucleotide sequences for the mite-transmitted wheat streak mosaic virus (WSMV) [44] and the fungal-transmitted barley yellow mosaic virus (BaYMV) [26] revealed strong similarities with the aphid-transmitted potyviruses including (i) the 3′ location of the coat-protein gene; (ii) the production of viral proteins by proteolytic processing from a large precursor; (iii) significant sequence identity (18–22%) with the coat proteins of other potyviruses [72] compared to very low sequence identity with the coat proteins of potexviruses, and no significant matches with the capsid proteins of other rod-shaped plant viruses; (iv) similar surface location of N- and C-terminal regions of the coat proteins; (v) similar hydrophilicity profiles for the BaYMV, WSMV, and potato virus Y (PVY) coat proteins; and (*vi*) similar morphology, size (226–227 residues) and sequence identity (22–27%) of the trypsin-treated core peptides. These genotypic and phenotypic characteristics, coupled with the ability of these viruses to form typical pinwheel cytoplasmic inclusions, provide strong evidence that these non-aphid-transmitted viruses are closely related to the potyviruses. The presence of a bipartite genome and consequently two modal lengths of virus particles should not disqualify the fungal-transmitted viruses from being grouped with other potyviruses.

Surprisingly, the recent molecular data has not changed the potyvirus group status of individual viruses apart from confirming the membership of the non-aphid-transmitted viruses, and providing the basis for the development of group-specific serology [25, 59]. In contrast, the molecular studies have had a dramatic effect on the second problem of potyvirus taxonomy, the development of criteria to distinguish "viruses" from "strains." These new developments and their impact on this very important aspect of potyvirus taxonomy are discussed in this paper. There are more detailed reviews available [53, 54, 72].

Traditional approaches to discriminating
viruses and strains

Characteristics used for distinguishing viruses and strains in the past were: host range, symptomatology, cross-protection, morphology of cytoplasmic inclusions, and serology. While these characteristics have played a significant role in the delineation of many potyviruses and their strains, they have not provided a workable solution for the potyvirus group as a whole because of the large size and the extensive biological and antigenic variation within the group and because of inadequacies in these classical methodologies.

Reliance on host range and symptomatology has created confusion in the identification of potyviruses infecting the Leguminosae [1, 5, 7, 24, 29, 69] and Gramineae [60]. Recently symptom phenotypes of tobacco mosaic virus [27] and cucumber mosaic virus [51] were dramatically altered by single point mutations indicating symptomatology is not a reliable marker of genetic relatedness.

Cross-protection tests were originally given considerable weight in establishing the virus/strain status of many plant viruses but doubt about the value of cross-protection arose when data emerged that conflicted with assignments based on other properties [33]. Some of the conflicting results in cross-protection experiments with potyviruses may be due to technical problems [54] but many may be attributed to misidentification of the viruses and strains used. For example, some strains of the sugarcane mosaic virus (SCMV) subgroup are known to cross protect while others such as strains A and B of maize dwarf mosaic virus (MDMV) do not [45,50,68]. Since the strains of SCMV were allocated to four distinct potyviruses instead of one, the reported cross-protection results with these strains conform to their assignments as four distinct viruses [60]. Other viruses such as soybean mosaic virus (SMV) [52], BYMV [18, 66], and bean common mosaic virus (BCMV) [63] each consist of more than one distinct virus, so much of the published information on unexpected cross-protection involving these viruses needs re-examination, and cross-protection may prove to be very useful as a taxonomic criterion for potyviruses once the assignment of the viruses and strains compared to date has been corrected.

Cytoplasmic inclusion (CI) morphology has no value in establishing hierarchical arrangements within the potyvirus group, since the four inclusion subgroupings [11] do not correlate with the four transmission mechanisms nor with major subgroupings by sequence identity. However it does have value in the identification of particular potyviruses and in signalling the existence of more than one virus in certain disease syndromes.

Finally, serology, which successfully differentiates viruses in other plant virus groups, has proven unsatisfactory when applied to potyviruses where the serological relationships between related strains and distinct members are complex and inconsistent [3, 5, 13, 20, 22, 42]. The location of the virus-specific, immunodominant coat-protein epitopes on the virion surface [57, 64] and the presence of highly conserved sequences in the coat-protein cores [53] suggest that much of the contradictory information on serological relationships among potyviruses can be attributed to the presence of variable proportions of cross-reacting core-targeted antibodies. Shukla et al. [58] used this information to develop a simple affinity chromatographic procedure to obtain virus-specific antibodies from polyclonal antisera by the selective removal of these cross-reacting antibodies by cross absorption. They used this strategy to show that 17 potyvirus strains infecting Gramineae belonged to four distinct potyviruses, johnsongrass mosaic virus (JGMV), sorghum mosaic virus (SrMV), MDMV and SCMV [60].

Sequence relationships between distinct potyviruses and strains

Coat-protein amino acid sequence data can be used to identify and differentiate distinct potyviruses and their strains [52, 53]. Analysis of the 136 possible pairings of the complete coat-protein amino acid sequences from 17 strains of eight distinct potyviruses revealed a bimodal distribution of sequence identity (Fig. 1). In this analysis the sequence identity between distinct members ranged from 38 to 71% (average 54%) while that between strains of the one virus ranged from 90 to 99% (average 95%). The only exceptions to this pattern were pepper mottle virus (PepMoV), which was as closely related to the four strains of PVY as were known strains of other potyviruses to each other, and the two SbMV isolates which were as different from each other as they were from other distinct members (Figs. 2 and 3). The very close structural relationship be-

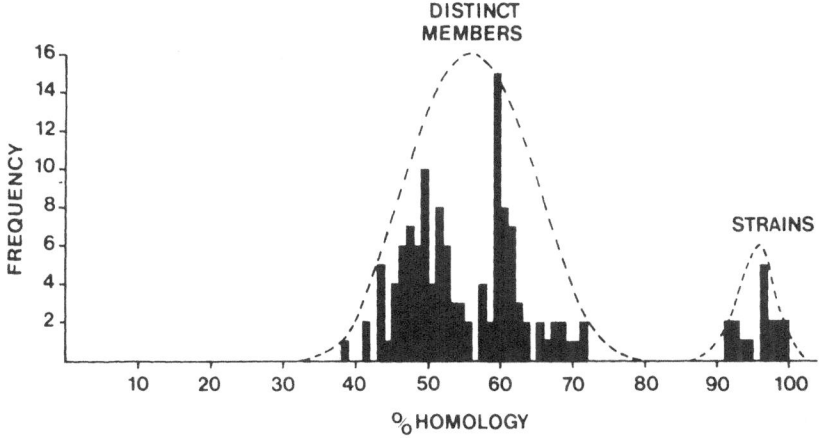

Fig. 1. Frequency distribution of amino acid sequence homologies for the coat proteins from 17 strains of eight distinct potyviruses. Reproduced from [52]

tween the coat proteins of PepMoV and the four strains of PVY (Fig. 2) compared to the much lower sequence identity between distinct members of the potyvirus group suggests that PepMoV could be considered a strain of PVY [55]. This analysis also indicates that SbMV-N and SbMV-V, with a sequence identity of 58%, should be considered as two distinct potyviruses rather than strains of SbMV [52].

This clear demarcation of sequence identity between distinct potyviruses and strains was not consistent with the "continuum" hypothesis [5, 32] proposed to explain the unsatisfactory taxonomy of potyviruses. The sequence identity between distinct members was little affected by the choice of strain used to make the comparison, indicating that, at least for the viruses examined, the boundaries between the peripheral virus strains were clear [52].

Since that report [52] coat-protein sequences of a further 23 aphid-transmitted potyviruses have been determined [72]. This new sequence data conforms to the same pattern and shows that:

 i PPV-R is a strain of plum pox virus (PPV);
 ii PVY-I and PVY-N are strains of PVY;
 iii BYMV-GDD, BYMV-CS and BYMV-S are strains of bean yellow mosaic virus (BYMV) but BYMV-30 is a strain of clover yellow vein virus (ClYVV) [66, 70];
 iv watermelon mosaic virus 1 (WMV 1) is not related to WMV 2 but is a strain of papaya ringspot virus (PRSV) as suggested by other data [48];
 v WMV 2 and SMV-N are closely related and appear to be strains of the same virus;
 vi zucchini yellow mosaic virus-C (ZYMV-C); ZYMV-NAT and ZYMV-F are all strains of ZYMV and are distinct from PRSV (WMV 1) and WMV 2;

vii in agreement with the recent serological findings [60], MDMV-A, MDMV-B, MDMV-O, and SCMV-H are strains of four distinct viruses named MDMV, SCMV, JGMV, and SrMV, respectively;

viii MDMV-O, MDMV-KS1, and SCMV-JG are all strains of JGMV;

ix MDMV-B, SCMV-SC, SCMV-BC, and SCMV-Sabi are strains of SCMV.

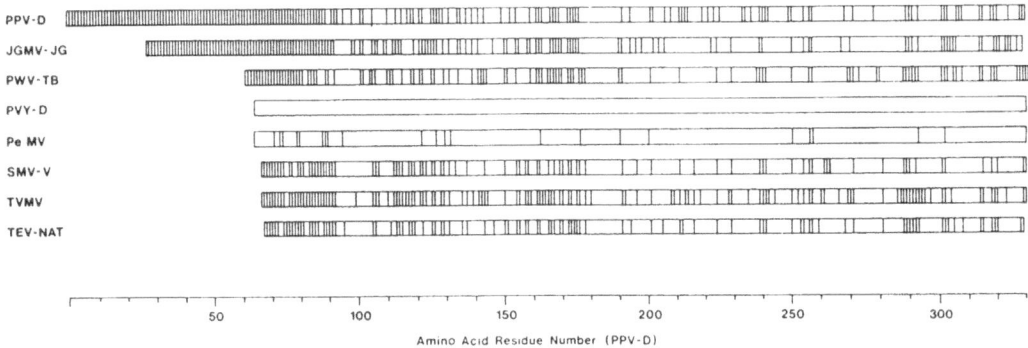

Fig. 2. Schematic diagram showing the location of sequence differences between distinct members of the genus *Potyvirus*. The sequences are compared with PVY-D, the type member. Reproduced from [52]. PeMV pepper mottle virus

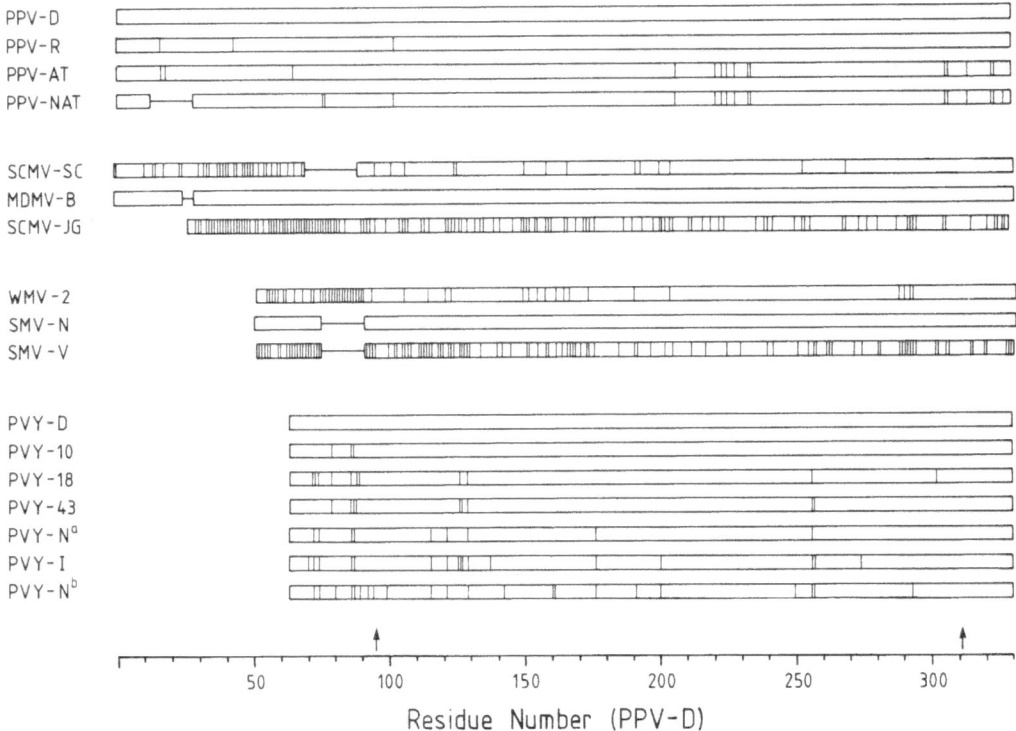

Fig. 3. Schematic diagram showing the location of sequence differences between reported strains of PPV, SCMV, SbMV and WMV2. Arrows delimit the core protein region. SCMV-JG is no longer considered a strain of SCMV, and is now recognized as a distinct potyvirus JGMV-JG

The sequence data for the non-aphid-transmitted potyviruses WSMV [44] and BaYMV [26] revealed that vector specificity correlates with a third level of sequence diversity. Figure 4 summarizes the evolutionary relationship among potyviruses based on coat-protein sequences. Here the potyvirus group is considered a family as suggested by Matthews [35]; vector transmission (and the highest level of sequence diversity) defines four genera, as suggested for geminiviruses [35]; distinct potyviruses (the middle level of sequence diver-

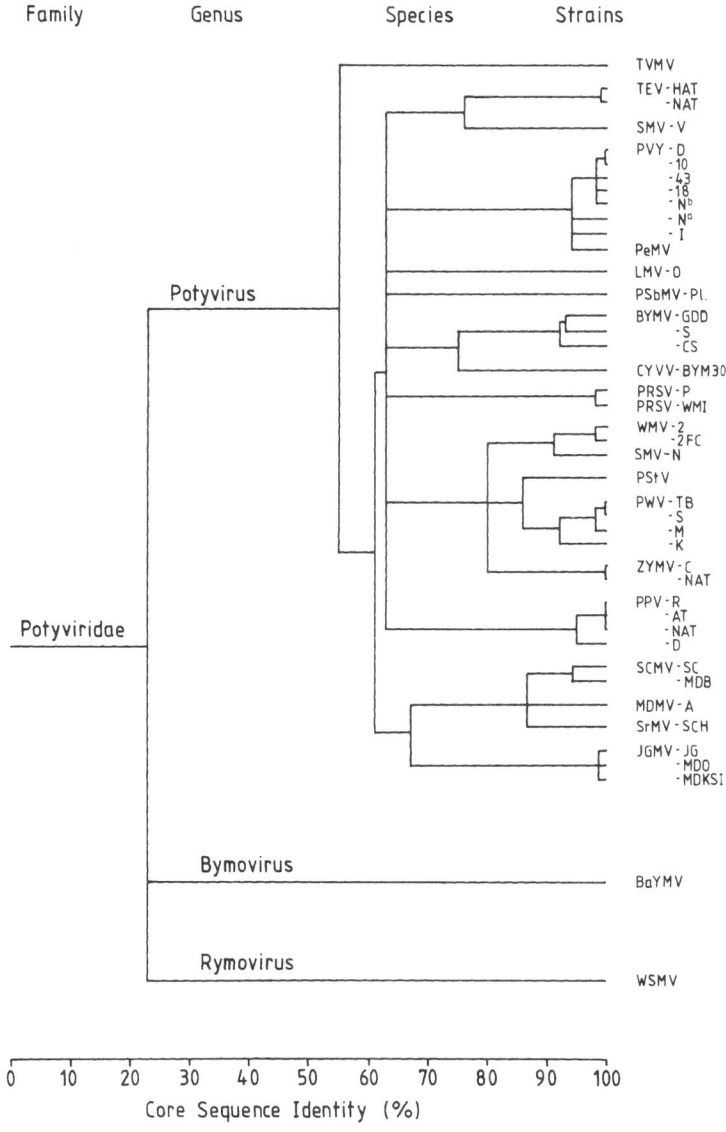

Fig. 4. Family tree showing phylogentic relationships within the family *Potyviridae*. The dendrogram reflects sequence identities between the core regions (D_{33}-R_{248} in PVY) of the potyvirus coat protein. The PStV sequence is from [38]; all other sequence sources and virus acronyms are listed in [72]. The fourth possible genus *Ipomovirus* (whitefly vector) is not shown in this dendrogram as no sequence data is available for the coat protein of the type member of this genus, sweet potato mild mottle virus. PeMV pepper mottle virus

sity) correspond to species; and variants with the lowest level of sequence diversity correspond to strains.

The sequence data (Fig. 4) has also revealed the existence of subgroups of distinct potyviruses, composed of closely related species within a genus such as:

i   SbMV-V/ tobacco etch virus (TEV),
ii BYMV/ClYVV,
iii ZYMV/WMV 2/ peanut stripe virus (PStV)[38]/passionfruit woodiness virus (PWV),
iv SCMV/MDMV/SrMV/JGMV.

How valid are such assignments based on the coat-protein core region, when it accounts for only 7% of the viral genome? Complete sequences are available for TEV [2], tobacco vein mottling virus (TVMV) [8], PVY [49], and three strains of PPV [28, 31, 67]. Pairwise comparisons, as illustrated in Fig. 5, show that strains exhibit high sequence identity irrespective of the gene product being considered while distinct viruses have significantly lower levels of identity between their equivalent gene products. Thus, conclusions about virus/strain status from coat-protein sequences are valid since coat-protein sequence identities are representative of sequence identities of the whole genome. In addition, an optimal protein sequence alignment of the polyproteins of TEV [2], TVMV [8], PVY [49], and PPV-R [28] revealed that the first protein (P1), the third protein (P3), and the N-terminal region of the coat

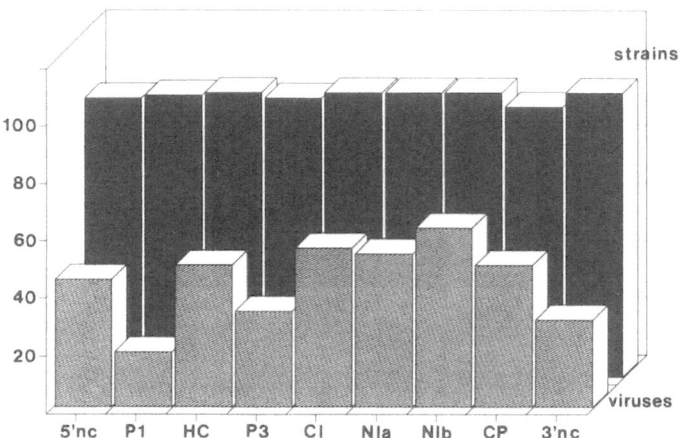

Fig. 5. Sequence identities between different regions of the genomes of distinct viruses and strains. The virus comparisons are PVY-N [49] and PPV-NAT [31]; the strain comparisons are PPV-NAT [31] and PPV-R [28]. The 5' and 3' non-coding regions are nucleotide sequence identities. The coding region products are amino acid sequence identities. The junction between P1 and HC is not known and is taken as the RG bond in the conserved FIVRG [72]

protein are the most variable regions in the viral genome [9, 62] as shown in Fig. 6. For this reason we have opted to use the coat-protein core region (equivalent to Asp_{33}–Arg_{248} in PVY) not the total coat protein, to establish taxonomic relationships within the potyvirus family (Fig. 4), because it more closely reflects the bulk of the coding and non-coding regions of the potyvirus genome.

The available data indicate that in most instances gene sequences or coat-protein sequences can readily establish the virus/strain status of a particular isolate because of the significant difference in the sequence identities found between viruses and strains. Because of this distinction, simpler techniques such as nucleic acid hybridization [21], HPLC peptide profiling [37, 39, 40, 56, 63] or N-terminal targeted serology [57, 58, 60] can be used as cruder measures of genetic relatedness. For most virus isolates, such approaches will be sufficient to decide if they are distinct viruses or related strains.

However, there are some instances where a more detailed analysis of the sequence data is required and where other characteristics will need to be considered. Major changes in the N-terminal region of related strains could be caused by cleavage site mutations, major deletions, or frameshift mutations [53]. N-terminal deletions have been reported for PPV-NAT [31] and WMV 2/SMV-N [15] but no evidence for the other two types of variation has yet been

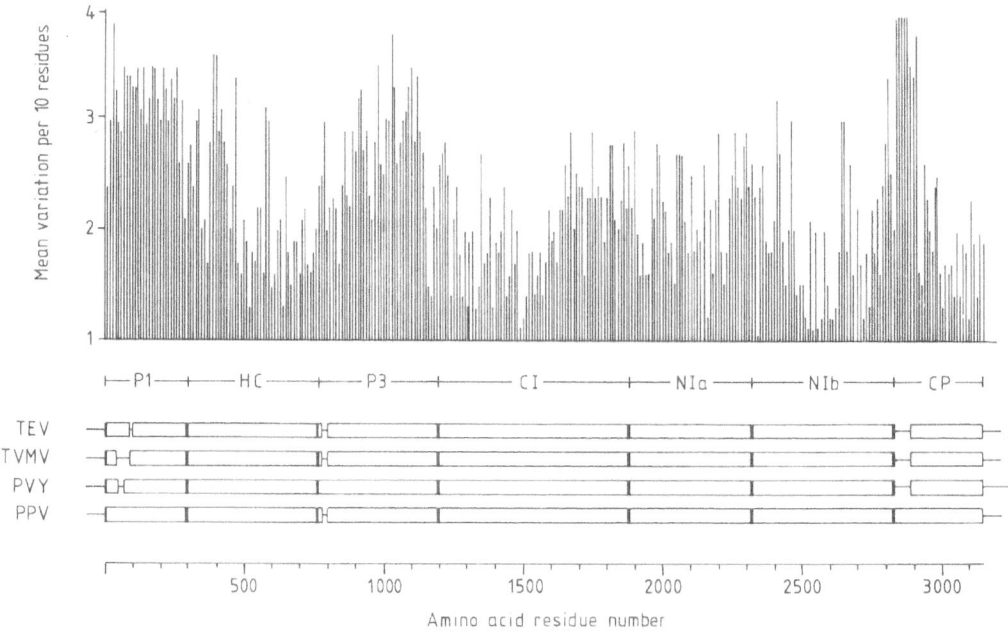

Fig. 6. Potyvirus genome variation. Each data entry represents the number of different amino acid residues found at each position in the polyprotein sequences of TEV-HAT [2], TVMV [8], PVY-N [49], and PPV-R [28] averaged for groups of 10 consecutive residues and plotted against the number of the first residue in each 10 residue sequence. Reproduced from [62]

reported. The recent data for the SCMV-SC/MDMV-B strain pair [16] and the MDMV/SCMV/SrMV subgroup [23] suggest other mechanisms of variation such as gene duplication or recombination [17] are possible. In such cases, more detailed analysis of the coat-protein/gene sequences, including the sequence of the 3' non-coding region [15] and upstream regions of the genome should allow the degree of genetic relatedness to be assessed more accurately.

The PepMoV/PVY relationship is another case where additional information is required. The coat-protein and 3' non-coding sequence data indicate they are related strains [10, 49, 55, 71]; their differences in host range and symptomatology are within the accepted limits for strains [43] and their CI proteins have similar gross morphology [46]. On the other hand the differential serological reactions of their coat proteins and CI proteins in agar gel diffusion tests; the lack of cross-protection; differences in the fine morphology of the CI proteins; and differential symptoms on tabasco pepper have been used to support the view that they are distinct viruses [43, 46, 47]. The availability of cDNA clones covering the complete genome of PVY [49] could be used to ascertain whether the high level of sequence homology seen in the coat protein and 3' non-coding regions is maintained over the entire genome, or whether some of the contrasting properties of PepMoV and PVY are due to a recombination event where the 5' end of their genomes are substantially different.

While sequence data are an excellent indication of genetic relatedness there is going to be some difficulty in deciding whether it is more appropriate to consider a borderline isolate a closely related but distinct virus rather than a distantly related strain. The PepMoV/PVY relationship is one example where opinions differ as to the most appropriate classification. Another borderline example is WMV 2. Should it be considered a distinct virus that is closely related to SbMV or should it be considered an SbMV strain [15] in the same way that WMV 1 is accepted to be a strain of PRSV [48]. Viruses are evolving constantly during each round of infection and the time frame for such evolution and diversification is short for viruses with single stranded RNA genomes [4]. Thus one would expect examples to be found where raw sequence identity scores lie on the border between distinct viruses and strains. For these isolates, phenotypic properties such as disease symptoms, host range, vector specificity, and cross-protection will have to be considered. It is important however to understand the molecular basis of specific phenotypic characters since only those that genuinely reflect significant genotype differences will be of value. If a single point mutation can result in a dramatic change to a different phenotypic subset then that phenotypic parameter has no value in establishing hierarchical relationships within a plant virus group.

Conclusions

In a review of the historical development of general taxonomic principles, Mayr [36] drew attention to Darwin's comment that taxonomy reflects propinquity of descent and that all true classifications are genealogical. Mayr [36] also traced the changes in the criteria used for general taxonomy from the use of descriptive morphological characters to the application of biochemical techniques that characterize the variation and evolution of molecules. Plant virus taxonomy is going through similar phases of development from the initial reliance on morphological, biological, and serological properties to assignments based on coat-protein and nucleic acid sequences. Thus, of the physical, biological and chemical properties used to classify potyviruses, protein and genome sequence information should represent the ultimate criteria. It is encouraging that the molecular data has allowed the relationships between members of the potyvirus group to be so finely established and provided assignments that are in general agreement with cross-protection, and CI morphology data. Previous examination of biological, physical, and serological properties had led to the conclusion that an effective taxonomy of potyviruses may be impossible as, in some instances, a continuum of variants or strains seemed to exist [5, 20]. Molecular data has overcome these problems, offering a sound basis for establishing a workable taxonomy for potyviruses and providing an explanation for failed cross-protection data and inconsistent serology, which had plagued earlier attempts at classification.

Acknowledgements

We are grateful to J. M. Jilka (U.S.A.), K. H. Gough (Australia), J. A. Garcia (Spain), C. L. Niblett (U.S.A.), S. A. Tolin (U.S.A.), I. Uyeda (Japan) and Miss S. L. Tracy (Australia) for generously providing gene sequences of potyviruses and preprint of papers prior to publication.

References

1. Abu-Samah N, Randles JW (1983) A comparison of Australian bean yellow mosaic virus isolates using molecular hybridization analysis. Ann Appl Biol 103: 97–107
2. Allison RF, Johnston RE, Dougherty WG (1986) The nucleotide sequence of the coding region of tobacco etch virus genomic RNA: evidence for the synthesis of a single polyprotein. Virology 154: 9–20
3. Barnett OW, Randles JW, Burrows PM (1987) Relationships among Australian and North American isolates of the bean yellow mosaic potyvirus subgroup. Phytopathology 77: 791–799
4. Bishop DHL (1985) The genetic basis for describing viruses as species. Intervirology 24: 79–93
5. Bos L (1970) The identification of three new viruses isolated from *Wisteria* and *Pisum* in the Netherlands, and the problem of variation within the potato virus Y group. Neth J Plant Pathol 76: 8–46
6. Brandes J, Berks R (1965) Gross morphology and serology as a basis for classification of elongated plant viruses. Adv Virus Res 11: 1–24
7. Dijkstra J, Bos L, Bouwmeester HJ, Hadiastono T, Lohuis H (1987) Identification of blackeye cowpea mosaic virus from germ plasm of yard-long bean and from soybean, and the relationships between blackeye cowpea mosaic virus and cowpea aphid-borne mosaic virus. Neth J Plant Pathol 93: 115–133

8. Domier LL, Franklin KF, Shahabuddin M, Hellman GM, Overmeyer JH, Hiremath ST, Siaw MFE, Lomonossoff GP, Shaw JG, Rhoads RE (1986) The nucleotide sequence of tobacco vein mottling virus RNA. Nucleic Acids Res 14: 5417–5430

9. Domier LL, Shaw JG, Rhoads RE (1987) Poytviral proteins share amino acid sequence homology with picorna-, como-, and caulimoviral proteins. Virology 158: 20–27

10. Dougherty WG, Allison RF, Parks TD, Johnston RE, Feild MJ, Armstrong FB (1985) Nucleotide sequence at the 3′-terminus of pepper mottle virus genomic RNA: evidence for an alternative mode of capsid protein gene organization. Virology 146: 282–291

11. Edwardson JR, Christie RG, Ko NJ (1984) Potyvirus cylindrical inclusions – subdivision IV. Phytopathology 74: 1111–1114

12. Fauquet C, Dejardin J, Thouvenel JC (1986) Evidence that the amino acid composition of the particle proteins of plant viruses is characteristic of the virus group. 2. Discriminant analysis according to structural biological and classification properties of plant viruses. Intervirology 25: 190–200

13. Francki RIB (1983) Current problems in plant virus taxonomy. In: Matthews REF (ed) A critical appraisal of viral taxonomy. CRC Press, Boca Raton, pp 63–104

14. Francki RIB, Milne RG, Hatta T (1985) Atlas of plant viruses, vol 2. CRC Press, Boca Raton, pp 183–217

15. Frenkel MJ, Ward CW, Shukla DD (1989) The use of 3′-noncoding nucleotide sequence in the taxonomy of potyviruses: application to watermelon mosaic virus 2 and soybean mosaic virus. J Gen Virol 70: 2775–2783

16. Frenkel MJ, Jilka JM, McKern NM, Strike PM, Clark, JM, Shukla DD, Ward CW (1991) Unexpected sequence diversity in the amino terminal ends of the coat proteins of strains of sugarcane mosaic virus. J Gen Virol 72: 237–242

17. Goldbach R., Wellink J (1988) Evolution of plus-stranded RNA viruses. Intervirology 29: 260–267

18. Hammond J, Hammond RW (1989) Molecular cloning, sequencing, and expression in *Escherichia coli* of the bean yellow mosaic virus coat protein gene. J Gen Virol 70: 1961–1974

19. Harrison BD (1985) Usefulness and limitations of the species concept for plant viruses. Intervirology 25: 71–78

20. Hollings M, Brunt AA (1981) Potyviruses. In: Kurstak E (ed) Handbook of plant virus infections: comparative diagnosis. Elsevier/North Holland, Amsterdam, pp 731–807

21. Hull R (1984) Rapid diagnosis of plant virus infections by spot hybridization. Trends Biotechnol 2: 88–91

22. Jarjees MM, Uyemoto JK (1984) Serological relatedness of strains of maize dwarf mosaic and sugar-cane mosaic viruses as determined by microprecipitin and enzyme-linked immunosorbent assays. Ann Appl Biol 104: 497–502

23. Jilka J (1990) Cloning and characterization of the 3′-terminal regions of RNA from select strains of maize dwarf mosaic virus and sugarcane mosaic virus. PhD Thesis, University of Illinois, Urbana

24. Jones RT, Diachun S (1977) Serologically and biologically distinct bean yellow mosaic virus strains. Phytopathology 67: 831–838

25. Jordan R and Hammond J (1991) Comparison and differentiation of potyvirus isolates and identification of strain-, virus-, subgroup-specific and potyvirus group-common epitopes using monoclonal antibodies. J Gen Virol 72: 25–36

26. Kashiwazaki S, Minobe Y, Omura T, Hibino H, (1990) Nucleotide sequence of barley yellow mosaic virus RNA 1: a close evolutionary relationship with potyviruses. J Gen Virol 71: 2781–2790

27. Knorr DN, Dawson WO (1988) A point mutation in the tobacco mosaic virus capsid protein gene induces hyposensitivity in *Nicotiana sylvestris*. Proc Natl Acad Sci USA 85: 170–174

28. Lain S, Reichmann JI, Garcia JA (1989) The complete nucleotide sequence of plum pox potyvirus RNA. Virus Res 13: 157–172

29. Lana AF, Lohuis H, Bos L, Dijkstra J (1988) Relationships among strains of bean common mosaic virus and blackeye cowpea mosaic virus – members of the potyvirus group. Ann Appl Biol 113: 493–505

30. Lesemann D E (1988) Cytopathology. In: Milne RG (ed) The plant viruses, vol 4, the filamentous plant viruses. Plenum, New York, pp 179–235

31. Maiss E, Timpe U, Brisske A, Jelkmann W, Casper R, Himmler G, Mattanovich D, Katinger HWD (1989) The complete nucleotide sequence of plum pox virus RNA. J Gen Virol 70: 513-524

32. Matthews REF (1979) Classification and nomenclature of viruses. Third report of the International Committee of Taxonomy of Viruses. Intervirology 12: 131-296

33. Matthews REF (1981) Plant virology, 2nd edn. Academic Press, New York

34. Matthews REF (1982) Classification and nomenclature of viruses. Fourth report of the International Committee on Taxonomy of Viruses. Intervirology 17: 1–199

35. Matthews REF (1985) Viral taxonomy for the non-virologist. Annu Rev Microbial 39: 451–474

36. Mayr E (1982) The growth of biological thought: diversity, evolution and inheritance. Harvard University Press, Cambridge

37. McKern NM, Shukla DD, Barnett OW, Vetten HJ, Dijkstra J, Whittaker LA, Ward CW (1991) Coat protein properties suggest that azuki bean mosaic virus, blackeye cowpea mosaic virus, peanut stripe virus and three isolates from soybean are all strains of the same potyvirus. Intervirology 33: 121–134

38. McKern NM, Edskes HK, Ward CW, Strike PM, Barnett OW, Shukla DD (1991) Coat protein of potyviruses 7. Amino acid sequence of peanut stripe virus. Arch Virol 119: 25–35

39. McKern NM, Whittaker LA, Strike PM, Ford RE, Jensen SG, Shukla DD (1990) Coat protein properties indicate that maize dwarf mosaic virus KS1 is a strain of Johnsongrass mosaic virus. Phytopathology 80: 907–912

40. McKern NM, Shukla DD, Toler RW, Jensen SG, Tosic M, Ford RE, Leon O, Ward CW (1991) Confirmation that sugarcane mosaic subgroup consists of four distinct potyviruses by using peptide profiles of coat proteins. Phytopathology 81: 1025–1029

41. Milne RG (1988) Taxonomy of rod-shaped filamentous viruses. In: Milne RG (ed) The plant viruses, vol 4, the filamentous plant viruses. Plenum, New York, pp 3–50

42. Moghal SM, Franki RIB (1976) Towards a system for the identification and classification of potyviruses. I. Serology and amino acid composition of six distinct viruses. Virology 73: 350–362

43. Nelson MR, Wheeler RE (1978) Biological and serological characterisation and separation of potyviruses that infect peppers. Phytopathology 68: 979–984

44. Niblett CI, Zagula KR, Calvert LA, Kendall TL, Stark DM, Smith CE, Beachy RN, Lommel SA (1991) cDNA cloning and nucleotide sequence of the wheat streak mosaic virus capsid protein gene. J Gen Virol 72: 499–504

45. Paulsen AR, Sill WH (1970) Absence of cross-protection between maize dwarf mosaic virus strains A and B in grain sorghum. Plant Dis Rep 54: 627–629

46. Purcifull DE, Hiebert E, McDonald JG (1973) Immunochemical specificity of cytoplasmic inclusions induced by viruses of the potato Y group. Virology 55: 275–279

47. Purcifull DE, Zitter TA, Hiebert E (1975) Morphology, host range and serological relationships of pepper mottle virus. Phytopathology 65: 559–562

48. Quemada H, L'Hostis B, Gonsalves D, Reardon IM, Heinrikson R, Hienert EL, Sieu LC, Slightom JL (1990) The nucleotide sequences of the 3'-terminal regions of papaya ringspot strains W and P. J Gen Virol 71: 203–210

49. Robaglia C, Durand-Tardif M, Tronchet M, Boudazin G, Astier-Manifacier S, Casse-Delbart F (1989) Nucleotide sequence of potato virus Y (N strain) genomic RNA. J Gen Virol 70: 935–947

50. Shepherd RJ (1965) Properties of a mosaic virus of corn and Johnsongrass and its relation to the sugarcane mosaic virus. Phytopathology 55: 1250–1256

51. Shinlaku M, Palukaitis PF (1990) Genetic mapping of cucumber mosaic virus. In: Pirone TP, Shaw JG (eds) Viral genes and plant pathogenesis. Springer, Berlin Heidelberg New York Tokyo, pp 156–164

52. Shukla DD, Ward CW (1988) Amino acid sequence homology of coat proteins as a basis for identification and classification of the potyvirus group. J Gen Virol 69: 2703–2710

53. Shukla DD, Ward CW (1989) Structure of potyvirus coat proteins and its application in the taxonomy of the potyvirus group. Adv Virus Res 36: 273–314

54. Shukla DD, Ward CW (1989) Identification and classification of potyviruses on the basis of coat protein sequence data and serology. Arch Virol 106: 171–200

55. Shukla DD, Thomas JE, McKern NM, Tracy SL, Ward CW (1988) Coat protein of potyviruses. 4. Comparison of biological properties, serological relationships, and coat protein amino acid sequences of four strains of potato virus Y. Arch Virol 102: 207–219

56. Shukla DD, McKern NM, Gough KH, Tracy SL, Letho SG (1988) Differentiation of potyviruses and their strains by high-performance liquid chromatographic peptide profiling of coat proteins. J Gen Virol 69: 493–502

57. Shukla DD, Strike PM, Tracy SL, Gough KH, Ward CW (1988) The N and C termini of the coat proteins of potyviruses are surface-located and the N terminus contains the major virus-specific epitopes. J Gen Virol 69: 1497–1508

58. Shukla DD, Jilka K, Tosic M, Ford RE (1989) A novel approach to the serology of potyviruses involving affinity purified polyclonal antibodies directed towards virus-specific N termini of coat proteins. J Gen Virol 70: 13–23

59. Shukla DD, Ford RE, Tosic M, Jilka J, Ward CW (1989) Possible members of the potyvirus group transmitted by mites or whiteflies share epitopes with aphid-transmitted definitive members of the group. Arch Virol 105: 143–151

60. Shukla DD, Tosic M, Jilka J, Ford RE, Toler RW, Langham MAC (1989) Taxonomy of potyviruses infecting maize, sorghum and sugarcane in Australia and the United States as determined by reactivities of polyclonal antibodies directed towards virus-specific N-termini of coat proteins. Phytopathology 79: 223–229

61. Shukla DD, Tribbick G, Mason TJ, Hewish DR, Geysen HM, Ward CW (1989) Localisation of virus-specific and group-specific epitopes of plant viruses by systematic immunochemical analysis of overlapping peptide fragments. Proc Natl Acad Sci USA 86: 8192–8196

62. Shukla DD, Frenkel MJ, Ward CW (1991) Structure and function of the potyvirus genome with special reference to the coat protein coding region. Can J Plant Pathol 13: 178–191

63. Shukla DD, McKern NM, Ward CW, Ford RE (1991) Molecular parameters suggest that cowpea aphid-borne mosaic virus is a distinct potyvirus and that bean common mosaic virus consists of at least three distinct potyviruses. Phytopathology 81: 1166

64. Shukla DD, Lauricella R, Ward CW (1992) Serology of potyvirus: current problems and some solutions. In: Barnett OW (ed) Potyvirus taxonomy. Springer, Wien New York, pp 57–69 (Arch Virol [Suppl] 5)

65. Stanarius A, Proeseler G, Richter J (1989) Immunelektronenmikroskopische Untersuchungen zur serologischen Verwandtschaft des Gerstengelbmosaik-Virus (barley yellow mosaic virus) mit anderen gestreckten Viren. Arch Phytopathol Pflanzenschutz 25: 303–307

66. Takahashi T, Uyeda I, Shikata E (1990) Nucleotide sequence of the capsid protein gene of bean yellow mosaic virus chlorotic spot strain. J Fac Agric Hokkaido Univ Sapporo 64: 152–163
67. Teycheney PY, Tavert G, Delbos R, Ravelondandro M, Dunez J (1989) The complete nucleotide sequence of plum pox virus RNA (strain D). Nucleic Acids Res 23: 10115–10116
68. Tosic M (1981) Cross protection among some strains of sugarcane mosaic virus and maize dwarf mosaic virus. Agronomie 1: 83–85
69. Tsuchizaki T, Omura T (1987) Relationships among bean common mosaic virus, blackeye cowpea mosaic virus, azuki bean mosaic virus, and soybean mosaic virus. Ann Phytopathol Soc Jpn 53: 478–488
70. Uyeda I, Takahashi T, Shikata E (1991) Relatedness of nucleotide sequence of the 3'-terminal region of clover yellow vein potyvirus RNA to bean yellow mosaic potyvirus RNA. Intervirology 32: 234–245
71. Van der Vlugt R, Allefs S, De Haan P, Goldbach R (1989) Nucleotide sequence of the 3'-terminus of potato virus Y^N RNA. J Gen Virol 70: 229–233
72. Ward CW, Shukla DD (1991) Taxonomy of potyviruses: current problems and some solutions. Intervirology 32: 269–296

Authors' address: C.W. Ward, CSIRO, Division of Biomolecular Engineering, 343 Royal Parade, Parkville, Vic. 3052, Australia.

Arch Virol (1992) [Suppl 5]: 299–304

The recombinative nature of potyviruses: implications for setting up true phylogenetic taxonomy

R. Goldbach

Department of Virology, Agricultural University, Wageningen, The Netherlands

Summary. Sequence comparisons reveal that positive-strand RNA viruses not only evolve by divergence from common ancestors but also by interviral recombination. A considerable number of these viruses, exemplified by the family *Potyviridae*, can in fact, be regarded as successful products of a number of recombination events. It is concluded that the recombinative character of RNA viruses will hamper any attempt to set up a true phylogenetic taxonomy. It is advisable, therefore, to avoid the introduction of any taxon higher-than-family in virus taxonomy.

Introduction

The proposed family *Potyviridae* represents the largest taxonomic group of plant viruses. Potyviruses have flexuous, filamentous particles which contain a positive-stranded RNA genome. The potyviral genome is characterized by having a VPg (Viral Protein genome-linked) at the 5' terminus and a poly(A) tail at the 3' terminus. Nucleotide sequence determination of the genome of a growing number of potyviruses, including potato virus Y [17], plum pox virus [16], tobacco vein mottling virus (TVMV) [3], tobacco etch virus (TEV) [1], all of the genus *Potyvirus*, and barley yellow mosaic virus (BaYMV) [2, 12, 13], belonging to the genus *Bymovirus*, has enabled us to compare the potyviruses with other viruses in terms of RNA and protein sequences. Such comparisons have revealed a number of remarkable genetic relationships between the *Potyviridae* and other groups and families of plant- and animal-infecting viruses. This short review will focus on the interviral relationships elucidated so far and will discuss the (im)possibilities of using these relationships to create taxa that rank higher than the family level.

Position of potyviruses within the supergroup of picorna-like viruses

As discussed previously in a number of reviews [6–8, 20], a large number of plant- and animal-infecting viruses can be placed into a limited number of "supergroups" or "superfamilies" (i.e., informal clusters of virus groups and families that may be considered as higher taxa). Thus the potyviruses have been placed into the supergroup of "picorna-like" viruses. Other viruses included in this supergroup are, in addition to the picornaviruses, the como- and nepoviruses (Fig. 1). The genetic interrelationship among these viruses is demonstrated by the following shared properties.

- *i* They all have positive-stranded RNA genomes, with a VPg and poly(A) tail.
- *ii* Their translation strategy is similar, involving the synthesis of poly-proteins (one or two, depending on the genome being split or not), from which the functional proteins are generated by proteolytic cleavages.
- *iii* They encode a number of non-structural proteins that exhibit sequence homology, i.e., the viral polymerase, a proteinase, and a (putative) helicase.
- *iv* Moreover, these conserved proteins are, together with VPg, encoded by a similarly arranged gene set (5'-helicase, VPg, proteinase, polymerase-3') located in the 3' halves of their respective genomes (Fig. 1).
- *v* Since these conserved non-structural proteins have all been demonstrated or suggested to be involved in the RNA replication process [6, 7], all viruses belonging to the picorna-like supergroup will share, in principle, a similar RNA replication strategy. This statement is further strengthened by the fact that the genomes of these viruses have similar terminal structures (VPg, poly(A) tail).

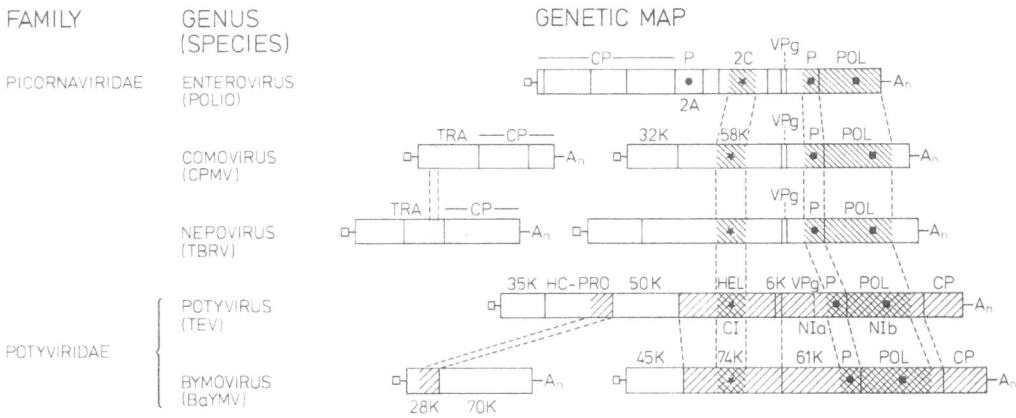

Fig. 1. Supergroup of picorna-like viruses. Genetic maps of poliovirus (*Picornaviridae*), cowpea mosaic virus (CPMV, comovirus), tomato black ring virus (TBRV, nepovirus), tobacco etch virus (TEV, genus *Potyvirus*, family *Potyviridae*) and barley yellow mosaic virus (BaYMV, genus *Bymovirus*, family *Potyviridae*). It is anticipated that comovirus and nepovirus groups will obtain the genus status. Coding regions in the genomes are indicated as open bars; regions of amino acid sequence homology in the gene products are indicated by similar shading. □ VPg; A_n poly(A) tail; *CP* coat protein(s); *TRA* transport protein; *HEL* helicase; *P* proteinase; *POL* polymerase; ★ NTP motif; ● cysteine proteinase motif; ■ polymerase motif

Differences between potyviruses and the other picorna-like viruses

Despite the fact that all viruses belonging to the supergroup of picorna-like viruses are very similar in genome structure, gene arrangement (3' half) and in the two basic processes of protein and RNA synthesis, they can be very different in other properties, like cell-to-cell movement, (vector) transmission, host range, and disease symptom expression. Moreover, potyviruses are very distinct from the other picorna-like viruses in having a rod-shaped particle morphology instead of having a pseudo T = 3 isometric particle. On the genetic level these differences are illustrated by the linkage of unique genes to the set of conserved genes (Fig. 1). This linkage of unique and conserved genes [5, 7, 21], together with other lines of evidence, demonstrates that gene shuffling by interviral recombination has been a major mechanism upon which RNA virus evolution is based. This delineation implies that potyviruses will not only be evolutionarily related by the conserved replicatory protein genes to the other picorna-like viruses (como-, nepo- and picornaviruses) but also (by their additional genes) to other viruses, not belonging to this supergroup. These additional interviral relationships are discussed.

Affinities between potyviruses and viruses not belonging to the picornavirus supergroup

Upstream of the replication gene module (helicase, VPg/proteinase, and polymerase genes) the genome of the potyviruses belonging to the genus *Potyvirus* contains three genes for which there seem to be no counterparts present in the genome of the other members of the picornavirus supergroup (Fig. 1). The 5' proximate encoded protein has been suggested to represent the cell-to-cell movement protein since for TVMV this protein (28 kDa in size) has some sequence homology to the TMV 30 kDa movement protein, though the TEV-encoded analogous protein (35 kDa) (Fig. 1) apparently does not show this homology [4]. Thus the conclusion may be drawn that, via the 5' proximate gene, the genus *Potyvirus* might be distantly related to tobamoviruses. It should be noted, however, that this gene seems to be lacking in the bipartite genome of BaYMV (Fig. 1).

The second unique gene of potyviruses encodes the helper component (HC-PRO), a two-domain protein required for aphid transmission (HC) and possessing proteolytic activity (PRO). This protein exhibits some sequence homology to the aphid transmission factor (gene II product) of caulimoviruses [4], which suggests a genetic interrelationship between poty- and caulimoviruses. BaYMV is transmitted by a fungus (*Polymyxa graminis*) and not by aphids. In concert with this, in the BaYMV genome only the PRO-domain is conserved, while the HC domain is lacking (Fig. 1).

The protein encoded by the third unique gene of potyviruses (i.e., the 50 kDa protein of TEV) does not show homology to any other known viral or cellular protein. A 45k cistron is found in the BaYMV genome (RNA-1), upstream of the helicase gene, which might be the bymoviral counterpart of the 50 kDa gene of TEV (Fig. 1).

In addition to the three unique genes in the 5' terminal part of the potyviral genome (of which two may be located in the RNA-2 of bymoviruses), a fourth gene is exclusively found in members of the *Potyviridae*, i.e., the coat-protein gene located at the 3' end of the genome. While picorna- and comoviruses (and probably also nepoviruses) encode coat proteins that are folded into three globular domains, or "β-barrels" [18], building up isometric virions with the same basic geometry, members of the *Potyviridae* encode a single capsid-protein species (30–35 kDa in size) that folds into a core, similar to the TMV coat protein, with extended N- and C-termini, and builds a tobamo-like rod-shaped particle [19]. Both the structure of the coat protein, the resulting rod-shaped architecture of the potyviral particle, and the deviant position of the coat-protein gene within the genome set the *Potyviridae* distinctly apart from the other picorna-like viruses, and links this virus family to rod-shaped plant viruses like the tobamoviruses.

Last but not least, there is a genetic link to the animal virus family *Flaviviridae*. Though the helicase (CI protein) of members of the *Potyviridae* shows significant sequence homology to the (putative) helicases of the como-, nepo- and picornaviruses (Fig. 1), this protein is even more closely related to the nucleotide-binding motif (NTP motif) containing protein (NS3) of *Flaviviridae* [9, 10, 14, 15]. On the basis of sequence homology, the NTP-motif containing proteins encoded by a considerable number of positive-strand RNA viruses can be placed into three main groups ("alphavirus-like," "picornavirus-like" and "poty/flavivirus-like"), each of them revealing closer relationships to distinct groups of cellular NTP-binding motif-containing proteins (both pro- and eukaryotic) than to the NTP-binding proteins of the other viral groups [9, 15].

Conclusion

Potyviruses share a considerable number of features with the other members of the picornavirus supergroup (como-, nepo-, and picornaviruses). They are all related in terms of genome structure, replication strategy, and mode of translation, and they encode a number of conserved proteins. Despite these shared characters, the *Potyviridae* are different from other picorna-like viruses in having a number of unique genes and in being a rod-shaped instead of a spherical virus. Comparisons with other viruses allow us to conclude that members of the *Potyviridae* can be regarded as (successful) products of a number of interviral recombination events. As a result of such recombination events potyviruses appear to possess genes that originate from at least four different genetic sources, i.e., the NIa (proteinase) and NIb (polymerase) genes which are both related to genes of como-, nepo- and picornaviruses, the CI (helicase) gene most closely related to the NS3 gene of flaviviruses, the HC-PRO gene distantly related to gene II of caulimoviruses, and a coat-protein gene related to those of other rod-shaped viruses. Due to this recombinative character of potyviruses (and many more viruses), the establishment of a true, phylogenetic virus taxonomy, that gives full account of all interviral relationships, will be very difficult, if not impossible. For instance, placing the *Potyviridae*, together with the como-, nepo- and picornaviruses, into a higher-than-family-level taxon (e.g., order *Picornavirales*) would on one hand emphasize the genetic relationships between these viruses, but, on the other hand, ignore other genetic relationships (to, e.g., *Flaviviridae*) as well as the tobamovirus-like particle morphology. Since virus evolution is not only a matter of common ancestry but also involves recombinational events, the obvious and less debatable solution is therefore not to introduce any higher-than-family-rank taxon in the taxonomy of viruses.

Acknowledgements

The author is grateful to Dr. Davidson and Dr. Kashiwazaki for providing results on BaYMV prior to publication, and to Dr. J. Dijkstra for critical reading of the text.

References

1. Allison RF, Johnston RE, Dougherty WG (1986) The nucleotide sequence of the coding region of tobacco etch virus genomic RNA: evidence for the synthesis of a single polyprotein. Virology 154: 9–20
2. Davidson, AD, Prols M, Schell J, Steinbiss H-H (1991) The nucleotide sequence of RNA 2 of barley yellow mosaic virus. J Gen Virol 72: 989–993
3. Domier LL, Franklin KM, Shahabuddin N, Hellmann GM, Overmeyer JH, Hiremath ST, Siaw MFE, Lomonossoff GP, Shaw JG, Rhoads RE (1986) The nucleotide sequence of tobacco vein mottling virus RNA. Nucleic Acids Res 14: 5417–5430
4. Domier LL, Shaw JG, Rhoads RE (1987) Potyviral proteins share amino acid sequence homology with picorna-, como-, and caulimoviral proteins. Virology 158: 20–27
5. Gibbs A (1987) Molecular evolution of viruses: 'trees,' 'clocks' and 'modules.' J Cell Sci [Suppl] 7: 319–337
6. Goldbach R (1986) Molecular evolution of plant RNA viruses. Annu Rev Phytopathol 24: 289–310
7. Goldbach R (1987) Genome similarities between plant and animal RNA viruses. Microbiol Sci 4: 197–202
8. Goldbach R, Le Gall O, Wellink J (1991) Alpha-like viruses in plants. Semin Virol 2: 19–25
9. Gorbalenya AE, Koonin EV, Donchenko AP, Blinov VM (1988) A novel superfamily of nucleoside triphosphate-binding motif containing proteins which are probably involved in duplex unwinding in DNA and RNA replication and recombination. FEBS Lett 235: 16–24
10. Hodgman TC (1988) A new superfamily of replicative proteins. Nature 333: 578
11. Hull R (1989) Movement of viruses in plants. Annu Rev Phytopathol 27: 213–240
12. Kashiwazaki S, Minobe Y, Hibino H (1991) Nucleotide sequence of barley yellow mosaic virus RNA 2. J Gen Virol 72: 995–999

13. Kashiwazaki S, Minobe Y, Omura T, Hibino H (1990) Nucleotide sequence of barley yellow mosaic virus RNA 1: a close evolutionary relationship with potyviruses. J Gen Virol 71: 2781–2790
14. Lain S, Riechmann JL, Martin MT, Garcia JA (1991) RNA helicase: a novel activity associated with a protein encoded by a positive strand RNA virus. Nucleic Acids Res 18: 7003–7006
15. Lain S, Riechmann JL, Martin MT, Garcia JA (1989) Homologous potyvirus and flavivirus proteins belonging to a superfamily of helicase-like proteins. Gene 82: 357–362
16. Maiss E, Timpe U, Brisske A, Jelkmann W, Casper R, Himmler G, Mattanovich D, Katinger HWD (1989) The complete nucleotide sequence of plum pox virus RNA. J Gen Virol 70: 513–524
17. Robaglia C, Durand-Tardif M, Tronchet M, Boudazin G, Astier-Manifacier S, Casse-Delbart F (1989) Nucleotide sequence of potato virus Y (N strain) genomic RNA. J Gen Virol 70: 935–947
18. Rossmann MG, Johnson JE (1989) Icosahedral RNA virus structure. Annu Rev Biochem 58: 533–573
19. Shukla DD, Strike PM, Tracy SL, Gough KH, Ward CW (1988) The N and C termini of the coat proteins of potyviruses are surface located and the N terminus contains major virus-specific epitopes. J Gen Virol 69: 1497–1508
20. Strauss JH, Strauss EG (1988) Evolution of RNA viruses. Annu Rev Microbiol 42: 657–683
21. Zimmern D (1987) Evolution of RNA viruses. In: Holland J, Domingo E, Ahlquist P (eds) RNA genetics. CRC Press, Boca Raton, pp 211–240

Author's address: R. Goldbach, Department of Virology, Agricultural University, P.O.B. 8045, NL-6700 EM Wageningen, The Netherlands.

Virus relationships

Arch Virol (1992) [Suppl 5]: 307–310
© by Springer-Verlag 1992

Nomenclature and relationships of some Brazilian leguminous potyviruses related to bean common mosaic and/or passionfruit woodiness viruses

O. Lovisolo[1] and **E. W. Kitajima**[2]

[1] Istituto di Fitovirologia Applicata del C.N.R., Torino, Italy
[2] Departamento de Biologia Celular, Universidade de Brasilia, Brasilia, Brazil

Summary. The main Brazilian literature of the last 10 years on potyviruses of leguminous plants related to bean common mosaic virus (BCMV) and/or to passionfruit woodiness virus (PWV) is discussed and summarized. The viruses dealt with are canavalia acronecrosis, mosaico de canavalia, cassia yellow spot, cowpea green vein-banding, cowpea rugose mosaic and cowpea severe mottle. The viruses have similar biological properties, such as a host range restricted mainly to the Leguminosae, aphid transmission, seed transmission in leguminous plants, and various degrees of serological relationships with BCMV and PWV.

Introduction

Several leguminous potyviruses serologically related to bean common mosaic virus (BCMV) and/or passionfruit woodiness virus (PWV) have been described in Brazil during the last 10 years. These viruses have similar biological properties, mainly host range, seed transmission in leguminous plants, aphid transmission, and various degrees of serological relationships with BCMV and/or PWV. Probably they are in an intermediate position between BCMV and PWV.

BCMV has serological relationships with 17 potyviruses [14], but many of these relationships are distant and not correlated with biological properties. Such a situation is common among potyviruses as recently discussed by Shukla and Ward [22].

Brazilian PWV isolates tested to date are closely related serologically to canavalia acronecrosis virus. PWV is also distantly related serologically to BCMV, blackeye cowpea mosaic (BlCMV), carnation vein mottle (CVMV), cassia yellow spot (CYSV), cowpea green vein-banding (CGVBV), cowpea rugose mosaic (CRMV), papaya ringspot-W (PRSV-W), soybean mosaic

(SbMV), watermelon mosaic 2 (WMV-2), and wisteria vein mosaic (WVMV) viruses [5, 11, 12].

PWV systemically infects *Canavalia ensiformis*, *C. brasiliensis*, and other Leguminosae, including French bean, in which the virus is seed-transmitted at the rate of 26.6% [1].

Canavalia acronecrosis virus (CanAV) and other canavalia potyviruses

Two strains of CanAV which infect *Canavalia* spp. but with different geographic distributions found in Pernambuco (CanAV-Pe) and Para (CanAV-Pa) states are known in Brazil. Both infect several Leguminosae including French bean, soybean, and cowpea, and cause top necrosis in *Canavalia ensiformis*, *C. brasiliensis,* and *C. gladiata* [3]. CanAV-Pa and CanAV-Pe strains are closely related serologically to PWV and have various degrees of relationship to CGVBV, CRMV, CYSV, cowpea severe mottle virus (CSMtV), WVMV, CVMV, BCMV, BlCMV, SbMV, and PRSV-W [2, 3; Lovisolo and Kitajima, unpubl.].

Some other potyviruses have been isolated from *Canavalia* spp.: (*i*) *Canavalia maritima* mosaic virus found in Puerto Rico in *C. maritima*, which has a host range wider than CanAV and is not related serologically to BCMV [16]; (*ii*) canavalia mosaic virus found in the Ivory Coast and Nigeria infecting *C. ensiformis*, which has a range of serological relationships that seem different from those of CanAV [6]; (*iii*) sword bean distortion mosaic virus isolated from *C. ensiformis* in India, with a host range restricted mostly to Leguminosae and a few species in the Aizoaceae and Chenopodiaceae, which shares a few antigenic determinants with BYMV and SbMV [13]; (*iv*) virus do mosaico da Canavalia found in Brazil (Rio de Janeiro), distantly related serologically to CanAV, BCMV, WVMV, and CVMV [3, 21; Lovisolo and Kitajima, unpubl.].

Cassia yellow spot virus (CYSV)

This virus was first found in Pernambuco state in *Cassia hoffmanseggi* and called "cassia yellow blotches virus" [15]. A similar name was used by Dale et al. [4] for a new bromovirus. Souto et al. [25] gave the Brazilian potyvirus the new name CYSV. The virus has a host range mainly restricted to Leguminosae, but infects some plants in other families. CYSV is serologically related to BCMV, CanAV-Pa, CanAV-Pe, CVMV, and PWV [3, 12, 24; Souto et al., unpubl.]; it was not related to BlCMV, an isolate of BCMV, PRSV, or WMV-2 [15]. CYSV was seed-transmitted in *C. occidentalis* at the rate of 12% [23].

Cowpea green vein-banding virus (CGVBV)

This virus was isolated from cowpea in the state of Goias. It infects mainly Leguminosae, but also some Chenopodiaceae and Amaranthaceae [8]. It is related serologically to CRMV, PWV, CanAV-Pa, and CanAV-Pe [11,12,20]. It is not related to BlCMV, cowpea aphid-borne mosaic virus (CABMV), BCMV, SbMV, or bean yellow mosaic virus [7,9].

Cowpea rugose mosaic virus (CRMV) and cowpea severe mottle virus (CSMtV)

These viruses were isolated from cowpea in the state of Piaui in 1979 [18]; CSMtV was isolated from soybean [17]. Their host ranges are mainly restricted to Leguminosae, but they also infect a few Chenopodiaceae and Amaranthaceae. CRMV is related serologically and very similar to CSMtV: the main differences are in the host range, vectors, and sources of resistance. CRMV is also distantly related serologically to two isolates of BCMV, CanAV-Pa, CanAV-Pe, CGVBV, and PWV. CRMV and CSMtV are not related to BlCMV, an isolate of BCMV, BYMV, SbMV, PRSV-W, or turnip mosaic virus [10–12, 19,20; Lovisolo and Kitajima, unpubl.].

Conclusion

These virus isolates from Brazil are all serologically related to BCMV or PWV and their biological properties are similar but each virus has a few unique characteristics. It would be useful to apply some of the techniques related to protein or nucleic acid sequences to these viruses.

References

1. Costa AF (1985) Contribuição ao estudo da epidemiologia e controle do virus do endurecimento dos frutos do maracuja. Fitopatol Brasil 10: 310
2. Costa CL, Costa AF, Marinho, VLA, Kitajima, EW, Lin MT, Avila AC, Brioso, PST, Albuquerque FC (1984) Potyvirus isolados de feijão de porco (*Canavalia* spp). Fitopatol Brasil 9: 400
3. Costa CL, Kitajima EW, Marinho VLA (1989) Caracterização do virus da *Acronecrose* da Canavalia. Fitopatol Brasil 14: 115
4. Dale JL, Gibbs AJ, Behncken GM (1984) Cassia yellow blotch: a new bromovirus from an Australian native legume, *Cassia pleurocarpa*. J Gen Virol 65: 281–288
5. De Sa PB, Lovisolo O, Kitajima EW (1989) Dados complementares a caracterrização de um isolado do virus do mosaico de melancia 2 (WMV-2). Fitopatol Brasil 14: 114
6. Fauquet C, Thouvenel JC (1980) Maladies virales des plantes cultivees en Cote d'Ivoire. ORSTOM, Paris, pp 1–128
7. Lin MT (1979) Purification and serology of legume and corn viruses in Brazil. Fitopatol Brasil 4: 203–213
8. Lin MT, Anjos JRN, Kitajima EW, Rios GP (1979) Um novo potyvirus isolado do caupi e potencialmente importante para a cultura do feijão no Brasil. Fitopatol Brasil 4: 120–121
9. Lin MT, Kitajima EW, Rios GP (1981) Serological identification of several cowpea viruses in Central Brazil. Fitopatol Brasil 6: 73–85

10. Lin MT, Rios GP (1985) Cowpea diseases and their prevalence in Latin America. In: Singh SR, Rachie KO (eds) Cowpea research, production and utilization. Wiley, New York, pp 199–204

11. Lovisolo O, Marinho VLA, Lin MT, Kitajima EW (1986) Relaçoes sorologicas do virus do endurecimento dos frutos do maracujazeiro ("passionfruit woodiness virus") com outros potyvirus. Fitopatol Brasil 11: 358

12. Lovisolo O, Marinho FLA, Souto E, Costa CL, Kitajima EW (1989) Investigation on passionfruit woodiness and legume potyviruses serologically related with bean common mosaic virus. Fitopatol Brasil 14: 119

13. Mali VR, Nirmal DD, Mundhe GE, Vyanjane NT, Raut KG (1985) Purification and some properties of a virus causing distortion mosaic of sword bean (*Canavalia ensiformis*). Indian Phytopathol 38: 282–285

14. Morales FJ, Bos L (1988) Bean common mosaic virus. AAB Descriptions of Plant Viruses, no 337

15. Paguio OR, Kitajima EW (1981) Isolamento de um potyvirus de *Cassia hoffmannseggii* Mart. Fitopatol Brasil 6: 187–191

16. Rodriguez RL, Bird J, Monllor AC, Waterworth HE, Kimura M, Maramorosch K (1975) A mosaic virus of *Canavalia maritima* (Bay-bean) in Puerto Rico. In: Bird J, Maramorosch K (eds) Tropical diseases of legumes. Academic Press, New York, pp 91–101

17. Santos AA (1985) Infecção natural em soja do virus do mosqueado severo do caupi. Fitopatol Brasil 10: 311

18. Santos AA, Lin MT, Costa CL, Anjos JRN (1980) Diferenciação de tres virus alongados isolados de caupi (*Vigna unguiculata*). Fitopatol Brasil 5: 457

19. Santos AA, Lin MT, Kitajima EW (1981) Properties of cowpea rugose mosaic virus. Phytopathology 71: 890

20. Santos AA, Lin MT, Kitajima EW (1984) Caracterização de dois potyvirus isolados de caupi (*Vigna unguiculata*) no estado do Piaui. Fitopatol Brasil 9: 567–582

21. Santos OR, Costa CL, Kitajima EW, Meyer MC, Ramagem RD (1990) Propriedades fisicas e biologicas do virus do mosaico da *Canavalia*. Fitopatol Brasil 15: 132

22. Shukla DD, Ward CW (1989) Structure of potyvirus coat proteins and its application in the taxonomy of the potyvirus group. Adv Virus Res 36: 273–314

23. Souto ER, Costa CL, Kitajima EW (1990) Transmissão do virus das manchas amarelas da cassia por afideos e por sementes. Fitopatol Brasil 15: 136

24. Souto ER, Kitajima EW, Marinho VLA, Costa AF (1988) Estudos adicionais sobre o virus de manchas amarelas da cassia. Fitopatol Brasil 13: 145

25. Souto ER, Marinho VLA, Oliveira CRB, Kitajima EW (1989) Caracterização de um isolado do virus de manchas amarelas da cassia. Fitopatol Brasil 14: 114

26. Trindade DR, Costa CL, Kitajima EW, Lin MT (1984) Identificação e caracterização de estirpes do virus do mosaico comun de feijoeiro no Brasil. Fitopatol Brasil 9: 1–12

Authors' address: O. Lovisolo, Istituto di Fitovirologia Applicata del C.N.R., Strada delle Cacce 73, I-10135 Torino, Italy.

Arch Virol (1992) [Suppl 5]: 311–316

Ecology and taxonomy of some European potyviruses

O. Lovisolo

Istituto di Fitovirologia Applicata del C.N.R., Torino, Italy

Summary. Observations on the ecology and taxonomy of henbane mosaic virus, amaranthus leaf mottle virus, European viruses of the sugarcane mosaic virus cluster, and some related potyviruses are given.

Introduction

The author and coworkers have investigated several potyviruses over the course of the last 30 years. This report consists of observations and considerations about these potyviruses which have ecological implications connected with taxonomy. The viruses dealt with are: henbane mosaic virus and possible related viruses; amaranthus leaf mottle virus and the bean yellow mosaic virus cluster; and European viruses of the sugarcane mosaic virus subgroup.

Henbane mosaic virus: a possible distinct potyvirus cluster

Henbane mosaic virus (HMV) was first found in 1932 in England in cultivated henbane (*Hyoscyamus niger*) [15], and later was infrequently reported in that plant in Europe. According to Varma [47], HMV is common on the Indian subcontinent in henbane grown as a medicinal plant. HMV was also found in another medicinal plant, *Atropa belladonna*, in England [43] and in West Germany [3].

In Europe, HMV seems to be common in wild plants. It has been found in *Datura stramonium* in England [43], in Italy [26], and in Hungary [39]. In northwest Italy the strain, alkekengi (HMV-A) is quite common in wild *Physalis alkekengi* [26; Lovisolo, unpubl.].

HMV is distantly related serologically to potato virus Y (PVY) [11], Colombian datura virus (CDV) [18, 26] and pokeweed mosaic virus (PkMV) [11].

An interesting feature of HMV is that it has a modal length greater than all other potyviruses: Bode et al. [3] found a length of about 900 nm for the atropa strain, Harrison and Roberts [16] about 925 nm for the same atropa strain, and Lovisolo and Bartels [26] about 850 nm for the alkekengi strain.

The modal lengths of potyviruses cover a wide range (680–900 nm), but generally have no taxonomic value for individual members. Moghal and Francki [33] showed that closely related potyviruses may have different particle lengths and concluded that such measurements should not be used for subdividing or identifying viruses within the group.

In some potyviruses the particle length depends on the composition of the suspending medium, as first found by Govier and Woods [12]; particles are longer and straighter in the presence of Ca and Mg cations, and shorter in the presence of EDTA. HMV has a modal length of about 800 nm in the absence of Mg and of about 900 nm in the presence of Mg, but this is not true of all potyviruses [2, 17].

Interestingly, HMV infections may cause mitochondrial aggregates in *Datura stramonium* leaves [19]. Such effects are known in only a few potyviruses: PVY [7], iris fulva mosaic virus [2], zucchini yellow fleck virus (ZYFV) [31], and an unidentified isolate related to tobacco etch virus (TEV) [20].

HMV, PkMV, and PVY belong to subdivision III on the basis of cylindrical inclusion morphology while datura 437 virus belongs to subdivision IV [7, 8].

Because of these features, in particular the modal length, the genome or the amino acid composition of HMV should be investigated. HMV could be compared with the viruses that have some serological relationship with it (CDV, PkMV, and PVY), and relationships between HMV and the following possible potyvirus members developed: datura 437 virus [6], datura distortion mosaic virus [30], datura mosaic virus [37] and Hungarian datura innoxia mosaic virus [34].

Amaranthus leaf mottle virus (AmLMV) and the bean yellow mosaic virus cluster

AmLMV is a potyvirus widespread in *Amaranthus deflexus* in the Mediterranean region [24, 27, 28]. A strain of AmLMV also has been found in *Cirsium arvense* [4].

The experimental host range is largely confined to members of the families Amaranthaceae and Chenopodiaceae, but some members of the families Leguminosae, Solanaceae, and Asteraceae also are infected, mainly in the case of the *Cirsium* strain.

AmLMV has never been found in cultivated plants, and the *Amaranthus* isolates seem to still be in a phase of easy mutability in regard to host range. Variation in host range is particularly evident during subculturing [28].

Serologically, the type AmLMV is distantly related to bean yellow mosaic virus (BYMV) [28], peanut mottle virus [38], zucchini yellow mosaic virus [22], plum pox virus [28] and a possible new virus of Lisianthus [21].

AmLMV is related serologically only to members of the BYMV cluster. The main biological difference between AmLMV and the viruses closest to

BYMV is that AmLMV does not infect French bean systemically, whereas it does infect *Chenopodium amaranticolor* and *C. quinoa* systemically.

European viruses of the sugarcane mosaic virus (SCMV) subgroup

Typical SCMV isolates were rarely reported in Europe until recently. Matz [32] identified a Spanish isolate from sugarcane, found to be very similar to another from Puerto Rico and thought to be of West Indian origin. Signoret [42] identified an isolate as maize dwarf mosaic virus-B, in fields for the production of hybrid maize seed near Versailles. Haeni [14] identified as SCMV an isolate from maize found in the Canton of Ticino, Switzerland. Recently Fuchs et al. [9] reported an increase in SCMV affecting maize in the German Democratic Republic.

Much more common, especially in Italy, Yugoslavia, France and some other regions of Europe, is the disease first called "arrossamento striato del sorgo." This disease has been known in Italy since the 1930's [10]. Grancini [13] and Lovisolo [23] studied this disease and a mosaic of maize, generally mild without dwarfing, and found both were caused by a virus of the SCMV complex, the sorghum red stripe virus (SRSV). Gracini [13] and Lovisolo and Acimovic [25] found that in nature SRSV infects not only grain sorghum (*Sorghum bicolor*) and maize, but also Johnsongrass (*S. halepense*). Another important feature is that SRSV does not infect sugarcane systemically but causes only local chlorotic lesions on the inoculated leaves [23].

We now know that some of these features (systemic mosaic in maize and Johnsongrass; no systemic infection in sugarcane) are typical of maize dwarf mosaic virus strain A, or SCMV-Jg.

The taxonomy of SCMV, MDMV and related viruses is rather confused. Until the 1950's SCMV was considered to be a single virus of worldwide distribution, composed of several strains [36]. In 1963 Dale [5] isolated a virus similar to SCMV associated with an extremely dangerous disease of maize, and found that it also infected Johnsongrass systemically; this virus was considered by Shepherd [40] to be a new strain of SCMV (Jg strain), while Williams and Alexander [48] considered it to be a new entity, maize dwarf mosaic virus.

In 1968 Taylor and Pares [44] in Australia described two closely related viruses in maize and Johnsongrass, and called them the maize and Johnsongrass isolates of MDMV.

In Venezuela [45] a virus that could infect Johnsongrass, considered to be a variant of MDMV-A, was isolated from grain sorghum.

Tosic et al. [46] compared two Italian isolates from maize with a Yugoslavian isolate and several strains of SCMV and MDMV from the U.S.A.. The Italian isolates were similar to those described by Grancini [13] and Lovisolo [23], to the Yugoslavian isolates, and to MDMV-A and SCMV-Jg.

Persley et al. [35] confirmed that the Johnsongrass infecting isolates from maize and sorghum in Europe (Italy, Yugoslavia, and France) were very similar to North American MDMV-A, while the Venezuelan and Australian isolates were different.

Shukla et al. [41] compared 17 SCMV/MDMV strains from Australia and the U.S.A. by electroblot immunoassay with cross-absorbed polyclonal antibodies directed toward surface-located N-termini of the coat proteins. They suggested that four distinct viruses were involved, for which the names johnsongrass mosaic, maize dwarf mosaic, sugarcane mosaic and sorghum mosaic virus were proposed.

Such work should be extended to SRSV and similar viruses from India [29] and Israel [1]. If SRSV were confirmed as being close to MDMV, the name "sorghum red stripe virus" should be given priority.

References

1. Antignus Y (1987) Comparative study of two maize dwarf virus strains infecting corn and Johnsongrass in Israel. Plant Dis 71: 687–791
2. Barnett OW, Alper M (1977) Characterization of iris fulva mosaic virus. Phytopathology 67: 448-454
3. Bode O, Brandes J, Paul HL (1969) Untersuchungen über ein neues, langgestrecktes Virus aus *Atropa belladonna*. BBA für Land- und Forstwirtschaft in Berlin und Braunschweig, Jahresbericht 1968, A61
4. Casetta A, D'Agostino G, Conti M (1986) Isolamento di "amaranthus leaf mottle virus" (ALMV) da *Cirsium arvense* Scop. Info Fitopatol 36: 43–46
5. Dale JL (1964) Isolation of a mechanically transmissible virus from corn in Arkansas. Plant Dis Rep 48: 661–663
6. Damsteegt VD (1974) Transmission, host range, and physical properties of a virus infecting *Datura candida*. Proc Am Phytopathol Soc 1: 50
7. Edwardson JR (1974) Some properties of the potato virus Y-group. Fla Agric Exp Stat Monogr Ser, no 4
8. Edwardson, JR, Christie RG, Ko NJ (1984) Potyvirus cylindrical inclusions – subdivision IV. Phytopathology 74: 1111–1114
9. Fuchs E, Gruentzig M, Bedri A (1990) On the ecology of potyviruses affecting maize in the German Democratic Republic. Arch Phytopathol Pflanzenschutz 26: 329–335
10. Goidanich G (1939) Ricerche sul deperimento del sorgo zuccherino verificatosi in Italia nella primavera del 1938. Boll R Staz Patol Veg Roma NS 19: 261–321
11. Govier DA, Plumb RT (1972) Henbane mosaic virus. CMI/AAB Descriptions of Plant Viruses, no 95
12. Govier DA, Woods RD (1971) Changes induced by magnesium ions in the morphology of some plant viruses with filamentous particles. J Gen Virol 13: 127–132
13. Grancini P (1957) Un mosaico del mais e del sorgo in Italia. Maydica 2: 83–104
14. Haeni A (1989) Identification of two potyviruses on maize in Ticino. Landwirtschaft Schweiz 2: 593–598
15. Hamilton MA (1932) On three new virus diseases of *Hyoscyamus niger*. Ann Appl Biol 19: 550–567
16. Harrison BD, Roberts IM (1971) Pinwheels and crystalline structures induced by Atropa mild mosaic virus, a plant virus with particles 925 nm long. J Gen Virol 10: 71–78

17. Hollings M, Brunt AA (1981) Potyvirus group. CMI/AAB Descriptions of Plant Viruses, no 245

18. Kahn RP, Bartels R (1968) The Colombian Datura virus – a new virus in the potato virus Y group. Phytopathology 58: 587–592

19. Kitajima EW, Lovisolo O (1972) Mitochondrial aggregates in *Datura* leaf cells infected with henbane mosaic virus. J Gen Virol 16: 265–271

20. Lesemann D-E (1988) Cytopathology. In: Milne RG (ed) The plant viruses, vol 4, the filamentous plant viruses. Plenum, New York, pp 179–235

21. Lisa V, Dellavalle G, Vaira AM, Milne RG, Masenga V, Boccardo G, D'Aquilio M (1990) Update on virus diseases of lisianthus. In: Proceedings of the XXIIIrd International Horticultural Congress, Firenze, 537

22. Lisa V, Lecoq H (1984) Zucchini yellow mosaic virus. CMI/AAB Descriptions of Plant Viruses, no 282

23. Lovisolo O (1957) Contributo sperimentale alla conoscenza ed alla determinazione del virus agente dell'arrossamento striato del sorgo e di un mosaico del mais. Boll Staz Patologia Veg Roma 14: 261–321

24. Lovisolo O (1989–1990) Centri di origine e di diversificazione delle virosi delle plante ed altri aspetti di ecologia virale. Ann Accad Agric Torino 132: 65–103

25. Lovisolo O, Acimovic M (1961) Sorghum red stripe disease in Yugoslavia. FAO Plant Protect Bull 9: 99–102

26. Lovisolo O, Bartels R (1970) On a new strain of henbane mosaic virus from *Physalis alkekengi*. Phytopathol Z 69: 189–201

27. Lovisolo O, Lisa V (1976) Characterization of a virus isolated from *Amaranthus deflexus*, serologically related to bean yellow mosaic virus. Agric Consp Scient 39: 553–559

28. Lovisolo O, Lisa V (1979) Studies on Amaranthus leaf mottle virus (ALMV) in the Mediterranean region. Phytopathol Medit 18: 89–93

29. Mali VR, Garud TB (1978) Sorghum red stripe – a Johnsongrass strain of sugarcane mosaic virus. FAO Plant Protect Bull 26: 28–29

30. Mali VR, Nirmal DD, Patel KV, Vyanjane NT (1985) Studies on a virus causing distortion mosaic of *Datura fastuosa*. Indian Phytopathol 38: 413–417

31. Martelli GP, Russo M, Vovlas C (1981) Ultrastructure of zucchini yellow fleck virus infections. Phytopathol Medit 20: 193–196

32. Matz J (1939) Comparative study of sugarcane mosaic from different countries. In: Proceedings of the 6th Congress International Society for Sugar Cane Technology, pp 572–580

33. Moghal SM, Francki RIB (1981) Towards a system for the identification and classification of potyviruses. II. Virus particle length, symptomatology and cytopathology of six distinct viruses. Virology 112: 210–216

34. Peralta E-L, Beczner L, Dezséry M (1981) Characterization of the Hungarian Datura innoxia mosaic virus. Acta Phytopathol Acad Sci Hung 16: 85–96

35. Persley DM, Henzell RG, Greber RS, Teakle DS, Toler RW (1985) Use of a set of differential sorghum inbred lines to compare isolates of sugarcane mosaic virus from *Sorghum* and maize in nine countries. Plant Dis 69: 1046–1049

36. Pirone TP (1972) Sugarcane mosaic virus. CMI/AAB Descriptions of Plant Viruses, no 88

37. Qureshi S, Mahmood K (1978) Purification and properties of Datura mosaic virus. Phytopathol Z 93: 113–119

38. Rajeshwari R, Iizuka N, Nolt BL, Reddy DVR (1983) Purification, serology and physico-chemical properties of a peanut mottle virus isolate from India. Plant Pathol 32: 197–205

39. Salamon P, Dezsery M (1983) Viruses pathogenic to solanaceous plants in Hungary. Novènyvèdelmi Tudomanyos Napok, Budapest

40. Shepherd, RJ (1965) Properties of a mosaic virus of corn and Johnsongrass and its relation to the sugarcane mosaic virus. Phytopathology 55: 1250–1256
41. Shukla DD, Tosic M, Jilka J, Ford RE, Toler RW, Langham MAC (1989) Taxonomy of potyviruses infecting maize, sorghum, and sugarcane in Australia and the United States as determined by reactivities of polyclonal antibodies directed towards virus-specific N-termini of coat proteins. Phytopathology 79: 223–229
42. Signoret PA (1974) Les maladies a virus des graminees dans le Midi de la France. Acta Biol Iugoslav Ser B 11: 115–120
43. Smith KM (1972) A textbook of plant virus diseases, 3rd edn. Longman, London
44. Taylor RH, Pares RD (1968) The relationship between sugarcane mosaic virus and mosaic viruses of maize and Johnsongrass in Australia. Aust J Agric Res 19: 767–773
45. Toler RW, Rosenow DT, Riccelli M, Mena HA (1982) Variability of Venezuelan isolate of maize dwarf mosaic virus in sorghum. Plant Dis 66: 849–850
46. Tosic M, Benetti MP, Conti M (1977) Studies on sugarcane mosaic virus (SCMV) isolates from northern and central Italy. Ann Phytopathol 9: 387–393
47. Varma A (1988) The Indian subcontinent. In: Milne RG (ed) The plant viruses, vol 4, the filamentous plant viruses. Plenum, New York, pp 371–378
48. Williams LE, Alexander LJ (1965) Maize dwarf mosaic, a new corn disease. Phytopathology 55: 802–804

Author's address: O. Lovisolo, Istituto di Fitovirologia Applicata del C.N.R., Strada delle Cacce 73, I-10135 Torino, Italy.

Arch Virol (1992) [Suppl 5]: 317–318

Relationships among iris severe mosaic virus (ISMV) isolates

C. I. M. van der Vlugt[1, 2], **R. Goldbach**[1], and **A. F. L. M. Derks**[2]

[1] Agricultural University, Wageningen
[2] Bulb Research Centre, Lisse, The Netherlands

Summary. The ISMV isolates studied so far, have been found to be indistinguishable both by serological and hybridization assays. Therefore, the different symptoms induced by these isolates may be based on only minor molecular changes.

*

To develop a test for the detection of iris severe mosaic virus [1, 2] in iris bulbs, it is necessary to investigate whether different isolates of this virus react in the same manner.

For this, we used serological assays as well as nucleic acid hybridization and tested six isolates of iris severe mosaic virus [3]: (*i*) four isolates which caused mild to moderate chlorotic symptoms in leaves of *Iris* × *hollandica* cultivar Professor Blaauw; two of which originated from bulbous iris, one from crocus and one from rhizomatous iris; (*ii*) one isolate from bulbous *Iris bucharica* which caused very severe symptoms in iris cultivar Professor Blaauw leading to premature death of the plants when grown under field conditions; (*iii*) one isolate from rhizomatous iris, previously described as bearded iris mosaic virus (BIMV) but now recognized as a strain of ISMV [1, 3]. This isolate was obtained from O. W. Barnett (Clemson, SC, U.S.A.).

Antisera were produced against a crocus isolate of ISMV [2] and against BIMV. All isolates could be detected by the antiserum against ISMV from crocus, as well as with the BIMV antiserum [3]. (The severe strain has not been tested yet with the BIMV antiserum.) Iris mild mosaic virus (IMMV), another potyvirus frequently present in irises [4], did not react with these antisera.

The crocus isolate was also used for cloning. A cDNA clone was obtained, 820 nucleotides long and comprising part of the polymerase gene NIb. This clone was used as a probe in hybridization assays.

The four mild isolates and the severe isolate all reacted positively with this probe when the isolates were transferred to the iris cultivar Professor Blaauw. So far we have not tested the BIMV isolate by hybridization assay as transmission to this iris cultivar has not been successful. With random cDNA hybridization O. W. Barnett [1] found a difference in cross hybridization between one ISMV isolate and a BIMV isolate. This needs further investigation. In our hybridization assays IMMV never reacted with the 820 nt probe.

From tests to date, we conclude that the isolates investigated, though differing considerably in expression of symptoms in the same host plant, are not distinguishable either in a serological assay (using antibodies directed against whole virus particles) nor by hybridization assays using a probe corresponding to part of the polymerase gene. These results question whether the previous division of the isolates into at least three different strains [1] is valid or emphasize that strains may be the result of minor molecular changes.

References

1. Brunt AA, Derks AFLM, Barnett OW (1988) Iris severe mosaic virus. AAB Descriptions of Plant Viruses, no 338
2. van der Vlugt CIM, Derks AFLM, Dijkstra J, Goldbach R (1988) Towards a rapid and reliable detection method for iris severe mosaic virus in iris bulbs. Acta Hortic 234: 191–198
3. Derks AFLM, Hollinger TC (1986) Similarities of and differences between potyviruses from bulbous and rhizomatous irises. Acta Hortic 177: 555–561
4. Asjes CJ (1979) Viruses and virus diseases in Dutch bulbous irises (*Iris hollandica*) in the Netherlands. Neth J Plant Pathol 85: 269–279

Authors' address: C. I. M. van der Vlugt, Bulb Research Centre, Postbus 85, Vennestraat 22, NL-2160 AB Lisse, The Netherlands.

Virus relationships – PVY subgroup

Arch Virol (1992) [Suppl 5]: 321–326

A comparison of pepper mottle virus with potato virus Y and evidence for their distinction

E. Hiebert and **D. E. Purcifull**

Department of Plant Pathology, University of Florida, Gainesville, Florida, U.S.A.

Summary. Pepper mottle virus (PepMOV) was identified as a distinct potyvirus infecting peppers in Arizona and Florida in the 1970's. The distinction of PepMoV from potato virus Y (PVY) has recently been challenged on the basis of sequence comparisons of the coat proteins and of the 3′ nontranslated regions of the viral RNAs. We summarize the biological, cytological, serological, and in vitro translational studies which compare the apparent differences, and also similarities, between PepMoV and PVY. We conclude that although PepMoV may be more closely related to PVY than to other known potyvirus, PepMoV should be maintained as a separate virus on the basis of its distinctive characteristics.

Introduction

Pepper mottle virus (PepMoV) was described in the early 1970's in Florida when sweet peppers resistant to common strains of potato virus Y (PVY) and tobacco etch virus (TEV) developed virus symptoms [22]. At the same time, Nelson and Wheeler [11] reported a new virus, serologically unrelated to PVY and TEV, infecting chili peppers in Arizona. These two pepper-infecting viruses were shown to be closely related on the basis of their reactivities with antisera to virions and cylindrical inclusions (CI) of the Florida isolate [16]. Since then the virus has been found to be a persistent problem in the major pepper growing areas in the Southern U.S.A., Mexico, and Central America [1, 11, 12, 14]. The causal agent was initially identified as a PVY strain but later was named as a new potyvirus based on a number of distinct properties [16]. PepMoV can be distinguished from PVY by pathogenicity studies, by cytological differences in their CI (mapped as protein #4 on the potyviral genome, P4) and the presence of prominent amorphous inclusions (AI; P2) in PepMoV infections, and by serology of their capsid protein, helper component (HC; P2)-AI protein, and CI protein. Recently, comparisons of the sequence from the 3′ end of a putative PepMoV isolate [3] with PVY isolates have been the basis for claiming that PepMoV does not warrant a distinction from PVY isolates [4, 5, 17–19, 21]. Here we summarize some of the data for the comparison of PepMoV with PVY and justify the distinction of these two viruses. The Florida PepMoV isolate was obtained from Zitter [22] and maintained by mechanical inoculation in our virus culture facilities. The culture was transferred by aphids for studies reported by de Mejia et al. [3]. A PVY isolate was obtained from G. V. Gooding, Jr. (North Carolina State University, Raleigh, NC).

Comparisons

Cytological properties

In infected tissue samples, PepMoV is readily distinguished from PVY on the basis of the formation of long, thin CI (pinwheel inclusions), which appear prominently and more abundantly at the cell wall during the early stages of PepMoV infections (Figs. 17 and 23 in [2]). The CI produced by PVY infections appear to be much shorter and are not as readily resolved at the cell wall (Fig. 17 in [2]). In later stages of infection when the CI accumulate in the cytoplasm, the CI appear much larger for PepMoV infections than those seen for PVY infections [2]. Purified PepMoV CI, unlike PVY CI, appear differently when negatively stained with uranyl acetate as compared with phosphotungstate negative stain [6]. The purified PepMoV CI (scrolls) stained with uranyl acetate appear rolled up tightly (long and slender) whereas in phosphotungstate the CI appear less tightly rolled (a decrease in their length/width ratio). Presumably this sensitivity to the negative stains represents a difference in the chemical composition of the PepMoV CI compared to the PVY CI. The AI (also known as irregular inclusions) [2] formed in PepMoV-infected tissues are stained much more intensely with Azure A or orange-green combination, and are more prominent in terms of size and frequency of appearance compared to those seen in PVY-infected tissues (Figs. 21, 23, 26 in [2]). In the electron microscope, the AI associated with PVY appear vacuolate and heterogeneous compared to the homogeneous and electron dense appearance of the AI in PepMoV-infected tissues (Fig. 26 in [2]; Fig. 2A in [3]). These differences can also be resolved by light microscopy [2].

Biological properties

PepMoV infects pepper cultivars which show resistance to common strains of PVY and tobacco etch virus (TEV) [16, 22]. Nelson and Wheeler [12] compared five PepMoV isolates (from Arizona, California, New Mexico, Florida, and North Carolina) with PVY and TEV for host range, cross-protection and serology. PepMoV was found in mixed infections with PVY and TEV in the field (no apparent cross-protection) [12]. In greenhouse experiments PVY did not protect tabasco pepper (*Capsicum frutescens*) against subsequent inoculations with PepMoV [12]. The Arizona and California isolates of PepMoV were identical, the Florida isolate and New Mexican isolates were somewhat different, and the North Carolina isolate was intermediate between PepMoV Arizona and PVY on the differential pepper cultivar tabasco. However, the PepMoV North Carolina isolate was serologically indistinguishable from the other PepMoV isolates. Nelson and Wheeler [13] identified a chili pepper line as a useful indicator host that results in a severe reaction to PVY infections while PepMoV infections produce mild mosaic symptoms.

Serological properties

PepMoV can be distinguished serologically from PVY by SDS-immuno-diffusion [11, 12, 16], by ELISA [1], and by ISEM [14] using polyclonal antisera. In a study of PepMoV isolates, Nelson and Wheeler [12] found that all of the PepMoV isolates were serologically identical when tested with PepMoV sera prepared to the Florida and Arizona isolates. The PepMoV isolates did not react specifically to antisera of PVY or TEV. Jordan and Hammond [9] studied 13 monoclonal antibodies to potyviruses and reported that only three epitopes were shared between PepMoV and PVY whereas 10 monoclonal antibodies differentiated the two viruses. The CI protein for the two viruses can also be serologically distinguished on the basis of reciprocal tests in SDS-immuno-diffusion, although their CI proteins also have some common epitopes [15]. An epitope difference between the P2 proteins (HC) for the two viruses is indicated since PVY HC antiserum failed to react with a PepMoV in vitro translation product (P1-P2 product) which was reactive with PepMoV AI antiserum [3, 7]. The P2 proteins for the two viruses are related because PepMoV AI antiserum reacted with PVY HC protein but not with tobacco vein mottling virus or papaya ringspot virus-W proteins in Western blot analysis [3].

Genomic properties

In vitro translation analysis of the PepMoV genome (Fig. 1 in [3]) reveals a product pattern similar to that for PVY (Fig. 1 in [7]) except that the estimated size of the P1-P2 polyprotein produced in the rabbit reticulocyte lysate (RRL) system for PVY is about 2 kDa larger than that determined for PepMoV (80 kDa vs. 78 kDa) (Fig. 1). Trace amounts of the putative P1 translation product for the two viruses are visible in the RRL system (Fig. 1). In the wheat germ (WG) system where the P1-P2 is processed efficiently [2], estimated size for the PepMoV P1 is 30 kDa while the PVY product is 32 kDa (data not shown). The P1 coding region has been shown to be the most variable in terms of size [8] and sequence of all the coding regions identified for the potyviral genome [18]. Another distinctive feature of the PepMoV RNA template in the RRL system is the lack of large polyproteins immunoprecipable with coat-protein antiserum (Fig. 1, lane 4 in [3] compared with Fig. 1, lane c in [7]). This pre-

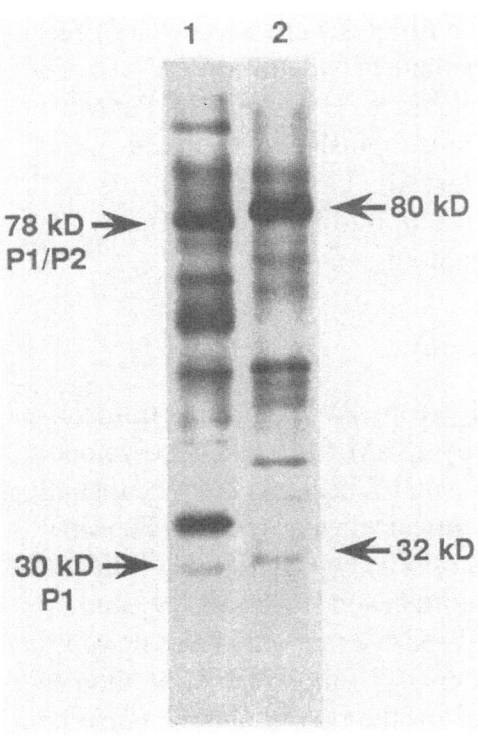

Fig. 1. Polyacrylamide gel electrophoresis of the total products of in vitro translations of PepMoV (*1*) and PVY (*2*) RNA in the rabbit reticulocyte lysate system. The sizes of the products (*P1/P2* and *P1*) were estimated by the use of MW markers in an adjacent lane (not shown). Note the size differences in the PepMoV and PVY P1/P2 and P1 products. The [^{35}S]methionine-labeled products were detected by fluorography. In vitro translation was done as described previously [3]

sumably is due to more efficient processing of the polyprotein by the small nuclear inclusion protein (P6) of PepMoV compared to PVY translations under similar in vitro conditions. The P2 products (helper components) for both viruses have a similar size as estimated on SDS-PAGE [3]. The other products, P3–P8 for both viruses, also cannot be distinguished on the basis of estimated sizes in SDS-PAGE.

Discussion

PepMoV has been disputed as being a distinct potyvirus on the basis of sequence comparisons of its coat protein and 3' nontranslated region with the sequences reported for PVY isolates [4, 5, 10, 17, 18, 21]. Now new sequence information for a PepMoV isolate from California, presumably similar to the Florida isolate used in our characterization studies, supports the distinction of PepMoV from PVY [20]. At this stage, the basis for the sequence discrepancies between the two presumed PepMoV isolates is subject to speculation. We believe that this PepMoV-PVY controversy illustrates the hazards of assessing viral relationships and viral taxonomy on the basis of a single property or criterion. This also is an example that the identification of viruses by classical methods (biological, cytological, and serological techniques) still is important in viral classification and taxonomy.

PepMoV has been a useful "model" potyvirus in many of our teaching and research projects. The PepMoV-CI and -AI associated with tobacco infections are resolved readily with the light microscope. The virions, CI and AI, can be purified efficiently from infected tobacco with good yields. The PepMoV RNA is a very active template in in vitro translations with both the RRL and WG systems. The translational product resolution on SDS-PAGE is better than that obtained with any other potyvirus we have tested to date. The products mapped to the 3' end of the potyviral genome are processed more efficiently during in vitro translations with PepMoV than those seen for any other potyviral translations.

We have summarized some of the important and apparent differences, and also similarities, between PepMoV and PVY. On the basis of these comparisons between PepMoV and PVY, and the fact that PepMoV has maintained an important niche in commercial pepper production in the Americas, we believe that PepMoV should be considered a potyvirus more closely related to PVY than to other known potyviruses, but PepMoV should maintain its separate status on the basis of its distinctive biological and other properties.

References

1. Abdalla OA, Desjardins PR, Dodds JA (1991) Identification, disease incidence, and distribution of viruses infecting peppers in California. Plant Dis 75: 1019–1023
2. Christie RG, Edwardson JR (1977) Light and electron microscopy of plant virus inclusions. Fla Agric Exp Stat Monogr Ser, no 9
3. de Mejia MVG, Hiebert E, Purcifull DE, Thornbury DW, Pirone TP (1985) Identification of potyviral amorphous inclusion protein as a nonstructural, virus-specific product related to helper component. Virology 142: 34–43
4. Dougherty WG, Allison RF, Parks TD, Johnston RE, Feild MJ, Armstrong FB, (1985) Nucleotide sequence at the 3' terminus of pepper mottle virus genomic RNA: evidence for an alternative mode of potyvirus capsid protein gene organization. Virology 146: 282–291
5. Hay JM, Fellowes AP, Timmerman GM (1989) Nucleotide sequence of the coat protein gene of a necrotic strain of potato virus Y from New Zealand. Arch Virol 107: 111–122
6. Hiebert E, McDonald JG (1973) Characterization of some proteins associated with viruses in the potato virus Y group. Virology 56: 349–361
7. Hiebert E, Thornbury DW, Pirone TP (1984) Immunoprecipitation analysis of potyviral in vitro translation products using antisera to helper component of tobacco vein mottling virus and potato virus Y. Virology 135: 1–9
8. Hiebert E, Dougherty WG (1988) Organization and expression of the viral genomes. In: Milne B (ed) The plant viruses, vol 4, the filamentous plant viruses. Plenum, New York, pp 155–178
9. Jordan RL, Hammond, J (1991) Comparison and differentiation of potyvirus isolates and identification of strain-, virus-, subgroup-specific and potyvirus group-common epitopes using monoclonal antibodies. J Gen Virol 72: 25–36
10. Matthews REF (1991) Plant virology, 3rd edn. Academic Press, New York, pp 1–835
11. Nelson MR, Wheeler RE (1972) A new virus disease of pepper in Arizona. Plant Dis Rep 56: 731–735

12. Nelson MR, Wheeler RE (1978) Biological and serological characterization and separation of potyviruses that infect pepper. Phytopathology 68: 979–984
13. Nelson MR, Wheeler RE (1981) A local lesion host for potato virus Y. Phytopathology 71: 241
14. Nelson MR, Wheeler RE, Zitter TA (1982) Pepper mottle virus. CMI/AAB Descriptions of Plant Viruses, no 253
15. Purcifull DE, Hiebert E, McDonald JG (1973) Immunochemical specificity of cytoplasmic inclusions induced by viruses in the potato virus Y group. Virology 55: 275–279
16. Purcifull DE, Zitter TA, Hiebert E (1975) Morphology, host range, and serological relationships of pepper mottle virus. Phytopathology 65:559–562
17. Shukla DD, Ward CW (1988) Amino acid sequence homology of coat proteins as a basis for identification and classification of the potyvirus group. J Gen Virol 69: 2703–2710
18. Shukla DD, Ward CW (1989) Identification and classification of potyviruses on the basis of coat protein sequence data and serology. Arch Virol 106: 171–200
19. Shukla DD, Frenkel MJ, Ward CW (1991) Structure and function of the potyvirus genome with special reference to the coat protein region. Can J Plant Pathol 13: 178–191
20. Vance VB, Jordan R, Edwardson JR, Christie R, Purcifull DE, Turpen T, Falk B (1992) Evidence that pepper mottle and potato virus Y are distinct viruses: analyses of the coat protein and 3′ untranslated sequences of a California isolate of pepper mottle virus. In: Barnett OW (ed) Potyvirus taxonomy. Springer, Wien New York, pp 337–345 (Arch Virol [Suppl] 5)
21. Wefels E, Sommer H, Salamini F, Rohde W (1989) Cloning of the potato virus Y genes encoding the capsid protein CP and the nuclear inclusion protein NIb. Arch Virol 107: 123–134
22. Zitter TA (1972) Naturally occurring pepper virus strains in Florida. Plant Dis Rep 56: 586–590

Authors' address: E. Hiebert, Department of Plant Pathology, University of Florida, Gainesville, FL 32611, U.S.A.

Arch Virol (1992) [Suppl 5]: 327–335

Is pepper mottle virus a strain of potato virus Y?

R. van der Vlugt

Department of Virology, Agricultural University, Wageningen, The Netherlands

Summary. On the basis of serological properties and host plant reactions pepper mottle virus (PepMoV) has been classified as a potyvirus related to, but distinct from, other pepper-infecting potyviruses, potato virus Y (PVY) and tobacco etch virus (TEV).

Recent amino acid and nucleotide sequence data show that PepMoV is more closely related to PVY than previously assumed. PepMoV shows a high degree of homology to various PVY strains in both the coat protein and the 3′ non-translated sequences, while unrelated potyviruses are generally less homologous in these regions. Detailed coat-protein amino acid sequence and 3′ non-translated region (3′ NTR) nucleotide sequence comparisons described in this paper confirm the close relationship between PepMoV and PVY and it is concluded that the isolate sequenced indeed represents a strain of PVY. Sequence data for several strains of PVY gave two groups with closer relationships among strains in a group than between groups.

Introduction

The potyvirus known as pepper mottle virus (PepMoV) was first described in 1972 by Zitter [27] as a new atypical pepper isolate of PVY from Florida, U.S.A., on the basis of differential host reactions and it was named PVY-speckling or PVY-S. Also in 1972 Nelson and Wheeler [14] described a new potyvirus isolated from chili pepper in Arizona. This virus was serologically and biologically unrelated to PVY and TEV and was named Arizona pepper virus. Both PVY-S and Arizona pepper virus differed from PVY and TEV, the only other North American potyviruses known to infect pepper, in that they caused characteristic necrotic primary infection spots on, and subsequent systemic necrosis and premature death of, the chili pepper, *Capsicum frutescens* L. 'Tabasco'. Purcifull et al. in 1973 [16] tentatively renamed PVY-S to PepMoV after finding its cytoplasmic inclusion protein to be serologically distinct from, though related to, PVY. Zitter and Cook [28] also referred to PVY-S as PepMoV. They observed that Brazilian bell pepper (*Capsicum annuum*) cultivar Avelar progeny plants are tolerant for PepMoV but resistant to infection by TEV or PVY. The authors considered this additional evidence that PepMoV is distinct from PVY. In 1975 PepMoV was more closely investigated by Purcifull et al. [17]. Immunodiffusion tests using coat-protein specific antisera and an antiserum specific for PepMoV cytoplasmic cylindrical inclusions showed that PepMoV is distinct from PVY, TEV and pepper veinal mottle virus (PVMV) from Ghana, the latter being the only non-American potyvirus known to infect pepper. In addition, the authors identified the Arizona pepper virus as a closely related strain of PepMoV since both viruses, and their induced lamellar inclusions, are serologically identical but appear to differ somewhat in host range [14]. This was the first report which identified two distinct strains of PepMoV. In 1978 Nelson and Wheeler [15] compared PVY, TEV, five isolates of PepMoV, and an unidentified pepper virus from North Carolina, named NC, by host range, cross-protection, and serology. The authors reported that only the PepMoV strains and the NC virus were serologically related, indicating a close relationship between NC and PepMoV. However, NC did not induce the primary infection spots on *Capsicum frutescens* 'Tabasco' characteristic for PepMoV isolates but symptoms were more intermediate between PVY and PepMoV (mosaic with occasional systemic necrosis occurs but no primary necrotic spots). Cross-protection tests between NC and PepMoV also did not indicate a relationship. Interestingly, clear biological differences occurred on different host plants allowing distinction between the PepMoV isolates which indicated the existence of PepMoV strains.

The biological data above do indicate a close relationship between PepMoV and PVY but are they sufficient to determine if PepMoV and PVY are distinct viruses or merely strains of the same virus?

Recently, amino acid and nucleotide sequence determinations have greatly added to our knowledge about the classification of potyviruses [24, 21, 8]. Most sequence data have been obtained from the coat-protein region, obviously because of its genomic localization and the application of the coat-protein genes in genetically engineered protection. Many amino acid sequence comparisons revealed extensive homologies among different potyviruses especially in the central and C-terminal part of the coat protein. Major differences, however, are found in length and sequence of their N-termini. Distinct potyviruses all differ in the number of amino acid residues in this region and only share a DAG motif which might be involved in aphid transmission [9, 3]. In contrast, different strains of a distinct virus all show very homologous N-terminal sequences. Coat-protein comparisons by Shukla and Ward [21] showed a bimodal distribution of sequence homologies with distinct viruses having 38 to 71% (average 54%) homology and established virus strains showing 90 to 99% homology. From this the authors concluded that amino acid data from coat proteins can reliably be used for identification and classification of potyviruses.

For many potyviruses the nucleotide sequence of the 3' non-translated region (3'NTR) has been determined. In this region a high sequence variability is observed between different viruses. Distinct viruses generally share 35–50% sequence homology in contrast to 85–100% for virus strains. This led to the conclusion that the potyviral 3'NTR sequences also are useful in the classification of potyviruses [8].

The first nucleotide sequence information of the 3' terminal region of a PVY-strain displayed [24] a strong homology to the 3'NTR of PepMoV. A high amino acid sequence homology between the coat proteins of PepMoV and PVY strains was reported previously [22]. From the degree of homology observed between the 3'NTRs and the coat-protein regions it was concluded that PepMoV should be regarded a strain of PVY [22, 24].

In this paper I report a more extensive comparison between PepMoV and PVY strains on the amino acid and nucleotide sequence level.

Results

To further elucidate the relationship between PepMoV and PVY the coat-protein amino acid sequence of PepMoV [7] was compared to those of 15 different PVY strains using the GAP program from the University of Wisconsin GCG package while additional alignments were made by hand. Figure 1 shows the alignment of the coat-protein sequences and a derived consensus sequence for the PVY subgroup. The PepMoV coat-protein sequence is similar in length to PVY coat proteins (267 amino acids) and shows a similar degree of homology to the consensus sequence (95.5%) as the other PVY coat-protein sequences (94.8–98.1%). This level of homology falls within the range of homologies (90.6–99.3%) found between strains of other potyviruses [21].

Also from Fig. 1 it can be observed that certain strains show identical differences from the consensus sequence and are therefore likely to be more closely related to each other than to the other PVY strains. Figure 1 shows a clear relationship between strains Ne [24], GO16 [25], Nz [11], Jp [10], and H [5] as well as between strains 10, 18, 43, and D [22].

In analogy with the coat-protein encoding region a comparison was made among the 3'NTRs of PepMoV and eight PVY strains for which the sequence of this region is available (Fig. 2). The regions are very similar in length (330–336 nt) and show a high degree of homology (82–98%) (Table 1). In contrast the 3'NTRs of other potyviruses are variable in length (ranging from 168 to 474 nt) and show only 38 to 48% homology to the PVY 3'NTRs (results not shown). Interestingly, the observation that certain PVY strains show identical differences from the derived coat-protein consensus sequence can also be made for the same strains in their 3'NTR (Fig. 2). This is supported by Table 1 in which two main clusters of PVY strains can be distinguished. Strains Fr [19], Ne [24], GO16 [25] and Nz [11] show 92–98% homology to each other while this group shows only 82–88% homology to the cluster containing

```
         1                                                           60
PVY-Fr               n              p                                  t
PVY-Ne     g             t      q        l     e     v     v
PVY-GO16   g             t      q        l     e           v                 k
PVY-Nz     ...           t      q        f     e           v
PepMoV         t   n         v           ss                          a
PVY-I          d         r               p.
PVY-O              n n              s     s ls             v                    r
PVY-Kg         en                        r p
PVY-Ru     g             t      q   v     l     e     e               r
PVY-10               e                    p                           a       r
PVY-18               dn  k                va.                          a       r
PVY-43               e                    ap                          a       r
PVY-D                e           r        gv p                        a       r
PVY-Ch             v d   n                s p                         p
PVY-Jp     g                     q        sl    e     e     v
PVY-H      g             t       q        l  n  e     v
Consensus  ANDTIDAGGS SKKDAKPEQG SIQPN-NKGK DKDVNAGTSG THTVPRIKAI TSKMRMPKSK

         61                                                          120
PVY-Fr
PVY-Ne                                                q
PVY-GO16                                              q
PVY-Nz       ia                                       q
PepMoV       a   k                                     v
PVY-I        vaa              t
PVY-O        vaa                                             q
PVY-Kg       ta                                     v
PVY-Ru                                            e
PVY-10           h                                          d
PVY-18       s                                              d
PVY-43       aa                                             d
PVY-D            h                                          d
PVY-Ch                               .
PVY-Jp                                      k   ql
PVY-H        a   k                              ql
Consensus  GATVLNLEHL LEYAPQQIDI SNTRATQSQF DTWYEAVRMA YDIGETEMPT VMNGLMVWCI

         121                                                         180
PVY-Fr
PVY-Ne         i            d
PVY-GO16       i            k
PVY-Nz         i            d
PepMoV         i            s
PVY-I                       s
PVY-O          i
PVY-Kg          l           s          s
PVY-Ru
PVY-10
PVY-18
PVY-43
PVY-D
PVY-Ch                          g
PVY-Jp         i            d
PVY-H          i            d
Consensus  ENGTSPNVNG VWVMMDGNEQ VEYPLKPIVE NAKPTLRQIM AHFSDVAEAY IEMRNKKEPY

         181                                                         240
PVY-Fr              mg
PVY-Ne                                                s
PVY-GO16                                              s
PVY-Nz                                                s
PepMoV             a                                  s
PVY-I              i                        e
PVY-O              i
PVY-Kg                                                s            v
PVY-Ru             vg
PVY-10             vg
PVY-18              g
PVY-43
PVY-D              vg
PVY-Ch             vg
PVY-Jp        v                                       s
PVY-H         v                                       s
Consensus  MPRYGLIRNL RDGSLARYAF DFYEVTSRTP VRAREAHIQM KAAALKSAQP RLFGLDGGIS
```

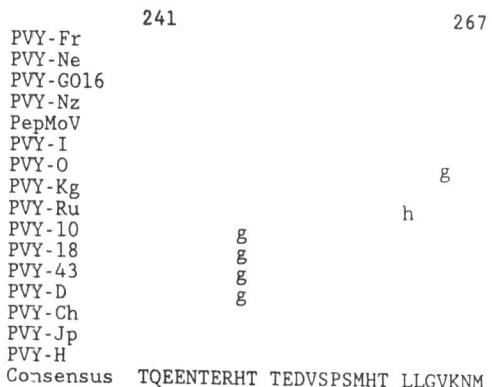

```
                241                          267
PVY-Fr
PVY-Ne
PVY-GO16
PVY-Nz
PepMoV
PVY-I
PVY-O                                  g
PVY-Kg
PVY-Ru                             h
PVY-10           g
PVY-18           g
PVY-43           g
PVY-D            g
PVY-Ch
PVY-Jp
PVY-H
Consensus  TQEENTERHT TEDVSPSMHT LLGVKNM
```

Fig. 1. Alignment of the coat-protein sequences of PepMoV and 15 PVY strains. The sources of sequence data are: PVY-Fr [19], PVY-Ne [24], PVY-GO16 [25], PVY-Nz [11], PepMoV [7], PVY-I [20], PVY-O [4], PVY-Kg (unpubl. results), PVY-Ru [18], PVY-10, PVY-18, PVY-43 and PVY-D [22], PVY-Ch [26], PVY-Jp [10], and PVY-H [5]. The bottom sequence represents the consensus derived from all 16 sequences

PepMoV [7], I [20], O [4], Ru [18] and Kg (J.E. Pot, unpubl. results). The strains in this second cluster also show an average homology of over 91%.

Discussion

From earlier biological data it was concluded that PepMoV was related to, but distinct from, PVY [15, 16, 18, 19, 30, 31]. Recently obtained sequence data on the coat protein and 3′NTR of PepMoV and an increasing number of different PVY strains indicate a close relationship [24, 26]. The high level of homology between the viruses in the 3′NTR suggests a very close relationship, because variability in the 3′NTR length and significantly lower homologies occur in this region between unrelated potyviruses. On the basis of the 3′NTR and coat-protein sequence data, this PepMoV isolate should therefore be considered a variant of PVY.

The coat protein and 3′NTR compose about one-third of the potyvirus genome. Of the potyviruses whose complete sequence has been determined, sequence comparisons of non-structural proteins corroborate conclusions from the 3′ portion of the genome. For instance, high levels of homology are observed in the non-structural proteins of potyvirus strains whose complete sequences have been determined. Homologies observed within two strains of TEV, TEV [2] and TEV-HAT [1; R.F. Allison, EMBL Accession no M11458]; three strains of plum pox virus, PPV-D [23], PPV-R [12] and PPV-NAT [13]; and two strains of PVY[N] [19; R.A.A. van der Vlugt, unpubl.] all exceed 85% while comparison of non-structural proteins of unrelated viruses only shows homologies ranging from 35 to 70%. This indicates that biological differences observed between strains are not likely to be caused by large differences in their non-structural proteins. However recombination events in the genome

```
               1                              .  .                                        .
Fr             .                              .. t               ga        tg t           .
Ne             .                a             .. t               a         tg t         .  .
GO16                            gc  cc         t                 ga        tg t                t
Nz                             ga                                          tg t                t
PepMoV         .                                                          g                .
I              .                              .. t                                             .
O                              ga                                                          .
Kg                                            ..
Ru

Consensus    TGATTGTAGT  G..TCTCTCC  GGACGATATA  TAAGTATTTA  CATATGCAGT  AAGTA.TTTT

               61                             a              .                  ggc
Fr             .                                                               ggc
Ne             .                                   c                   t       g c
GO16           .                              a              .                 g c
Nz             .                              t    c    g                    t c    a
PepMoV         g                              t c  c                         t a    a
I              .                              a                              a
O              .                              t    c                         t a
Kg             .                              a                              a
Ru

Consensus    .GGCTTTTCC  TGTACTACTT  TTATCGTAAT  TAATAATCAG  TTTGAATATT  ACTVATAGAT

               121      g     t    ta c         g    t a    ct  g       cgg
Fr             g     tg   ta c         g    .ca    ct  g       cg
Ne             g     t    ta c         g    . a    ct  g       cg
GO16           g     t    ta c         g    . a    ct  g       cg
Nz                   .   .      a    t               .       t  g
PepMoV               .   .      a    t               .       t  t        c
I                              t               .                t a     c c
O                              t             c    .                 t     c
Kg                                     t             .             t a   c c
Ru

Consensus    AGAGGTGGCA  GGGTGATTCC  GTCATTG.TG  GTGACTCTAT  CTGTBATTTC  TGTATTATTA

               181       c t g        a  .              .    ac         t
Fr             t         g.           .  .                   c          t
Ne             t         g. g         .  .             a c          c
GO16           t         g.           .  .                   c          t
Nz                       c a          .  .        c         a g       aat
PepMoV         c   a  .         . c              a a    c a c
I                         t gg        ...  g          c      a          tc
O                         t           .  .            a g       a c
Kg                        c t a       .  .                   a          a  .
Ru

Consensus    AGTTTNATAT  AAAA.GT.GC  CGGGTTGTTG  TTGTTGTGGV  TGATCYATCG  ATTAGGTGAT

               241   c t           a                     t  g                    c
Fr                               .                        t  g                    c
Ne                               .                        t  g                    ac
GO16                   t  .                               t  g      g             c
Nz                     t        c a                    t     c              g    a
PepMoV                          c                         t  c              g
I                                                        t  c  g            g
O              c       t  c                               t  c  g           g
Kg                     t          a                   c   t  c              g
Ru                                                       t  c  g            g

Consensus    GTTGCGATTC  TGTCGTAGCA  GTGACTATGT  CTGGATCTAB  TTACTTGGGT  GATGTTGTGA

               301      t                                    340
Fr                 t
Ne                 t
GO16               t          t
NZ                 t
PepMoV         t            a
I                          a
O              t           a
Kg
Ru             t           a                    . .  . . . . . . . . .

Consensus    TTCCGTCATA  GCAGTGACTG  TAAACTTCAA  TCAGGAGACA_n
```

Fig. 2. Alignment of the 3′ NTRs of PepMoV and eight PVY strains. The sources of the sequence data are as indicated in Fig. 1. The bottom sequence represents the consensus derived from all nine sequences. Y=T/C, V=C/C/G, B=T/C/G, N=A/T/C/G. Dots indicate gaps introduced to maximize alignments

Table 1. Percentage of nucleotide sequence homologies among the 3' NTRs of PVY strains. The sources of sequence data are as indicated in Fig. 1.

	Fr	Ne	GO16	Nz	PepMoV	I	O	Kg	Ru
Fr	–	93.3	91.7	93.0	83.3	82.1	87.8	86.1	86.8
Ne		–	96.0	97.9	83.9	83.5	84.2	86.1	85.5
GO16			–	95.7	83.0	83.8	84.5	87.0	84.9
Nz				–	83.9	83.8	84.2	86.4	85.5
PepMoV					–	91.6	87.6	93.4	89.4
I						–	89.4	94.0	90.9
O							–	90.9	96.9
Kg								–	92.8
Ru									–

structure of potyviruses cannot be ruled out as a cause of dramatic biological differences.

In view of the above it can be concluded that amino acid and nucleotide sequence comparisons, preferably of complete genomes as suggested by Dijkstra [6], form the only reliable means for the distinction between different potyviruses and should therefore form the basis of potyvirus taxonomy. Biological criteria however can be very useful tools in distinguishing between different virus strains.

It would be interesting to obtain more sequence data of the PepMoV genome and additional strains. In this respect the NC strain of PepMoV described by Nelson and Wheeler [15] is particularly interesting since on the basis of symptomatology it appears to be an intermediate between PepMoV and PVY. Furthermore it might be worthwhile to reinvestigate the precise biological differences that appear to exist between PepMoV and other PVY strains. Experiments whereby PVY^O, PVY^N, PepMoV, and PVY-Kg, a sweet pepper variant of PVY^N, are compared for their mutual biological relatedness are currently underway in our department.

Acknowledgements

I thank J. E. Pot and Dr. P. Vos of Keygene nv. Wageningen for providing the PVY-Kg and its sequence, Dr. R. Goldbach for critical reading of the manuscript and Dr. J. Dijkstra for stimulating discussions.

References

1. Allison RF, Dougherty WG, Parks TD, Johnston RE, Kelly ME, Armstrong FB (1985) Biochemical analysis of the capsid protein gene and capsid protein of tobacco etch virus: N-terminal amino acids are located on the virion's surface. Virology 147:309–316

334   R. van der Vlugt

2. Allison RF, Johnston RE, Dougherty WG (1986) The nucleotide sequence of the coding region of tobacco etch virus genomic RNA: evidence for the synthesis of a single polyprotein. Virology 154: 9–20

3. Atreya CD, Raccah B, Pirone TP (1990) A point mutation in the coat protein abolishes aphid transmissibility of a potyvirus. Virology 178: 161–165

4. Bravo-Almonacid FF, Mentaberry AN (1989) Nucleotide cDNA coding for the PVY° coat protein. Nucleic Acids Res 17: 4401

5. Dalmay T, Balázs E (1990) Nucleotide sequence of an altered virulence potato virus Y coat protein gene (PVY^H). Nucleic Acids Res 18: 6721

6. Dijkstra J (1992) Importance of host ranges and other biological properties for the taxonomy of plant viruses. In: Barnett OW (ed) Potyvirus taxonomy. Springer, Wien New York, pp 173–176 (Arch Virol [Suppl] 5)

7. Dougherty WG, Allison RF, Parks TD, Johnston RE, Feild MJ, Armstrong FB (1985) Nucleotide sequence at the 3' terminus of pepper mottle virus genomic RNA: evidence for an alternative mode of potyvirus capsid protein gene organization. Virology 146: 282–291

8. Frenkel MJ, Ward CW, Shukla DD (1989) The use of 3' NTR sequences in the taxonomy of potyviruses: application to watermelon mosaic virus 2 and soybean mosaic virus-N. J Gen Virol 70: 2775–2783

9. Harrison BD Robinson, DJ (1988) Molecular variation in vector-borne plant viruses: epidemiological significance. Philos Trans R Soc Lond [Biol] 321: 447–462

10. Hataya T, Sano T, Ohshima K, Shikata E (1990) Polymerase chain reaction-mediated cloning and expression of the coat protein gene of potato virus Y in *Escherichia coli*. Virus Genes 4: 339–350

11. Hay JM, Fellowes AP, Timmerman GM (1989) Nucleotide sequence of the coat protein gene of a necrotic strain of potato virus Y from New Zealand. Arch Virol 107: 11–122

12. Laín S, Riechmann JL, García JA (1989) The complete nucleotide sequence of plum pox virus potyvirus RNA. Virus Res 13: 157–172

13. Maiss E, Timpe U, Brisske A, Jelkmann W, Casper R, Himmler G, Mattanovich D, Katinger HWD (1989) The complete nucleotide sequence of plum pox virus RNA. J Gen Virol 70: 513–524

14. Nelson MR, Wheeler RE (1972) A new virus disease of pepper in Arizona. Plant Dis Rep 56: 731–735

15. Nelson MR, Wheeler RE (1978) Biological and serological characterization and separation of potyviruses that infect peppers. Phytopathology 68: 979–984

16. Purcifull DE, Hiebert E, McDonald JG (1973) Immunochemical specificity of cytoplasmic inclusions induced by viruses in the potato virus Y group. Virology 55: 275–279

17. Purcifull DE, Zitter TA, Hiebert E (1975) Morphology, host range and serological relationships of pepper mottle virus. Phytopathology 65: 559–562

18. Puurand Ü, Saarma M (1990) Cloning and sequencing of the 3' terminal region of potato virus Y^N (Russian isolate) RNA genome. Nucleic Acids Res 18: 6694

19. Robaglia C, Durand-Tardif M, Tronchet M, Boudazin G, Astier-Manifacier S, Casse-Delbart F (1989) Nucleotide sequence of potato virus Y (N strain) genomic RNA. J Gen Virol 70: 935–947

20. Rosner A, Raccah B (1988) Nucleotide sequence of the capsid protein gene of potato virus Y (PVY). Virus Genes 1: 255–260

21. Shukla DD, Ward CW (1988) Amino acid sequence homology of coat proteins as a basis for identification and classification of the potyvirus group. J Gen Virol 69: 2703–2710

22. Shukla DD, Thomas JE, McKern NM, Tracy SL, Ward CW (1988) Coat protein of potyviruses 4. Comparison of biological properties, serological relationships, and coat protein amino acid sequences of four strains of potato virus Y. Arch Virol 102: 207–219

23. Teycheney PY, Taven G, Delbos R, Ravelonandro M, Dunez J (1989) The complete nucleotide sequence of plum pox virus RNA (strain D). Nucleic Acids Res 17: 10115–10116
24. van der Vlugt RAA, Allefs S, De Haan PT, Goldbach RW (1989) Nucleotide sequence of the 3' terminal region of potato virus Y^N RNA. J Gen Virol 70: 229–233
25. Wefels E, Sommer H, Salamini F, Rohde W (1989) Cloning of the potato virus Y genes encoding the capsid protein CP and the nuclear inclusion protein NIb. Arch Virol 107: 123–134
26. Zhou X–R, Fang R-X, Wang C-Q, Mang K-Q (1990) cDNA sequence of the 3'-coding region of PVY genome (the Chinese isolate). Nucleic Acids Res 18: 5554
27. Zitter TA (1972) Naturally occurring pepper virus strains in Florida. Plant Dis Rep 56: 586–590
28. Zitter TA, Cook AA (1973) Inheritance of tolerance to a pepper virus in Florida. Phytopathology 63: 1211–1212

Author's address: R. van der Vlugt, Department of Virology, Agricultural University, Binnenhaven 11, NL-6709 PD Wageningen, The Netherlands.

Arch Virol (1992) [Suppl 5]: 337–345

Evidence that pepper mottle virus and potato virus Y are distinct viruses: analyses of the coat protein and 3' untranslated sequence of a California isolate of pepper mottle virus

Vicki Bowman Vance[1], R. Jordan[2], J. R. Edwardson[3], R. Christie[3], D. E. Purcifull[4], T. Turpen[5], and B. Falk[6]

[1] Department of Biological Sciences, University of South Carolina, Columbia, South Carolina
[2] U.S. Department of Agriculture, Agricultural Research Service, BARC-W, Beltsville, Maryland
Departments of [3] Agronomy and [4] Plant Pathology, University of Florida, Gainesville, Florida
[5] Biosource Genetics Corporation, Vacaville, California
[6] Department of Plant Pathology, University of California, Davis, California, U.S.A.

Summary. Pepper mottle virus (PepMoV) is a member of the large and complex genus *Potyvirus*, and is classically distinguished from other members of the genus by differential host range and cytopathology as well as serology of the coat protein and cytoplasmic inclusion body proteins. Here we report the deduced amino acid sequence of the coat protein of a California potyvirus identified by a variety of classical methods as PepMoV (PepMoV C). Comparison of the 3' untranslated nucleic acid sequence and the deduced coat-protein amino acid sequence of the PepMoV C isolate with those of PVY and other potyviruses indicates that PepMoV C is sufficiently diverged to be considered a distinct virus species. Thus, comparative sequence analyses of the PepMoV C isolate support earlier serological and biological evidence that PepMoV and PVY are distinct viruses.

Introduction

The potyvirus now known as pepper mottle virus (PepMoV) was first described in 1972 as an atypical pepper isolate of potato virus Y (PVY) from Florida, U.S.A., [27] and a similar virus from Arizona was described the same year [16]. Further investigations determined that these two isolates, as well as a third isolate from North Carolina, were strains of the same virus [20]. The PepMoV strains are distinct from PVY strains based on differential host range and cytopathology of the infection, as well as serological differences in both the coat and cytoplasmic inclusion body proteins [1, 5, 17, 19].

Within the genus *Potyvirus*, the borders between distinct viruses are blurred and it remains difficult to determine if an individual potyvirus isolate is a strain of another potyvirus or is a distinct virus. With the advent of cDNA cloning and DNA sequencing, the 3' untranslated and coat-protein sequences for numerous potyviruses have become available. Comparison of these regions of the potyvirus genome has become the major taxonomic tool to establish the virus/strain status of a particular potyviral isolate. A partial sequence of the genomic RNA of a PepMoV isolate of unreported origin has been published [7], and comparison with other potyviral sequences indicates that this particular PepMoV isolate is not a distinct virus, but a strain of PVY [23, 26].

We have recently cloned and sequenced the entire genome of a California isolate of a potyvirus identified by classical means as PepMoV. Comparison of the 3' untranslated region and the deduced coat-protein sequence of this PepMoV isolate (PepMoV C) with those of PVY and other related potyviruses indicates that the PepMoV C isolate is distinct from PVY.

Materials and methods

PepMoV C isolate

The isolate of PepMoV used in this work was originally isolated in 1974 from field grown pepper in California and identified at that time as PVY. The original collection was by Dr. A. O. Paulus, Department of Plant Pathology, University of California, Riverside, California, and the isolate was subsequently propagated by Dr. L. G. Weathers (UC, Riverside) by single aphid transmission. The RNA genome of this putative PVY isolate was molecularly cloned and a partial sequence (5' and 3' untranslated regions) was published in 1989 [25]. Sequence data reported here was derived from the same cDNA clones reported at that time. For the present study, a number of procedures were used to definitively determine that this potyviral isolate is not PVY but a strain of PepMoV.

Nucleic acid sequencing

The nucleotide sequence of PepMoV C deletion subclones was determined by dideoxy-nucleotide chain termination reactions on denatured double stranded templates using a genetically engineered form of T7 DNA polymerase exactly as indicated by the manufacturer (US Biochemical, Cleveland, Ohio). Internal sequencing primers were synthesized and used as needed to complete sequence determination for both strands of the cDNA.

Computer analyses

All the computer analyses were performed using the Sequence Analysis Software Package (GCG package, version 7) by Genetics Computer, Inc. (Madison, Wisconsin). The alignment of coat-protein amino acid sequences was created using the program "PileUp" which uses a simplification of the progressive alignment method of Feng and Doolittle [9].

Cytology

Epidermal strips from PepMoV C-infected tobacco were examined for inclusions in the O-G stain combination, "Calcomine orange 2RS-Luxol brilliant green BL" [4].

Host range analysis

The PepMoV C isolate was mechanically inoculated to a series of pepper cultivars which had shown differential susceptibility to potato virus Y, pepper mottle virus, and tobacco etch viruses. These cultivars included: Calwonder 300; Yolo Y; VR2; Agronomico 10; VR4; Del Ray Bell; PI 152225; PI 159236, and Greenleaf tobasco. Results were assessed by symptom development and ELISA one month after inoculation.

SDS-immunodiffusion

Immunodiffusion tests were conducted in agar plates containing sodium dodecyl sulfate (SDS) using polyclonal antibodies to virus and cylindrical inclusion proteins of California potyviruses, essentially as described [19]. Antigens were prepared by grinding infected leaf material in H_2O (1 g/ml) and adding SDS [19].

Antigen coated plate (ACP)-ELISA

Virus samples were evaluated using the PTY MAbs in an indirect ACP-ELISA as described [14]. Purified viruses were diluted to 2 µg/ml in 0.05 M sodium carbonate-bicarbonate buffer (CB) and dispensed to duplicate wells of Nunc MaxiSorp polystyrene ELISA plates. Plant samples were prepared as described [14], except that 0.2% sodium diethyldithiocarbamate was added to the CB/2% PVP extraction buffer.

Purified preparations of bean yellow mosaic virus (BYMV GDD), pepper mottle virus (PepMoV NC165), potato virus Y (PVY 3), and tobacco etch virus (TEV NAT), and BYMV-infected *Nicotiana benthamiana*, PVY 3-infected potato, and PepMoV NC165- and TEV PV69-infected *N. tabacum* cultivar Burley 21 were prepared as previously described [14]. PepMoV C was supplied as purified virus [25] or as systemically infected *N. tabacum* cultivar Xanthi leaves.

Monoclonal antibodies

Twenty-five monoclonal antibodies (MAbs) from a previous study [14], designated "PTY" MAbs, were used to compare PepMoV C to four other potyviruses. Members of this panel of MAbs recognize epitopes that are (*i*) specific to BYMV strains, (*ii*) common to members of the BYMV subgroup, (*iii*) distinctive for unique potyviruses, or (*iv*) common to many distinct potyviruses.

Results

Identification of the virus

Cytopathology of PepMoV C-infected tobacco

Epidermal peels of PepMoV C-infected tobacco leaves were examined by light microscopy for the presence of inclusion bodies. Long, very sharp-pointed cylindrical inclusions near the cell walls and irregularly shaped, cytoplasmic "amorphous" inclusions were observed. These inclusions are diagnostic of PepMoV-infected tissues and are readily distinguished from those induced by PVY [8].

Host range analysis

PepMoV C gave systemic symptoms on five of five plants of cultivars Calwonder 300 and Yolo Y, and none on any plants of the other cultivars tested. ELISA analysis confirmed the infections in Calwonder 300 and Yolo Y but also showed that AG10 and VR2 were symptomless hosts for PepMoV C. The resistance of the other symptomless cultivars to infection with this viral isolate was confirmed by ELISA analyses. Calwonder 300 is rated as universally susceptible to pepper potyviruses, while Yolo Y is rated as a cultivar resistant to PVY [22]. AG10 and VR2 are pepper cultivars with resistance to PVY isolates, but generally are susceptible to PepMoV isolates [18]. Thus the host range for PepMoV C is consistent with previous observations with California PepMoV isolates.

Serological analysis of the cytoplasmic inclusion (CI) body protein

Extracts of PepMoV C-infected tobacco leaves reacted positively with antisera to virions of a California PepMoV, and to CI proteins of California and Florida PepMoV isolates. However, no positive reactions for PepMoV C were obtained when tested with CI and virion antisera to a California TEV or with virion antiserum to a California PVY isolate.

Comparative serological analysis of PepMoV C and other potyviruses using monoclonal antibodies

The reactivity patterns of PepMoV C with a panel of 26 virus-specific and group-cross reactive MAbs were compared with those of four other potyviruses in ACP-ELISA. This panel of MAbs has been useful in examining the intra-virus, inter-virus, and intra-group serological relationships among/between diverse potyviruses [14, 12, 13]. Purified virions and plant sap extracts of all viruses tested exhibited the same reactivity patterns with the panel of MAbs.

The reactivity patterns of PepMoV C, PVY 3, PepMoV NC165, TEV PV69, and BYMV GDD with 11 of the MAbs are shown in Table 1. In this study, eight of the 26 PTY MAbs detected their respective epitopes on PepMoV C; these eight MAbs are among the 11 MAbs illustrated in Table 1. The 11 selected MAbs define 11 different potyviral epitopes [14]. MAbs PTY 1 through 12 each define unique potyvirus group-common epitopes present on 9 to 33 of 33 distinct potyviruses. MAbs 21, 30, and 33 define three unique

Table 1. Differential reactivity of PTY monoclonal antibodies with PepMoV C and four other potyviruses in ACP-ELISA[a]

Potyvirus[b]	PTY monoclonal antibody										
	1	2	3	4	8	10	12	21	30	33	35
BYMV GDD	+++	+++	+++	+++	++	(+)	+	+++	+++	+++	+++
PepMoV NC165	+++	–	+++	+	–	+++	++	+++	+	++	–
PepMoV c	+++	–	+++	+	–	+++	++	+++	+	++	–
PVY 3	+++	+++	–	+++	++	+++	–	–	–	–	–
TEV PV69	+++	+++	–	–	++	++	+	–	–	–	–

[a] Indirect ELISA using 2 µg/ml purified potyvirus diluted in carbonate buffer and antigen coating of Nunc ELISA A_{405} values (1 to 3 h substrate incubation): +++ ≥ 1.2; ++ 0.6 to 1.2; + 0.2 to 0.6; (+) ≥ 0.5 after 6 to 16 h substrate incubation; and negative controls < 0.1; – < 0.05

[b] Virus acronyms and sources are in text

epitopes shared between specific isolates of BYMV and PepMoV NC165, and now PepMoV C. MAb PTY 35 defines an epitope shared between six of nine BYMV isolates. The reaction profile of PepMoV C is indistinguishable from that of PepMoV NC165 and distinct from any of the other tested potyviruses. The entire reaction profile of PepMoV C against the panel of 26 MAbs (data not shown) is also identical to PepMoV NC165, and distinct from the other four potyviruses tested here. These data indicate that PepMoV C is indistinguishable from the previously identified PepMoV NC165 but distinct from PVY based on serology of the coat protein.

Homologies of PepMoV C sequences with those of the other potyviruses

PepMoV C 3′ untranslated region homologies

Analyses of potyviral 3′ untranslated sequences from a large number of viral isolates have suggested that homologies in this region may be used to differentiate distinct viral species from strains of the same virus [11]. Strains of the same virus are observed to share 83–99% homology within this region, while distinct viral species are dramatically more diverged sharing only 39–53% homology. Comparison of the previously reported 3′ untranslated region of PepMoV C [25] with that of several other potyviruses [2, 6, 15, 21] revealed homologies ranging from 32–44% (Table 2). These data suggest that the PepMoV C 3′ untranslated region is sufficiently diverged from that of other potyviruses so that this isolate should be considered a distinct virus species.

Table 2. Percent homology of PepMoV C 3′ untranslated nucleic acid sequence with those of other potyviruses

	PVY Nª	PepMoV	PepMoV C	TEV	TVMV	PPV
PVY N	100.0	83.2	36.4	36.7	37.7	33.0
PepMoV		100.0	37.2	38.3	38.1	33.9
PepMoV C			100.0	44.2	31.7	33.0
TEV				100.0	36.2	30.3
TVMV					100.0	36.6
PPV						100.0

ª Sources of viral sequence are as follows: PVY N [21]; PepMoV [7]; TEV [2]; TVMV [6]; and PPV [15]

Table 3. Percent similarity of PepMoV C coat protein amino acid sequence with those of other potyviruses

	PepMoV C	PVY Nª	PVY O	PepMoV	TEV	TVMV
PepMoV C	100.0	80.2	80.9	78.0	76.0	71.3
PVY N		100.0	95.9	94.4	76.4	69.4
PVY O			100.0	94.0	76.4	69.8
PepMoV				100.0	76.8	69.1
TEV					100.0	72.6
TVMV						100.0

ª Sources of viral coat protein sequence are as follows: PVY N [21]; PVY O [3]; PepMoV [7]; TEV [2]; and TVMV [6]

PepMoV C coat-protein sequence homologies

Like 3′ untranslated sequences, the sequences of potyviral coat proteins have been used to differentiate viral species from viral strains [24]. In this case, known strains of the same potyvirus are 90–99% homologous, while distinct potyviral species are again more diverged from one another, showing homologies ranging from 38–71%. The coat-protein sequence of PepMoV C was deduced from the cDNA sequence and compared with that of several other potyviruses including another isolate of PepMoV and two strains of PVY (Table 3). The PepMoV C coat-protein sequence is approximately 80% similar to that of the two strains of PVY and 78% similar to that of the other sequenced isolate of PepMoV [7]. These data suggest that PepMoV C is sufficiently diverged to be considered a distinct virus. In contrast, the previously reported coat-protein sequence of the other sequenced PepMoV isolate and those of both PVY strains have homologies ranging from 94–95%, suggesting that they are strains of the same viral species. The deduced PepMoV C coat-protein sequence is shown in Fig. 1 in alignment with those of several other potyviruses. As has been demonstrated with other members of the group [23], the major regions of divergence are located within the amino terminal region of the coat-protein sequence.

```
          1                                                        50
PepMoV C  SSSRSDTLDA GEEKKKNKEV ATVSDGMGKK EVESTRDSDV NAGTVGTFTI
PepMoV    ...an--i-t -gns--d... vkpeq-siqp ssnkgk-k-- ----s--h--
PVY N     ...an--i-- -gsn--d... -kpeq-siqp npnkgk-k-- ----s--h--
PVY O     ...an--i-- -gnn--d... -kpeqssiqs nlskgk-k-- -v--s--h--
TEV       ....gg-v-- sa-vg-.... .kkdqkdd-v aeqask-r-- ----s---s-
TVMV      ....---v-- -k--ardqkl -dkp....tl a--r-k-k-- -t--s---s-


          51                                                       100
PepMoV C  PRIKSITEKM RMPKQKRKGV LNLAHLLEYK PSQVDISNTR STQAFDNWY
PepMoV    ----a--a-- ----s-gaa- -k-d-----a -q-------- a--s---t--
PVY N     ----a--s-- ---ts-gat- p--e-----a -q-------- a--s---t--
PVY O     ----a--s-- ---rs-gvaa ---e-----a -q-------- a--s---t--
TEV       ---namat-- qy-rm-gev- v--n---g-- -q---l--a- a-he--aa-h
TVMV      --l-kaamn- k---vggss- v--d---t-- -a-ef-v--- a-hs--ka-h


          101                                                      150
PepMoV C  CEVMKAYDLQ EEAMGTVMNG LMVWCIENGT SPNISGTWTM MDGDEQVEFP
PepMoV    ea-rv---ig -te-p----- ---------- ----n-v-v- ---s------
PVY N     ea-rm---ig -te-p----- ---------- ----n-v-v- ---n------
PVY O     ea-rm---ig qte-p----- ---------- ----n-v-v- ---n------
TEV       qa--t--gvn --q-kil--- ---------- ---ln---v- -------s--
TVMV      tn--ael--n --q-ki---- ---------- ------v--- ----------


          151                                                      200
PepMoV C  LKPVIENAKP TFRQIMAHFS DVAEAYIEMR NKQEPYMPRY GLVRNLRDMG
PepMoV    ---------- ---------- ---------- --k------- --------as
PVY N     ---------- ---------- ---------- --k------- ----------
PVY O     ---------- ---------- ---------- --k------- -------is
TEV       ---m----q- ------t--- -l-------- -rer------ --q--it--s
TVMV      ie-m-kh-n- s-----k--- nl-----r-- -seqv-i--- --q-g-v-rn


          201                                                      250
PepMoV C  LARYAFDFYE VTSRTSTRAR EAHIQMKAAA LKSAQTRLFG LDGGIGTQGE
PepMoV    ---------- -----pv--- ---------- -----s---- ----s--e-
PVY N     ---------- -----pv--- ---------- -----p---- -----s--e-
PVY O     ---------- -----pv--- ---------- -----p---- -----s--e-
TEV       -s-------- l--k-pv--- ---m------ vrnsg----- ---n---ae-
TVMV      --p------- -nga-pv--- ---a------ -rns-q---c ---s-sg-e-


          251                273
PepMoV C  NTERHTTEDV SPDMHTLLGV REM
PepMoV    ---------- --s------- kn-
PVY N     ---------- --s------- kn-
PVY O     ---------- --s------g kn-
TEV       d-----ah-- nrn------- -q.
TVMV      ------v--- naq--h---- kgv
```

Fig. 1. Computer assisted alignment of coat-protein amino acid sequence of pepper mottle virus strain C with those of the other sequenced isolate of pepper mottle virus [7], potato virus Y strain N, potato virus Y strain O, tobacco etch virus, and tobacco vein mottling virus. A dash indicates that the amino acid in that position is identical to that in the PepMoV C sequence present. A dot indicates that no amino acid is present in the analogous position

Discussion

Based on biological and serological properties, PepMoV is currently classified as a distinct viral species, closely related to PVY. This taxonomic status as a distinct viral species has recently come into question based on sequence data from a single PepMoV isolate which showed high homology to numerous PVY strains [10]. We report the coat-protein and 3′ untranslated region sequence of a potyviral isolate identified as PepMoV by a number of classical means, including host range and cytopathology of infection as well as serology of coat and cytoplasmic inclusion body proteins. These data suggest that this California isolate of PepMoV is in fact distinct from PVY. The confusion in PVY/PepMoV taxonomic status points to the need for further sequence analyses using a variety of PepMoV isolates.

Acknowledgments

We wish to gratefully acknowledge and thank Mary Ann Guaragna for excellent technical assistance. We thank Dr. W. G. Dougherty for critical reading of the manuscript. The use of trade, firm, or corporation names in this article does not imply the endorsement or approval by the USDA-ARS of any product to the exclusion of others that may be suitable. This work was supported in part by Grant DBM-8817233 from the National Science Foundation.

References

1. Abdalla OA, Desjardins PR, Dodds JA (1991) Identification, disease incidence, and distribution of viruses infecting peppers in California. Plant Dis 75: 1019–1023
2. Allison R, Johnston RE, Dougherty WG (1986) The nucleotide sequence of the coding region of tobacco etch virus genomic RNA: evidence for the synthesis of a single polyprotein. Virology 154: 9–20
3. Bravo-Almonacid FF, Mentaberry AN (1989) Nucleotide cDNA sequence coding for the PVY O coat protein. Nucleic Acids Res 17: 4401
4. Christie RG, Edwardson JR (1977) Light and electron microscopy of plant virus inclusions. Fla Agric Exp Stat Monogr Ser, no 9
5. deMejia MVG, Hiebert E, Purcifull DE, Thornbury DW, Pirone TP (1985) Identification of potyviral amorphous inclusion protein as a nonstructural, virus-specific protein related to helper component. Virology 142: 34–43
6. Domier LL, Franklin KM, Shahabuddin M, Hellmann GM, Overmeyer JH, Hiremath ST, Siaw MFE, Lomonossoff GP, Shaw JG, Rhoads RE (1986) The nucleotide sequence of tobacco vein mottling virus RNA. Nucleic Acids Res 14: 5417–5430
7. Dougherty WG, Allison RF, Parks TD, Johnston RE, Feild MJ, Armstrong FB (1985) Nucleotide sequence at the 3′ terminus of pepper mottle virus genomic RNA: evidence for an alternative mode of potyvirus capsid protein gene organization. Virology 146: 282–291
8. Edwardson JR, Christie RG, Purcifull DE, Petersen MA (1992) Inclusions in diagnosing plant virus diseases. In: Matthews REF (ed) Diagnosis of plant virus diseases. CRC Press, Boca Raton (in press)
9. Feng D, Doolittle RF (1987) Progressive sequence alignment as a prerequisite to current phylogenetic trees. J Mol Evol 25: 351–360

10. Francki RIB, Fauquet CM, Knudson DL, Brown F (eds) (1991) Classification and nomenclature of viruses. Fifth report of the International Committee on Taxonomy of Viruses. Springer, Wien New York (Arch Virol [Suppl] 2)

11. Frenkel MJ, Ward CW, Shukla DD (1989) The use of 3′ NTR sequences on the taxonomy of potyviruses: application to watermelon mosaic virus 2 and soybean mosaic virus-N. J Gen Virol 70: 2775–2783

12. Hammond J, Jordan RL, Larsen RC, Moyer JW (1991) Serological relationships among three filamentous viruses of sweet potato examined using polyclonal and monoclonal antibodies. Phytopathology 81: 1217

13. Hampton RO, Shukla DD, Jordan RL (1992) White lupin mosaic virus: comparative host range, serology, and coat protein peptide profiles. Phytopathology 82: 566–571

14. Jordan RL, Hammond J (1991) Comparison and differentiation of potyvirus isolates and identification of strain-, virus-, subgroup-specific and potyvirus group-common epitopes using monoclonal antibodies. J Gen Virol 72: 25–36

15. Maiss E, Timpe U, Brisske A, Jelkmann W, Casper R, Himmler G, Mattanovich D, Katinger HWD (1989) The complete nucleotide sequence of plum pox virus RNA. J Gen Virol 70: 513–524

16. Nelson MR, Wheeler RE (1972) A new virus disease of pepper in Arizona. Plant Dis Rep 56: 731–735

17. Nelson MR, Wheeler RE (1978) Biological and serological charcterization and separation of potyviruses that infect peppers. Phytopathology 68: 979–984

18. Nelson MR, Wheeler RE, Zitter TA (1982) Pepper mottle virus. CMI/AAB Descriptions of Plant Viruses, no 253

19. Purcifull DE, Hiebert E, McDonald JG (1973) Immunochemical specificity of cytoplasmic inclusions induced by viruses in the potato Y group. Virology 55: 275–279

20. Purcifull DE, Zitter TA, Hiebert E (1975) Morphology, host range and serological relationships of pepper mottle virus. Phytopathology 65: 559–562

21. Robaglia C, Durand-Tardif M, Tronchet M, Boudazin G, Astier-Manifacier S, Casse-Delbart F (1989) Nucleotide sequence of potato virus Y (N strain) genomic RNA. J Gen Virol 70: 935–947

22. Shifriss C, Marco S (1980) Partial dominance of resistance to potato virus Y in *Capsicum*. Plant Dis 64: 57–59

23. Shukla DD, Ward CW (1988) Amino acid sequence homology of coat proteins as a basis for identification and classification of the potyvirus group. J Gen Virol 69: 2703–2710

24. Shukla DD, Ward CW (1989) Identification and classification of potyviruses on the basis of coat protein sequence data and serology. Arch Virol 106: 171–200

25. Turpen T (1989) Molecular cloning of a potato virus Y genome: nucleotide sequence homology in non-coding regions of potyviruses. J Gen Virol 70: 1951–1960

26. van der Vlugt RAA, Allefs S, DeHaan PT, Goldbach RW (1989) Nucleotide sequence of the 3′ terminal region of potato virus Y^N RNA. J Gen Virol 70: 229–233

27. Zitter TA (1972) Naturally occurring pepper virus strains in Florida. Plant Dis Rep 56: 586–590

Authors' address: Vicki Bowman Vance, Department of Biological Sciences, University of South Carolina, Columbia, SC 29208, U.S.A.

Virus relationships – SCMV subgroup

Arch Virol (1992) [Suppl 5]: 349–351

A viewpoint on the taxonomy of potyviruses infecting sugarcane, maize, and sorghum

S. G. Jensen

U.S. Department of Agriculture, Agricultural Research Service,
Department of Plant Pathology, University of Nebraska, Lincoln, Nebraska, U.S.A

Summary. The value of taxonomy lies in its ability to communicate concepts and relationships. Our present concepts of the poaceous potyviruses, based on their biology, serology, and biochemistry, identify four viruses that can be distinguished by each characteristic. Identifying these as four distinct viruses has important implications for disciplines such as epidemiology, plant breeding, and diagnostics.

*

I recently bought a book, *Extraordinary People*, by Dr. Darold A. Trefert. The first line in this book has particular reference to the subject at hand: "The beginning of wisdom is to call things by their right name."

For years we have spoken of maize dwarf mosaic virus (MDMV) and then apologized because, we said, it was really just a strain of sugarcane mosaic virus (SCMV) with the wrong name. There were also many strains of SCMV, some that were so unique that we wondered where the missing links between them and other members of the SCMV group might be found. The whole system of names and strains grew with each new description with no scientific basis.

In the last five years new information and new techniques have begun to establish some semblance of order. In my own lab I had made antisera to the inclusion proteins of several of the members of this potyvirus complex in an attempt to use a different method to show relationships among members of this group. We concluded that there were at least four distinct groups of 'strains' in this complex [3]. In addition, these groups were really quite different from each other in a number of other ways.

Then Shukla et al. [7] described four different viruses based on the serology of the coat protein done by improved methodology. My work fits perfectly with their results and I support their conclusions completely.

This concept of four different viruses led to an immediate breakthrough. A potyvirus isolate from Kansas, U.S.A., called KS-1, had a number of characteristics different from MDMV, SCMV, or any other maize potyvirus. KS-1 inclusion body proteins had serological similarities [3] to a strain found in Texas, described as MDMV-O [4]. Thus, KS-1 should infect oats like MDMV-O and it did. Shukla et al. [6] had placed MDMV-O with johnsongrass mosaic virus (JGMV strain MD-O). If that was true then JGMV should also infect oats; in subsequent research it was found that JGMV does infect oats [8]. In cooperative work with the Australian group [5] we have shown that these three isolates have many other similarities and are undoubtedly isolates of the same virus from three distant parts of the world.

Based on both the coat proteins and the inclusion proteins, identifying and naming four distinct viruses seems valid biochemically, serologically, and biologically.

In addition to the scientific validity in separating these isolates into four groups and giving each a separate name there are some good practical reasons. Breeders have done a good job of keeping the virus diseases under control. However, they were regularly surprised and disappointed when some of their 'resistance' broke down under a new 'strain' of the virus. Our work clearly shows that the fault has been with the pathologists for not providing a clear understanding of which were 'viruses' and which were 'strains.' Now we can look past strains and identify genes that confer resistance to each of the viruses.

It is not only useful knowledge for the breeders but also for the epidemiologists and the diagnosticians. In the U.S.A. the common wisdom was that MDMV-A is in the south and SCMV-MD-B (previously MDMV-B) in the north because "A" can infect Johnsongrass while "B" cannot and Johnsongrass is found only in the south. But, there is more to the story. "A" is also more virulent than "B" on the northern grasses [1] and should be more common. Yet "B" predominates in the north for reasons unknown. Different viruses would not necessarily compete for the same niche.

JGMV-MD-O was found on the Texas coast [4] while a nearly identical isolate, JGMV-KS-1, is found only around Hays, Kansas, U.S.A. [2], and the original JGMV is from Australia [7]. Is there some reason why the JGMV isolates are found in these few areas thousands of miles apart? Perhaps now that we know how JGMV differs from SCMV and MDMV, JGMV may be found all over the world.

Sorghum mosaic virus (SrMV), including strains SC-H, SC-I, and SC-M, occurs only in the deep south in the U.S.A. [7]. However, the serious concern is, will it move north? If we knew its biology and its potential for expansion we would know if the breeders in Kansas and Nebraska should brace for a new threat or just ignore it. Now that SrMV is known to be different from SCMV, about which a great deal of epidemiological information has been collected, it is recognized that more needs to be learned about SrMV.

Less than 10 years ago while working with an international sorghum project on the sorghum diseases of the developing countries of Africa, I was told that they really had no virus problems. Now, with better education and broader experience they are recognizing viruses everywhere. The key to helping them advance is to call their viruses 'by the right name'.

So now, what is 'the right name' for these potyviruses? We recognize that humans are all one species but we have other 'names', based upon ethnic background, race, or gender, each of which conveys different meanings. Should we drop these names and just call everyone 'human'? I think not. Certainly the potyviruses, like humans, are a large group, but properly chosen names can help us sort them out and identify them. Putting the various maize, sorghum, and sugarcane virus strains that have been distinguished into the proper virus classification, MDMV, SCMV, JGMV or SrMV, seems justified by all criteria used. For now, the four viruses identified seem solid and well defined from each other on the basis of their differences in biochemistry, biology, and serology. Their similarities simply identify them as members of the genus *Potyvirus*. In the future we may want to add other names to the list as new viruses with new properties are found. For now each strain or isolate that has been examined carefully, in all of its properties, has fit nicely into one of the four named viruses.

References

1. Jensen SG, Palomar MK, Gorz HJ, Haskins FA (1986) Epidemiology of maize dwarf mosaic virus in the northern United States. In: Proceedings of the Workshop on Epidemiology of Plant Virus Diseases, Orlando, Florida, Sec I-21
2. Jensen, SG, Staudinger, JL (1989) A Kansas isolate of MDMV (KS-1) is serologically similar to MDMV-O. Phytopathology 79: 1004
3. Jensen SG, Staudinger JL (1989) Serological grouping of the cytoplasmic inclusions of 6 strains of sugarcane mosaic virus. Phytopathology 79: 1215
4. McDaniel LL, Gordon DT (1985) Identification of a new strain of maize dwarf mosaic virus. Plant Dis 69: 602–607
5. McKern NM, Shukla DD, Toler RW, Jensen SG, Tosic M, Ford RE, Leon O, Ward CW (1991) Confirmation that the sugarcane mosaic virus subgroup consists of four distinct potyviruses by using peptide profiles of coat proteins. Phytopathology 81: 1025–1029
6. McKern NM, Whittaker LA, Strike PM, Ford RE, Jensen JG, Shukla DD (1990) Coat protein properties indicate that maize dwarf mosaic virus-KS1 is a strain of Johnsongrass mosaic virus. Phytopathology 80: 907–912
7. Shukla DD, Tosic M, Jilka J, Ford RE, Toler RW, Langham MAC (1989) Taxonomy of potyviruses infecting maize, sorghum, and sugarcane in Australia and the United States as determined by reactivities of polyclonal antibodies directed towards virus-specific N-termini of coat proteins. Phytopathology 79: 223–229
8. Tosic M, Ford RE, Shukla DD, Jilka J (1990) Differentiation of sugarcane, maize dwarf, Johnsongrass, and sorghum mosaic viruses based on reactions of oat and some sorghum cultivars. Plant Dis 74: 549–552

Author's address: S.G. Jensen, USDA/ARS, Department of Plant Pathology, University of Nebraska, Lincoln, NE 68583-0722, U.S.A.

Arch Virol (1992) [Suppl 5]: 353–361

Differentiation of the four viruses of the sugarcane mosaic virus subgroup based on cytopathology

D.-E. Lesemann[1], D. D. Shukla[2], M. Tosic[3], and W. Huth[1]

[1] Biologische Bundesanstalt für Land- und Forstwirtschaft, Institut für Biochemie und Pflanzenvirologie, Braunschweig, Federal Republic of Germany
[2] CSIRO, Division of Biomolecular Engineering, Parkville, Victoria, Australia
[3] Institute of Plant and Food Protection, Faculty of Agriculture, University of Belgrade, Belgrade, Yugoslavia

Summary. A cytological comparison has been made of representative isolates of johnsongrass mosaic (JGMV), maize dwarf mosaic (MDMV), sorghum mosaic (SrMV) and sugarcane mosaic (SCMV) viruses. These four viruses now encompass the complex of virus strains which were formerly considered as strains of sugarcane mosaic and/or maize dwarf mosaic viruses. The structure of the cytoplasmic cylindrical inclusions induced by these viruses, together with other cytological alterations, allow the four viruses to be distinguished. Pinwheels, scrolls and laminated aggregates were produced only by SCMV whereas JGMV, MDMV, and SrMV produced only pinwheels and scrolls. SrMV produced amorphous cytoplasmic inclusions which are not produced by JGMV and MDMV. The latter two were rather similar in cytological effects except that the SCMV-JG (U.S.A.) isolate of MDMV produced aggregates of needle-like structures in the cytoplasm which were not found with JGMV and the other MDMV isolates. The specific cytological effects induced by these viruses thus corroborate the recent classification of these viruses based mainly on the properties of the coat-protein gene, the 3′ noncoding nucleotide sequences, and host reactions.

Introduction

Potyviruses infecting sugarcane and maize (and other Gramineae) were described long ago and many isolates have been studied, but their definitive taxonomic status remained uncertain. The heterogeneity in cytopathology of the viruses assigned to the sugarcane mosaic virus (SCMV) complex of strains was recognized early. Edwardson [2] noted that different strains of SCMV induced different types of cylindrical inclusions (CI). The general experience obtained before and after Edwardson's 1974 report was that the cytopathology of cells infected with different virus strains is similar [6]. The same subdivision type of CI is generally detected with different isolates and even strains of one potyvirus, although there have been a few exceptions [1, 7].

Comparative studies on antigenic properties of the coat protein, on amino acid sequence and peptide profiling of coat protein, on nucleotide sequences of the 3′ noncoding region and molecular hybridization using probes of the 3′ noncoding region, and on reactions of differential host plants have recently revealed that a clear taxonomic order can be achieved by grouping the strains of SCMV into four different categories. These appear sufficiently separate to be considered as separate viruses, namely johnsongrass mosaic (JGMV), maize dwarf mosaic (MDMV), sorghum mosaic (SrMV) and SCMV [for review, see 8]. We report here results of a comparative study of the cytological alterations induced by isolates representing these four viruses.

Material and methods

Most of the isolates studied (Table 1) were sent by M. Tosic to Braunschweig as fresh maize leaves infected with the individual isolates and subcultivated by mechanical inoculation onto maize seedlings. Two isolates originated from maize in Germany [4]. Symptomatic parts of systemically infected, intermediately aged maize leaves were embedded in Epon as described before [5].

Results

The results (Table 1) were obtained from 36 different tissue samples comprising 15 different virus isolates. With four additional isolates studied (MDMV-A, Roma, Italy; SCMV-A, U.S.A.; SCMV-B, U.S.A.; SCMV-SC, Australia) no information on cytopathology was obtained due to insufficient tissue samples. Information was considered representative for given isolates only if tissue blocks displayed well-developed cellular inclusions. This was important because in tissues from leaves which were too young or too old the cylindrical cytoplasmic inclusions and other inclusions were sometimes incompletely developed or already degraded, respectively, so that a definitive description of relevant structures was not possible.

All isolates produced typical cytoplasmic CI, although in two different modifications (scrolls and laminated aggregates) [2]. It was not clear whether the CI induced by MDMV and by SrMV isolates were of the short curved laminated aggregate type [3]. Virus particles were inconspicuous in almost all samples studied. Sometimes small bundles of virus-like particles were seen in the cytoplasm or monolayers of virions were seen associated with the tonoplast. Possibly the virions were poorly preserved by the fixation method. As is typical for many potyviruses, almost all of the isolates often induced the formation of large clusters of small vesicles (Figs. 1A, 3C, and 4A). These accumulated in the cytoplasmic inclusion complexes which also contained the

Table 1. Cytological alterations induced by the four viruses of the sugarcane mosaic virus complex[a]

Cytological alterations[d]	JGMV[b]	MDMV	SCMV	SrMV
	SCMV-JG(AU)[c], MDMV-O(US)	MDMV-A(US), SCMV-JG(US), MDMV-481(GER), MDMV-A(I) MDMV-A(YU)	MDMV-B(US), SCMV-D(US), SCMV-E(US), SCMV-BC(AU), SCMV-SABI(AU), SCMV-522(GER)	SCMV-H(US) SCMV-I(US)
Scrolls only	+	+	−	+
Scrolls + laminated aggregates	−	−	+	−
Amorphous inclusions	−	−	−	+
Needle-like inclusions	−	(+)[e]	−	−
Cytoplasmic crystals	+	(+)[f]	+	+
Vesicle accumulation	+	+	+	+

[a] Virus naming and assignment of strains to the four viruses follows Shukla et al. [8]
[b] JGMV johnsongrass mosaic virus; MDMV maize dwarf mosaic virus; SCMV sugarcane mosaic virus; SrMV sorghum mosaic virus
[c] *AU* Australia; *US* United States; *GER* Germany; *I* Bergamo, Italy; *YU* Yugoslavia
[d] For explanations see text
[e] With strain SCMV-JG (US) only
[f] With MDMV-481 (GER) only

CI. Such vesicles sometimes showed a fibrous content reminiscent of ds-RNA (Fig. 1 A, inset). With seven of the isolates (representing all four viruses) crystalline inclusions were found in the cytoplasm (Figs. 1 C, inset, 3 B, C, and 4 C) which were different from crystalline inclusions of peroxisomes as they were not bounded by a membrane. These crystals were rather rare and hence it could not be clearly decided whether they were specific for the virus infections. Similar crystals were, however, not found in maize tissues infected by other viruses (Lesemann, unpubl.).

The six isolates of SCMV were clearly distinct from those of the other three viruses in respect to the structure of the CI which comprised laminated aggregates in addition to pinwheels and scrolls (Fig. 1A–C). These inclusions were abundant in infected tissues.

The five isolates of MDMV, two of JGMV and two of SrMV were characterized by the production of pinwheels and scrolls without laminated aggregates (Figs. 2A, B, 3C, and 4A). Again many infected cells often had large accumulations of CI.

With four isolates of MDMV no other cytological alterations than pinwheel and scroll inclusions were visualized (Table 1). However, in three different embeddings of the fifth isolate, the MDMV-strain SCMV-Jg (U.S.A.), conspicuous accumulations of dark staining needle-shaped material in the cytoplasm (Fig. 2C) as well as small bundles of filamentous structures in the nuclei and cytoplasm (Fig. 3A) were consistently observed in addition to CI.

Johnsongrass mosaic virus isolates did not reveal specific inclusions apart from CI (Fig. 3C). However, both isolates of SrMV consistently induced rounded accumulations of dark staining amorphous-granular material in the cytoplasm (Fig. 4B).

Discussion

The CI structure allowed the clear differentiation of SCMV from the other three viruses. SrMV differed from MDMV and JGMV in the production of amorphous inclusions. MDMV and JGMV showed no cytological differentiation except for one strain of MDMV. In general, the cytological alterations corroborated a subdivision of the SCMV complex into the four viruses as proposed by Shukla et al. [8]. The CI are products translated from parts of the viral genome other than the coat-protein gene, hence our results substantially strengthen the distinction between SCMV and the other three viruses which was based mainly on differences in the coat-protein gene. The significance of the other cellular inclusions (needle-like, amorphous, crystals, filamentous) is not known. It is suggested, however, that they may represent products of still other genome parts and in this case they again substantiate the distinction of these four viruses.

Three of the four viruses appear homogeneous in respect to the cytological morphology of inclusions. However, isolate SCMV-Jg (U.S.A.) of MDMV appeared different from the other MDMV isolates, as well as from the other viruses. This may indicate the need for further differentiation of isolates now attributed to MDMV.

Fig. 1. Cytological alterations induced by strains of sugarcane mosaic virus. Bars: 1 μm; for insets, 500 nm. **A** Cylindrical inclusions comprising pinwheel, scroll and laminated aggregate structures, and accumulation of vesicles induced by SCMV-D (U.S.A.). Inset shows vesicles containing fibrillar material at higher magnification. **B** and **C** Pinwheel, scroll and laminated aggregate inclusions produced by SCMV-BC (Australia) and SCMV 552 (Germany), respectively. Inset in **C** shows a cytoplasmic crystal in a cell infected by SCMV-BC

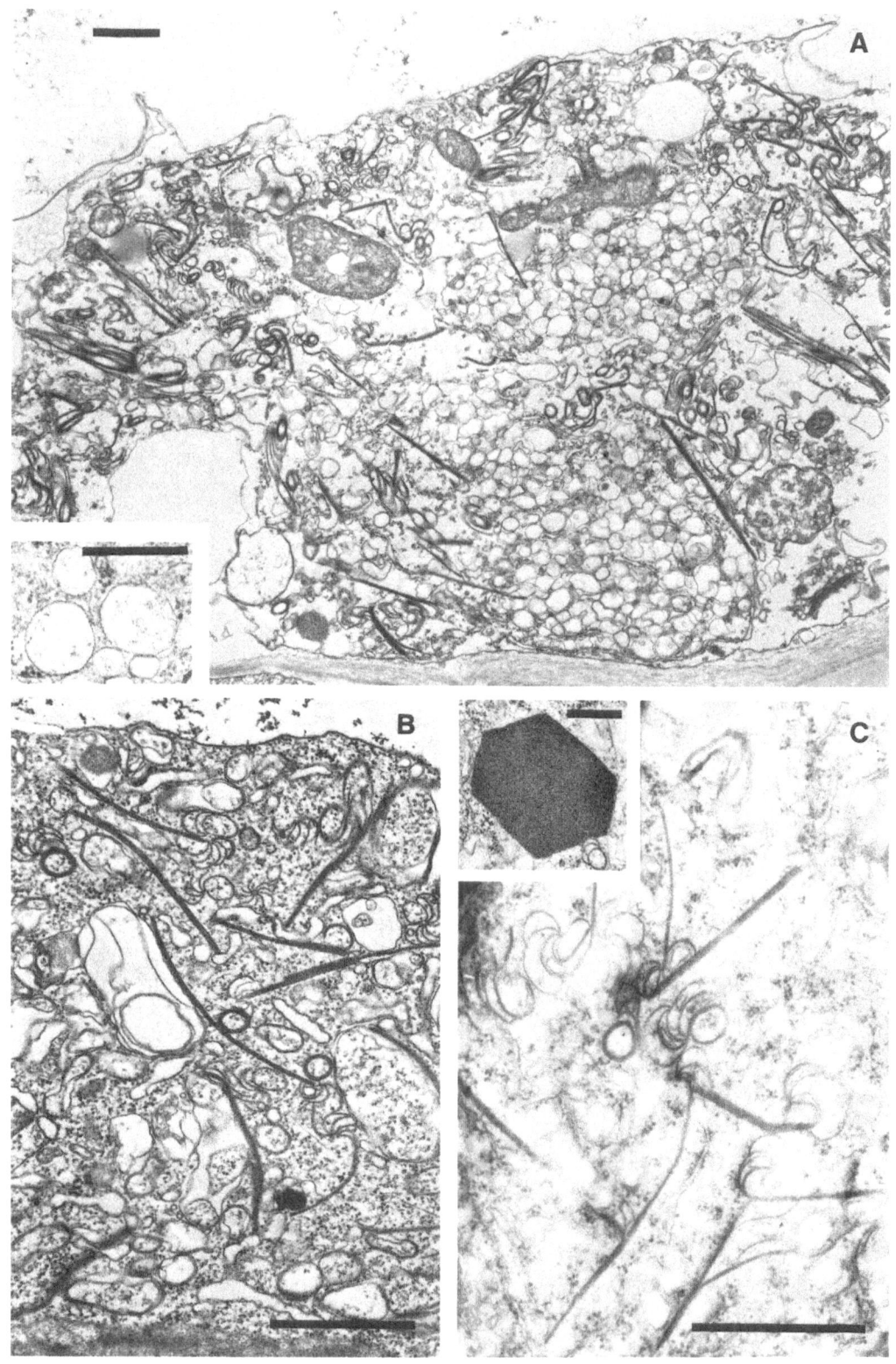

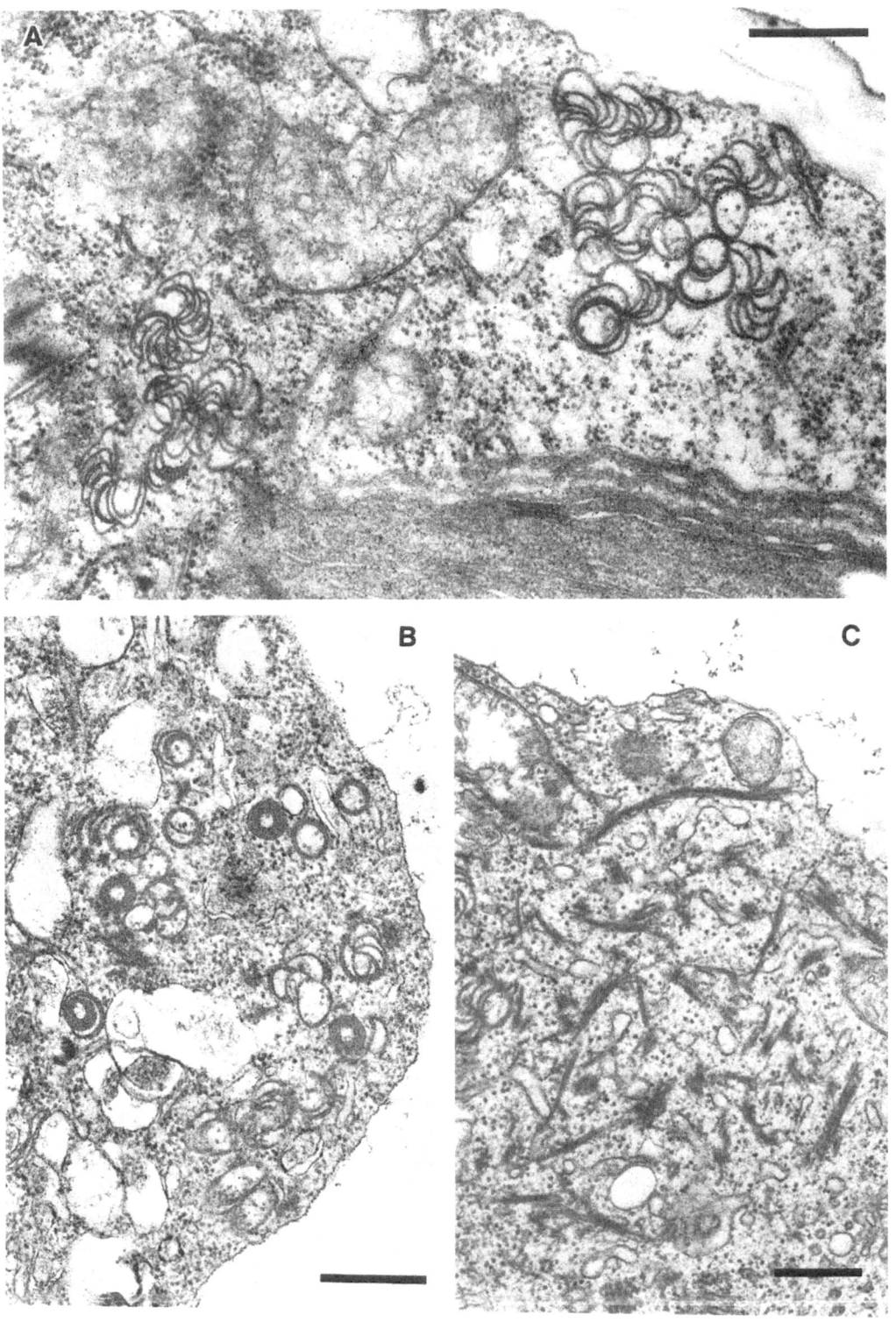

Fig. 2. Cytoplasmic inclusions produced by MDMV. Bars: 500 nm. **A** and **B** Pinwheel and scroll inclusions of MDMV-A (U.S.A.) and SCMV-JG (U.S.A.), respectively. **C** Needle-like cytoplasmic inclusions of SCMV-JG (U.S.A.)

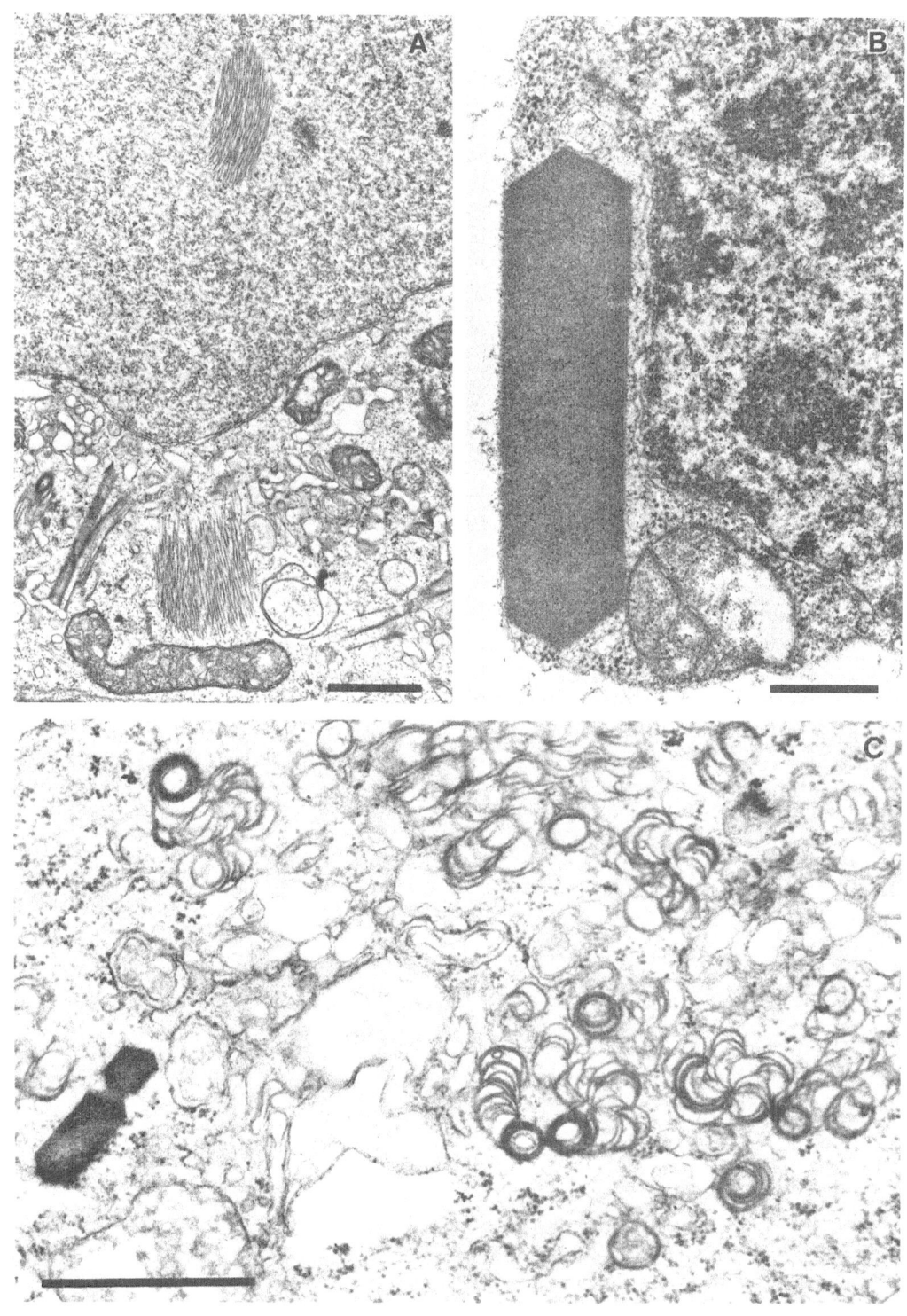

Fig.3. A and **B** Inclusions of strain SCMV-JG (U.S.A.) of MDMV. **A** Aggregates of filamentous structures in nucleus and cytoplasm. **B** Cytoplasmic crystal. **C** Pinwheel and scroll inclusions and cytoplasmic crystal of JGMV strain SCMV-JG (Australia). Bars: A and C, 1 μm; B, 500 nm

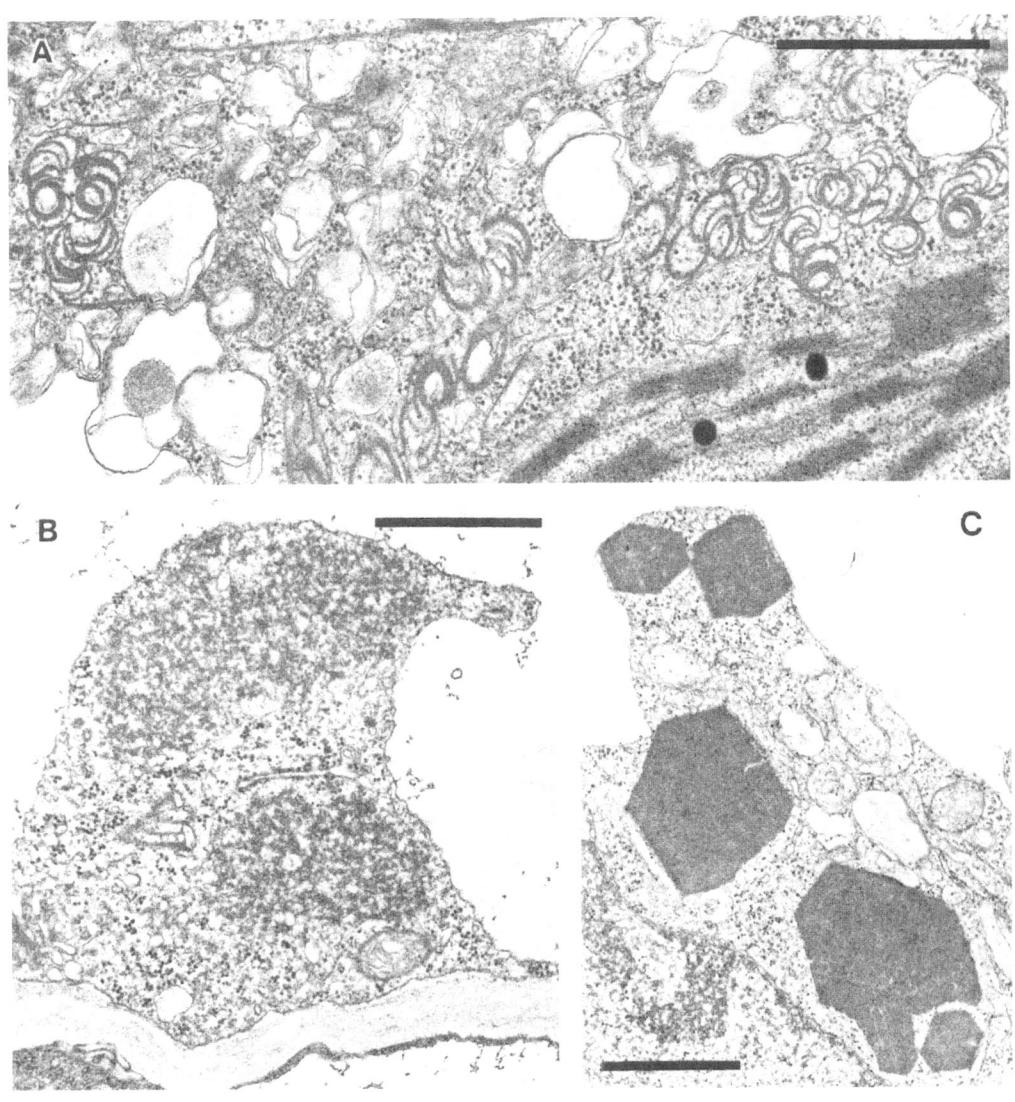

Fig. 4. Cytoplasmic inclusions induced by SrMV strain SCMV-H. Bars: 1 μm. **A** Pinwheel and scroll inclusions. **B** Granular amorphous inclusions. **C** Cytoplasmic crystals

Acknowledgements

We thank Cora Broistedt, Christina Maaß and Sabine Schuhmann for excellent technical assistance.

References

1. Chamberlain JA, Catherall PL, Jellings AJ (1977) Symptoms and electronmicroscopy of ryegrass mosaic virus in different grass species. J Gen Virol 36: 297–306

2. Edwardson JR (1974) Some properties of the potato virus Y group. Fla Agric Exp Stat Monogr Ser, no 4
3. Edwardson JR, Christie RC, Ko NJ (1984) Potyvirus cylindrical inclusions – subdivision IV. Phytopathology 74: 1111–1114
4. Huth W, Lesemann D-E (1991) Detection of maize dwarf mosaic and sugarcane mosaic viruses in the Federal Republic of Germany. In: Proceedings of the 5th Conference on Virus Diseases of Gramineae in Europe, 24–27 May 1988. Acta Phytopathol Entomol Hung 26: 125–130
5. Koenig R, Lesemann D-E (1985) Plant viruses in German rivers and lakes. I Tombusviruses, a potexvirus and carnation mottle virus. J Phytopathol 118: 105–116
6. Lesemann D-E (1988) Cytopathology. In: Milne RG (ed) The plant viruses, vol 4, the filamentous plant viruses. Plenum, New York, pp 179–235
7. McDonald JG, Hiebert E (1975) Characterization of the capsid and cylindrical inclusion proteins of three strains of turnip mosaic virus. Virology 63: 295–303
8. Shukla DD, Frenckel MJ, McKern NM, Ward CW, Jilka J, Ford RE (1992) Present status of the sugarcane mosaic potyvirus subgroup. In: Barnett O W (ed) Potyvirus taxonomy. Springer, Wien New York, pp 363–373 (Arch Virol [Suppl] 5)

Authors' address: D.-E. Lesemann, Biologische Bundesanstalt für Land- und Forstwirtschaft, Institut für Biochemie und Pflanzenvirologie, Messeweg 11–12, D-W-3300 Braunschweig, Federal Republic of Germany.

Arch Virol (1992) [Suppl 5]: 363–373

Present status of the sugarcane mosaic subgroup of potyviruses

D. D. Shukla[1], M. J. Frenkel[1], N. M. McKern[1], C. W. Ward[1], J. Jilka[2], M. Tosic[3], and R. E. Ford[4]

[1] CSIRO, Division of Biomolecular Engineering, Parkville, Victoria, Australia
[2] Monsanto Company, St. Louis, Missouri, U.S.A.
[3] Faculty of Agriculture, University of Belgrade, Belgrade, Yugoslavia
[4] Department of Plant Pathology, University of Illinois, Urbana, Illinois, U.S.A.

Summary. Until recently, sugarcane mosaic virus (SCMV) was believed to be a single potyvirus consisting of a large number of strains, differing from each other in certain biological and antigenic properties. The use of affinity-purified polyclonal antibodies directed towards the surface-located, virus-specific amino termini of the coat proteins showed that 17 strains from Australia and the United States represented four distinct potyviruses, namely johnsongrass mosaic virus (JGMV), maize dwarf mosaic virus (MDMV), sorghum mosaic virus (SrMV) and SCMV. Comparisons of strains from each of these four viruses on the basis of reactions on differential sorghum and oat cultivars, cell-free translation of RNAs, morphology and serology of cytoplasmic cylindrical inclusions, amino acid sequence and peptide profiling of coat proteins, 3′ non-coding nucleotide sequences, and molecular hybridization with probes corresponding to the 3′ non-coding regions, resulted in exactly the same taxonomic assignments as obtained using amino-terminal serology. These results further confirm that the former sugarcane mosaic virus actually consists of four distinct viruses and show that MDMV, SrMV, and SCMV are more closely related to each other than they are to JGMV. Because these four viruses are closely related but distinct, formation of a sugarcane mosaic subgroup in the genus *Potyvirus* would be appropriate.

Introduction

Sugarcane mosaic virus (SCMV), a definitive member of the *Potyvirus* genus, infects maize, sorghum, sugarcane, and other poaceous plant species throughout the world [24, 37]. Traditionally, isolates originating in sugarcane were designated as strains of SCMV [1] and those originating in maize as strains of maize dwarf mosaic virus (MDMV) [18]. However, strains of SCMV and

MDMV share many common properties [14, 25, 36], and MDMV has been considered a strain of SCMV [24]. The main reasons for assigning all these isolates to SCMV were that they: (*i*) had very similar host ranges, (*ii*) produced similar symptoms in many hosts, (*iii*) had common aphid vectors, and (*iv*) were interrelated serologically [24]. However, questions about the homogeneity of SCMV arose when some strains: (*i*) did not cross protect against each other [23], (*ii*) were assigned to different cytoplasmic inclusion morphology subdivisions [3], (*iii*) did not cross react in some serological tests [9, 12, 14, 19, 26, 34], and (*iv*) were found to have different coat-protein structures [10, 31]. Based on these observations, Francki et al. [5] noted that some of the SCMV strains differed so much from each other that they merited the status of distinct viruses.

Classification of the SCMV subgroup into four distinct viruses

Amino-terminal serology

Recently, Shukla et al. [35] compared 17 SCMV strains from Australia and the United States on the basis of their reactivities in electro-blot immunoassay (Fig. 1) with cross-absorbed polyclonal antibodies directed towards surface-located, virus-specific amino termini of coat proteins [32]. Their results demonstrated that the 17 SCMV strains did not represent one potyvirus but in fact belonged to four distinct potyviruses, johnsongrass mosaic virus (JGMV), MDMV, sorghum mosaic virus (SrMV) and SCMV [35]. Strains of each virus reacted with virus-specific antibodies produced by an isolate of the respective virus but not with virus-specific antibodies produced against the other three viruses.

This assignment of SCMV strains into four distinct potyviruses, based on serology of the coat-protein amino terminus, has now been supported by other biological and biochemical data.

Reactivities on differential oat and sorghum cultivars

When 15 strains of the four viruses were tested on 11 sorghum inbred-lines and one oat cultivar, they were separated into four groups based on symptoms [38], giving the same assignments for the strains as obtained with amino-terminal serology [35]. For example: (*i*) the sorghum line TX2786 is infected by the strains of JGMV, MDMV, and SCMV but not by strains of SrMV; (*ii*) JGMV and SrMV strains cause necrosis in the sorghum line Trudex but MDMV does not; and (*iii*) the oat cultivar Clintland is infected only by JGMV [38]. MDMV-O and MDMV-KS1, which are now considered to be strains of JGMV, also infect oat cultivars [15, 19].

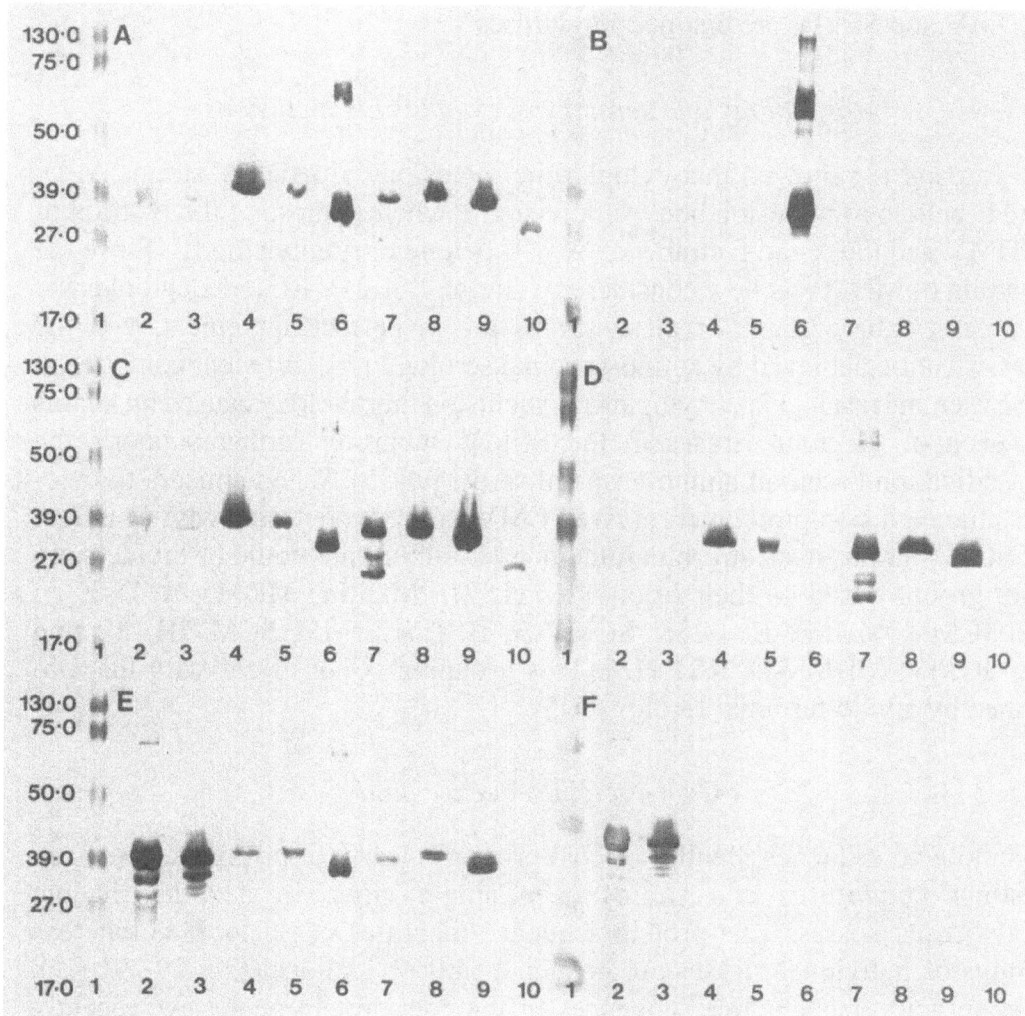

Fig. 1. Electro-blot immunoassay of strains of JGMV, MDMV and SCMV. *1* Bio-Rad (**A–E**) or BRL (**F**) prestained standards. *2–10* JGMV-JG, JGMV-MDO, SCMV-SC, SCMV-E, MDMV-A, SCMV-MDB, SCMV-BC, SCMV-Sabi and tobacco etch virus, respectively. **A**, **C**, and **E** probed with unfractionated antisera to MDMV-A, SCMV-MDB and JGMV-MDO, respectively. **B**, **D**, and **F** probed with affinity purified, virus-specific antisera directed to N terminus of coat proteins of MDMV-A, SCMV-MDB and JGMV-MDO. Reproduced from [35]

Cell-free translation of RNAs

A comparative analysis of the translation products directed by the RNAs of MDMV-A (type strain of MDMV) and MDMV-B (a strain of SCMV) in the rabbit reticulocyte cell-free system showed distinct protein profiles. The four major polypeptide products in MDMV-A-directed translation were of M_r 57k, 91k, 100k and 117k, whereas in the MDMV-B-directed translation the four major products had M_r of 39k, 55k, 86k and 121k [2], again suggesting that MDMV and SCMV are distinct potyviruses.

Morphology and serology of cytoplasmic inclusions

Edwardson [3] showed that cytoplasmic inclusions of MDMV strains A, D, and E belong to inclusion body subdivision I, whereas those of the B strain of MDMV and the A and E strains of SCMV belong to subdivision III. Since the B strain of MDMV is now considered a strain of SCMV [35], the cytoplasmic inclusion subgroupings of these strains are in perfect agreement with the classification achieved by amino-terminal serology [35]. A recent comparison by Lesemann et al. [17] of cytoplasmic inclusion morphology caused by strains of each of the four viruses in the SCMV subgroup further supports the classification based on amino-terminal serology [35]. When antisera to cyto-plasmic inclusion proteins from five SCMV strains were tested with a number of SCMV strains in electro-blot immunoassay, the strains could be divided into four groups based on their reactivities viz.: (*i*) MDMV; MDMV-A, D, E, F; (*ii*) SCMV; MDMV-B, I-188, SCMV-A, B; (*iii*) SrMV; SCMV-H, M; and (*iv*) JGMV; MDMV-O, KS1 [15]. This grouping is consistent with that ob-tained by amino-terminal serology [35].

Coat-protein sequence data

Amino acid sequence identity of coat proteins clearly discriminates between distinct potyviruses and strains of a single virus. In general, distinct potyviruses possess coat-protein sequence identities of 38 to 71% whereas strains of individual viruses are greater than 90% identical [28, 29, 30, 39]. Amino acid or nucleotide sequences of the coat-protein gene have recently been determined for three strains of JGMV [11, 16, 31], two strains of SCMV [6, 16] and one strain each of MDMV and SrMV [16]. Comparison of these sequences revealed sequence identities ranging from 93 to 99% for the three JGMV strains, and 79% for the two SCMV strains but only 51 to 71% among JGMV, MDMV, SrMV and SCMV (Table 1). These values conform to se-quence identity scores generally observed between distinct potyviruses and their strains [39] except for the low sequence identity of 79% between the two SCMV strains, SCMV-SC and MDMV-B [6]. This low sequence identity between SCMV-SC and MDMV-B is due to an unexpected sequence diversity in the amino-terminal regions of the two coat proteins spanning amino acid residues 27 to 70 in SCMV-SC [6]. This diverse region of SCMV-SC is smaller (44 residues) than the equivalent region in MDMV-B (59 residues) and shows only 22% identity to the MDMV-B sequence. The origin of this diversity is unknown and may be the result of recombination. Sequence analysis of the corresponding regions of other strains assigned by serology to SCMV [35] will establish whether one of the two sequence types predominates. Despite this

Table 1. Percent amino acid sequence identity between the coat proteins from strains of viruses in the SCMV subgroup[a]

Virus strain	JGMV-JG	JGMV-MDKS1	JGMV-MDO	SCMV-SC	SCMV-MDB	MDMV-A	SrMV-SCH
JGMV-JG		94	93	56	55	57	52
JGMV-MDKS1	99		99	55	52	55	51
JGMV-MDO	99	99		55	52	55	51
SCMV-SC	67	68	68		79	7·1	67
SCMV-MDB	67	68	68	94		66	65
MDMV-A	68	68	68	88	86		70
SrMV-SCH	66	66	66	85	84	88	

[a] The sources of sequence data were from [6, 16, 31]. Sequence identities were calculated as described previously [28]. Comparisons above the diagonal refer to total coat protein; figures below the diagonal are from the coat protein core region (equivalent to Asp_{70}–Arg_{285} in JGMV-JG [31]).

large change in which two-thirds of the N-terminal region is altered, SCMV-SC and MDMV-B retain strong virus-specific serological cross-reactivity [35] and induce similar symptoms on most sorghum lines [38]. However, SCMV-SC and MDMV-B do induce distinct symptoms in some sorghum lines [38], and it will be of interest to see if this change has any effect on other biological properties of these two SCMV strains such as aphid transmissibility and cross-protection [6].

When the amino acid sequences from the core regions (devoid of N and C termini) of the coat proteins from strains of the four viruses were compared (Table 1), the three JGMV strains displayed sequence identity of 99% and the two SCMV strains of 94%. The sequence identity between the JGMV strains and MDMV, SCMV and SrMV in the coat-protein core regions ranged from 66 to 68% and that among the latter three viruses from 84 to 88%, suggesting that MDMV, SCMV and SrMV are more closely related to each other than they are to JGMV. This relationship is shown graphically in the phylogenetic tree constructed from the sequence identities of the core regions of potyvirus coat proteins [39].

Coat-protein peptide mapping by high performance liquid chromatography (HPLC)

Comparison of HPLC peptide profiles of coat-protein tryptic digests from six strains of SCMV, three strains of JGMV, and two strains each of MDMV and SrMV revealed that the peptide profiles could be divided into four distinct groups (Fig. 2) [21, 22]. This technique reflects the extent of sequence identity between proteins and has been shown to clearly differentiate between distinct potyviruses and their strains [13, 20, 33].

3' non-coding regions

A comparison of 3' non-coding sequences from 14 strains of seven distinct potyviruses by Frenkel et al. [8] revealed that this region in strains of the same potyvirus is greater than 80% identical and is approximately the same length, while distinct potyviruses have 3' non-coding sequences generally less than 50% identical and can be quite different in length. Analysis of 3' non-coding regions showed that the three JGMV strains have sequence identities of 96 to 98%, the two SCMV strains of 88%, and the sequence identities among the 3' non-coding regions of JGMV, MDMV, SCMV, and SrMV ranges from 44 to 61% (Table 2).

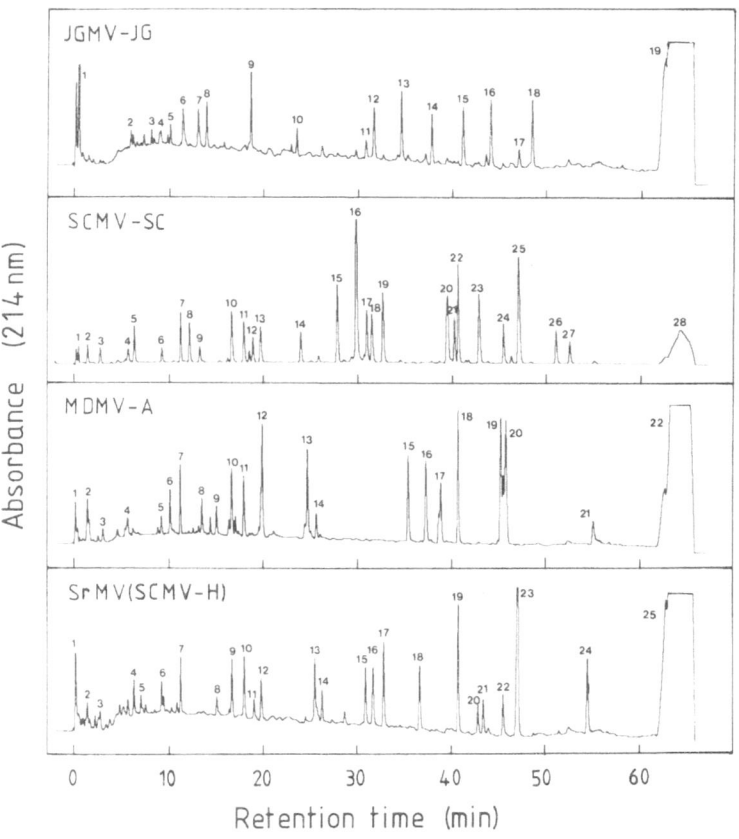

Fig. 2. HPLC peptide profiles of tryptic digests of coat proteins of the four viruses in the SCMV subgroup. Reproduced from [21]

A nucleic acid probe (amplified by the polymerase chain reaction and radiolabeled) corresponding to the 3′ non-coding region of the genome of the SC strain of SCMV hybridized only with the strains of SCMV (Fig. 3) and not with JGMV, MDMV or SrMV [7], thus further confirming the assignments made on the basis of N-terminal serology [35].

Table 2. Percent nucleotide sequence identity between the 3′ non-coding regions from strains of viruses in the SCMV subgroup[a]

Virus strain	JGMV-JG	JGMV-MDKS1	JGMV-MDO	SCMV-SC	SCMV-MDB	MDMV-A	SrMV-SCH
JGMV-JG							
JGMV-MDKS1	97						
JGMV-MDO	96	98					
SCMV-SC	44	46	46				
SCMV-MDB	48	47	45	88			
MDMV-A	44	44	44	54	54		
SrMV-SCH	45	48	47	61	59	57	

[a] The sources of sequence data were from [6, 11, 16]. Sequence identities were calculated as described previously [8]

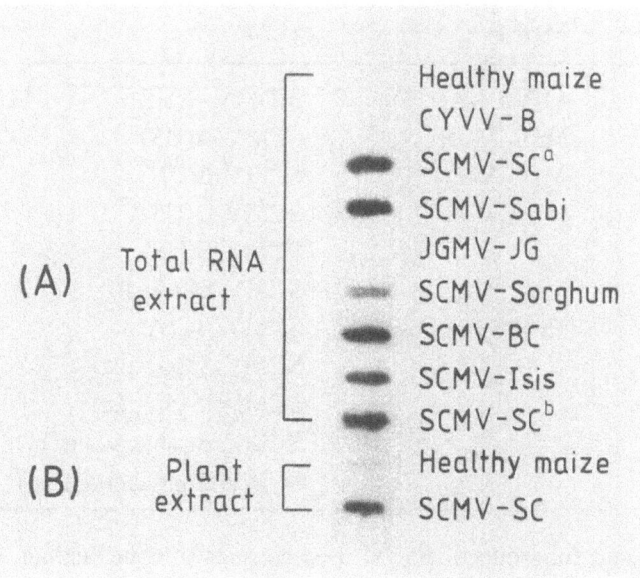

Fig. 3. Hybridization of purified RNA isolated from healthy and infected tissue (**A**) and of crude plant extract (**B**) with the PCR-amplified, radiolabeled 3′ non-coding region of SCMV-SC as the probe. CYVV Clover yellow vein virus. Reproduced from [7]

Conclusions

The SCMV subgroup contains four distinct potyviruses: JGMV, MDMV, SCMV, and SrMV (Table 3). These viruses were separated first by amino-terminal serology [35] which is now supported by the following lines of evidence:

  i   reactions on selected sorghum and oat cultivars,
  ii   the nature of their cell-free translation products,
  iii the morphology and serology of their cytoplasmic inclusions,
  iv   cross-protection,
  v   coat-protein sequence,
  vi   coat-protein HPLC peptide profiles,
  vii 3′ non-coding sequence identity and hybridization.

In addition, the coat-protein core sequence data has revealed that MDMV, SCMV, and SrMV form a closely related subset of potyvirus species with JGMV being more distantly related. This classification has now been accepted by the editors of the AAB Descriptions of Plant Viruses. Consequently, the description for SCMV has been revised [37], new descriptions for JGMV [27] and MDMV [4] published, and the description for SrMV is in preparation (DD Shukla, RW Toler, and SG Jensen, unpubl.).

Table 3. Grouping of virus strains in the SCMV subgroup

JGMV	MDMV	SCMV	SrMV
SCMV-JG[a] (Aust)	MDMV-A(US)	MDMV-B(US)	SCMV-H(US)
MDMV-O(US)	MDMV-D(US)	SCMV-A(US)	SCMV-I(US)
MDMV-KS1(US)	MDMV-E(US)	SCMV-B(US)	SCMV-M(US)
	MDMV-F(US)	SCMV-D(US)	
		SCMV-E(US)	
		SCMV-SC(Aust)	
		SCMV-BC(Aust)	
		SCMV-Sabi(Aust)	
		SCMV-ISIS(Aust)	
		SCMV-Brisbane(Aust)	
		SCMV-Bundaberg(Aust)	

Sugarcane mosaic subgroup of potyviruses consists of four distinct species. JGMV johnsongrass mosaic virus, MDMV maize dwarf mosaic virus, SCMV sugarcane mosaic virus, SrMV sorghum mosaic virus
[a] Original names of the strains
Aust Australia, *US* United States

In a previous review [30], we drew attention to the need to give careful thought to the new nomenclature that will be required following the re-assignment of viruses and strains to new classifications. We believe that the new name should not only reflect the new assignment but also permit ready connection to the past literature. Thus MDMV-O and MDMV-KS1 which are both strains of JGMV could be renamed JGMV-MDO and JGMV-MDKS1, i.e., the maize dwarf O or maize dwarf KS1 strains of JGMV, without any necessity to repeat the symbols for mosaic or virus. Similarly, MDMV-B becomes SCMV-MDB, and SCMV-H, SCMV-I, and SCMV-M are renamed SrMV-SCH, SrMV-SCI, and SrMV-SCM.

References

1. Abbott EV, Tippet RL (1966) Strains of sugarcane mosaic virus. US Dept Agric Res Serv Tech Bull 1340

2. Berger PH, Luciano CS, Thornbury DW, Benner HI, Hill JH, Zeyen RJ (1989) Properties and in vitro translation of maize dwarf mosaic virus RNA. J Gen Virol 70: 1845–1851

3. Edwardson JR (1974) Some properties of the potato virus Y-group. Fla Agric Exp Stat Monogr Ser, no 4

4. Ford RE, Tosic M, Shukla DD (1989) Maize dwarf mosaic virus. AAB Descriptions of Plant Viuses, no. 341

5. Francki RIB, Milne RG, Hatta T (1985) Atlas of plant viruses, vol 2. CRC Press, Boca Raton,

6. Frenkel MJ, Jilka J, McKern NM, Strike PM, Clark Jr JM, Shukla DD, Ward CW (1991) Unexpected sequence diversity in the amino-terminal ends of the coat proteins of strains of sugarcane mosaic virus. J Gen Virol 72: 237–242

7. Frenkel MJ, Jilka JM, Shukla DD, Ward CW (1991) Differentiation of potyviruses and their strains by hybridization with the 3′ non-coding region of the viral genome. J Virol Methods 36: 51–62

8. Frenkel MJ, Ward CW, Shukla DD (1989) The use of 3′ non-coding nucleotide sequences in the taxonomy of potyviruses: application to watermelon mosaic virus 2 and soybean mosaic virus-N. J Gen Virol 70: 2775–2783

9. Giorda LM, Toler RW, Miller FR (1986) Identification of sugarcane mosaic virus strain H isolate in commercial grain sorghum. Plant Dis 70: 624–628

10. Gough KH, Shukla DD (1981) Coat protein of potyviruses. I. Comparison of four Australian strains of sugarcane mosaic virus. Virology 111: 455–462

11. Gough KH, Azad AA, Hanna PJ, Shukla DD (1987) Nucleotide sequence of the capsid protein and nuclear inclusion protein genes from the Johnsongrass strain of sugarcane mosaic virus RNA. J Gen Virol 68: 297–304

12. Hewish DR, Shukla DD, Gough KH (1986) The use of biotin conjugated antisera in immunoassays for plant viruses. J Virol Methods 13: 79–85

13. Jain RK, McKern NM, Tolin SA, Hill JH, Barnett OW, Tosic M, Ford RE, Beachy RN, Yu MH, Ward CW, Shukla DD (1992) Confirmation that fourteen potyvirus isolates from soybean are strains of one virus by comparing coat protein peptide profiles. Phytopathology 82: 294–299

14. Jarjees MM, Uyemoto JK (1984) Serological relatedness of strains of maize dwarf mosaic and sugaracne mosaic viruses as determined by microprecipitin and enzyme-linked immunosorbent assays. Ann Appl Biol 104: 497–501

15. Jensen SG, Staudinger JL (1989) Serological grouping of the cycloplasmic inclusions of six strains of sugarcane mosaic virus. Phytopathology 79: 1215

16. Jilka JM (1990) Cloning and characterization of the 3' terminal regions of RNA from select strains of maize dwarf mosaic virus and sugarcane mosaic virus. PhD Thesis, University of Illinois, Urbana, Illinois

17. Lesemann DE, Shukla DD, Tosic M, Huth W (1992) Differentiation of the four viruses of the sugarcane mosaic virus subgroup based on cytopathology. In: Barnett OW (ed) Potyvirus taxonomy. Springer, Wien New York, pp 353–361 (Arch Virol [Suppl] 5)

18. Louie R, Knoke JK (1975) Strains of maize dwarf mosaic virus. Plant Dis Rep 59: 518–522

19. McDaniel LL, Gordon DT (1985) Identification of a new strain of maize dwarf mosaic virus. Plant Dis 69: 602–607

20. McKern NM, Shukla DD, Barnett OW, Vetten HJ, Dijkstra J, Whittaker LW, Ward CW (1992) Coat protein properties suggest that azuki bean mosaic virus, blackeye cowpea mosaic virus, peanut stripe virus and three isolates from soybean are all strains of the same potyvirus. Intervirology 33: 121–134

21. McKern NM, Shukla DD, Toler RW, Jensen SG, Tosic M, Ford RE, Leon O, Ward CW (1991) Confirmation that the sugarcane mosaic virus subgroup consists of four distinct potyviruses by using peptide profiles of coat proteins. Phytopathology 81: 1025–1029

22. McKern NM, Whittaker LA, Strike PM, Ford RE, Jensen SG, Shukla DD (1990) Coat protein properties indicate that maize dwarf mosaic virus-KS1 is a strain of Johnsongrass mosaic virus. Phytopathology 80: 907–912

23. Paulsen AR, Sill WH (1970) Absence of cross-protection between maize dwarf mosaic virus strain A and B in grain sorghum. Plant Dis Rep 54: 627–629

24. Pirone TP (1972) Sugarcane mosaic virus. CMI/AAB Descriptions of Plant Viruses, no 88

25. Shepherd RJ (1965) Properties of a mosaic virus of corn and Johnsongrass and its relation to the sugarcane mosaic virus. Phytopathology 55: 1250–1256

26. Shukla DD, Gough KH (1984) Serological relationships among four Australian strains of sugarcane mosaic virus as determined by immune electron microscopy. Plant Dis 68: 204–206

27. Shukla DD, Teakle DS (1989) Johnsongrass mosaic virus. AAB Descriptions of Plant Viruses, no. 340

28. Shukla DD, Ward CW (1988) Amino acid sequence homology of coat proteins as a basis for identification and classification of the potyvirus group. J Gen Virol 69: 2703–2710

29. Shukla DD, Ward CW (1989) Structure of potyvirus coat proteins and its application in the taxonomy of the potyvirus group. Adv Virus Res 36: 272–314

30. Shukla DD, Ward CW (1989) Identification and classification of potyviruses on the basis of coat protein sequence data and serology. Arch Virol 106: 171–200

31. Shukla DD, Gough KH, Ward CW (1987) Coat protein of potyviruses. 3. Comparison of amino acid sequences of the coat protein of four Australian strains of sugarcane mosaic virus. Arch Virol 96: 59–74

32. Shukla DD, Jilka J, Tosic M, Ford RE (1989) A novel approach to the serology of potyviruses involving affinity purified polyclonal antibodies directed towards virus-specific N termini of coat proteins. J Gen Virol 70: 13–23

33. Shukla DD, McKern NM, Gough KH, Tracy SL, Letho SG (1988) Differentiation of potyviruses and their strains by high performance liquid chromatographic peptide profiling of coat proteins. J Gen Virol 69: 493–502

34. Shukla DD, O'Donnell IJ, Gough KH (1983) Characteristics of the electro-blot radioimmunoassay (EBRIA) in relation to the identification of plant viruses. Acta Phytopathol Acad Sci Hung 18: 79–84

35. Shukla DD, Tosic M, Jilka J, Ford RE, Toler RW, Langham MAC (1989) Taxonomy of potyviruses infecting maize, sorghum and sugarcane in Australia and the United States as determined by reactivities of polyclonal antibodies directed towards virus-specific N termini of coat proteins. Phytopathology 79: 223–229

36. Snazelle TE, Bancroft JB, Ullstrup AJ (1971) Purification and serology of maize dwarf mosaic and sugarcane mosaic viruses. Phytopathology 61: 1059–1063

37. Teakle DS, Shukla DD, Ford RE (1989) Sugarcane mosaic virus. AAB Descriptions of Plant Viruses, no 342 (no 88 revised)

38. Tosic M, Ford RE, Shukla DD, Jilka J (1990) Differentiation of sugarcane, maize dwarf, Johnsongrass, and sorghum mosaic viruses based on reactions of oat and some sorghum cultivars. Plant Dis 74: 549–552

39. Ward CW, Shukla DD (1991) Taxonomy of potyviruses: current problems and some solutions. Intervirology 32: 269–296

Authors' address: D. D. Shukla, CSIRO, Division of Biomolecular Engineering, 343 Royal Parade, Parkville, Vic. 3052, Australia.

Virus relationships – BYMV subgroup

Arch Virol (1992) [Suppl 5]: 377–385

Bean yellow mosaic virus subgroup; search for the group specific sequences in the 3' terminal region of the genome

I. Uyeda

Department of Botany, Faculty of Agriculture, Hokkaido University, Sapporo, Japan

Summary. In order to examine relationships among viruses of the bean yellow mosaic subgroup of the *Potyvirus* genus, several isolates of bean yellow mosaic virus (BYMV) and clover yellow vein virus (ClYVV) were compared by amino acid sequence of the coat protein and nucleotide sequence of the 3' terminal non-coding region. The sequence comparisons showed that BYMV and ClYVV were distinct viruses but had close affinity to each other (85–95% homology among isolates of a virus but 70–77% homology between viruses), justifying establishment of the BYMV subgroup. There was an oligonucleotide consensus sequence present in the 3' terminal non-coding region of all potyviruses examined. This consensus sequence divided the potyviruses into three groups whose significance is not clear.

Introduction

The bean yellow mosaic virus subgroup of the genus *Potyvirus* [3, 19] includes bean yellow mosaic virus (BYMV), clover yellow vein virus (ClYVV), pea mosaic virus (PMosV), and sweet pea mosaic virus. Host ranges and symptoms produced by these viruses are distinct but similar to each other. They are serologically related in varying degrees. Several attempts have been made to classify the related viruses using host range, symptomatology, serology, direct enzyme-linked immunosorbent assay, and nucleic acid hybridization [3, 4, 13, 20]. These studies clearly show that all viruses in the subgroup are related but that differentiation of the viruses within the subgroup is difficult due to lack of a criterion for unambiguous identification [19]. Identification of the viruses is confusing since treating these viruses as strains of BYMV or distinct viruses requires interpretation of the techniques utilized.

Recent proposals by Shukla and Ward [21, 23] classified potyviruses on the basis of amino acid sequences of the coat-protein gene. In this communication, I attempt to clarify the relationship among strains of BYMV and ClYVV by comparing amino acid sequences of the coat-protein gene and 3' non-coding nucleotide sequences. These criteria are straightforward and practical for the taxonomy of potyviruses, because the genome nucleotide sequence is an intrinsic property of the virus and is easily accessed once it is deposited in a data base. Use of sequence information makes classification more reliable than that based on properties such as host range, serology, and molecular hybridization. For example, host range and symptomatology are difficult to standardize because test plants may be grown under different conditions and different cultivars may be tested by different researchers which results in different host reactions. Serological comparison of the viruses shows that they could be grouped differently with different antisera [3]. Based on the sequence comparison presented in this communication, it is proposed that gene diagnosis may be the choice for rapid and accurate classification and perhaps even for identification.

These sequence comparisons reveal how closely two viruses or strains of the same virus are related, assuming there is a hierarchical relationship among these viruses. This technique might even allow classification of a virus isolate which has an amino acid or nucleotide sequence homology intermediate between two viruses, if such an isolate were found. Sequence relationships are important characteristics for any discussion of the taxonomy of potyviruses.

Amino acid sequence of the coat protein

Accumulation of nucleotide sequence data of BYMV [5, 11, 26, 27] and ClYVV [6, 27, 29] has made possible examination of molecular relationships between these viruses. Amino acid sequence homology of the coat protein among four isolates of BYMV was from 87 to 94%, whereas that of BYMV to ClYVV isolates was from 70 to 77% (Table 1 and Fig. 1). The homology

Table 1. Amino acid sequence homology of coat proteins among strains of BYMV and ClYVV[a]

		BYMV				ClYVV		
		CS	Danish	GDD	S	30	B	NZ
BYMV	CS	–	89[b]	88	87	73	77	77
	Danish		–	94	89	71	76	73
	GDD			–	89	70	74	73
	S				–	71	74	73
ClYVV	30					–	92	88
	B						–	93

[a] Amino acid sequences were taken from the following: BYMV-CS [26], -Danish [5], -GDD [11], -S [27], and ClYVV-30 [29], -B [27], -NZ [6]

[b] Sequence homology was calculated by a program of GENETYX

```
CS       1   SDQEKLNASEKKKDKDKKVEDQSTKESEGQSSKQIIPDRDVNAGTTGTFSVPRLKKIAGK
GDD      1         Q    G E     R   N GNPN D        VR  V              V
DANISH   1         Q    G E     K   N ENPD N      N R  V              V          K
S        1         P    G            I NPS D D       RR  V     I T   V     I
30       1     K      VG QQ S     ESRQYEIL EV E  NR           I    S          K       S
B        1     K       G QQ F     E-PR RDQ G -N NR           I              K       S
NZ       1   GK Q      G QQ PR    D-PK REQ P A- N             I                      S

CS      61   LHIPKVNGKIVLNLDHLLEYNPSQDDISNTIATDEQFKAWYNGVKQAYEVEDSQMSIILN
GDD     61   N   IG    F        K    P      V    QA  E                  R G
DANISH  61   N   IG              P  G  V    QV  E   S                   G
S       61   N T IG             D    P      QA  E          HE    N        D   G
30      61   SL  IK  GL       V V N   L  N     Q  LE  HE    N      D Q   E    C
B       59   SL  IK  GL       V V N      N     Q  LE  HE    N      D Q   E    C
NZ      59   SL  IK  GL       L V N      N     Q  LE  HE    N D D Q   E    C

CS     121   GLMVWCIENGTSGDLQGEWTMMDGDEQVTYPLKPILDNAKPTFRQIMSHFSQVAEAYIEK
GDD    121                         E                             E
DANISH 121                         E                             E
S      121                         E                       N  E
30     121                         EK    F         F     L     A    S     F
B      119                         EK    F         F     L     A    S     F
NZ     119                         K     F   V     F     L     A    S     F

CS     181   RNATERYMPRYGLQRNLTDYGLARYAFDFYRLTSRTPVRAREAHMQMKAAAIRGKSNRLF
GDD    181                 E                      E                 V A   T
DANISH 181                              · K                      V    T
S      181             T                      K                   V
30     181   KKLNRV                          K  A       IE          N  HM
B      179     S                             K  A       E             HM
NZ     179     C                             K  T                      HM

CS     241   GLDGNVGTDEENTERHTAGDVNRDMHTMLGVRI
GDD    241
DANISH 241
S      241                 N                 V
30     241              N     N   HIA A F
B      239              N     N   HIA A F
NZ     239              N     N   HIA A F
```

Fig. 1. Alignment of coat protein amino acid sequences of the virus isolates in the BYMV subgroup. Blanks, amino acids identical to BYMV CS strain. Dashes, gaps needed for optimal alignment. The sequence data were taken from those listed in Table 1

among three isolates of ClYVV was from 88 to 93%. Clearly, BYMV and ClYVV should be treated as distinct species [27, 29] based on the criteria proposed by Shukla and Ward [21, 23]. A close comparison of the sequences revealed highly conserved regions (nearly 100%) [29] among isolates of these two viruses. When these regions were compared with the comparable positions of the sequences in other potyviruses, BYMV subgroup-specific sequences were found. These are -GDLQGEWT-, -TDYG-, and -NVGTDE- in the regions II, III, and IV, respectively, as defined by Uyeda et al. [29] (Fig. 2). Particularly, G at a position 133 from the N-terminus was present only in viruses in this subgroup, where P was invariably present in other potyviruses. Similarly, I at position 91 was BYMV subgroup-specific whereas R was present in all other potyviruses (Fig. 2).

The amino acid sequence, -DAG-, in the N-terminal region of the coat protein is required for aphid transmission of tobacco vein mottling virus (TVMV) and tobacco etch virus (TEV) [2, 12]. This sequence is present in all

Region	2 I	II	III	IV	V
	1 15	88 *93	120 * 144	187 210	240 266
BYMV-CS	SDQEKLNASEKKKDK	SNTIAT	NGLMVWCIENGTSGDLQGEWTMMDG	YMPRYGLQRNLTDYGLARYAFDFY	FGLDGNVGTDEENTERHTAGDVNRDMH
BYMV-DANISH	SDQEQLNAGEEKKDK	SNVIAT	NGLMVWCIENGTSGDLQGEWTMMDG	YMPRYELQRNLTDYGLARYAFDFY	FGLDGNVGTDEENTERHTAGDVNRDMH
BYMV-GDD	SDQEQLNAGERKKDK	SNVIAT	NGLMVWCIENGTSGDLQGEWTMMDG	YMPRYELQRNLTDYGLARYAFDFY	FGLDGNVGTDEENTERHTAGDVNRDMH
BYMV-S	SDQEKPNAGEKKKDK	SNTIAT	NGLMVWCIENGTSGDLQGEWTMMDG	YMTRYGLQRNLTDYGLARYAFDFY	FGLDGNVGTDEENTERHTAGDVNRDMH
ClYVV-30	SDKEKLNVGEQQKSK	SNNIAT	NGLMVWCIENGTSGDLQGEWTMMDG	YMPRYGLQRNLTDYGLARYAFDFY	FGLDGNVGTDEENTERHTANDVNRNMH
ClYVV-B	SDKEKLNAGEQQKFK	SNNIAT	NGLMVWCIENGTSGDLQGEWTMMDG	YMPRYGLQRNLTDYGLARYAFDFY	FGLDGNVGTDEENTERHTANDVNRNMH
ClYVV-NZ	SGKEQLNAGEQQKPR	SNNIAT	NGLMVWCIENGTSGDLQGEWTMMDG	YMPRYGLQRNLTDYGLARYAFDFY	FGLDGNVGTDEENTERHTANDVNRNMH
PWV-TB	KDEIIDVGADGKKV	FNTRAT	NPFMVWCIENGTSPDINGVWVMMDG	YMPRYGLLRNLRDKNLARYAFDFY	FGLDGNVATISEDTERHTARDVNQNMH
SbMV-N	SGKEKEGDMDADKDPKKS	FNTRAT	NGFMVWCIDNGTSPDANGVWVMMDG	YMPRYGLLRNLRDRELARYAFDFY	FGLDGNISTNSENTGRHTARDVNQNMH
WMV 2	SGKETVENLDAGKESKKD	FNTRAT	NGFMVWCIDNGTSPDVNGVWVMMDG	YMPRYGLLRNLRDRELARYAFDFY	FGLDGNISTNSENTGRHTARDVNQNMH
ZYMV-F	SGTQPTVADARVTKKDK	YNTRAS	NGFMVWCIENGTSPDINGVWFMMDG	YMPRYGLLRNLRDRSLARYAFDFY	FGLDGNVATTSEDSERHTARDVNRNMH
PVY-O	ANDTIDAVEINKKE	SNTRAT	NGLMVWCIENGTSPNVNGVWVMMDG	YMPRYGLIRNLRDVGLARYAFDFY	FGLDGGISTQEENTERHTTEDVSPSMH
PVY-T	GNDTIDAGGSTKKD	SNTRAT	NGLMVWCIENGTSPNINGVWVMMDG	YMPRYGLVRNLRDGSLARYAFDFY	FGLDGGISTQEENTERHTTEDVSPSMH
TEV	SGTVDAGADAGKK	SNARAT	NGFMVWCIENGTSPNLNGTWVMMDG	YMPRYGLQRNITDMSLSRYAFDFY	FGLDGNVGTAEEDTERHTAHDVNRNMH
TVMV	SDTVDAGKDKARD	VNTRAT	NGFMIWCIENGTSPNISGVWTMMDG	YIPRYGLQRGLVDRNLAPFAFDFF	FCLDGSVSGQEENTERHTVDDVNAQMH
JGMV	SGNEDAGKQKSAT	SNARAT	NGLMVWCIENGTSPDINGYWTMVDG	YMPRYGLLRNLNDKSLARYAFDFY	FGLDGIVGESSENTERHTAADVSRNVH
SCMV-SC	AGTVDAGAQGGGG	SNTRAT	SGLMVWCIENGCSPNISGSWTMMDG	YMPRYGLQRNLTDYSLARYAFDFY	FGLDGNVGETQENTERHTAGDVSRNMH
MDMV-B	SGRVDAGAQGGSG	SNTRAT	SGLMVWCIENGCSPNINGNWTMMDK	YMPRYGLQRNISDYSLARYAFDFY	FGLDGNVGETQENTERHTAGDVSRNMH
TuMV	AGETLDADLTEEQK	SNTRST	NGLRVWCIENGTSPNINGMWVMMDG	YMPRYGLQRNLTDMSLARYAFDFY	FGLDGNVGTTVENTERHTTEDVIRNMH
PRSV-W	SKNEAVDTGLNEKFK	SNTRAT	NGLMVWCIENGTSPDISGVWVMMDG	YMPQYGIKRNLTDISLARYAFDFY	FGIDGSVTNKEENTERHTVEDVNRDMH
PPV-SP	DEREDEEEVDAGKPSVVT	SNTRAP	NGLMVWCIENGTSPNINGMWVMMDG	YMPRYGIQRNLTDYSLARYAFDFY	FGLDGNVGTQEEDTERHTAGDVNRNMH

Fig. 2. Comparison of coat-protein amino acid sequences of potyviruses in specified regions. Proposed BYMV subgroup specific sequences are underlined. Asterisks, amino acids conserved only in the BYMV subgroup. Position of the amino acid from the N-terminus of BYMV-CS was shown at the top line. N-terminal sequence of PWV was not available. Sequences were taken from following: PWV [22], SbMV [8], WMV [18], ZYMV [18], PVY-O and -T [16],TEV [1],TVMV [7], JGMV [10], SCMV [9], MDMV [9], TuMV [28], PRSV [17], PPV [15]

potyviruses, except for the BYMV subgroup. Interestingly the amino acid sequence of BYMV and ClYVV at the comparable position is -NAG- rather than -DAG- (Fig. 2) [24]. So far this sequence also separates the BYMV subgroup from other potyviruses.

3′ non-coding sequence

The nucleotide sequence homology among BYMV isolates ranged from 86 to 94%. Significant homologies (71 to 77%) were also detected between isolates of BYMV and ClYVV (Table 2, Fig. 3). The relatedness of sequences between the two viruses was similar to that detected in amino acid sequences of the coat protein.

There is virtually no similarity to other potyviruses in the length or sequence of the 3′ non-coding region of BYMV or ClYVV.

There was an imperfect direct repeat in the 3′ non-coding region (Fig. 3) of both BYMV and ClYVV. A search for the similar sequence in other potyviruses revealed the consensus sequence

-AG-GAGG - - - - - CCUCc- or -AG-GUGG - - - - - cCACC-

(Fig. 4). Although the consensus sequence was in the region with the direct repeat in the BYMV subgroup, a similar repeat was not always present in other potyviruses. These consensus sequences separated the potyviruses into three

groups. The first group included BYMV, ClYVV, plum pox virus (PPV), TVMV, TEV, sugarcane mosaic virus (SCMV), and turnip mosaic virus with the
-AG-GAGG - - - - - CCUCc-
sequence. The second group included soybean mosaic virus (SbMV), water melon mosaic virus 2 (WMV 2), and johnsongrass mosaic virus (JGMV) with the
-AG-GUGG - - - - - cCACC-
sequence. The third contains only potato virus Y (PVY) without either sequence. Papaya ringspot virus (PRSV) had both consensus sequences in the imperfect direct repeat region. An exception, zucchini yellow mosaic virus (ZYMV), had the sequence
-AG-GUGG-CCUCC-.

Discussion

Amino acid sequence comparisons of the coat proteins of BYMV and ClYVV show that they should be treated as distinct viruses [27, 29] in the potyvirus taxonomy scheme [21, 23].

Nevertheless the sequence comparisons of BYMV and ClYVV to other potyviruses clearly show that the two viruses are closely related at the molecular level. First, the number of amino acids in the coat proteins of the subgroup is nearly identical (271 or 273). Second, the non-coding regions of the two viruses are similar in both length and sequence, whereas those of different potyviruses differ significantly in these characteristics. These properties justify the formation of the BYMV subgroup to include these related viruses. A close comparative examination of amino acid sequences of different potyviruses reveals several short stretches of subgroup specific sequences in the coat-

Table 2. Nucleotide sequence homology of non-coding 3′ terminal regions between BYMV and ClYVV isolates[a]

		BYMV				ClYVV		
		CS	Danish	GDD	S	30	B	NZ
BYMV	CS	–	94[b]	86	90	75	74	73
	Danish		–	92	94	77	76	74
	GDD			–	88	73	74	71
	S				–	74	73	72
ClYVV	30					–	98	92
	B						–	92

[a] Nucleotide sequences were taken from the following: BYMV-CS [26], -Danish [5], -GDD [11], and -S [27], and ClYVV-30 [29], -B [27], -NZ [6]

[b] Sequence homology was calculated by a program of GENETYX

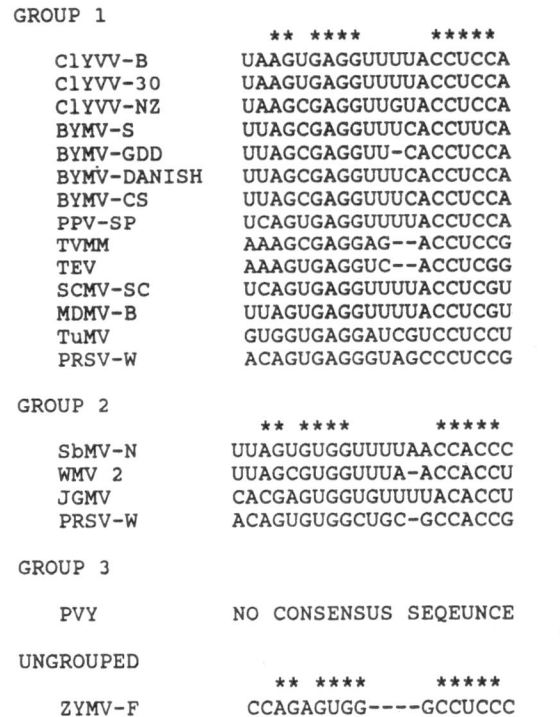

```
NZ       1            AA          A AU                    C
B        1     AGUAUCCGUCUUUAAAUUCUCCGUUAAUUUCGAAGUUUUACUAUUAUAGCACUAUGU-UA
30       1     AGUAUCCGUCUUUAAAUUCUCCGUUAAUUUCGAAGUUUUACUAUUAUAGCACUAUGU-UA
S        1     A              UA-   C - GC  AC U  C -  AU      G U
GDD      1                    UA-   - -C -AAC U  C -  AU -AC AGA U
DANISH   1                    UA-   - GC  AC U  C -  AU      A U
CS       1         A          U -A  - GC  AC U  C -  AU      A U


NZ      60     C      G              U          G
B       60     AGUGAGGUUUUACCUCCAUUUCACUUUAAGUAUAAAUAGUAAUCCAUUCUCUCUAUUCUG                    C
30      60     AGUGAGGUUUUACCUCCAUUUCACUUUAAGUAUAAAUAGUAAUCCAUUCUCUCUAUUCUG
S       58     C      C    U  --A  U    A UC G      GUAU              C -
GDD     55           - C       --G  U    A UC G      GUAU            - C GA
DANISH  58            C         --G  U    A UC G      GUAU              C
CS      58     C               --G  U    A UC G  A   GU U              C


NZ     120                               U        CA
B      120                               C
30     120     ACAGAGU-AGC--UAAGUGAGGUUAUACCUCGUUGUGAAUCUGAUCUUUAUAGAGCGAG
S      115       AG A   AG U       --     -   GCU        G
GDD    111     U  G   A  AG U      --G     -    GC        G
DANISH 116        G   A  AG U      --G     -    GC        G
CS     116        G   A  UG U      --U     -A  GGC        G
```

Fig. 3. Alignment of 3'-terminal non-coding nucleotide sequences of BYMV and ClYVV isolates with respect to ClYVV 30. The sequences were aligned manually for optimum homology. Blanks, a nucleotide identical to ClYVV 30. Dashes, gaps introduced for optimal alignment. Imperfect direct repeat is underlined. The sequence data were taken from those listed in Table 1

```
GROUP 1
                              ** ****    *****
          ClYVV-B      UAAGUGAGGUUUUACCUCCA
          ClYVV-30     UAAGUGAGGUUUUACCUCCA
          ClYVV-NZ     UAAGCGAGGUUGUACCUCCA
          BYMV-S       UUAGCGAGGUUUCACCUUCA
          BYMV-GDD     UUAGCGAGGUU-CACCUCCA
          BYMV-DANISH  UUAGCGAGGUUUCACCUCCA
          BYMV-CS      UUAGCGAGGUUUCACCUCCA
          PPV-SP       UCAGUGAGGUUUUACCUCCA
          TVMM         AAAGCGAGGAG--ACCUCCG
          TEV          AAAGUGAGGUC--ACCUCGG
          SCMV-SC      UCAGUGAGGUUUUACCUCGU
          MDMV-B       UUAGUGAGGUUUUACCUCGU
          TuMV         GUGGUGAGGAUCGUCCUCCU
          PRSV-W       ACAGUGAGGGUAGCCCUCCG


GROUP 2
                              ** ****    *****
          SbMV-N       UUAGUGUGGUUUUAACCACCC
          WMV 2        UUAGCGUGGUUUA-ACCACCU
          JGMV         CACGAGUGGUGUUUUACACCU
          PRSV-W       ACAGUGUGGCUGC-GCCACCG


GROUP 3

          PVY          NO CONSENSUS SEQEUNCE


UNGROUPED
                              ** ****    *****
          ZYMV-F       CCAGAGUGG----GCCUCCC
```

Fig. 4. Grouping of potyviruses based on the consensus nucleotide sequence in the 3'-terminal non-coding region. The sequence data were taken from those listed in Fig. 2. Asterisks, the proposed consensus

protein gene. These regions may help differentiate other potyviruses from the BYMV subgroup.

Since the nucleotide sequences of only BYMV and ClYVV are available at present, sequence relationships of PMosV and sweet pea mosaic virus to BYMV and ClYVV are not yet known. Molecular hybridization experiments using amplified DNA probes corresponding to the 3' non-coding regions of BYMV and ClYVV show that hybridization occurs only with their homologous viruses, not with PMosV [27]. This suggests that PMosV is a distinct member of the subgroup as was proposed by previous studies based on other properties [3]. The conclusive answer awaits determination of the nucleotide sequences of PMosV and sweet pea mosaic virus.

An additional rapid and accurate method for practical and routine identification of the viruses in the subgroup is needed. Previous studies on host range, serology, and molecular hybridization using cDNA probes made by random priming show that it is difficult to interpret results in order to clarify the relationships among viruses and strains or isolates within the BYMV subgroup. Relationships of the viruses based on serology or molecular hybridization are often inconsistent because of different antisera or cDNA probes [3, 20] and interpretation tends to be arbitrary. Use of monoclonal antibodies directed against the subgroup-specific and virus-specific epitopes may allow rapid identification and classification [14]. Alternatively, molecular hybridization using synthetic oligonucleotide probes complementary to the subgroup or virus-specific nucleotide sequences [25], or PCR amplification [27] and sequencing of a particular region of the genome is promising. Takahashi [25] designed synthetic 20 mer oligonucleotide DNA probes complementary to ClYVV-30 and BYMV-CS coat-protein gene regions where there were two nucleotide mismatches. The two viruses were successfully differentiated by a dot blot hybridization using these probes. Sequence information of the viruses so far available suggests that strains of a single virus (BYMV or ClYVV) from different countries and hosts share highly conserved sequences. Thus probes designed on the basis of these sequences will be able to detect and discriminate the viruses in the BYMV subgroup.

There are oligonucleotide consensus sequences in the non-coding region of most potyviruses. The grouping based on this consensus is consistent in most part with that proposed by Quemada et al. [18] using nucleotide sequence data of the coat-protein gene. Their grouping of BYMV, TEV, TVMV, PPV, and PVY is consistent with the one presented here except for PVY. Potato virus Y does not have either consensus sequence and was placed in a third group in this paper. Both grouped JGMV, SbMV, and WMV-2 together. ZYMV cannot be assigned based on the consensus sequences here. Papaya ringspot virus is intermediate between the two groups having both consensus sequences and this is in reasonably good agreement with that of Quemada et al. [18]. Although a function of this sequence is not known, it can be used as one criterion for grouping potyviruses.

Acknowledgements

The author would like to thank Dr. Eishiro Shikata for critical reading of this manuscript. Thanks are extended to Drs. D. D. Shukla, G. T. Bryan, and R. L. S. Forster for providing gene sequences of bean yellow and clover yellow vein viruses prior to publication.

References

1. Allison RF, Sorenson JC, Kelly ME, Armstrong FB, Dougherty WG (1985) Sequence determination of the capsid protein gene and flanking regions of tobacco etch virus: evidence for synthesis and processing of a polyprotein in potyvirus genome expression. Proc Natl Acad Sci USA 82: 3969–3972
2. Atreya CD, Raccah B, Pirone TP (1990) A point mutation in the coat protein abolishes aphid transmissibility of a potyvirus. Virology 178: 161–165
3. Barnett OW, Randles JW, Burrows PM (1987) Relationships among Australian and North American isolates of the bean yellow mosaic potyvirus subgroup. Phytopathology 77: 791–799
4. Bos L, Kowalska CZ, Maat DZ (1974) The identification of bean mosaic, pea yellow mosaic and pea necrosis strains of bean yellow mosaic virus. Neth J Plant Pathol 80: 173–191
5. Boye K, Jensen PE, Stummann BM, Henningsen KW (1990) Nucleotide sequence of cDNA encoding the BYMV coat protein gene. Nucleic Acids Res 18: 4926
6. Bryan GT, Gardner RC, Forster RLS (1992) Nucleotide sequence of the coat protein gene of a strain of clover yellow vein virus from New Zealand: conservation of a stemloop structure in the 3' region of potyviruses. Arch Virol 124: 133–146
7. Domier LL, Franklin KM, Shahabuddin M, Hellmann GM, Overmeyer JH, Siaw MFE, Lomonossoff GP, Shaw JG, Rhoads RE (1986) The nucleotide sequence of tobacco vein mottling virus RNA. Nucleic Acids Res 14: 5417–5430
8. Eggenberger AL, Stark DM, Beachy RN (1989) The nucleotide sequence of a soybean mosaic virus coat protein-coding region and its expression in *Escherichia coli*, *Agrobacterium tumefaciens*, and tobacco callus. J Gen Virol 70: 1853–1860
9. Frenkel MJ, Jilka JM, McKern NM, Strike PM, Clark Jr JM, Shukla DD, Ward CW (1991) Unexpected sequence diversity in the amino-terminal ends of the coat proteins of strains of sugarcane mosaic virus. J Gen Virol 72: 237–242
10. Gough KH, Azad AA, Hanna PJ, Shukla DD (1987) Nucleotide sequence of the capsid and nuclear inclusion protein genes from the Johnson grass strain of sugarcane mosaic virus RNA. J Gen Virol 68: 297–304
11. Hammond J, Hammond RW (1989) Molecular cloning, sequencing and expression in *Escherichia coli* of the bean yellow mosaic virus coat protein gene. J Gen Virol 70: 1961–1974
12. Harrison BD, Robinson DJ (1988) Molecular variation in vector-borne plant viruses: epidemiological significance. Philos Trans R Soc Lond [Biol] 321: 447–462 (Virology & AIDS Abstracts 22: 121, 1989)
13. Jones RT, Diachun S (1977) Serologically and biologically distinct bean yellow mosaic virus strains. Phytopathology 67: 831–838
14. Jordan R, Hammond J (1991) Comparison and differentiation of potyvirus isolates and identification of strain-, virus-, subgroup-specific and potyvirus group-common epitopes using monoclonal antibodies. J Gen Virol 72: 25–36
15. Lain S, Riechmann JL, Mendez E, Garcia JA (1988) Nucleotide sequence of the 3' terminal region of plum pox potyvirus RNA. Virus Res 10: 325–342
16. Ohshima K, Hataya T, Sano T, Inoue AK, Shikata E (1991) Comparison of amino acid sequences, biological properties and serological characteristics between potato virus Y ordinary strain and necrotic strain. Ann Phytopathol Soc Jpn 57: 615–622

17. Quemada H, L'Hostis B, Gonsalves D, Reardon IM, Heinrikson R, Hiebert EL, Sieu LC, Slightom JL (1990) The nucleotide sequence of the 3'-terminal regions of papaya ringspot virus strains W and P. J Gen Virol 71: 203–210
18. Quemada H, Sieu LC, Siemieniak DR, Gonsalves D, Slightom JL (1990) Watermelon mosaic virus II and zucchini yellow mosaic virus: cloning of 3'-terminal regions, nucleotide sequences, and phylogenetic comparisons. J Gen Virol 71: 1451–1460
19. Randles JW, Davies C, Gibbs AJ, Hatta T (1980) Amino acid composition of capsid protein as a taxonomic criterion for classifying the atypical S strain of bean yellow mosaic virus. Aust J Biol Sci 33: 245–254
20. Reddick BB, Barnett OW (1983) A comparison of three potyviruses by direct hybridization analysis. Phytopathology 73: 1506–1510
21. Shukla DD, Ward CW (1988) Amino acid sequence homology of coat proteins as a basis for identification and classification of the potyvirus group. J Gen Virol 69: 2703–2710
22. Shukla DD, McKern NM, Ward CW (1988) Coat protein of potyviruses 5. Symptomatology, serology, and coat protein sequences of three strains of passionfruit woodiness virus. Arch Virol 102: 221–232
23. Shukla DD, Ward CW (1989) Structure of potyvirus coat proteins and its application in the taxonomy of the potyvirus group. Adv Virus Res 36: 272–314
24. Shukla DD, Frenkel MJ, Ward CW (1991) Structure and function of the potyvirus genome with special reference to the coat protein coding region. Can J Plant Pathol 13: 178–191
25. Takahashi T (1990) Studies on gene structure of bean yellow mosaic virus and clover yellow vein virus. PhD thesis, Hokkaido University, Sapporo
26. Takahashi T, Uyeda I, Ohshima K, Shikata E (1990) Nucleotide sequence of the capsid protein gene of bean yellow mosaic virus chlorotic spot strain. J Fac Agric Hokkaido Univ 64: 152–163
27. Tracy SL, Frenkel MJ, Gough KH, Hanna PJ, Shukla DD (1992) Bean yellow mosaic, clover yellow vein and pea mosaic are distinct potyviruses: evidence from coat protein gene sequence and molecular hybridization involving the 3' non-coding regions. Arch Virol 122: 249–261
28. Tremblay M-F, Nicolas O, Sinha RC, Lazure C, Laliberte J-F (1990) Sequence of the 3'-terminal region of turnip mosaic virus RNA and the capsid protein gene. J Gen Virol 71: 2769–2772
29. Uyeda I, Takahashi T, Shikata E (1991) Relatedness of the nucleotide sequence of the 3'-terminal region of clover yellow vein potyvirus RNA to bean yellow mosaic virus potyvirus RNA. Intervirology 32: 234–245

Author's address: I. Uyeda, Department of Botany, Faculty of Agriculture, Hokkaido University, Sapporo 060, Japan.

Virus relationships – BCMV subgroup

Arch Virol (1992) [Suppl 5]: 389–395

A proposal for a bean common mosaic subgroup
of potyviruses

Jeanne Dijkstra and **J. A. Khan**

Department of Virology, Agricultural University, Wageningen, The Netherlands

Summary. In order to elucidate the taxonomic positions of bean common mosaic virus (BCMV) and blackeye cowpea mosaic virus (BlCMV), several strains of these viruses were compared on the basis of host ranges, antigenic properties established with antisera to virions and to N-terminal peptide domains of their coat proteins, and high performance liquid chromatographic peptide profiles. The comparison includes three strains of BCMV, viz. NL1, NL3 and NY15, four strains of BlCMV, viz. Fla, Ind, NR, and W, and the Moroccan isolate (Mor) of cowpea aphid-borne mosaic virus (CABMV), formerly designated as BlCMV-Mor. Based on these parameters, Fla, NR, and W are strains of one virus, whereas NL3, Ind and CABMV-Mor (and possibly NL1 and NY15) are separate viruses. In view of these characteristics which allow similar viruses to be separated, we propose that these viruses be included in a bean common mosaic subgroup of the genus *Potyvirus*.

Introduction

A cluster composed of viruses closely related to bean yellow mosaic virus was proposed in 1980 [6]. This idea has been expanded in a number of subsequent studies and a bean yellow mosaic subgroup of potyviruses accepted. Uyeda [10] found a sequence common to viruses in this subgroup.

The genus *Potyvirus* contains a large number of member and possible member viruses. While subgroups have no formal taxonomic status with the International Committee on Taxonomy of Viruses, a subgroup makes dealing with closely related viruses easier. For some purposes, diagnosis to the subgroup level will be enough. Once a diagnosis to the subgroup level is made, other diagnostic techniques can be used to differentiate individual species and strains. We are proposing a bean common mosaic subgroup here.

Bean common mosaic virus (BCMV) and blackeye cowpea mosaic virus (BlCMV) are two major viruses of French bean (*Phaseolus vulgaris*) and cowpea (*Vigna unguiculata*), respectively, each with a large number of strains.

The BCMV strains have been distinguished mainly on the basis of their genetic interaction with cultivars of French bean [2], and host ranges and serological properties [4, 11, 12].

Strains of BCMV have been arranged in three main groups according to symptoms in bean cultivars:(*i*) strains that never induce systemic necrosis, but mosaic;(*ii*) strains that induce systemic necrosis in cultivars of some II resistance groups, according to temperature (temperature-dependent, necrosis-inducing strains); (*iii*) strains inducing local and systemic necrosis at all temperatures in II genotypes susceptible to the strain concerned (temperature-independent, necrosis-inducing strains) [2]. Also, serologically, groups 1 and 2 on the one hand, and group 3 on the other hand, can be distinguished and have been designated serotype B and A, respectively [11, 12].

Distinctions between BlCMV strains are also based on host ranges and antigenic properties [1, 8, 9]. However, a lack of clear differences between these viruses and their strains has made it obvious that their taxonomic status could not be determined by biological and conventional serological criteria alone [4, 7].

Recently, more information on the structure of their coat protein has led to a better distinction between the different strains. Use of antibodies to the N-terminal parts of coat proteins makes it possible to distinguish between a number of strains of BCMV and BlCMV [3]. Peptide profiling of coat-protein tryptic digests by high performance liquid chromatography (HPLC) has also contributed to further classification of strains of both BCMV and BlCMV [5; D. D. Shukla, pers. comm.].

To assign a definite taxonomic status to a virus species, knowledge of the complete nucleotide sequence of its genome is desirable. In the absence of the latter, a combination of other parameters, such as host ranges, conventional serology with antibodies to virions, N-terminal serology, and HPLC might be used to distinguish between the viruses and their strains. On the basis of these parameters, an effort is made in this paper to tentatively classify the BCMV strains NL1 and NY15 (both belonging to serotype B), NL3 (a serotype A strain), the BlCMV strains Fla, Ind, NR, W, and the Moroccan isolate (Mor) of cowpea aphid-borne mosaic virus (CABMV), earlier described as BlCMV-Mor [1, 4] and henceforth referred to as Mor.

Parameters

Host range

All the bean cultivars in Table 1 were infected by NL3 and most of them by NL1 and NY15. Resulting infection was usually systemic, with or without symptoms. Strain W induced mostly local symptoms or did not infect some cultivars at all. This response proved true for Mor, as well as Fla, Ind, and NR, but the number of bean cultivars tested with these strains was too small to justify such a conclusion.

The cowpea genotypes inoculated with NL1 and NL3 were either uninfected or had mostly symptomless infection. NY15 induced clear mosaic symptoms in 'California Blackeye' and a number of TVu lines, as did all the BlCMV strains and Mor (Table 2).

Of the non-legumes, *Chenopodium amaranticolor* and *C. quinoa* were not infected by NL1 and NL3, but reacted with local lesions after infection with NY15, all BlCMV strains, and Mor. *Nicotiana benthamiana* was symptomlessly infected by NL1, not infected by NL3, but showed symptoms with NY15, all strains of BlCMV and also Mor.

Table 1. Reactions of bean cultivars to bean common mosaic virus (BCMV) strains NL1, NL3, NY15, and to blackeye cowpea mosaic virus (BlCMV) strains Fla, Ind, NR, W, and cowpea aphid-borne mosaic virus (CABMV) isolate Morocco (Mor)

Host group[a]	Differential bean cultivar	Strains of BCMV			Strains of BlCMV				Mor
		NL1	NL3	NY15	Fla	Ind	NR	W	
1.	Dubbele Witte	S	S	S	S	S,SN	–	s,S	L
	Stringless Green Refugee	S	S	S	.	.	.	L	.
	Bountiful	S	S	S	.	.	.	S	.
	Saxa	S	S,SN	S	S	S,SN	S	L	L
2.	Redlands Greenleaf C	L	s,S	S	.	.	.	L	.
	Puregold Wax	–	s,S	S	.	.	.	L	.
	Imuna	L,s	s,S	–	.	.	.	–	.
	Bataaf	S	S,SN	S	L	L	L	L	L
3.	Redlands Greenleaf B	L,s	S	L	.	.	.	L	.
	Great Northern UI 123	L,s	s	–	.	.	.	–	.
4.	Sanilac	L	S	S	.	.	.	–	.
	Michelite	s,S	S	S	.	.	.	L	.
	Red Mexican UI 34	L,s	S	S	.	.	.	L	.
5.	Pinto	L,s	S	S	L	L	L	L	L
6.	Monroe	L,s	L	L	.	.	.	L	.
	Great Northern UI 31	L	s	L,s	.	.	.	–	.
	Red Mexican UI 35	L	S	L	.	.	.	L	.
8.	Widusa	–,s	SN	–	.	.	.	–	.
	Black Turtle Soup	–	SN	–	.	.	.	–	.
9a.	Jubila	–	SN	L	.	.	.	–	.
9b.	Topcrop	s,S	SN	–	.	.	.	–	.
	Improved Tendergreen	–	SN	–	.	.	.	–	.

Adapted from [1] and [4]

[a] Host group based on resistance to BCMV

L Local symptoms; *S* systemic, non-necrotic symptoms; *SN* systemic, necrotic symptoms; *s* symptomless infection or very weak symptoms

– No infection; dots, not tested

392 Jeanne Dijkstra and J. A. Khan

Table 2. Reactions of cowpea cultivars and lines to bean common mosaic virus (BCMV) strains NL1, NL3, NY15, and to blackeye cowpea mosaic virus (BlCMV) strains Fla, Ind, NR, W, and cowpea aphid-borne mosaic virus (CABMV) isolate Morocco (Mor)

Cowpea cultivars and lines	Strains of BCMV			Strains of BlCMV				Mor
	NL1	NL3	NY15	Fla	Ind	NR	W	
California Blackeye	S	–	S	S	S	S	S	S
Early Red	L	–	–	–	–	–	–	S
IITA TVu 196	1	–	S	S	s,S	S	S	s,S
IITA TVu 401	–	–	s	–,S	–,S	–,S	–,S	–,S
IITA TVu 1582	s	s	s,S	S	s,S	S	S	–
IITA TVu 1593	s	s	s,S	–,S	L,s	S	S	–
IITA TVu 2460	–	–	–	.	SN	.	S	.
IITA TVu 2657	s	–	–	–	s,S	–	–,S	–,S
IITA TVu 2740	s	–	–	–,S	1,L	–	–,S	s
IITA TVu 2845	S	S	S	S	S	S	S	S
IITA TVu 3270	–	–	–	–,S	S	–	–,S	S
IITA TVu 3433	s	s	–	S	–	–,S	–	S

Adapted from [1] and [4]

1 Symptomless local infection. For other legends, see Table 1

Conventional serology

Homologous reactions among NL1, NY15, Fla, NR, and W were recorded in reciprocal SDS-immunodiffusion tests, but not in direct DAS-ELISA. The relationship between NL3, Ind, and Mor, as opposed to the other strains of BCMV and BlCMV, was non-reciprocal (Table 3).

N-terminal peptide domains of the coat proteins

When using N-terminal specific antibodies in electroblot immunoassay and direct DAS-ELISA, NL1 reacted reciprocally with NY15 and W, but NY15 and W did not react with each other's antiserum, and NL3 showed reactions only with homologous antiserum [3]. Mor also did not react with any of the N-terminal specific antibodies to the other strains (the reciprocal test has not been carried out).

High performance liquid chromatography

It has been reported [5] that the coat proteins of Fla and W have similar amino acid compositions. The peptide profiles and amino acid compositions

Table 3. Reactions of bean common mosaic virus (BCMV) strains NL1, NL3, NY15, and of blackeye cowpea mosaic virus (BlCMV) strains Fla, Ind, NR, W, and cowpea aphidborne mosaic virus (CABMV) isolate Morocco (Mor) to homologous and heterologous antisera in SDS-immunodiffusion tests (*D*) and direct ELISA (*EL*)

Antisera	Antigens															
	Strains of BCMV						Strains of BlCMV									
	NL1		NL3		NY15		Fla		Ind		NR		W		Mor	
	D	EL[a]	D	EL	D	EL	D	EL	D	EL	D	EL	D	EL	D	EL
NL1	+++	3	++	3	+++	3	+++	2	+++	1	+++	2	+++	2	+++	2
NL3	++	1	+++	3	++	1	++	1	++	1	++	1	++	1	++	1
NY15	+++	2	+	2	+++	2	+++	2	++	1	+++	2	+++	2	++	2
Fla	+++	2	–	1	++	2	+++	3	++	1	+++	3	+++	3	+++	3
Ind	+	1	+	1	+	1	++	1	+++	3	++	1	++	1	++	1
NR	+++	3	++	3	++	3	+++	3	+	2	+++	3	+++	3	+++	3
W	++	3	++	2	++	3	+++	3	++	1	+++	3	+++	3	+++	3
Mor	+	1	–	1	+	1	+	1	+	1	+	1	+	1	+++	3

Adapted from [4]

[a] Absorbance values at 405 nm as percentage of that of the homologous reaction arranged in three groups, viz. *1* (1–15%), *2* (15–50%), *3* (50–100%)

+++ Reaction of homology or identity; ++ strong heterologous reaction with spur formation; + weak heterologous reaction with spur formation; – no reaction

of some peptides of BCMV and BlCMV strains revealed a great similarity between NY15, Fla, and W; the peptide profiles of NL1, NL3, and Mor differed greatly from each other and from NY15, Fla and W (D.D. Shukla, pers. comm.).

Generalizations

Results from use of these four parameters allow the following generalizations. Strains NL1, NL3, and NY15 usually induce distinct systemic symptoms in susceptible bean cultivars and latent infections in a number of cowpea genotypes. However, NY15 causes mosaic symptoms in the latter, thus resembling in this respect Fla, Ind, NR, W, and Mor. In SDS-immunodiffusion tests and ELISA, NL1 and NY15 are closely related to each other, and to Fla, NR, and W, but there is a non-reciprocal relationship to NL3, Ind, and Mor. In N-terminal serology, NL1 and W cross react with each other, but not with NL3 and Mor. However, there is no reaction between NY15 and W. HPLC results show great similarity between NY15, Fla, and W, but not between these strains and NL1, NL3, and Mor, and also not among the latter three themselves.

Conclusions

The above findings suggest Fla, NR, and W are strains of one virus and NL3, Ind, and Mor need designations of their own. In conventional and N-terminal serology NL1 resembles NY15 and W, but not in HPLC analysis. On the basis of conventional serology and HPLC profiles of coat proteins, NY15 and W are strains of one virus, but not in N-terminal serology. Therefore, for the time being, it is advisable to also give both NL1 and NY15 taxonomic positions of their own. Since these viruses are closely related biologically and by serological, chemical and physical properties and yet can be distinguished, we propose that these viruses be placed in a subgroup of the potyvirus genus. While a subgroup has no official taxonomic status, its use recognizes the similarities among the virus so grouped.

References

1. Dijkstra J, Bos L, Bouwmeester HJ, Hadiastono T, Lohuis H (1987) Identification of blackeye cowpea mosaic virus from germplasm of yardlong bean and from soybean, and the relationship between cowpea mosaic virus and cowpea aphid-borne mosaic virus. Neth J Plant Pathol 93: 115–133
2. Drijfhout E (1978) Genetic interaction between *Phaseolus vulgaris* and bean common mosaic virus with implications for strain identification and breeding for resistance. Agricultural Research Reports 872. Pudoc, Wageningen
3. Khan JA, Lohuis H, Goldbach RW, Dijkstra J (1990) Distinction of strains of bean common mosaic virus and blackeye cowpea mosaic virus using antibodies to N- and C- or N-terminal peptide domains of coat protein. Ann Appl Biol 117: 583–593
4. Lana AF, Lohuis H, Bos L, Dijkstra J (1988) Relationships among strains of bean common mosaic virus and blackeye cowpea mosaic virus – members of the potyvirus group. Ann Appl Biol 113: 493–505
5. McKern NM, Shukla DD, Barnett OW, Vetten HJ, Dijkstra J, Whittaker LA, Ward CW (1992) Structural properties of coat proteins suggest that azuki bean mosaic virus, blackeye cowpea mosaic virus, peanut stripe virus and three potyvirus isolates from soybean are all strains of the same potyvirus. Intervirology 33: 121–134
6. Randles JW, Davies C, Gibbs AJ, Hatta T (1980) Amino acid composition of capsid protein as a taxonomic criterion for classifying the atypical S strain of bean yellow mosaic virus. Austr J Biol Sci 33: 245–254
7. Shukla DD, McKern NM, Barnett OW, Ward CW (1990) Identification and classification of potyviruses infecting tropical legumes. In: VIIIth International Congress of Virology, Berlin, 1990, abstract W87-8
8. Taiwo MA, Gonsalves D (1982) Serological grouping of isolates of blackeye cowpea mosaic virus and cowpea aphid-borne mosaic viruses. Phytopathology 72: 583–589
9. Taiwo MA, Gonsalves D, Provvidenti R, Thurston HD (1982) Partial characterization and grouping of isolates of blackeye cowpea mosaic and cowpea aphid-borne mosaic viruses. Phytopathology 72: 590–596
10. Uyeda I (1992) Bean yellow mosaic virus subgroup: search for the group specific sequences in the 3′ terminal region of the genome. In: Barnett OW (ed) Potyvirus taxonomy. Springer, Wien New York, pp 377–385 (Arch Virol [Suppl] 5)
11. Wang W-Y (1983) Serology of bean common mosaic virus strains. MSc Thesis, Washington State University, Pullman, Washington

12. Wang W-Y (1985) Production and characterization of hybridoma cell lines and a broad spectrum monoclonal antibody against bean common mosaic virus. PhD Thesis, Washington State University, Pullman, Washington

Authors' address: Jeanne Dijkstra, Department of Virology, Agricultural University, P.O. Box 8045, NL-6700 EM Wageningen, The Netherlands.

Arch Virol (1992) [Suppl 5]: 397–406

Serological and biological relationships among viruses in the bean common mosaic virus subgroup

G. I. Mink[1] and **M. J. Silbernagel**[2]

[1] Washington State University Irrigated Agriculture Research and Extension Center, Prosser, Washington
[2] U.S. Department of Agriculture, Agricultural Research Service, Irrigated Agriculture Research and Extension Center, Prosser, Washington, U.S.A.

Summary. Bean common mosaic virus (BCMV), blackeye cowpea mosaic virus (BlCMV), cowpea aphid-borne mosaic virus (CABMV), azuki bean mosaic virus (AzMV), and peanut stripe virus (PStV) are five species of the genus *Potyvirus*, family *Potyviridae* which are seed-transmitted in beans or cowpeas. Eighteen isolates of BCMV, five isolates of BlCMV, four isolates of CABMV, and one isolate each of AzMV, and PStV were compared serologically using a panel of 13 monoclonal antibodies (MAbs) raised against BCMV, BlCMV, CABMV, or PStV in indirect enzyme-linked immunosorbent assay (ELISA). Four MAbs detected all virus isolates; one detected all isolates except those of CABMV. Three MAbs were specific only for serotype A isolates of BCMV. Four MAbs detected all serotype B isolates of BCMV plus all isolates of BlCMV, AzMV, and PStV. None of the antibodies distinguished among these four viruses. However, in biological tests with 11 bean cultivars selected for differentiating BCMV pathotypes, all isolates of BlCMV, AzMV, and PStV could be differentiated from the BCMV serotype B isolates by their reactions on a few bean cultivars in host group I and the cowpea cultivar California Blackeye #5. Potential problems that can arise from the use of nonauthenticated isolates are also discussed.

Introduction

At least five species of the genus *Potyvirus*, family *Potyviridae* are seed-borne in beans or cowpeas. These are azuki bean mosaic virus (AzMV), bean common mosaic virus (BCMV), blackeye cowpea mosaic virus (BlCMV), cowpea aphid-borne mosaic virus (CABMV), and peanut stripe virus (PStV). While there are close biological and serological relationships among strains of these five viruses, it has been difficult to precisely define characteristics that identify the individual viruses [1, 7, 8]. As a consequence, they are often referred to informally as the bean common mosaic subgroup of the genus *Potyvirus*.

Each of the viruses in this subgroup consists of numerous isolates with similar, recognizable biological characters which collectively are referred to as "strains." The identity and early description of many BCMV strains was reviewed by Drijfhout [4]. The identity and biological behavior of an international collection of 22 BCMV strains was described by Drijfhout et al. [5]. Strain relationships among BlCMV, CABMV and AzMV have been examined by Taiwo et al. [13, 14] and Tsuchizaki and Onura [15]. Some relationships among strains of BCMV and BlCMV were examined by Lana et al. [10]. Despite the plethora of described strains and extensive investigation of serological and biological properties, there are few, if any, characteristics that can be used to distinguish unequivocally among viruses in this subgroup.

The term strain is widely used by virologists to identify a group of isolates with a similar, identifiable biological character. However, Drijfhout [4] introduced the term pathogenicity group (or pathogroup now more commonly termed pathotype) to clarify breeding for BCMV resistance in *Phaseolus* bean. The gene-for-gene concept established pathotypes which included BCMV strains having the same pathogenicity spectrum as defined by a series of differential cultivars. These cultivars were selected on the basis of their resistance genes. While this grouping of variants simplified development of BCMV-resistant bean cultivars, it did little to clarify relationships among BCMV strains and related viruses. In fact, the pathotyping system which has been so successful for plant breeders seems to have exacerbated difficulties in defining BCMV variants. Variants were assigned to pathotypes entirely on their ability or inability to systemically infect cultivars with specifically defined resistance genes. This system ignores entirely the biological variability that occurs in symptoms induced on inoculated leaves of any cultivar and places low priority on variations in the type or intensity of systemic symptoms. Consequently, isolates that differ markedly in symptomatology are frequently placed in the same pathotype simply because they systemically invade specific *Phaseolus* genotypes. One result of this, as we will illustrate, is that multiple isolates of some BCMV strains now exist that vary considerably in their biological properties, but carry the same strain designation. This will have profound effects on interpretation of results from non-biological techniques. Wang [17] demonstrated that BCMV strains could be divided into two serologically distinct groups; serotype A which included strains NL-3, NL-5, NL-8, and TN-1 and serotype B which included most other strains tested. Recently, monoclonal antibodies (MAbs) have been raised against strains of BCMV [18], BlCMV [16], CABMV (R. I. Hamilton, unpubl.), and PStV (J. Sherwood, unpubl.). In this study, we tested a selection of these MAbs against a panel of variants representing each of the viruses.

Methods and materials

Each of the biological variants used here (Table 1) was initially described as a distinct strain. However, many of them had been maintained in serial culture for several years and thus could represent selections that deviate in one or more characteristics from the original. Consequently, for the BCMV strains we obtained new isolates from infected seed lots and

Table 1. Source of virus isolates

Virus	Isolate	Source[a]	Authenticated biologically[b]
BCMV	NL-8	ED	Y
	NL-3	ED	Y
	NL-5	ED	Y
	TN-1	MJS	Y
	US-1	MJS	Y
	PR-1	RA	Y
	NL-1	ED	Y
	US-7	MJS	Y
	US-3	MJS	Y
	US-4	MJS	Y
	CH-2	MJS	Y
	NL-6	ED	Y
	NY-15 (P)	RP	
	NL-2	ED	Y
	US-6	MJS	Y
	NL-7	ED	Y
	US-5	MJS	Y
	NL-4	ED	Y
AzMV	MJS	MJS	Y
BlCMV	Type	OWB	N
	Fla	ROH	Y
	GA	JWD	N
	R03	ROH	Y
	R04	ROH	Y
CABMV	Mor	ROH, OWB	Y
	R05	ROH	Y
	R09	ROH	Y
	R010	ROH	Y
PStV	GA	JWD	N

[a] *ED* E. Drijfhout, IVT, Wageningen, The Netherlands
MJS M. J. Silbernagel, USDA-ARS, Prosser, WA
OWB O. W. Barnett, Clemson University, Clemson, SC
RA R. Alconero, USDA-ARS, Geneva, NY
ROH R. O. Hampton, USDA-ARS, Corvallis, OR
RP R. Provvidenti, Cornell University, Geneva, NY
JWD J. W. Demski, University of Georgia, Griffin, GA
[b] *Y* Yes; *N* no

Table 2. Source of monoclonal antibodies (MAb)

Immunizing virus	MAb ID	Source[a]
BCMV	I2	GIM
	I3	
	I59	
	II197	
	II463	
BlCMV	B1A4	HJV
	B5E5	
	B5B8	
	B6C5	
CABMV	16G5	RIH
	10G5	
PStV	7C14	JKS
PMV[b]	10D4	

[a] *GIM* G. I. Mink, WSU-Prosser IAREC, Prosser, WA

HJV H. J. Vetten, IBP, Braunschweig, Germany

RIH R. I. Hamilton, Ag Canada, Vancouver, BC, Canada

JKS J. K. Sherwood, Oklahoma State University, Stillwater, OK

[b] *PMV* Peanut mottle virus

authenticated their biological similarity to the original pathotype descriptions as recommended by Drijfhout et al. [5]. While no comparable scheme exists to authenticate isolates of AzMV, BlCMV, CABMV and PStV, we defined the pathogenicity of our isolates on the BCMV differential cultivars.

All isolates of BCMV and PStV were maintained in Dubble Witte bean; those of BlCMV and CABMV were maintained in cowpea cultivar California blackeye #5. AzMV was maintained in azuki bean cultivar Erimo. Comparative host ranges were performed in growth chambers using 16 h photoperiods at 24 °C.

The sources and identity of the monoclonal antibodies used are listed in Table 2. Serological evaluations were made by indirect, antigen coated enzyme-linked immunosorbent assay (ELISA). Fresh or air-dried leaf tissue was triturated 1:40 in pH 9.6 carbonate buffer [2] containing 2% egg albumin, 0.2% polyvinyl pyrollidone, and 0.45% sodium diethyldithiocarbamate. Antibodies were tested at dilutions between 1:250 and 1:3000.

Results

Four MAbs (BCMVII197, BlCMB1A4, CABMV16G5, and PStV7C14) were broad spectrum in that they detected all of the virus isolates used in this study (Table 3). One antibody (BCMVII463) detected all isolates of AzMV, BCMV, BlCMV, and PStV but failed to detect any of the four isolates of CABMV. Three antibodies (BCMVI2, I3, and I59) detected only those BCMV isolates previously described as serotype A isolates [17]. Four antibodies (BlCMB5E5,

B5B8, B6C5 and PMV 10D4) detected all BCMV isolates previously described as serotype B [17]. Although not shown in Table 3, MAb B5E5 exhibited some cross-reaction with serotype A isolate BCMV-NL3. The other three B serotype-specific antibodies did not cross react with serotype A isolates. All

Table 3. Serological grouping of virus isolates in the bean common mosaic virus subgroup using a panel of 13 monoclonal antibodies in indirect ELISA

Virus	Isolate	BCMV sero-type	BCMV patho-type	BCMV II 197, BlCM B1A4, CABMV 16G5, PStV 7C14	BCMV II463	BCMV I2 BCMV I3 BCMV I59	BlCM B5E5 BlCM B5B8 BlCM B6C5 PMV 10D4	CABM 10G5
CABM	Mor	.	(I)	+	−	−	−	−
CABM	ROH5	.	(I)	+	−	−	−	−
CABM	ROH9	.	(I)	+	−	−	−	−
CABM	ROH10	.	(I)	+	−	−	−	−
BCMV	NL-8	A	III	+	+	+	−	−
BCMV	NL-3	A	VI	+	+	+	−	−
BCMV	NL-5	A	VI	+	+	+	−	−
BCMV	TN-1	A	VI	+	+	+	−	−
BlCMV	Fla	(B)	(I)	+	+	−	+	+
BlCMV	GA	(B)	(I)	+	+	−	+	+
BlCMV	ROH3	(B)	(I)	+	+	−	+	+
BlCMV	ROH4	(B)	(I)	+	+	−	+	+
PStV	GA	(B)	(I)	+	+	−	+	+
BCMV	US-1	B	I	+	+	−	+	+
BCMV	PR-1	B	I	+	+	−	+	+
BCMV	NL-1	B	I	+	+	−	+	+
BCMV	US-7	B	II	+	+	−	+	+
BCMV	US-3	B	IV	+	+	−	+	+
BCMV	US-4	B	IV	+	+	−	+	+
BCMV	NL-6	B	IV	+	+	−	+	+
BCMV	NY-15 (P)	B	V	+	+	−	+	+
BCMV	NL-2	B	V	+	+	−	+	+
AzMV	MJS	(B)	(I)	+	+	−	+	−
BCMV	NL-7	B	II	+	+	−	+	−
BCMV	US-5	B	IV	+	+	−	+	−
BCMV	CH-2	B	IV	+	+	−	+	−
BCMV	NL-4	B	VII	+	+	−	+	−
BCMV	US-6	B	VII	+	+	−	+	−
Healthy	cowpea			−	−	−	−	−
Healthy	bean			−	−	−	−	−

four BCMV serotype B-specific antibodies also detected all isolates of AzMV, BlCMV, and PStV used. None of these four antibodies detected any of the four CABMV isolates tested. One antibody (CABM 10G5) detected BlCMV, PStV and all B serotype BCMV isolates except NL-7, US-5, CH-2, NL-4 and US-6; AzMV also was not detected. This antibody which was presumably raised against CABMV also failed to detect any of the CABMV isolates.

MAb CABM 10G5 distinguished AzMV from BlCMV and PStV in indirect ELISA; none of the MAbs evaluated in these tests were able to differentiate AzMV, BlCMV, or PStV from the serotype B isolates of BCMV. When inoculated to the 10 standard BCMV differential bean cultivars, the AzMV and BlCMV isolates (as well as CABMV) induced systemic mosaic symptoms in the host group 1 cultivar Dubble Witte but did not systemically infect cultivars representing the other 10 host groups (Table 4). In this respect, these isolates resembled those of BCMV pathogroup I. Unlike the BCMV isolates, however, all isolates of AzMV, BlCMV and CABMV systemically infected cowpea cultivar California Blackeye #5. When inoculated to an expanded list of BCMV host group 1 cultivars, all isolates of BlCMV could be distinguished from BCMV isolates by the severe reactions they induced on cultivars Black Turtle II, Sutter Pink, and Sierra and by their failure to infect Stringless Green Refugee (Table 5). All isolates of CABMV could be distinguished from either BlCMV or BCMV by their reactions on host group 1 cultivars as well as by their serololgical distinctiveness.

Table 4. Infection of bean common mosaic virus (BCMV) differential cultivars by four viruses in the BCMV subgroup

Virus	Isolate	Host group												
		1	2	3	4	5	6	7	8	9	10	11		
		DW	RGC	RGB	SAN	114	GN31	7214	BTl	TC	AMA	7233	AZ	CP
BCMV	US-1	+++[a]	+––	++–	+––	–––	+––	–––	–––	–––	–––	–––	+++	+––
	NL-1	+++	+––	++–	+––	–––	+––	–––	–––	–––	–––	–––	+++	+––
	PR-1	+++	+––	++–	+––	+––	+––	–––	–––	–––	–––	–––	+++	+––
AzMV	MJS	+++	+––	+––	–––	–––	–––	–––	–––	–––	–––	–––	+++	+++
BlCMV	Fla	+++	+––	+––	+––	+––	+––	–––	+––	–––	–––	–––	–+–	+++
	ROH3	+++	+––	+––	+––	+––	+––	–––	–––	–––	–––	–––	–+–	+++
	ROH4	+++	+––	+––	+––	+––	+––	–––	+––	–––	–––	–––	–+–	+++
CABMV	Mor	+++	+––	+––	–––	+––	+––	–––	–––	–––	–––	–––	·	+++
	ROH9	+++	+––	+––	–––	+––	+––	–––	–––	–––	–––	–––	·	+++
	ROH10	+++	+––	+––	–––	+––	+––	–––	–––	–––	–––	–––	·	+++

[a] ELISA results four weeks after inoculation using primary leaf, first trifoliate leaflet, tip trifoliate leaflet, respectively. +++ All three tissues positive, ––– all three tissues negative; dots, not tested

Table 5. Behavior of BCMV-NY15(P), three BlCMV isolates, and two CABMV isolates on BCMV host group I cultivars

Cultivar	Isolate					
	BCMV NY-15 (P)	BlCMV Fla	BlCMV R03	BlCMV R04	CABMV Mor	CABMV R05
Dubble Witte	CS/CS,Mo,LR[a]	–/Mo,LR,D	CS/EP,Mo,D	–/Mo,LR	NS/LR,D	NS/Mo
Black Turtle II	CS/CS,Mo,St	Ns/EP,St,D	NS/EP,St,D	–/EP,St,D	NS/D	NS/Mo
Sutter Pink	NL/Mo,LR,St	–/EP,St,D	NS/EP,St,D	–/EP,St,D	NS/D	CS/Mo
Sierra	–/Mo,Bl,LR	–/EP,Mo,D	CS/Mo,St,D	–/EP,Mo,St	NS/–	NS/–
Long Tom	–/CR,Mo,LR	–/Mo	CS/Mo,LR	–/Mo	–/Mo	–/Mo
Beautiful Stringless	–/GB,Mo,LR	CS/CS	CS/CS	–/CS	–/–	–/–
Green Refugee	–/Mo,LR,St	–/–	–/–	–/–	CS/Mo	CS/Mo

[a] Symptoms on inoculated primary leaves/trifoliate leaves: *CS* chlorotic spots; *Mo* mosaic; *LR* leafroll; *NS* necrotic spots; *EP* leaf epinasty; *St* stunt; *CR* chlorotic rings; *Bl* blisters; *D* death

Our authenticated isolate of BCMV strain NL-1 exhibited the same pathogenicity spectrum as the NL-1 isolate described by Drijfhout in 1978 [4]. It systemically infected only host group 1 cultivars (Table 6). In contrast, an isolate designated NL-1 reported by Lana et al. 10 years later [10] not only infected host group 1 but also infected cultivars in host groups 3, 5 and 9

Table 6. Inconsistencies among isolates of BCMV strain NL-1

Host group	Differential cultivar	Prosser 1991	Wageningen		
			[4]	[5]	[6]
1	DW	+	+	+	+
2	RGC	–	–	–	–
3	RGB	–	–	+(L)	+(L)
4	SAN	–	–	–	–
5	P114	–	–	+(L)	+(L)
6	GN31	–	–	–	–
7	7214	–	–	–	–
8	BTS	–	–	.	–
9	TC	–	–	+(L)	+(S)
10	AMA	–	–	–	–
11	7233	–	–	.	.

L Latent, *S* symptoms, + infected, – no infection; dots, not tested

without symptoms. Three years later, an isolate from the same laboratory [6] with the same designation not only infected host groups 3, 5, and 9 but produced strong mosaic symptoms in the host group 9 cultivar Top Crop. This type of variability among different isolates of a described strain may not be unusual for BCMV and related viruses. We currently maintain three biologically distinct isolates (variants) all described initially as BCMV strain NY-15 [3, 9, 12]. While all infect the same differential cultivars, they differ markedly in the intensity of symptoms induced on various hosts.

Discussion

Antibodies (BCMVII197, BICMVB1A4, CABMV16G5 and PStV7C14) raised against four different "viruses" detected all 28 virus isolates suggesting their coat proteins have at least one epitope in common. It has not yet been determined if all four MAbs recognize the same or different epitopes. However, unpublished results in our laboratory suggest that the epitope recognized by BCMV II 197 is different from that recognized by the POLY-1 McAb available from Agdia, Inc. (Mishawaka, IN). Since the latter McAb also recognized all 28 isolates (data not presented) it is likely that two or more distinct epitopes can be found on the coat proteins of all 28 isolates.

During this study more than 20 different MAbs were evaluated. Table 3 provides data for the 13 which produced consistent results in all tests. Several other antibodies produced highly variable results from test to test for reasons that have not yet been determined. Nevertheless, none of the MAbs used in our tests recognized epitopes specific to isolates of CABMV. This includes two antibodies prepared against an isolate presumed to be CABMV.

Despite considerable work from several different laboratories, confusion still exists in labeling isolates or strains within the AzMV-BCMV-BICMV-CABM cluster of viruses. One major problem involves the mentally taxing activity of trying to assign a given isolate to a fixed position within what has been described as a continuum of variants that has no well defined parameters for individual viruses [1, 7, 8, 10]. A more mundane but no less frustrating problem is that of maintaining isolate fidelity over time. The two situations we presented here demonstrate that variants with the same strain designation can exhibit great biological diversity even when obtained from the same laboratory.

It is not surprising that isolates of the same virus which are maintained at different locations or by different protocols may evolve differently with time. However, as increasing emphasis is given to molecularly oriented techniques that characterize the non-biological properties of viruses, seemingly less emphasis is placed on the biological fidelity of the isolates used. While these newer techniques may provide additional parameters for characterizing virus isolates, the results obtained will be of limited value in clarifying an already confused situation if the biological characteristics of the individual isolates used are not provided in each report. This seems to us to be of paramount importance whenever attempts are made to rename strains or to reclassify strains from one virus to another on the basis of non-biological properties [11].

Acknowledgement

PPNS Paper No. 0108. Project No. 1719. Washington State University, College of Agriculture and Home Economics Agricultural Research Center, Pullman, WA 99164-6240.

References

1. Bos L (1970) The identification of three new viruses isolated from *Wisteria* and *Pisum* in the Netherlands, and the problem of variation within the potato virus Y group. Neth J Plant Pathol 76: 8–46
2. Clark MF, Adams AN (1977) Characteristics of the microplate method of enzyme-linked immunosorbent assay for the detection of plant viruses. J Gen Virol 34: 475-483
3. Dean LL, Hungerford CW (1946) A new bean mosaic virus in Idaho. Phytopathology 36: 324–326
4. Drijfhout E (1978) Genetic interaction between *Phaseolus vulgaris* and bean common mosaic virus with implications for strain identification and breeding for resistance. Agriculture Research Reports, no 872. Pudoc, Wageningen
5. Drijfhout E, Silbernagel MJ, Burke DW (1978) Differentiation of strains of bean common mosaic virus. Neth J Plant Pathol 84: 13–26
6. Dykstra J, Khan JA (1992) A proposal for a bean common mosaic subgroup of potyviruses. In: Barnett OW (ed) Potyvirus taxonomy. Springer, Wien New York, pp 389–395 (Arch Virol [Suppl] 5)
7. Harrison B (1985) Usefulness and limitations of the species concept for plant viruses. Intervirology 24: 71–78
8. Hollings M, Brunt AA (1981) Potyviruses. In: Kurstak E (ed) Handbook of plant virus infections. Comparative diagnosis. Elsevier/North Holland, Amsterdam, pp 731–807
9. Kyle MM, Provvidenti R (1987) A severe isolate of bean common mosaic virus NY15. Ann Rep Bean Improv Coop 30: 87–88
10. Lana AF, Lohuis H, Bos L, Dykstra J (1988) Relationships among strains of bean common mosaic virus and blackeye cowpea mosaic virus – members of the potyvirus group. Ann Appl Biol 13: 493–505
11. McKern NM, Ward CW, Shukla DD (1992) Strains of bean common mosaic virus consist of at least two distinct potyviruses. In: Barnett OW (ed) Potyvirus taxonomy. Springer, Wien New York, pp 407–414 (Arch Virol [Suppl] 5)
12. Richards BL, Burkholder WH (1943) A new mosaic of beans. Phytopathology 33: 1215–1216
13. Taiwo MA, Gonsalves D (1982) Serological grouping of isolates of blackeye cowpea mosaic virus and cowpea aphidborne mosaic viruses. Phytopathology 72: 583–589
14. Taiwo MA, Gonsalves D, Provvidenti R, Thurston HD (1982) Partial characterization and grouping of isolates of blackeye cowpea mosaic and cowpea aphid borne mosaic viruses. Phytopathology 72: 590–596
15. Tsuchizaki T, Omura T (1987) Relationships among bean common mosaic virus, blackeye cowpea mosaic virus, azuki bean mosaic virus and soybean mosaic virus. Ann Phytopathol Soc Jpn 53: 478–488
16. Vetten HJ, Lesemann DE, Maiss E (1992) Serotype A and B strains of bean common mosaic virus are two distinct potyviruses. In: Barnett OW (ed) Potyvirus taxonomy. Springer, Wien New York, pp 415–431 (Arch Virol [Suppl] 5)

17. Wang W-Y (1983) Serology of bean common mosaic virus strains. MSc Thesis, Washington State University, Pullman, Washington
18. Wang W-Y (1985) Production and characterization of hybridoma cell lines and a broad spectrum monoclonal antibody against bean common mosaic virus. PhD Thesis. Washington State University, Pullman, Washington

Authors' address: G. I. Mink, WSU-Prosser IAREC, Route 2, Box 2953-A, Prosser, WA 99350-9687, U.S.A.

Arch Virol (1992) [Suppl 5]: 407–414

Strains of bean common mosaic virus consist of at least two distinct potyviruses

N. M. McKern, C. W. Ward, and **D. D. Shukla**

CSIRO, Division of Biomolecular Engineering, Parkville, Victoria, Australia

Summary. Bean common mosaic virus (BCMV) consists of a large number of pathotypes and strains which have largely been identified by their characteristic interactions with a selected number of differential bean cultivars. The relationships among these strains and other potyviruses that infect legumes are complex, with indications that BCMV, blackeye cowpea mosaic virus (BlCMV) and azuki bean mosaic virus (AzMV) may be strains of the one virus. Using high performance liquid chromatographic peptide profiles of coat-protein digests, the NL3 and NY15 strains of BCMV were compared with each other, with the Type and W strains of BlCMV and with the mild mottle strain of peanut stripe virus (PStV). The results suggest that BCMV-NL3 and BCMV-NY15 are distinct potyviruses, not strains of the one virus, and that BCMV-NY15 is a strain of the same potyvirus that includes BlCMV, PStV, AzMV and three potyvirus isolates (74, PM, PN) from soybeans.

Introduction

Bean common mosaic virus (BCMV), a definitive member of the genus *Potyvirus* [12], is perhaps the most common and most destructive of the 34 viruses now known to naturally infect beans [11]. Strains of the virus cause either common mosaic symptoms which are usually associated with leaf malformation, or black root which is characterized by vascular necrosis and death of the plant. The nature and severity of symptoms depend on the bean cultivar, time of infection, environmental conditions, and the strains or pathotype of the virus [1, 17]. On the basis of characteristic interactions with a selected number of bean cultivars [4, 5], a large number of strains of BCMV have been reported from different parts of the world [1, 10, 11, 17, 28–30]. The interrelationships among these strains, as well as their relationship to other potyviruses infecting legumes, are complex, with some recent reports suggesting a close relationship between BCMV strains and other potyviruses infecting legumes. Following a comparison of host range, cross-protection, transmission, antigenic and coat-protein properties, Tsuchizaki and Omura [28] concluded that some isolates of BCMV, blackeye cowpea mosaic virus (BlCMV) and azuki bean mosaic virus (AzMV) were not distinct potyviruses but strains of the one virus. Similarly, Lana et al. [11] were unable to draw a clear line between some strains of BCMV, BlCMV and cowpea aphid-borne mosaic virus (CABMV) following a detailed comparison of biological and serological properties.

At least 15 distinct potyviruses are now reported to naturally infect leguminous plant species. Each of these viruses further contains a large number of variants which differ from each other in host range, symptomatology and in some instances serological properties. Clear cut taxonomic assignments for many of these variants have been difficult to achieve using biological and serological criteria [24].

Structural information from coat-protein or genomic sequences has recently been shown to clearly differentiate distinct potyviruses and their strains [20–22]. Use of such information for the potyvirus isolates infecting legumes should resolve their taxonomic status. The determination of complete sequences for a large number of isolates is impractical, but an effective alternative is the use of high performance liquid chromatographic (HPLC) peptide profiling of coat-protein tryptic peptides. This procedure gives a measure of the extent of sequence homology between two proteins and clearly differentiates between distinct potyviruses and their strains [25]. HPLC peptide profiling of coat proteins [15, 16] confirmed the previous classification of sugarcane mosaic virus isolates into four distinct potyviruses based on N-terminal serology [23–26]. It also showed that AzMV, BlCMV-Type and -W, the stripe, mild mottle, and blotch strains of peanut stripe virus (PStV), and the soybean isolates PM, PN, and 74, are all strains of the same potyvirus [13]. Furthermore, 14 isolates of soybean mosaic virus (SbMV) were shown to be closely related to SbMV-N and to be strains of the one potyvirus [10].

In this paper we have compared the HPLC peptide profiles of coat-protein digests from two strains of BCMV (NL3 and NY15), two strains of BlCMV (Type and W), and the mild mottle strain of PStV. Results have been analyzed in the light of previous information on biological, biochemical and serological properties of strains of these viruses. These data indicate that BCMV consists of at least two distinct potyviruses.

Materials and methods

Strains of the viruses investigated were: BCMV-NL3, -NY15 [30]; BlCMV-Type [18], -Wageningen (W) [4]; and PStV-mild mottle (MM) [2, 32]. BlCMV-W was purified according to the method of Dijkstra et al. [4] and BCMV-NL3, -NY15, BlCMV-Type, and PStV-MM according to Method-2 of Reddick and Barnett [19].

Enzyme digests were prepared by suspending 0.5 to 1.0 mg freeze-dried viral preparations in 250–500 µl of 0.05 M ammonium bicarbonate by sonication, followed by incubation overnight at 37°C with trypsin (TPCK-treated, Worthington, U.S.A.) at a 1:50

enzyme: protein ratio. Solutions were dried, vortexed with 250–500 µl 0.1% trifluoroacetic acid and centrifuged at 9000 g in a benchtop centrifuge. Soluble peptides were separated as described in Fig. 1 by reverse-phase chromatography using a 5 µ Vydac (California) C18 column connected to a Perkin Elmer (Norwalk, Conn., U.S.A.) Series 4 liquid chromatograph.

Results

Reverse-phase HPLC of tryptic digests of five coat proteins from these legume-infecting potyviruses are shown in Fig. 1. The overall elution pattern of peptides from BCMV-NL3 differs from that obtained from the four other profiles, including BCMV-NY15. Pairwise comparisons of retention times of

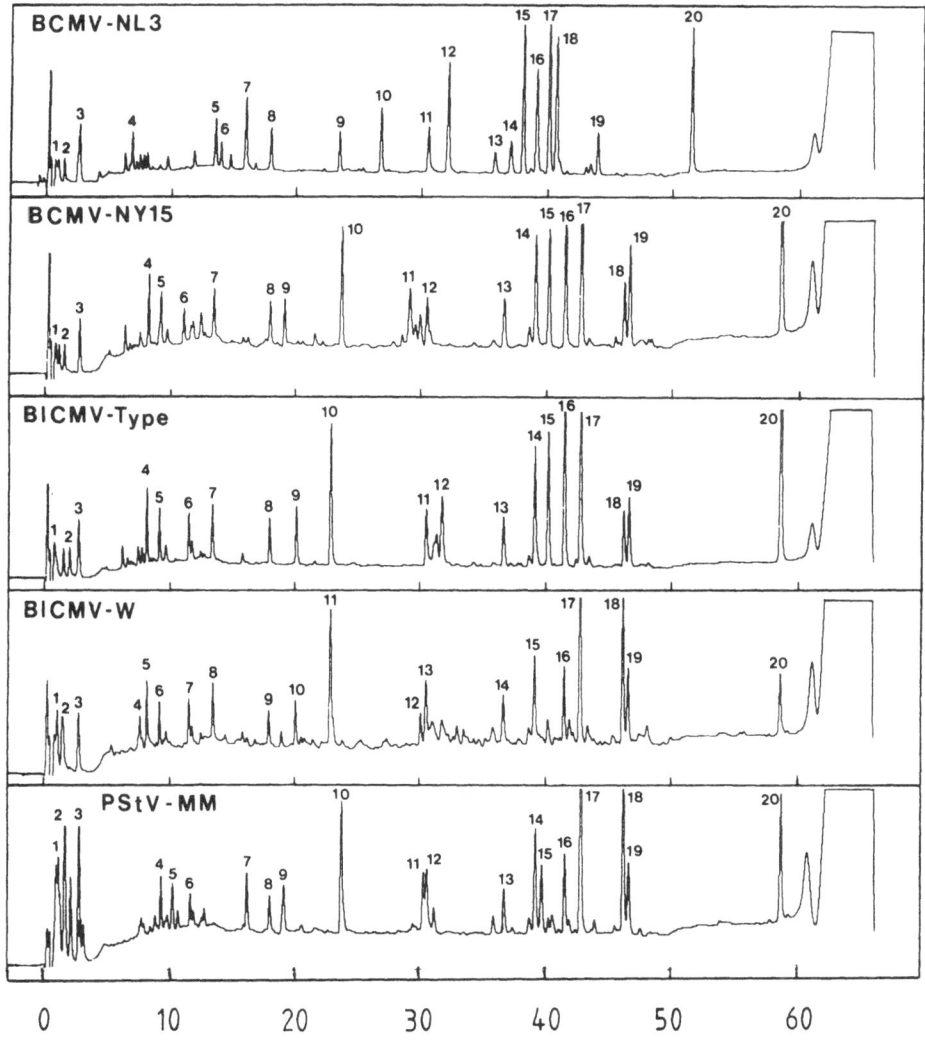

Fig. 1. Reverse-phase high performance liquid chromatography of tryptic digests of coat proteins from BCMV-NL3, BCMV-NY15, BICMV-Type, BICMV-W, and PStV-MM. Peptides bound to the column were eluted with a linear gradient of 0–33% acetonitrile in 0.1% aqueous trifluoroacetic acid over 60 min at a flow rate of 1 ml/min and column temperature of 45 °C. Numbered fractions were collected and in some cases analyzed

Table 1. Comparison of retention times of 17 major peaks from HPLC peptide profiles of tryptic digests of five coat proteins from legume-infecting potyviruses[a]

	Number of peaks (%) with common retention times[b]				
	BCMV-NL3	BCMV-NY15	BlCMV-Type	BlCMV-W	PStV-MM
BCMV-NL3	–	35	35	29	29
BCMV-NY15		–	76	71	71
BlCMV-Type			–	88	59
BlCMV-W				–	65

[a] Comparison based on Fig. 1. The twenty major peaks (based on peak height) of each profile were compared, excluding peaks 1–3 which were common to all profiles and represented material not bound to the column

[b] Retention times were considered to be similar if they were within ~0.2 min of each other

major peaks from the BCMV-NL3 profile were made with each of the other profiles. Peaks 1–3 were not included in these comparisons, since they were common to all profiles and were therefore not useful as indicators of similarity of the coat proteins. A summary of the pairwise comparison of peaks is given in Table 1. Between 29% and 35% of the 17 major, bound peaks of BCMV-NL3 have retention times co-incident with retention times of peaks from other profiles.

Unlike the profile of BCMV-NL3, the peptide profile of BCMV-NY15 showed substantial similarity to that of several of the other profiles in Fig. 1. Between 59% and 88% of the BCMV-NY15 peaks had retention times similar to those of peaks from BlCMV-Type, BlCMV-W and PStV-MM (Table 1).

Discussion

The finding that between 59% and 88% of the peaks in peptide profiles of coat protein from BCMV-NY15, BlCMV-Type and -W, and PStV-mild mottle have common retention times, indicates they have very similar coat proteins and suggests they are strains of the one virus.

In contrast, the peptide profile of the coat protein of BCMV-NL3 is quite distinct, less than 36% of the major peaks shared retention times in pairwise comparisons with the other profiles in Fig. 1.

HPLC peptide profiles of tryptic digests of coat proteins from a number of potyviruses are now available [9, 13–16, 24, 25]. Comparison of these profiles show that at least half of the major peaks from strains of one potyvirus have similar retention times, whereas those from distinct potyviruses share only a minority of peaks with common retention times. Whether these observations hold universally for tryptic digests of potyviruses remains to be determined, but at present peptide profiles of coat proteins provide a useful means for rapid examination of the relationships of potyvirus isolates.

In the context of these observations, the results of the present study suggest that BCMV-NY15, BlCMV-Type, BlCMV-W, and PStV-mild mottle are strains of one potyvirus, the few observed differences such as peak 11 of BCMV-NY15 (with a different amino acid composition), probably reflect differences in the amino termini of these coat proteins. Coat proteins of recognized strains of other potyviruses show significant variation at the amino termini, within otherwise closely matching sequences [21]. In contrast, the relatively low identity of peak retention times between BCMV-NL3 and BCMV-NY15 suggests they represent two distinct potyviruses. It is interesting to observe that the above assignments generally correlate well with previous information on biological, serological, and coat-protein properties from strains of these viruses. In a detailed study on host range and symptomatology [11], BCMV-NL3 and -NY15 induced distinct symptoms in susceptible bean cultivars and differed in the symptoms (latent versus necrosis) in cowpea cultivars. Only BCMV-NL3 caused temperature-insensive necrosis in bean cultivars with dominant I gene. On the other hand, BCMV-NY15 and BlCMV-W induced similar but severe symptoms in cowpea cultivars and only weak or no symptoms (BlCMV-W in particular) in bean cultivars. These workers concluded that there was a close biological and serological relationship between BlCMV and the non-necrosis inducing strains of BCMV, particularly -NL1 and -NY15.

The serological studies showed that BCMV-NY15 and BlCMV-W were closely related to each other and to other BlCMV strains (namely BlCMV-Fla, -Type and -NR) in double antibody sandwich enzyme-linked immunosorbent assay (DAS-ELISA) [11]. In contrast, BCMV-NL3 showed only a very distant relationship in DAS-ELISA with BCMV-NL1, -NY15, BlCMV-Fla, -NR, and -W [11]. Such distant serological relationships are likely to be due to antibodies directed to the core region of the coat proteins, which show high sequence homology throughout the potyvirus group [21-22, 26, 31]. When core antibodies were removed by cross absorption, virus-specific antibodies directed to the N terminus [23] of BCMV-NL3 did not react with BCMV-NL1, -NY15 and BlCMV-W [10], suggesting that BCMV-NL3 is distinct from BlCMV and these other BCMV strains.

On the basis of reactivities of several monoclonal antibodies (MAbs) and polyclonal antisera, BCMV strains have recently been grouped into two serogroups, A and B [29; G.I. Mink in 4; G.I. Mink, unpubl. results]. Serogroup A includes strains BCMV-NL3, -NL5, -NL8 and -TN1 whereas serogroup B contains BCMV-NL1, -NL2, -NL4, -NL6, -NL7, -NY15, -CH1, -CH2, -CR, -IR1, -PR1, -US1, -US2, -US3, -US4, -US-5, -US6, -US7 and -US9 [29] (G.I. Mink, unpubl. results). Members of each of these serogroups display close serological relationships among themselves but none, or only distant serological relationships with members of the other serogroup. For instance, in

DAS-ELISA polyclonal antisera to members of serogoup A recognized only members of this serogroup but not those in serogroup B. Similarly, MAbs BCI2 and BCI3 reacted only with members in serogroup A with exactly the same degree of specificity, whereas MAbs BCII134(1) and BCII134(2) gave varying degrees of specificity only with members in serogroup B and did not recognize members in serogroup A (W. Y. Wang and G. I. Mink, unpubl. results). This grouping also correlates well with the grouping based on pathotypes of the BCMV strains. For example, only the members of serogroup A, but not B, cause the necrosis reaction in bean cultivars with the dominant I gene (G. I. Mink, unpubl. results).

A detailed comparison of coat-protein HPLC peptide profiles has recently demonstrated that AzMV, three strains of PStV (stripe, blotch and mild mottle), two strains of BlCMV (Type and W) and three potyvirus isolates from soybean in Taiwan are all strains of the same virus [12], a finding consistent with earlier observations that the strains of BlCMV and PStV are very closely related serologically [3,8]. The peptide profiles in Fig. 1 clearly show that BCMV-NY15 belongs to this large group of viruses.

Nucleotide sequences of the coat-protein coding regions of two BCMV strains, NL4 (serogroup B) and NL8 (serogroup A) [29], showed that the coat proteins of BCMV-NL4 and -NL8 are different in size (287 residues, M_r 32,489 and 261 residues, M_r 29,662, respectively) and sequence (72% identity). This confirms that they are two distinct potyviruses [31] as suggested here by HPLC profiling of a serogroup A (BCMV-NL3) and serogroup B (BCMV-NY15) isolate. Of great interest is the observation that the BCMV-NL4 coat protein is the same size (287 amino acid residues) and has high sequence identity (93%) with the coat protein of PStV [14], as indicated here by HPLC for BCMV-NY15.

On the basis of coat-protein peptide profiling results presented in this paper and previous information on biological, serological and molecular properties, BCMV strains appear to represent at least two distinct potyviruses. The first consists of the necrosis-inducing, serogroup A isolates represented by BCMV-NL3 and -NL8 and the second by the serogroup B strains BCMV-NL4 and -NY15, which also includes BlCMV-Type and -W, PStV-stripe, -blotch and -mild mottle, AzMV and the soybean isolates -74, -PM, and -PN.

Verification of this conclusion will only be forthcoming when the large number of other BCMV strains in serogroups A and B are examined. Peptide profiling of their coat proteins or nucleotide probes involving 3' noncoding regions of viral RNA [6,7] should facilitate the assignment of these unclassified strains to either of the two types distinguished here, and establish whether the BCMV group of strains contains members of additional distinct potyviruses.

Acknowledgements

The authors are indebted to J. Dijkstra and O. W. Barnett for supplying samples of coat protein and to G.I. Mink for providing unpublished data. We thank Ms. Loyce Whittaker for excellent technical assistance, Mr. Nick Bartone for amino acid analysis and Mrs. Brenda Wood for typing the manuscript. This work was supported by the Rural Credits Development Fund of the Reserve Bank of Australia.

References

1. Bos L (1971) Bean common mosaic virus. CMI/AAB Descriptions of Plant Viruses, no 73
2. Demski JW, Lovell GR (1985) Peanut stripe virus and the distribution of peanut seeds. Plant Dis 69: 734–738
3. Demski JW, Reddy DVR, Wongkaew S, Kameya-iwaki M, Saleh N, Xu Z (1988) Naming of peanut stripe virus. Phytopathology 78: 631–632
4. Dijkstra J, Bos L, Bouwmeester HJ, Hadiastono T, Lohuis H (1987) Identification of blackeye cowpea mosaic virus from germplasm of yard-long bean and from soybean, and the relationships between blackeye cowpea mosaic virus and cowpea aphid-borne mosaic virus. Neth J Plant Pathol 93: 115–133
5. Drijfhout E (1978) Genetic interaction between *Phaseolus vulgaris* and bean common mosaic virus with implications for strain identification and breeding for resistance. Agricultural Research Reports 872. Pudoc, Wageningen
6. Frenkel MJ, Ward CW, Shukla DD (1989) The use of 3' non-coding nucleotide sequences in the taxonomy of potyviruses: application to watermelon mosaic virus 2 and soybean mosaic virus-N. J Gen Virol 70: 2775–2783
7. Frenkel MJ, Jilka JM, Shukla DD, Ward CW (1991) Differentiation of potyviruses and their strains by hybridization with the 3' non-coding region of the viral genome. J Virol Methods 36: 51–62
8. Green SK, Lee DR, Vetten HJ, Lesemann DE (1986) Occurrence of an unidentified potyvirus of soybean in Taiwan. Trop Agric Res Ser 19: 108–114
9. Jain RK, McKern NM, Tolin SA, Hill JH, Barnett OW, Tosic M, Ford RE, Beachy RN, Yu MH, Ward CW, Shukla DD (1992) Confirmation that fourteen potyvirus isolates from soybean are strains of one virus by comparing coat protein peptide profiles. Phytopathology 82: 294–299
10. Khan JA, Lohuis H, Goldbach RW, Dijkstra J (1990) Distinction of strains of bean common mosaic and blackeye cowpea mosaic virus using antibodies to N- and C- or N-terminal peptide domains of coat proteins. Ann Appl Biol 117: 583–593
11. Lana AF, Lohuis H, Bos L, Dijkstra J (1988) Relationships among strains of bean common mosaic virus and blackeye cowpea mosaic virus – members of the potyvirus group. Ann Appl Biol 113: 493-505
12. Matthews REF (1982) Classification and nomenclature of viruses. Intervirology 17: 1–199
13. McKern NM, Shukla DD, Barnett OW, Vetten HJ, Dijkstra J, Whittaker LA, Ward CW (1992) Coat protein properties suggest that azuki bean mosaic virus, blackeye cowpea mosaic virus, peanut stripe virus and three isolates from soybean are all strains of the same potyvirus. Intervirology 33: 121–134
14. McKern NM, Edskes HK, Ward CW, Strike PM, Barnett OW, Shukla DD (1991) Coat protein of potyviruses. 7. Amino acid sequence of peanut stripe virus. Arch Virol 119: 25–35
15. McKern NM, Shukla DD, Toler RW, Jensen SG, Tosic M, Ford RE, Leon O, Ward CW (1991) Confirmation that the sugarcane mosaic virus subgroup consists of four distinct potyviruses by using peptide profiles of coat proteins. Phytopathology 81: 1025–1029
16. McKern NM, Whittaker LA, Strike PM, Ford RE, Jensen SG, Shukla DD (1990) Coat protein properties indicate that maize dwarf mosaic virus-KS1 is a strain of Johnsongrass mosaic virus. Phytopathology 80: 907–912

17. Morales FJ, Bos L (1988) Bean common mosaic virus. AAB Descriptions of Plant Viruses, no 337 (no 73 revised)
18. Murphy JF, Barnett OW, Witcher W (1987) Characterization of a blackeye cowpea mosaic virus strain from South Carolina. Plant Dis 71: 243–248
19. Reddick BB, Barnett OW (1983) A comparison of three potyviruses by direct hybridization analysis. Phytopathology 73: 1500–1510
20. Shukla DD, Ward CW (1988) Amino acid sequence homology of coat proteins as a basis for identification and classification of the potyvirus group. J Gen Virol 69: 2703–2710
21. Shukla DD, Ward CW (1989) Structure of potyvirus coat proteins and its application in the taxonomy of the potyvirus group. Adv Virus Res 36: 273–314
22. Shukla DD, Ward CW (1989) Identification and classification of potyviruses on the basis of coat protein sequence data and serology. Arch Virol 106: 171–200
23. Shukla DD, Jilka J, Tosic M, Ford RE (1989) A novel approach to the serology of potyviruses involving affinity-purified polyclonal antibodies directed towards virus-specific N termini of coat proteins. J Gen Virol 70: 13–20
24. Shukla DD, McKern NM, Barnett OW, Ward CW (1990) Identification and classification of potyviruses infecting tropical legumes. In: VIIIth International Congress of Virology, Berlin, August 26–31 1990, W87-8
25. Shukla DD, McKern NM, Gough KH, Tracy SL, Letho SG (1988) Differentiation of potyviruses and their strains by high performance liquid chromatographic peptide profiling of coat proteins. J Gen Virol 69: 493–502
26. Shukla DD, Tosic M, Jilka J, Ford RE, Toler RW, Langham MAC (1989) Taxonomy of potyviruses infecting maize, sorghum and sugarcane in Australia and the United States as determined by reactivities of polyclonal antibodies directed towards virus-specific N-termini of coat proteins. Phytopathology 79: 223–229
27. Shukla DD, Tribbick G, Mason TJ, Hewish DR, Geysen HM, Ward CW (1989) Localization of virus-specific and group-specific epitopes of plant potyviruses by systematic immunochemical analysis of overlapping peptide fragments. Proc Natl Acad Sci USA 86: 8192–8196
28. Tsuchizaki T, Omura T (1987) Relationships among bean common mosaic virus, blackeye cowpea mosaic virus, azuki bean mosaic virus, and soybean mosaic virus. Ann Phytopathol Soc Jpn 53: 478–488
29. Vetten HJ, Lesemann DE, Maiss E (1990) Differentiation of bean common mosaic virus strains. In: VIIIth International Congress of Virology, Berlin, August 26–31 1990, P84-8
30. Wang WY, Mink GI, Silbernagel MJ (1982) Comparison of direct and indirect enzyme-linked immunosorbent assay (ELISA) in the detection of bean common mosaic virus. Phytopathology 72: 954
31. Ward CW, Shukla DD (1991) Taxonomy of potyviruses: current problems and some solutions. Intervirology 32: 269–296
32. Xu ZY, Yu ZL, Liu JL, Barnett OW (1983) A virus causing peanut mild mottle in Hubei Province, China. Plant Dis 67: 1029–1032

Authors' address: D. D. Shukla, CSIRO, Division of Biomolecular Engineering, 343 Royal Parade, Parkville, Vict. 3052, Australia.

Arch Virol (1992) [Suppl 5]: 415–431

Serotype A and B strains of bean common mosaic virus are two distinct potyviruses

H. J. Vetten, D.-E. Lesemann, and **E. Maiss**

Biologische Bundesanstalt für Land- und Forstwirtschaft, Institut für Biochemie und Pflanzenvirologie, Braunschweig, Federal Republic of Germany

Summary. The serological relationships among strains of bean common mosaic virus (BCMV) (genus *Potyvirus*, family *Potyviridae*) were investigated by testing 13 isolates of the 10 known BCMV pathotypes with two monoclonal antibodies and six antisera to BCMV strains. In addition, other properties of serologically distinct BCMV strains were compared. Two groups of BCMV strains were obtained by ELISA and Western blot serology: serotype A contained the BCMV strains NL 3, NL 5, and NL 8 and serotype B contained the BCMV strains NL 1, NL 2, NL 4, NL 6, US 4, NL 7, NY 15, and Fla. SDS polyacrylamide gel electrophoresis and Western blotting of freshly purified preparations, and of extracts from leaves infected with eleven BCMV strains showed that the apparent molecular mass of the capsid protein of the serotype A isolates NL 3, NL 5, and NL 8 are lower (about M_r 33,000) than those of the serotype B isolates (M_r 34,500 to 35,000). The normal lengths of the particles of the serotype A isolates were shorter (810–818 nm) than those of most isolates (except NL 6 and NY 15) of serotype B (847–886 nm). All isolates studied induced cytoplasmic pinwheel and scroll inclusions. Cells infected with serotype A isolates contained a specific type of proliferated endoplasmic reticulum which was never found in cells infected with serotype B isolates. The capsid protein gene of a representative member of each serotype was cloned and sequenced. Molecular mass calculations based upon nucleotide sequence-derived amino acid sequences yielded M_r of 29,662 and 32,489 for the capsid proteins of the serotype A isolate NL 8 and the serotype B isolate NL 4, respectively. Comparison of the coat-protein sequences showed considerable differences at the N-termini whereas the core regions and the C-termini appeared to be highly conserved. Marked differences were also observed within the 3' non-coding regions of cloned cDNAs of NL 4 and NL 8. The striking differences between the two serotypes of BCMV strongly suggest that they be classified as two distinct potyviruses which naturally infect *Phaseolus* beans.

Introduction

Bean common mosaic virus (BCMV), a species of the genus *Potyvirus*, is seed-borne in many cultivars of *Phaseolus* bean and hence occurs in all bean production areas [25]. BCMV strains have been grouped into pathotypes on the basis of the interactions between seven host genes for resistance and three BCMV genes for pathogenicity [4]. Symptomatologically unusual strains of BCMV, that cause a lethal systemic necrosis, commonly referred to as "black root," in bean genotypes possessing the dominant *I* gene, were described in Europe as early as 1963 and again in 1977 [5, 10] and several epidemic outbreaks of these strains have been reported from the U.S.A. in recent years [8, 12, 25]. Results of recent surveys in Africa suggest that these necrotic strains of BCMV predominate over most parts of eastern and southern Africa [24, 31].

Striking serological differences among BCMV strains [13, 16, 20, 34] led Wang [34] to distinguish between serotype A and serotype B strains. Using BCMV isolates of all known pathotypes as well as several monoclonal antibodies (MAb) and antisera (AS) to BCMV strains, we investigated the serological relationship among BCMV strains. Moreover, we compared the cytological effects, the particle dimensions, and apparent molecular mass (M_r) of the capsid protein of strains from all known BCMV pathotypes. For one representative member of each serotype of BCMV, the size, the nucleotide sequence-derived amino acid sequence of the coat protein and the nucleotide sequence of the 3' non-coding region were determined. Our data provide further evidence for separation of BCMV strains into two groups. A preliminary report of parts of this study was presented earlier [32].

Materials and methods

Source and maintenance of virus isolates

Isolates of bean common mosaic virus (BCMV) representing all known pathotypes of BCMV [4] were supplied as dried tissue or in seeds from various sources as follows: the BCMV isolates MSS 1 and 7 which resembled NL 3 in pathogenicity and serological properties (Vetten, unpubl.) were obtained from seeds of the local cultivar Misamfu Speckled Sugar collected in Zambia and supplied by D. J. Allen, Tanzania; isolate 'Persiel' originating from *Phaseolus* bean in Germany by F. Persiel, Germany; strain NY 15(P) by R. Provvidenti, U.S.A. [15]; and the strains NL 5(C), Fla and NY 15(C) by F. J. Morales, Colombia; strain NL 6 by D. G. A. Walkey, U.K.; strain NL 5(a) by L. Bos, The Netherlands; and the strains NL 1, 2, 3, 4, 7, 8 and US 4 by M. J. Silbernagel, U.S.A.

Isolates were re-activated from storage by growing out infected seeds or by mechanical inoculation of dried material onto *Phaseolus vulgaris* cultivars Dubbele Witte, Bountiful and Black Turtle II (BT II). The identity of the BCMV isolates obtained was subsequently verified by using the standard set of bean cultivars for differentiating BCMV strains as proposed by Drijfhout [4]. Each isolate was maintained in BT II kept in an insect-proof glasshouse and sprayed weekly with an aphicide.

Virus purification

All BCMV isolates were propagated in BT II and purified as described [20, 33]. At least three consecutive quasi-isopycnic centrifugations in 400 mg/ml CsCl were used for each preparation. Virus preparations were tested by SDS gel electrophoresis to assure that the coat protein was not degraded.

Antisera and monoclonal antibodies

Antisera to the BCMV isolates MSS 7, NL 5, NL 8, NY 15 (C), NY 15(P), and Fla were produced. In general, rabbits were given an intramuscular injection of 1 mg purified virus antigen emulsified in Freund's complete adjuvant (Difco). A second and third injection of about 1 mg antigen, but emulsified with incomplete adjuvant (Difco), were given one and eight weeks, respectively, after the first injection. Rabbits were bled weekly, starting one week after the second injection.

In addition to the antisera produced by ourselves, we also used the monoclonal antibodies (MAb) B-1-1G4 (specific for an epitope on the trypsin-resistant core of the capsid protein of several potyviruses (Vetten, unpubl. data) and the serotype B-specific MAb B-1-5E5, both from the stock of our laboratory and the serotype A-specific MAb bc-I-3 [35] from G. Mink, U.S.A.

Molecular mass determinations by SDS-PAGE and immunoblotting

Capsid-protein molecular masses were determined by sodium dodecyl sulphate-poly-acrylamide gel electrophoresis (SDS-PAGE) using a 4% stacking gel on a 12% separating gel and the buffer systems of Laemmli and Favre [17]. Purified virus preparations were mixed and boiled at a ratio of 4 : 1 with 5× sample buffer (SP) under reducing conditions and stored frozen until use. Extracts from leaves infected with the individual BCMV isolates were ground in liquid nitrogen, and 1 g of the resulting powder was immediately mixed with 5 × SP and heated at 100 °C for 5 min. Such extracts were clarified by low-speed centrifugation and stored frozen until use.

Appropriate protein concentrations of the denatured leaf extracts and purified virions were first determined by SDS-PAGE and Coomassie brilliant blue staining prior to Western blot analyses. The separated proteins were transferred to nitrocellulose using a semi-dry blotting apparatus (Biometra, Göttingen) and the buffer system of Kyhse-Andersen [14]. After blocking membranes with 3% (w/v) gelatin (Biorad) for 1 h, probing of immunoblots was done essentially as described [1] using the culture supernatant of the broad-spectrum MAb B-1-1G4 at a dilution of 1 : 50 for M_r determinations on Western blots as well as MAb bc-I-3 at 1 μg/ml and the culture supernatant of MAb B-1-5E5 at a dilution of 1 : 2 for serotyping of BCMV isolates. Detection of antibodies was with alkaline phosphatase labelled goat anti-mouse IgG (Dianova, Hamburg) at a dilution of 1 : 2500. Biotinylated marker proteins were used and visualized with streptavidin alkaline phosphatase complexes (Dakopatts, Hamburg) at a final dilution of 1 : 6000.

Trypsin treatment for epitope analysis by immunoblotting

For determining the location of the epitopes of the two serotype-specific MAbs, trypsin-treated extracts from leaves infected with three members of each serotype were prepared by mixing 0.9 ml of crude sap with 0.1 ml of 0.2 M Tris buffer, pH 8.3, and then adding trypsin to a final concentration of 4 μg/ml. After incubation for 30 min at room temperature, extracts were mixed at a ratio of 4:1 with 5 × SP, heated at 100°C for 5 min and stored frozen until use. These extracts were compared with untreated extracts using the two serotype-specific MAbs for immunoblotting.

Direct and indirect ELISA procedures

ELISA was done essentially as described [2]. For direct double-antibody sandwich ELISA (DAS-ELISA) wells were first coated with virus-specific immunoglobulins at 1 μg/ml for 4 h at 30°C; extracts of infected plants diluted 1:400 in 0.1 M sodium phosphate buffer, pH 7.8, containing 50 mM EDTA (PE) were incubated overnight; immunoglobulins labelled with alkaline phosphatase were used at a dilution of 1:1000 and incubated for 4 h.

For indirect ELISA with MAbs and antisera, wells were first coated overnight with extracts from virus-infected leaves diluted 1:20 and 1:100 in carbonate buffer, pH 9.6, and 1:100 in PE, respectively. After a blocking step using 1% non-fat milk powder in PBS, the MAb bc-I-3 and the culture supernatant B-1-5E5 were incubated for 2 h at a concentration of 2 μg/ml and at a dilution of 1:5 in conjugate buffer, respectively. Antisera to BCMV NL 5 and NY 15(C) were used as detecting antibodies at dilutions of 1:1000 and 1:5000 in conjugate buffer. Alkaline phosphatase labelled goat anti-mouse (for MAb) and goat anti-rabbit (for antisera) IgG (Dianova, Hamburg) were both used at a dilution of 1:2500.

After 60 and 90 min substrate hydrolysis for direct and indirect ELISA, respectively, the reaction was measured at 405 nm with a Titertek Multiscan ELISA reader. Reactions were considered positive when readings were higher than twice the value of the healthy control.

Electron microscopy

Morphology of negatively stained virions was determined on carbon-pioloform coated copper grids which had been floated for 5 min on crude extracts of virus-infected leaf tissue, washed with double-distilled water, and negatively stained with 1% aqueous uranyl acetate. Particles were measured directly in a Zeiss EM 10C electron microscope to which a Zeiss Morphomat 30 image analyzer was attached. The microscope magnification was calibrated using a diffraction grating replica with a periodicity of 463 nm.

Cytology of infected cells was studied in ultrathin sections of leaf tissues of BT II embedded in Epon [11]. Decoration titers were determined essentially as specified [23].

Cloning and sequencing procedures

Synthesis of cDNA was carried out as described [19]. Briefly, after first strand cDNA synthesis with AMV reverse transcriptase, the second strand was synthesized with RNase H and DNA polymerase I. The double-stranded cDNA was annealed without further modification to HincII-cut pT7T319U (Pharmacia). Sequencing of selected clones was performed as specified [27], using single-stranded DNA templates generated with the helper phage M13K07. Other recombinant DNA procedures were essentially as described [28]. Sequence analyses were done with the GCG program package [3].

Results

Serological relationship

When 13 isolates of BCMV representing all 10 known pathotypes of BCMV were examined by DAS-ELISA using one MAb and six antisera to BCMV strains (Table 1), BCMV strains NL 3, NL 5(a), NL 5(C), and NL 8 gave strong reactions only with MAb bc-I-3 and antisera to NL 5, NL 8 and MSS 7 but no

Table 1. Reaction of BCMV strains with various BCMV antisera and monoclonal antibodies in direct and indirect ELISA

| BCMV strain[a] | Direct DAS-ELISA[b] with | | | | | | | Indirect ELISA[c] with | | | | | |
| | MAb | Antisera to | | | | | | MAbs | | Antisera to | | | |
	bc-I-3	NL 5	NL 8	MSS 7	NY 15(C)	NY 15(P)	Fla	bc-I-3	B-1-5E5	NL 5 10^{-3}	NL 5 2×10^{-4}	NY 15(C) 10^{-3}	NY 15(C) 2×10^{-4}
NL 3	+++[d]	+++	+++	+++	0	0	0	+++	0	++	++	+	0
NL 5(a)	+++	+++	+++	+++	+	0	0	++	0	+++	++	+	0
NL 5(C)	+++	+++	+++	+++	0	0	0	+++	0	++	++	+	0
NL 8	+++	+++	+++	+++	+	+	0	+++	0	+++	+++	++	++
NL 1	0	0	0	0	++	++	+	0	+++	+	0	++	++
NL 2	0	0	+	0	++	++	++	0	+++	+	0	++	++
NL 4	0	0	+	+	+++	+++	+++	0	+++	++	0	+++	+++
NL 6	0	0	0	0	++	+	+	0	++	++	0	+++	+++
US 4	0	0	0	+	+++	++	++	0	+++	+	0	+++	+++
NL 7	0	0	0	0	++	++	++	0	+++	++	0	++	++
NY 15(P)	0	0	+	+	+++	+++	+++	0	++	+	0	+++	+++
NY 15(C)	0	0	0	+	+++	+++	+++	0	+++	++	0	+++	+++
Fla	0	0	+	+	+++	+++	+++	0	+++	++	+	+++	++

[a] Extracts of infected plants diluted 1:400 in 0.1 M sodium phosphate buffer containing 50 mM EDTA (PE), pH 7.8, and 1:20 in carbonate buffer, pH 9.6, for direct and indirect ELISA, respectively. For indirect ELISA with antisera, plant extracts were diluted 1:100 in PE
[b] Double-antibody sandwich ELISA (DAS-ELISA) using IgG at 1 µg/ml for coating and each conjugate at a dilution of 1:1000
[c] Indirect ELISA using bc-I-3 at 2 µg/ml and the culture supernatant of B-1-5E5 at a dilution of 1:5 as detecting antibody. Antisera were used as indicated in table. Goat anti-mouse (for MA) and goat anti-rabbit (for antisera) conjugates were both used at a dilution of 1:2500
[d] Absorption at 405 nm after a substrate reaction of 60 and 90 min for direct and indirect ELISA, respectively: $0 < 0.03$; $+ \, 0.03–0.099$; $++ \, 0.100–0.599$; $+++ \, > 0.600$

reaction or very weak reactions with antisera to strain Fla and to two symptomatically distinct strains of NY 15 [15]. Conversely, the other nine strains of BCMV reacted strongly with the three latter antisera but did not react or reacted only weakly with the MAb bc-I-3 or the NL 5, NL 8, and MSS 7 antisera. Using antisera to NL 5 and NY 15(C) at two different dilutions in indirect ELISA, a similar differentiation of NL 3 and NL 5 (and to a lesser extent NL 8) from the other BCMV isolates was possible particularly when

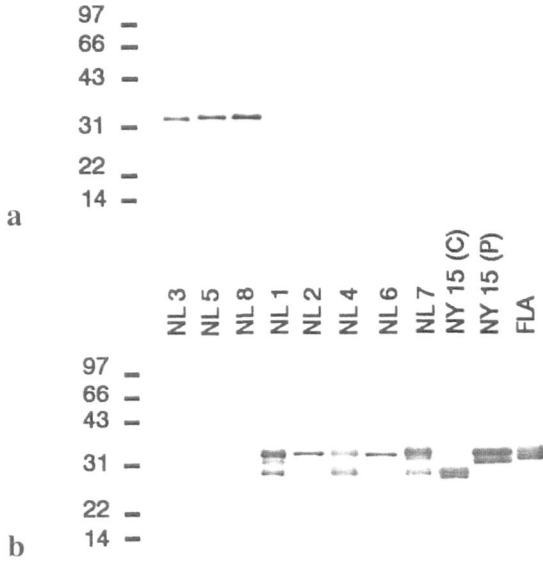

Fig. 1. Western blot of freshly purified preparations of eleven BCMV strains using the serotype A-specific MAb bc-I-3 (**a**) and the serotype B-specific MAb B-1-5E5 (**b**). The figures on the left indicate the positions of the molecular weight markers (in kilodaltons)

Table 2. Homologous and heterologous decoration titers[a] of the two serotype A strains NL 5 and NL 8 and the two serotype B strains NL 4 and NY 15(P) of BCMV with two antisera of each serotype

Antisera	BCMV strains			
	NL 5(C)	NL 8	NL 4	NY 15(P)
NL 5(a)	3200[b]	3200	400	800
NL 8	>6400	>6400	200	200
NY 15(C)	400	1600	>6400	>6400
Fla	200	200	3200	>6400

[a] End points of decoration reaction following incubation of adsorbed virus particles for 15 min with twofold dilutions of antisera [23]

[b] Reciprocal value of highest antiserum dilution yielding a reaction

antisera were diluted 1:5000 (Table 1). However, a clear-cut grouping of strains was obtained with the MAb bc-I-3 and B-1-5E5 in indirect ELISA (Table 1) and in Western blots (Fig. 1). The high specificity of the BCMV antisera and MAbs for certain BCMV strains in different ELISA procedures and immunoblotting experiments strongly suggests that there are two serotypes of BCMV. Therefore, in accordance with previous data [34] we group the BCMV strains NL 3, NL 5, and NL 8 in serotype A and the BCMV strains NL 1, NL 2, NL 4, NL 6, US 4, NL 7, NY 15, and Fla in serotype B of BCMV.

A clear distinction between two members of each serotype was also demonstrated on the basis of homologous and heterologous decoration titers (Table 2). Isolates of each serotype gave very similar and high titers with antisera to isolates of the same serotype whereas much lower titers were obtained with antisera to strains of the other serotype.

Specificity and location of serotype-specific epitopes

When the serotype-specific MAbs bc-I-3 and B-1-5E5 were tested in indirect ELISA against about 50 distinct potyviruses, MAb bc-I-3 only reacted with serotype A strains of BCMV whereas MAb B-1-5E5 reacted with azuki bean mosaic, blackeye cowpea mosaic (Florida isolate), peanut stripe, pepper veinal mottle, and soybean mosaic viruses (data not shown), in addition to all serotype B strains of BCMV. When trypsin-treated and untreated extracts of leaves infected with some members of each serotype were analyzed by immunoblotting using serotype-specific MAbs, the serotype A-specific epitope on the capsid protein of NL 3, NL 5, and NL 8 was destroyed in trypsin-treated extracts while the serotype B-specific epitope on the coat protein of NL 4, NL 7, and NY 15(C) remained intact (Fig. 2). Moreover, MAb bc-I-3 yielded similar reactions in direct and indirect ELISA procedures and densely deco-

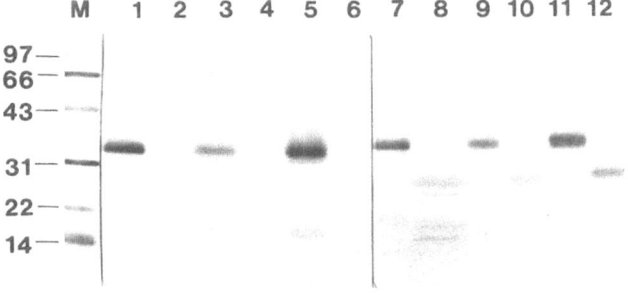

Fig. 2. Western blot of untreated extracts from leaves infected with NL 3 (*1*), NL 5 (*3*), NL 8 (*5*), NL 4 (*7*), NL 7 (*9*), NY 15C (*11*) and of trypsin-treated extracts from leaves infected with NL 3 (*2*), NL 5 (*4*), NL 8 (*6*), NL 4 (*8*), NL7 (*10*), NY 15C (*12*). *M* Biotinylated marker proteins. *1–6* reacted with MAb bc-1-3; *7–12* reacted with MAb B-1-5E5. For further details see Fig. 1

rated particles of serotype A isolates of BCMV in immunoelectron microscopy (IEM) whereas MAb B-1-5E5 did not decorate serotype B strains in IEM and gave high ELISA values only with antigen trapped on plates using carbonate buffer but not with antibody-trapped antigen (data not shown). These results suggest that the serotype A- and serotype B-specific epitopes of the capsid proteins of BCMV strains are located at the N- (or C) terminal end and within the trypsin-resistant core, respectively. The latter epitope seems to be inaccessible on intact particles of serotype B isolates but is exposed after alkaline degradation of virions and on immunoblots.

Apparent molecular mass (M_r) of capsid proteins

SDS-PAGE (data not shown) and Western blots (Fig. 3 a) of freshly purified preparations of 11 BCMV strains with the broad-spectrum MAb B-1-1G4 showed that some strains, in particular the serotype A isolates, yielded one major protein band but most of the other BCMV strains had several major and minor polypeptide bands indicating that the capsid proteins of many BCMV strains had been proteolytically degraded during virus purification although to a variable extent but in a fashion typical for potyviruses. To minimize proteolytic degradation as a result of the purification process, apparent mo-

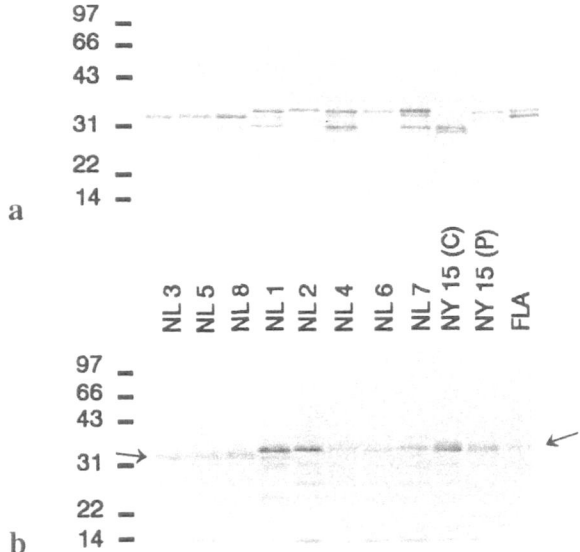

Fig. 3. Western blots of freshly purified preparations (**a**) and extracts (**b**) from leaves infected with various BCMV strains using the broad-spectrum MAb B-1-1G4. Arrows on the left and right show the positions of the major, not yet degraded capsid protein in freshly prepared extracts from leaves infected with serotype A and B isolates, respectively (**b**). The numerous minor bands which were visible with leaf extracts were also present in extracts from healthy leaves (not shown). For further details see Fig. 1

lecular masses (M_r) of the individual capsid proteins were determined using Western blots of extracts from BCMV-infected plants. As demonstrated in Fig. 3b, the M_r of the major protein band of the serotype A isolates NL 3, NL 5, and NL 8 are very similar (about M_r 33,000) but significantly lower than those of the serotype B isolates which range from M_r 34,500 to 35,000$_r$ (Table 3).

Immunoblotting experiments with extracts from leaves infected with several field isolates of BCMV which had been obtained from *Phaseolus* beans and wild leguminous weeds in Africa and identified as serotype A isolates with MAbs bc-I-3 [31; Vetten, unpubl. data], revealed that M_r of the capsid proteins of all these isolates were identical to those of NL 3, NL 5, and NL 8 (data not shown) indicating that the smaller capsid protein is a characteristic property of serotype A isolates.

Table 3. Properties differentiating strains of bean common mosaic virus (BCMV)

BCMV strain	Pathotype[a]	Particle length (nm)	Coat protein size (M_r)	Cytology		
				pinwheels and scrolls	finger-like ER- proliferations	accumulation of vesicles
Serotype A						
MSS 1	necrotic	n.d.[b]	n.d.	+[c]	+	−
NL 3	necrotic	818	32,500	+	+	−
NL 5	necrotic	815	33,000	+	+	−
NL 8	necrotic	810	33,000	+	+	−
Serotype B						
NL 1	mosaic	867	34,500	+	−	−
NL 2	(mosaic)	863	35,000	+	n.d.	n.d.
NL 4	mosaic	863	34,500	+	−	−
NL 6	(mosaic)	817	34,500	+	−	−
US 4	(mosaic)	847	n.d.	+	−	−
NL 7	mosaic	886	35,000	+	−	−
NY 15 (C)	(mosaic)	828	34,500	+	−	−
NY 15 (P)	(mosaic)	831	34,500	+	−	+
Fla	mosaic	853	35,000	n.d.	n.d.	n.d.

[a] "Necrotic" and "(mosaic)" pathotypes cause systemic vascular necrosis ("black root") at ambient temperatures or at > 32 °C, respectively, in bean cultivars possessing the dominant *I* gene whereas "mosaic" strains never induce necrosis but only mosaic in susceptible bean cultivars [4, 15]

[b] *n.d.* Not determined

[c] + Cytological alterations of this type were readily found in infected cells; − cytological alterations of this type were not found

Particle length

Normal length determinations of the particles of individual BCMV strains revealed that serotype A isolates have shorter particles (810–818 nm) than most isolates (except NL 6 and NY 15) of serotype B, which range from about 847 to 886 nm (Table 3).

Cytological effects

All isolates studied induced indistinguishable cytoplasmic cylindrical inclusions of the pinwheel and scroll type [6] as shown in Fig. 4. Cells infected with serotype A isolates consistently contained a specific type of proliferated endoplasmic reticulum with "finger-like" extensions (Fig. 5 a) [18], whereas such structures were never found in cells infected with serotype B isolates. Accumulation of vesicles similar to those found with other potyviruses [18] were inconspicuous with all BCMV isolates studied, except with NY 15(P), a serotype B isolate, which induced prominent vesicle clusters (Fig. 5 b and Table 3).

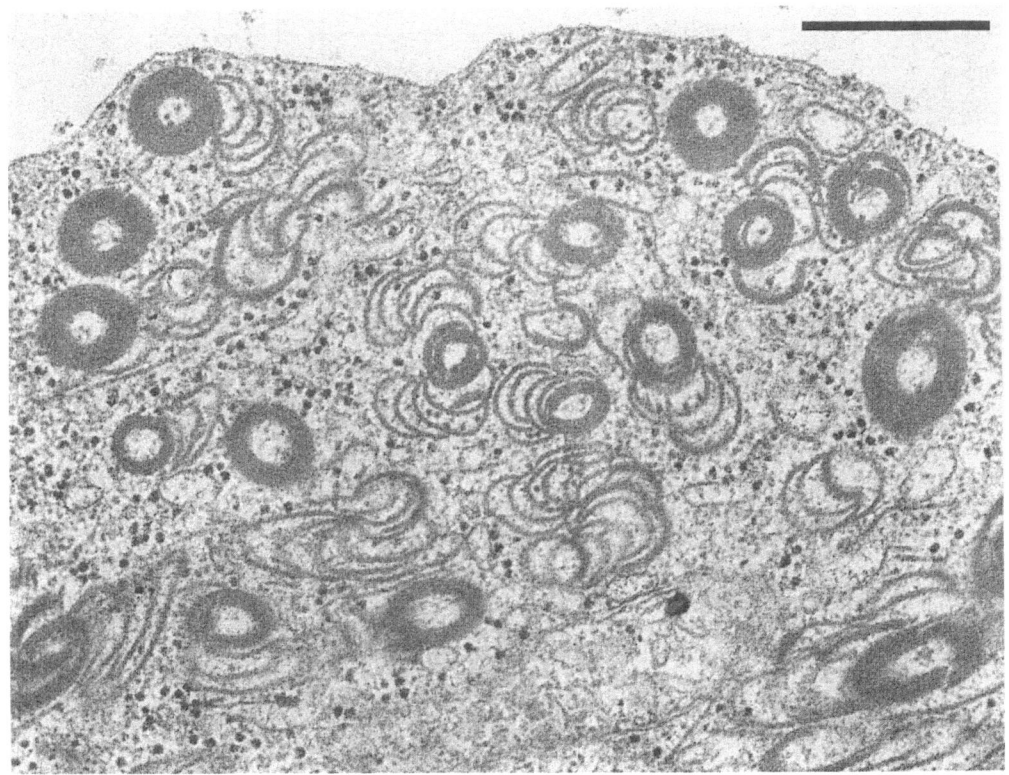

Fig. 4. Cylindrical inclusions, pinwheels and scrolls induced by all BCMV isolates studied. Photograph shows a section of bean tissue infected with BCMV isolate 'Persiel', a representative of serotype A. Bar: 500 nm

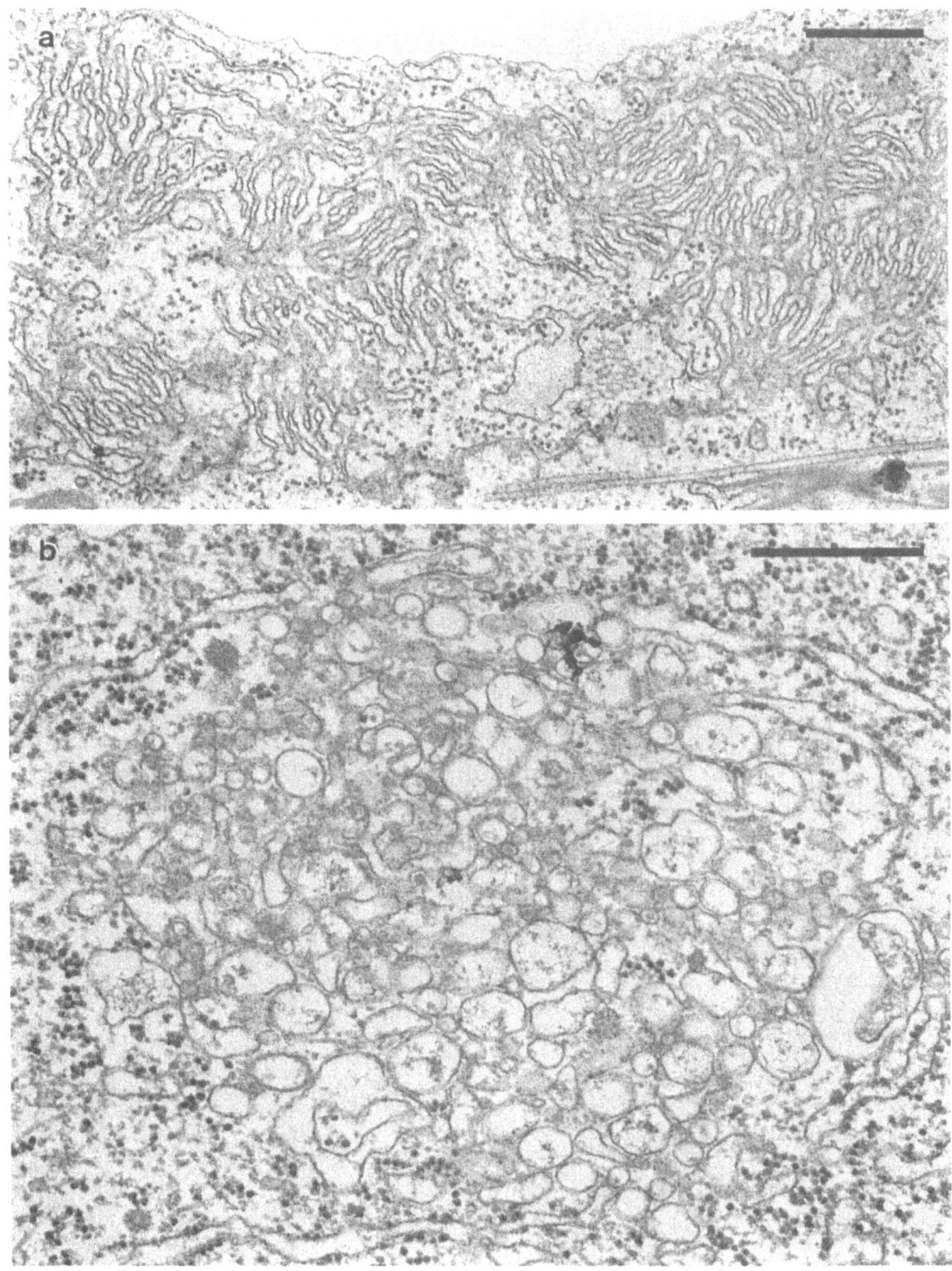

Fig. 5. Modifications of the cellular membrane system induced by BCMV strains. **a** Finger-like proliferations of the endoplasmic reticulum (ER) induced by strain MSS 1, a representative of the serotype A of BCMV. **b** Accumulation of cytoplasmic vesicles induced by BCMV strain NY15. Bars: 500 nm

Sequences of the capsid protein and 3' non-coding region

The capsid-protein gene of a representative member of each serotype was cloned and sequenced. The amino acid sequence and the putative cleavage site (ESVXXQ/S) of BCMV NIa protease located between the NIb and the coat protein were determined (Fig. 6). Molecular mass calculations based upon nucleotide sequence-derived amino acid sequences gave values of 29,662 and 32,489 for the capsid proteins of the serotype A isolate NL 8 and the serotype B isolate NL 4, respectively, confirming the above mentioned differences in M_r determined by SDS-PAGE.

Comparison of the amino acid sequences of the coat proteins showed considerable differences in the N-termini where two major deletions in the NL 8 isolate were observed. The core region and the C-termini appeared to be highly conserved. The degree of homology between the capsid proteins of the two BCMV strains was about 80% (Fig. 6). When the amino acid sequence of the capsid protein of NL 4 was compared with that of peanut stripe virus [21] to which NL 4 is serologically more closely related than to NL 8 [33; Vetten, unpubl.], a homology of approximately 90% was obtained.

```
  1 AESALKTLYTNKKTKIEELARYLEVLDFDYEVGCGESVHLQSGSGHPPLP 50
    ||.||:.||||:|..|:||:. ||. |:|: : :..|||  ||:.
 11 AETALRKLYTDKDAKMEEMQEYLKQLEFGSDDEVCESVSTQSSKK..... 55

 51 VVDAGVDTEKDKKDKSSRGKDPENKEETRNNSRGTENPTMRDKDVNAGSR 100
    :.: |.| ||::| .:||:|.                 ||||..||:
 56 ..EEEKDAEADKREK.DKGKGPA................DKDVGTGSK 84

101 GKVVPRLQRITKRMNLPMVKGNVILNLDHLLDYKPEQTDLFNTRATKMQF 150
    |||||||.:|||:||||||| |.:|||||||::|||||:||||||| ||
 85 GKVVPRLQKITKKMNLPMVGGRMILNLDHLIEYKPQQTDLYNTRATKAQF 134

151 EMWYNAVKGEYEIDDDQMAIIMNGFMVWCIDNGTSPDVNGTWVMMDGDEQ 200
    |.|:||||.||||.||||:::|||||||||||||||||||.||||||||
135 ERWYEAVKTEYELDDQQMGVVMNGFMVWCIDNGTSPDVNGVWVMMDGDEQ 184

201 VEYPLKPMVENAKPTLRQIMHHFSDAAEAYIEMRNSERPYMPRYGLLRNL 250
    :||||||||||||||||:||||||||||||||||||||.|||||||||||
185 IEYPLKPMVENAKPTLRQVMHHFSDAAEAYIEMRNSEGFYMPRYGLLRNL 234

251 RDKNLARYAFDFYEVTSKTSDRAREAVAQMKAAALSNVSSKLFGLDGNVA 300
    |||.|.|||||||||.|||||||||||||||||.||..:||||||||||
235 RDKSLARYAFDFYEVNSKTSDRAREAVAQMKAAALANVNTRLFGLDGNVA 284

301 TTSENTERHTARDVNQNMHTLLGMGPPQ 328
    ||||||||||||||||||| ||||...|
285 TTSENTERHTARDVNQNMHHLLGMTSGQ 312
```

Fig. 6. Comparison of the amino acid sequences of the coat proteins of NL 4 (top) and NL 8 (bottom). The putative cleavage site of the BCMV NIa protease is shaded. Gaps were introduced to make an optimal alignment. Dots above the line are for orientation and mark every tenth amino acid symbol. Strokes indicate identical amino acids; colons and dots indicate that the comparison value for two amino acids is greater than, or equal to, 0.5 and 0.1, respectively. Symbol comparison of the GCG program package is based upon the evolutionary distance of amino acids according to Dayhoff [28a] and normalized by Gribskov [7a]

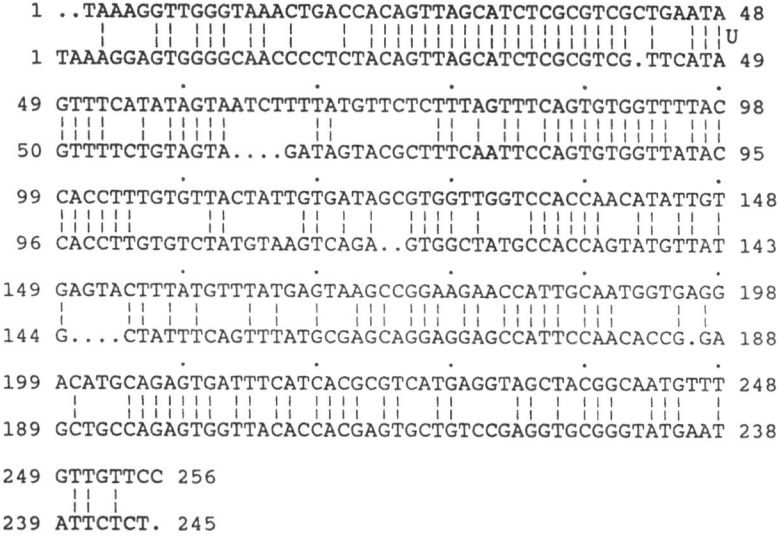

```
  1 ..TAAAGGTTGGGTAAACTGACCACAGTTAGCATCTCGCGTCGCTGAATA 48
     |  || ||| || |   | |||||||||||||||||| |   |||U
  1 TAAAGGAGTGGGGCAACCCCTCTACAGTTAGCATCTCGCGTCG.TTCATA 49

 49 GTTTCATATAGTAATCTTTTATGTTCTCTTTAGTTTCAGTGTGGTTTTAC 98
    ||||  | ||||||        ||     || | || |||||||||| |||
 50 GTTTTCTGTAGTA....GATAGTACGCTTTCAATTCCAGTGTGGTTATAC 95

 99 CACCTTTGTGTTACTATTGTGATAGCGTGGTTGGTCCACCAACATATTGT 148
    |||||| |   ||  |||| | |   |||| |  ||||||  || || |
 96 CACCTTGTGTCTATGTAAGTCAGA..GTGGCTATGCCACCAGTATGTTAT 143

149 GAGTACTTTATGTTTATGAGTAAGCCGGAAGAACCATTGCAATGGTGAGG 198
    |    ||| |   | |   ||  ||| ||| || |||| |||
144 G....CTATTTCAGTTTATGCGAGCAGGAGGAGCCATTCCAACACCG.GA 188

199 ACATGCAGAGTGATTTCATCACGCGTCATGAGGTAGCTACGGCAATGTTT 248
    |   ||||||| || || |||| || || |   |||  ||| |||| |
189 GCTGCCAGAGTGGTTACACCACGAGTGCTGTCCGAGGTGCGGGTATGAAT 238

249 GTTGTTCC 256
    || |
239 ATTCTCT. 245
```

Fig. 7. Alignment of the nucleotide sequences of the 3' non-coding regions of NL 4 (top) and NL 8 (bottom). Gaps were introduced for optimal alignment. Dots above the line mark every tenth nucleotide symbol

Optimal alignment of the 3' non-coding regions of NL 4 and NL 8 revealed only 68% homology between the two BCMV strains (Fig. 7).

Discussion

At temperatures as low as 20°C, certain isolates of BCMV induce systemic vascular necrosis in bean genotypes possessing the dominant *I* gene and are referred to as temperature-insensitive necrotic strains [4]. In previous serological investigations on BCMV strains [13, 16, 20] using only a limited number of BCMV strains and antisera as well as in a more extensive study by Wang [34] with a range of BCMV strains, some of these necrotic strains were shown to differ considerably from other BCMV strains in serological properties. This led Wang [34] to classify these necrotic strains as serotype A isolates of BCMV and the mosaic-inducing strains of BCMV as serotype B isolates. Results of our serological tests with two monoclonal antibodies and six antisera and with 13 isolates belonging to the ten known pathotypes of BCMV, substantiate, and provide a broader basis to, Wang's [34] serological classification. Even when using antisera in indirect ELISA and decoration-titer experiments, neither of which gives strain-specific reactions typical for DAS-ELISA, serotype A and serotype B isolates could be clearly distinguished in the present study. Therefore, we grouped the temperature-insensitive, necrotic strains NL 3, NL 5, and NL 8 as serotype A isolates and the mosaic-inducing strains NL 1, NL 2, NL 4, NL 6, US 4, NL 7, NY 15, and Fla as serotype B isolates. This serotyping of BCMV strains was most clearly accomplished with the MAbs, but may also be possible with polyclonal antibodies specific for the N-terminal part of the capsid protein [13].

By studying other properties of individual members of each BCMV serotype, we provided considerably more evidence for the distinction of necrotic strains from the remaining BCMV strains: necrotic strains had smaller capsid proteins, appeared to have shorter particles than serotype B isolates, and induced membrane proliferations in host cells not found with other BCMV strains (Table 3). Moreover, when comparing the coat-protein genes and the 3' non-coding regions of one representative member of each serotype, remarkable differences in amino acid and nucleotide sequences with homologies of 78% and 62%, respectively, were revealed. According to guidelines recently proposed for *Potyvirus* classification [7, 29], virus strains and distinct viruses are supposed to have homologies of more than 90% and in the range of 35 to 75%, respectively, for amino acid sequences of capsid proteins and the respective homologies for the 3' non-coding regions of potyviruses should range from 83 to 99% and from 39 to 53%, respectively. Although the degrees of homology between the two selected members of each BCMV-serotype were intermediate to the proposed ranges [7, 29], they are sufficient for considering NL 4 and NL 8 as distinct potyviruses. However, it needs to be determined if other BCMV strains give different sequence homologies. Based upon their specific pathogenic properties [4] and upon all the other differentiating properties determined in the present study, we regard the temperature-insensitive, necrotic strains as strains of a distinct virus, separate from the mosaic-inducing strains of BCMV. This distinction seems justified although BCMV strains are serologically related and all hitherto known BCMV strains, irrespective of serotype, cause very similar symptoms in bean genotypes lacking resistance genes [25].

Although serotype B isolates are very complex in their interactions with host genes for BCMV resistance [4] and appeared to vary in particle dimensions in our studies, they varied negligibly in cytopathology, had identical capsid-protein sizes and shared serological properties (Table 3). Alignments of the amino acid sequence of the serotype B strain NL 4 with that of peanut stripe virus [21] yielded a degree of homology of approximately 90% indicating a very close relationship between the two viruses and a closer relationship of BCMV NL 4 to peanut stripe virus (PStV) than to BCMV NL 8. This is supported by studies [13, 16, 30, 33] demonstrating a close serological relationship between certain serotype B isolates and viruses, such as peanut stripe virus (PStV), azuki bean mosaic virus (AzMV) and certain isolates of blackeye cowpea mosaic virus (BlCMV). This close relationship has been further supported by a recent study [22] comparing the molecular properties of the coat proteins of PStV, AzMV, and BlCMV and concluding these viruses are all strains of the same potyvirus. In addition to serological and molecular properties of the capsid proteins, there are also close similarities of serotype B strains to most of these other viruses in coat protein size [22, 33] and certain cytopathological features [33; Lesemann, unpubl.]. Only PStV induces cylindrical inclusions with scrolls and laminated aggregates and differs in this respect from AzMV, BlCMV, and serotype B isolates [33]. Therefore, we propose to group the serotype B isolates of BCMV with viruses, such as BlCMV, PStV, and AzMV, and to consider the serotype B isolates as strains belonging to this cluster but having a special adaptation to *Phaseolus* bean.

We have recently obtained potyvirus isolates from *Phaseolus* bean in Africa which are most closely related serologically to certain strains of BlCMV and cowpea aphid-borne mosaic virus (CABMV) and induce severe BCMV-like symptoms in many bean genotypes but are mildly pathogenic to cowpea genotypes known to be very susceptible to BlCMV and CABMV [31]. This observation exemplifies the notion of various researchers [9] that "the greater the number of isolates studied the more evident it becomes that sharply defined borderlines separating individual potyviruses cannot be drawn." Nevertheless, it appears necessary to reconsider the current classification of potyviruses by splitting and grouping strains of viruses wherever deviations and similarities in properties significant to virus classification are encountered.

This is the first known report comparing capsid-protein size, particle dimensions and cytopathology of various BCMV strains as well as providing amino acid and nucleotide sequence information on two BCMV strains. Based upon all the data presented, we propose to regard the temperature-independent, necrotic strains of BCMV, which were identified some decades later than the mosaic-inducing strains, in future as a distinct virus which should be given a new name: bean necrotic mosaic virus. For priority reasons, serotype B strains of BCMV should retain their name. Accumulating data [13, 16, 22, 31, 33] suggest that potyviruses, such as AzMV, BlCMV and PStV share many taxonomically significant properties with serotype B strains of BCMV. Although each of these viruses has a narrow, hardly overlapping host range with a special adaptation to its principal host [16, 25, 33] which has been a major criterion for distinction of potyviruses, they all appear to be members of a cluster of closely related viruses and hence should perhaps also be referred to as BCMV strains for priority reasons. At the present state of our knowledge it appears premature for making a final decision. Nucleotide sequence information on genome parts other than that of the coat protein and a better understanding of the significance of differences in nucleotide sequences may be required to address the question as to whether serotype B isolates of BCMV should be grouped separately or together with viruses such as AzMV, BlCMV, and PStV.

Acknowledgements

We are most grateful to Drs. F. J. Morales, R. Provvidenti, M. J. Silbernagel, and D. G. A. Walkey for providing BCMV isolates and to Dr. G. Mink for a generous gift of the monoclonal antibody bc-I-3. We thank F. J. Morales and M. J. Silbernagel for supplying seed of BCMV differential cultivars of bean and N. M. McKern and D. D. Shukla for their generous provision of sequence data of peanut stripe virus prior to publication and for providing us the correct amino acid sequence of residues 256 and 257 in the capsid protein of BCMV-NL4. We thank Mrs. Anke Brisske, Doris Hasselmann, Ina Kramer, Christina Maaß, Angelika Sieg-Müller, Sabine Schuhmann, and Susanne Steiner for expert technical assistance. Part of this study was supported by a grant from the Gesellschaft für Technische Zusammenarbeit (GTZ).

References

1. Burgermeister W, Koenig R (1984) Electroblot immunoassay – a means for studying serological relationships among plant viruses? Phytopathol Z 111: 15–25
2. Clark MF, Adams AN (1977) Characteristics of the microplate method of enzyme-linked immunosorbent assay for the detection of plant viruses. J Gen Virol 34: 475–483
3. Devereaux J, Haeberli P, Smithies O (1984) A comprehensive set of sequence analysis programs for the VAX. Nucleic Acids Res 12: 387–395
4. Drijfhout E (1978) Genetic interaction between *Phaseolus vulgaris* and bean common mosaic virus with implications for strain identification and breeding for resistance. Agricultural Research Reports 872. Pudoc, Wageningen
5. Drijfhout E, Bos L (1977) The identification of two new strains of bean common mosaic virus. Neth J Plant Pathol 83: 13–25
6. Edwardson JR (1974) Some properties of the potato virus Y-group. Fla Agric Exp Stat Monogr Ser, no 4
7. Frenkel MJ, Ward CW, Shukla DD (1989) The use of the 3'-non-coding nucleotide sequences in the taxonomy of potyviruses: application to watermelon mosaic virus 2 and soybean mosaic virus-N. J Gen Virol 70: 2775–2783
7a. Gribskov M, Burgess R (1986) Sigma factors of *E. coli, B. subtilis,* phage SPO1, and phage T4 are homologous proteins. Nucleic Acids Res 14: 6745–6763
8. Hampton RO, Silbernagel MJ, Burke DW (1983) Bean common mosaic virus strains associated with bean mosaic epidemics in the northwestern United States. Plant Dis 67: 658–661
9. Harrison BD (1985) Usefulness and limitations to the species concept for plant viruses. Intervirology 24: 71–78
10. Hubbeling N (1963) Complicaties bij de toetsing van bonerassen op resistentie tegen *Phaseolus* virus 1 ten gevolge van het voorkomen van afwijkende virusstammen. Meded Landbouwhogeschool Opzoekingsstat Staat Gent 28: 1025–1033
11. Jones RAC, Koenig R, Lesemann D-E (1980) Pepino mosaic virus, a new potexvirus from pepino (*Solanum muricatum*). Ann Appl Biol 94: 61–68
12. Kelly JD, Saettler AW, Morales M (1984) New necrotic strains of bean common mosaic virus in Michigan. Bean Improv Coop 27: 38–39
13. Khan JA, Lohuis H, Goldbach RW, Dijkstra J (1990) Distinction of strains of bean common mosaic virus and blackeye cowpea mosaic virus using antibodies to N- and C- or N-terminal peptide domains of coat proteins. Ann Appl Biol 117: 583–593
14. Kyhse-Andersen J (1984) Electroblotting of multiple gels: a simple apparatus without buffer tank for rapid transfer of proteins from polyacrylamide to nitrocellulose. J Biochem Biophys Methods 10: 203–209
15. Kyle MM, Provvidenti R (1988) A severe isolate of bean common mosaic virus NY 15. Bean Improv Coop 30: 87–88
16. Lana AF, Lohuis H, Bos L, Dijkstra L (1988) Relationships among strains of bean common mosaic virus and blackeye cowpea mosac virus – members of the potyvirus group. Ann Appl Biol 113: 493–505

17. Laemmli UK, Favre M (1973) Maturation of the head of bacteriophage T4. J Mol Biol 80: 575–599

18. Lesemann D-E (1988) Cytopathology. In: Milne RG (ed) The plant viruses, vol 4, the filamentous plant viruses. Plenum, New York, pp 179–235

19. Maiss E, Breyel E, Brisske A, Casper R (1988) Molecular cloning of cDNA complementary to the RNA genome of plum pox virus (PPV). J Phytopathol 122: 222–231

20. Makkouk KM, Lesemann D-E, Vetten HJ, Azzam OI (1986) Host range and serological properties of two potyviruses isolated from *Phaseolus vulgaris* in Lebanon. Trop Agric Res Serv 19: 187–194

21. McKern NM, Edskes HK, Ward CW, Strike PM, Barnett OW, Shukla DD (1991) Coat protein of potyviruses. 7. Amino acid sequence of peanut stripe virus. Arch Virol 119: 25–35

22. McKern NM, Shukla DD, Barnett OW, Vetten HJ, Dijkstra J, Whittaker LW, Ward CW (1992) Coat protein properties suggest that azuki bean mosaic virus, blackeye cowpea mosaic virus, peanut stripe virus and three isolates from soybean are all strains of the same potyvirus. Intervirology 33: 121–134

23. Milne RG (1984) Electron microscopy for the identification of plant viruses in in vitro preparations. In: Maramorosch K, Koprowski H (eds) Methods of virology, vol VII. Academic Press, Orlando, pp 87–120

24. Mink G (1985) Bean virus serological survey, Kenya and Tanzania. Washington State University, Prosser (unpubl. report)

25. Morales FJ, Bos L (1988) Bean common mosaic virus. AAB Descriptions of Plant Viruses, no 337

26. Provvidenti R, Silbernagel MJ, Wang W-Y (1984) Local epidemic of NL-8 strain of bean common mosaic virus in bean fields of western New York. Plant Dis 68: 1092–1094

27. Sanger F, Nicklen S, Coulson AR (1977) DNA sequencing with chain-terminating inhibitors. Proc Natl Acad Sci USA 74: 5463–5467

28. Sambrook J, Fritsch EF, Maniatis, T (1989) Molecular cloning: a laboratory manual, 2nd edn. Cold Spring Harbor Laboratory, New York

28a. Schwarz RM, Dayhoff MO (1979) In: Dayhoff MO (ed) Atlas of protein sequences and structure. National Biomedical Research Foundation, Washington DC, pp 353–358

29. Shukla DD, Ward CW (1989) Structure of potyvirus coat proteins and its application in the taxonomy of the potyvirus group. Adv Virus Res 36: 273–314

30. Tsuchizaki T, Omura T (1987) Relationships among bean common mosaic virus, blackeye cowpea mosaic virus, adzuki bean mosaic virus and soybean mosaic virus. Ann Phytopathol Soc Jpn 53: 478–488

31. Vetten HJ, Allen DJ (1991) Recent progress in the identification of viruses of *Phaseolus vulgaris* in Africa. Bean Improv Coop 34: 3–4

32. Vetten HJ, Lesemann D-E, Maiss E (1990) Differentiation of bean common mosaic virus strains. In: VIIIth International Congress of Virology, Berlin, 1990, P84-008

33. Vetten HJ, Green SK, Lesemann D-E (1992) Characterization of peanut stripe virus isolates from soybean in Taiwan. J Phytopathology 135 (in press)

34. Wang W-Y (1983) Serology of bean common mosaic virus strains. MSc Thesis, Washington State University, Pullman, Washington

35. Wang W-Y (1985) Production and characterization of hybridoma cell lines and a broad spectrum monoclonal antibody against bean common mosaic virus. PhD Thesis, Washington State University, Pullman, Washington

Authors' address: H. J. Vetten, Biologische Bundesanstalt für Land- und Forstwirtschaft, Institut für Biochemie und Pflanzenvirologie, Messeweg 11–12, D-W-3300 Braunschweig, Federal Republic of Germany.

Summary

Arch Virol (1992) [Suppl 5]: 435–444

A summary of potyvirus taxonomy and definitions

O. W. Barnett

Department of Plant Pathology and Physiology, Clemson University,
Clemson, South Carolina, U.S.A.

Summary. The current taxonomic status of the family *Potyviridae* is presented with suggestions for resolving some taxonomic problems. Terms such as strain, pathotype, serotype, variant, mutant, and isolate are discussed in relation to the family.

Introduction

Viruses of the family *Potyviridae* are numerous and diverse. They infect many wild and cultivated plant species often causing devastating diseases. Attempts to logically separate and classify these viruses have been only partially successful. More accurately, virologists have not reached a consensus about taxonomic placement because of the genetic and biological diversity of viruses in the family *Potyviridae* and because we have not been able to separate taxonomic and diagnostic characteristics in our own minds. However, the need for a useful and acceptable taxonomy requires that some conclusions be drawn, else the current situation will continue.

Part of the problem is the biological complexity of viruses in the *Potyviridae*. Virologists recognize that reference virus strains have been maintained for years and still give similar diagnostic reactions [7]. If a different host is used for propagation of the strain, or if other selection pressures are applied, isolates may adapt and may differ biologically [11, 14]. When only two viruses are known to infect species of a plant family they are easily recognized. As more isolates of each virus are characterized and compared, it becomes difficult to decide to which virus a new isolate belongs. As more isolates are obtained characteristics of the viruses seem to overlap or form a continuum [3].

Serology has been used to identify these viruses, but polyclonal antibodies often are not specific because of cross-reactivity among closely related viruses. Simple and accurate diagnosis of closely related viruses is difficult but this capability must be available for effective disease prediction and control.

Potyvirus isolates which infect sugarcane, maize, and sorghum were all assigned to a single virus, sugarcane mosaic virus (SCMV), because their host ranges were similar and all isolates were antigenically interrelated. The confusion developed as virologists realized that some groups of SCMV strains seemed to have no connecting strains [9]. This dilemma was solved when Shukla and co-workers [13, 18] separated the isolates of SCMV into four viruses. Thus, although many of the host reactions and overall host ranges of all strains were similar, the four viruses were different when compared by virus-specific amino terminal polyclonal antibodies, amino acid sequences of the coat protein, nucleotide sequences of the 3' non-translated region of genome RNA, peptide profiles of the coat protein, cell free translation products of the RNAs, size and serology of the cylindrical inclusion proteins, and inclusion body morphologies. Establishment of four viruses even allowed selection of differential host plants to aid in diagnosis and explained anomalies in earlier cross protection results.

Viruses in the family *Potyviridae* now can be differentiated accurately based on molecular properties. Biological and antigenic properties which agree with this classification can be used for diagnosis. Part of the confusion among potyvirologists has been with use of terms such as strain, pathotype, serotype, variant, mutant, and isolate. These terms will be discussed with reference to the family *Potyviridae*. The current taxonomic status of the family also will be presented with suggestions for resolving some taxonomic problems.

Definitions

The International Committee on Taxonomy of Viruses (ICTV) recently accepted an official definition for the term virus species: "A virus species is a polythetic class of viruses that constitutes a replicating lineage and occupies a particular ecological niche" [21]. The term polythetic means that all members need not possess a specific property. Thus, virus isolates that compose the species have many biological and physico-chemical properties in common but individual isolates may differ in one or more properties and no single property can define a species. Isolates of a species are similar because of the replicating lineage, but because the species occupies an ecological niche, isolates of the species from different times and places will vary within limits imposed by the selection pressure.

The strength of the polythetic species concept is that it does not depend upon strict definitions. Thus, there need not be common rules. ICTV Study Groups, such as the Potyvirus Study Group, can decide what is relevant to their particular viruses and devise criteria according to the peculiarities of those viruses. Taxonomic considerations at levels lower than virus species have not been addressed by the ICTV. This allows virologists an opportunity to provide definitions applicable to the potyviruses. I submit the following interpretative definitions for consideration.

Isolate

An isolate refers to a single isolation from an infected plant. A newly obtained isolate has no taxonomic implications because (*i*) it may contain particles of more than one virus species, (*ii*) it may be a member of a previously described strain (*iii*) it may be the first member of an undescribed strain of a known virus species, or (*iv*) it may be the first member of an undescribed virus species. An isolate must be characterized and compared with similar entities to be placed in one of the categories. It should then be qualified as an isolate of species X, strain Y, or pathotype Z once the identification has been made.

Strain

A strain is a collection of naturally occurring isolates of a virus species that all have the same recognizable property or properties not present in other isolates of the virus and which are "so similar to each other that it is useful to give them a ... name" [21].

Pathotype

A pathotype is a collection of isolates which cause a recognizable host response, such as host range or reaction to host resistance genes. Hampton and Provvidenti [7] illustrate this well; a "pathotype is ... controlled by a host gene specific for that entity. Isolates of a given pathotype may vary in virulence, however, as illustrated by isolates L (mild) and L_1 (severe) of pea seed-borne mosaic virus pathotype P-2. Although distinguishable in virulence on susceptible *Pisum* genotypes, they are both controlled by the same genes ..."

Serotype

A serotype is a collection of isolates which differs from another collection of isolates of the same virus in antigenic properties. Therefore, pathotypes and serotypes are specific types of strains based on host reactions or antigenic properties, respectively. For example, bean common mosaic virus isolates are subdivided into pathotypes based on reactions on a series of differential bean cultivars [14]. Strains may differ in other properties such as atypical vector relations and rates of seed transmission [11].

To exist in nature, strains must achieve some degree of ecological stability in a given niche. Often, isolates of a particular strain may occur in a given locale over a number of years. However, a strain may arise through mutation of a single nucleotide, for instance from aphid transmitted to non-aphid transmitted [2]. Because such small molecular changes can lead to significant biological changes which are prominent enough to warrant strain designation, there is no justification for giving strains the status of subspecies, a term with taxonomic implications (Van Regenmortel, pers. comm.). A single nucleotide change shows no temporal pattern of change useful for phylogenic relationships.

The requirement that strains be naturally occurring means that the recognized character responsible for a strain designation must have achieved some level of stability under various competitive pressures. Stability of the recognizable character must be proven by obtaining multiple isolates of a virus with this characteristic. So, all isolates of a virus cannot be different strains and strains are composed of a collection of isolates.

Variant

The term variant may be used for any novel or distinctive isolate which occurs naturally or which is man-made (Murant and Mayo, pers. comm.). If a number of variants are found with a common recognizable, stable property then they should be given a strain designation. However, variants also may have slight or minor changes in characters other than the stable characters used for strain determination, i.e., strains may include several variants. These minor changes may even be reversible or unstable. For instance, Lovisolo [12] notes that amaranthus leaf mottle virus is found only in wild plants and different isolates have various host ranges. This virus may be in a stage of easy mutability in the host range determinant. Tomato black ring nepovirus (TBRV) furnishes another example of the term variant (Murant and Mayo, pers. comm.). TBRV has two major serotypes (strains) which differ by two or three serological differentiation index (SDI) units and which have different nematode vectors. Within each strain there are serological variants which form weak spurs in gel-diffusion tests; other variants differ in virulence of symptoms; and other variants differ in efficiency of nematode transmission though not necessarily in virulence or serology. Variants and mutants may be considered synonymous but mutants are more generally regarded as products of treatments under laboratory conditions. Thus isolates, variants, and mutants refer to individually obtained isolations.

After characterization, isolates and variants may be included as members of a virus strain. However, many virologists equate a strain with a single, characterized isolate. These isolates may have been worked with for many years and begin to typify the strain so strongly in our thoughts that a single isolate in effect becomes the strain. We must not allow ourselves to forget that a strain comprises a collection of isolates each of which may have one or more minor differences while all have the character which defines the strain.

A strain is not a static entity as could be pictured by the narrow perception that a single isolate equals a strain, but is a variable, changeable collection of isolates with particles having the common bounds of the virus species and of the strain characteristics. Strain variability is illustrated by isolates of bean common mosaic virus [14]. In the BCMV pathotype V there are at least three isolates identified by different authors as NY-15, all of which infect the correct differential indicators which define pathotype V. These three isolates all react characteristically (and stably) on another set of differential hosts which separate NY-15 from other pathotype V isolates. Thus, NY-15 is a strain or pathotype which happens to be included in pathotype V. There is no reason to develop a progression of terms to fit smaller categories. These terms are flexible so that pathotypes and serotypes are strains, but pathotypes can encompass strains.

Variability must also be considered when storing and propagating isolates. An isolate may change after continuous passage in a single host [11]. One way to maintain an isolate with some level of genetic stability is to dry infected plant tissue because a large number of particles are retained and used as inoculum in the future with the same ratio of particle variability as was present at the time the tissue was fresh (assuming all particles in the tissue are equally stable to desiccation). What happens to isolates when they are stored in other manners? Bean common mosaic virus can be stored in infected seed. Sometimes isolates recovered from different seeds may possess different characters (Silbernagel, pers. comm.). Seed, aphid, and sap transmission impose different selection pressures upon isolates. Tests to verify biological and serological trueness-to-type of all new isolates are important for all virologists to conduct, regardless of whether the isolate is used for molecular or biological experimentation.

This discussion of variation leads to a concept which may be self-evident. All isolates, variants, and mutants consist of populations of virions [15]. Even if infection is established by a single virion, a population with differences among virions will soon develop. For any given propagation environment virions with different genomic sequences will establish a quasiequilibrium, usually with one genomic sequence predominating. Imposition of a different selection pressure upon the isolate could result in a shift in the equilibrium, perhaps leading to recognizable variants. However, in general, virus isolates are remarkably stable [7]. Perhaps viruses have some mechanism which contributes to genome stability. An example might be that aphid transmission is controlled by genes on either end of the genome [2]. This placement of genes might limit recombination events. A classification system requires some level of stability to be meaningful.

Current taxonomy of the family *Potyviridae*

Participants at the Potyvirus Taxonomy Workshop in Braunschweig in 1990 proposed that the definitive virus members and possible members of the potyvirus group be placed in a family of plant viruses called the *Potyviridae* [1]. All of these viruses have several common characteristics such as particle morphology, induction of pinwheel inclusions, genome organization, replication module, processing of a polyprotein, a 5'VPg, and a 3'poly A [23] (Fig. 1). Three genera and one possible genus were proposed. The Plant Virus Subcom-

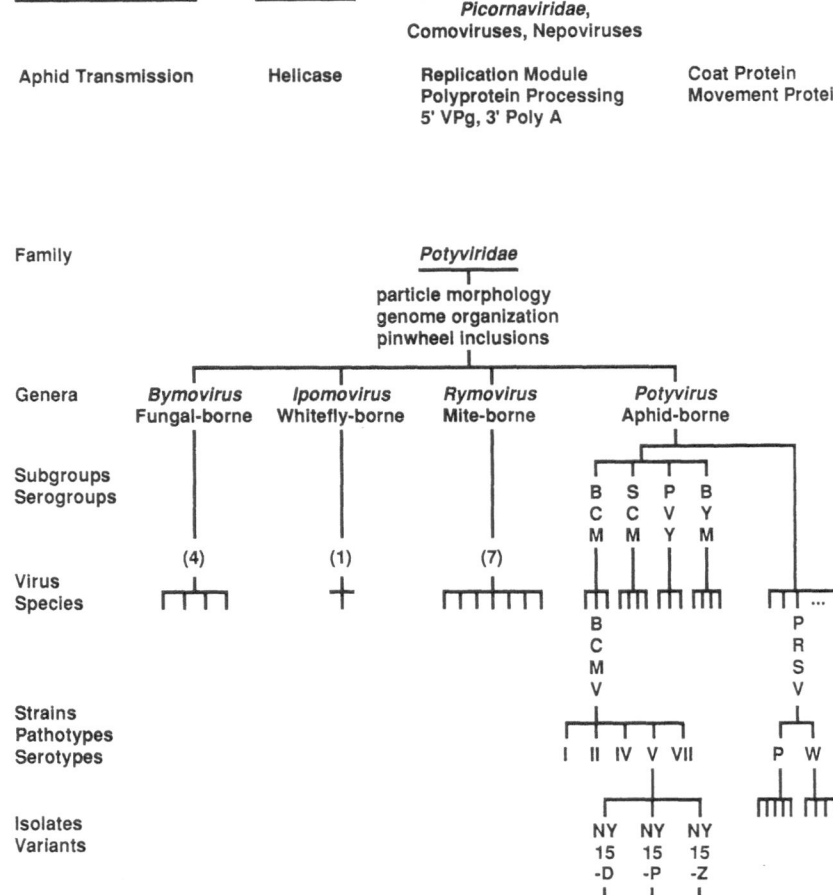

Fig. 1. Relationships of the family *Potyviridae*. At the top are displayed other virus groups and families along with characters which are similar to viruses of the *Potyviridae*. All *Potyviridae* species have similar particle morphology, genome organization, and produce pinwheel inclusions. The genera are separated based upon sequence information of the nucleotides or amino acid sequences of the coat proteins. Fortuitously, the different genera contain viruses with different vectors. Subgroups are a convenient collection of closely related species. Variation of the species is illustrated by the placement of strains (pathotypes or serotypes), isolates, or variants, all of which contain variability

mittee of the Committee on Taxonomy of Viruses approved these suggestions but modified two of the proposed genus names. The genus names approved by the Plant Virus Subcommittee were the genus *Potyvirus* composed of aphid-transmitted viruses with potato virus Y as the type species, the genus *Rymovirus* composed of mite-transmitted viruses with ryegrass mosaic virus as the type species, the genus *Bymovirus* composed of fungal-transmitted viruses with barley yellow mosaic virus as the type species, and the possible genus *Ipomovirus* composed of whitefly-transmitted viruses with sweet potato mild mottle virus the type species. Final approval of the family and genera requires ratification by the Executive Committee of ICTV and by delegates of the ICTV at the International Congress of Virology in 1993 at Glasgow, Scotland.

The family *Potyviridae* is related by nucleotide sequence homology of certain genes to several other virus families [6, 17] (Fig. 1). In spite of these sequence similarities to other families the *Potyviridae* species form a distinct cluster of genera. Rybicki and Shukla [17] call the family *Potyviridae* the "largest unique group of plant viruses evolved from a common ancestor without reorganization or addition of genes or modules." Due to the apparent modular evolution, selection of a single hierarchical relationship at levels higher than family seems impossible. Goldbach and Rybicki have made independent suggestions that taxa higher than that of family should not be erected [6, 17].

Below the family level all viruses which form cylindrical inclusions fit into one of the four genera with the exception of maclura mosaic and narcissus latent viruses. The particle lengths of these two viruses are shorter than those of members in the four genera and the coat protein molecular weight is larger. After further characterization these two viruses might form another genus of the *Potyviridae* [4].

A subgroup of legume viruses was proposed for viruses with many properties similar to bean yellow mosaic virus [19]. This BYMV subgroup has been confirmed by cluster analysis of the amino acid sequences of the coat proteins [17, 23]. Cluster analysis [1,23] also shows close relationships within groups of virus species such as sugarcane mosaic virus [9,13,18], potato virus Y [8,20,22], bean common mosaic virus [5,14,24], and watermelon mosaic virus [17,23]. These clusters of species should be given subgroup designations. Species in these subgroups often require careful discrimination for identification. Use of this subgroup category has proved convenient and should be encouraged.

A *subgroup* might be defined as a collection of virus species which are closely related and have many properties in common, making them difficult to separate. This collection of viruses is between the genus and species level. Recently the A and B 'serotypes' were shown to be separate virus species [24]. This means that BCMV isolates (see Table 3 in [14]) have several levels of antigenic relationships at the species level and above rather than below the species level. The term *serogroup* should be used above the species levels as a parallel to the subgroup term (Fig. 1). In fact, antigenic relationships of potyviruses seem more common at the serogroup than serotype level.

The International Committee on Taxonomy of Viruses defines and approves families, genera, and species. Subgroups, strains, and isolates are defined and recommended by study groups such as the Potyvirus Study Group of the Plant Virus Subcommittee. Current technology has helped resolve some of the questionable relationships among species of the *Potyviridae* and confirmed the relationship of the bipartite bymoviruses to the monopartite viruses in this family. The status of taxonomy in the *Potyviridae* is not a report of disarray and chaos. Scientists have been very careful in descriptions of new viruses and the Plant Virus Subcommittee has been conservative in assigning members to the former potyvirus group. Acceptance of the family with genera has removed one area of confusion (monopartite and bipartite genomes). The next task will be to determine if the possible members of the genus *Potyvirus* are different species or if they represent synonyms of currently recognized species. As new species are proposed or synonyms are found the Potyvirus Study Group should be involved so that a consensus for the new nomenclature can be developed among a wide range of virologists. The same procedure can be used for subgroup, strain, and pathogroup assignments.

The confusion virologists have had over the taxonomy of these viruses seems to result from different points of view. Molecularly-oriented scientists see the virus particle while pathologists see the disease. This volume presents both viewpoints. My conclusion from reading these chapters is that taxonomy must rely on molecular information derived directly from the sequence of the RNA genome or of the amino acid sequences of proteins derived from the genome. So far information at this level has allowed clear determination of species as evidenced by the PVY/PepMoV area of confusion which may be resolved [8, 20, 22]. It is encouraging that speciation by molecular properties places virus isolates with similar biological properties together. Thus in the sugarcane mosaic virus subgroup, speciation by end group specific serology, peptide profiles, and coat protein sequences defined viruses in which all isolates had the same cytological properties and agreed with positive cross protection tests. Biological properties are important from the disease viewpoint. Genetically controlled resistance in plants is often at the strain level. Characterization of isolates and variants of strains, that is variation within a strain, is important when working with resistance mechanisms.

Once species and strains have been separated and defined there is a need for rapid identification of these entities. Richter and Jordan recommend that certain laboratories be designated as repositories of reference diagnostic materials for species of the family *Potyviridae* [10, 16]. This might be a task for an international body such as the Potyvirus Study Group to organize.

Acknowledgements

Discussions with M. A. Mayo, G. I. Mink, A. F. Murant, and M. H. V. Van Regenmortel were most helpful in arriving at these definitions and C. H. Calisher critically read the draft, but the final choice of wording is my responsibility. Technical contribution no. 3275 of the South Carolina Agricultural Experiment Station, Clemson University.

References

1. Barnett OW (1991) *Potyviridae*, a proposed family of plant viruses. Arch Virol 118: 139–141
2. Berger PH (1992) The usefulness of aphid transmission as a taxonomic criterion for potyviruses. In: Barnett OW (ed) Potyvirus taxonomy. Springer, Wien New York, pp 251–255 (Arch Virol [Suppl] 5)

3. Bos L (1992) Potyviruses, chaos or order? In: Barnett OW (ed) Potyvirus taxonomy. Springer, Wien New York, pp 31–46 (Arch Virol [Suppl] 5)

4. Brunt AA (1992) The general properties of potyviruses. In: Barnett OW (ed) Potyvirus taxonomy. Springer, Wien New York, pp 3–16 (Arch Virol [Suppl] 5)

5. Dijkstra J, Kahn JA (1992) A proposal for a bean common mosaic subgroup of potyviruses. In: Barnett OW (ed) Potyvirus taxonomy. Springer, Wien New York, pp 389–395(Arch Virol [Suppl] 5)

6. Goldbach R (1992) The recombinative nature of potyviruses: implications for setting up a true phylogenetic taxonomy. In: Barnett OW (ed) Potyvirus taxonomy. Springer, Wien New York, pp 299–304 (Arch Virol [Suppl] 5)

7. Hampton RO, Provvidenti R (1992) Specific infectivity and host resistance have predicated potyviral and pathotype nomenclature but relate less to taxonomy. In: Barnett OW (ed) Potyvirus taxonomy. Springer, Wien New York, pp 183–187 (Arch Virol [Suppl] 5)

8. Hiebert E, Purcifull DE (1992) A comparison of pepper mottle virus with potato virus Y and evidence for their distinction. In: Barnett OW (ed) Potyvirus taxonomy. Springer, Wien New York, pp 321–326 (Arch Virol [Suppl] 5)

9. Jensen SG (1992) A viewpoint on the taxonomy of potyviruses infecting sugarcane, maize, and sorghum. In: Barnett OW (ed) Potyvirus taxonomy. Springer, Wien New York, pp 349–351 (Arch Virol [Suppl] 5)

10. Jordan R (1992) Potyviruses, monoclonal antibodies and antigenic sites. In: Barnett OW (ed) Potyvirus taxonomy. Springer, Wien New York, pp 81–95 (Arch Virol [Suppl] 5)

11. Lecoq H, Purcifull DE (1992) Biological variability of potyviruses, an example, zucchini yellow mosaic virus. In: Barnett OW (ed) Potyvirus taxonomy. Springer, Wien New York, pp 229–234 (Arch Virol [Suppl] 5)

12. Lovisolo O (1992) Ecology and taxonomy of some European potyviruses. In: Barnett OW (ed) Potyvirus taxonomy. Springer, Wien New York, pp 311–316 (Arch Virol [Suppl] 5)

13. Leseman D-E, Shukla DD, Tosic M, Huth W (1992) Differentiation of the four viruses of the sugarcane mosaic virus subgroup based on cytopathology. In: Barnett OW (ed) Potyvirus taxonomy. Springer, Wien New York, pp 353–361 (Arch Virol [Suppl] 5)

14. Mink GI, Silbernagel MJ (1992) Serological and biological relationships among viruses in the bean common mosaic virus subgroup. In: Barnett OW (ed) Potyvirus taxonomy. Springer, Wien New York, pp 397–406 (Arch Virol [Suppl] 5)

15. Ohki ST (1992) A potyvirus in nature: indistinct populations. In: Barnett OW (ed) Potyvirus taxonomy. Springer, Wien New York, pp 217–219 (Arch Virol [Suppl] 5)

16. Richter J (1992) Polyclonal reference antisera may be useful for the differentiation of potyvirus species. In: Barnett OW (ed) Potyvirus taxonomy. Springer, Wien New York, pp 71–74 (Arch Virol [Suppl] 5)

17. Rybicki EP, Shukla DD (1992) Coat protein phylogeny and systematics of potyviruses. In: Barnett OW (ed) Potyvirus taxonomy. Springer, Wien New York, pp 139–170 (Arch Virol [Suppl] 5)

18. Shukla DD, Frenkel MJ, McKern NM, Ward CW, Jilka J, Tosic M, Ford RE (1992) Present status of the sugarcane mosaic subgroup of potyviruses. In: Barnett OW (ed) Potyvirus taxonomy. Springer, Wien New York, pp 363–373 (Arch Virol [Suppl] 5)

19. Uyeda I (1992) Bean yellow mosaic virus subgroup: search for the group specific sequences in the 3' terminal region of the genome. In: Barnett OW (ed) Potyvirus taxonomy. Springer, Wien New York, pp 377–385 (Arch Virol [Suppl] 5)

20. van der Vlugt R (1992) Is pepper mottle virus a strain of potato virus Y? In: Barnett OW (ed) Potyvirus taxonomy. Springer, Wien New York, pp 327–335 (Arch Virol [Suppl] 5)

21. Van Regenmortel MHV (1992) What is a virus? In: Barnett OW (ed) Potyvirus taxonomy. Springer, Wien New York, pp 47–53 (Arch Virol [Suppl] 5)
22. Vance VB, Jordan R, Edwardson JR, Christie R, Purcifull DE, Turpen T, Falk B (1992) Evidence that pepper mottle virus and potato virus Y are distinct viruses: analyses of the coat protein and 3' untranslated sequences of a California isolate of pepper mottle virus. In: Barnett OW (ed) Potyvirus taxonomy. Springer, Wien New York, pp 337–345 (Arch Virol [Suppl] 5)
23. Ward CW, McKern NM, Frenkel MJ, Shukla DD (1992) Sequence data as the major criterion for potyvirus classification. In: Barnett OW (ed) Potyvirus taxonomy. Springer, Wien New York, pp 283–297 (Arch Virol [Suppl] 5)
24. Vetten HJ, Lesemann D-E, Maiss E (1992) Serotype A and B strains of bean common mosaic virus are two distinct potyviruses. In: Barnett OW (ed) Potyvirus taxonomy. Springer, Wien New York, pp 415–431 (Arch Virol [Suppl] 5)

Author's address: O. W. Barnett, Department of Plant Pathology and Physiology, Clemson University, Clemson, SC 29634-0377, U.S.A.

Subject index

R.I.B. Francki, C.M. Fauquet, D.L. Knudson, F. Brown (eds.)

Classification and Nomenclature of Viruses

Fifth Report of the International Committee on Taxonomy of Viruses

(Archives of Virology / Supplementum 2)

1991. IV, 450 pages.
Soft cover DM 110,-, öS 770,-
Reduced price for subscribers
to "Archives of Virology":
Soft cover DM 99,-, öS 693,-
ISBN 3-211-82286-0

Prices are subject to change without notice

The Fifth Report of the International Committee on Taxonomy of Viruses (ICTV), summarizes the proceedings and decisions reached by the ICTV at its meetings held at the International Congresses of Virology in Sendai (1984), Edmonton (1987), and Berlin (1990). The report places 2.430 viruses into 73 families and groups (including 19 new ones), according to the universal virus taxonomic system. There are many new tables, diagrams, and keys and there are expanded descriptions of virion properties, viral replication, and biological properties.

This report represents the work of many virologists from many countries, working to apply the explosion of new molecular biologic, physicochemical, serologic, and biologic data to the universal virus taxonomic system. New data and new viruses are well represented in the construction and refinement of virus families and genera. This volume will serve virologists well as they work to identify, characterize, and compare new isolates, and as they work to advance our understanding of the nature and properties of classic strains. This volume will serve virologists well as the basis for all future developments in virus classification, taxonomy, and nomenclature.

Springer-Verlag Wien New York

ISSN 0304-8608
Title No. 705

Archives of Virology

Official Journal of the Virology Division
of the International Union of Microbiological Societies

Archives of Virology publishes original contributions from all branches of research on viruses, viruslike agents, and virus infections of humans, animals, plants, insects, and bacteria. Coverage includes the broadest spectrum of topics, from initial descriptions of newly discovered viruses, to studies of virus structure, composition, and genetics, to studies of virus interactions with host cells, host organisms, and host populations. Multidisciplinary studies are particularly welcome, as are studies employing molecular biologic, molecular genetic, and modern immunologic and epidemiologic approaches. For example, studies on the molecular pathogenesis, pathophysiology, and genetics of virus infections in individual hosts, and studies on the molecular epidemiology of virus infections in populations, are encouraged. Studies involving applied research, such as diagnostic technology, development, monoclonal antibody panel development, vaccine development, and antiviral drug development, are also encouraged. However, such studies are often better presented in the context of a specific application or as they bear upon general principles of interest to many virologists. In all cases, it is the quality of the research work, its significance, and its originality which will decide acceptability.

As a new opportunity for publication of proceedings of meetings, treatises, and large reviews the series of Special Issues of *Archives of Virology* was initiated in 1990. Individuals who are organizing a meeting, symposium, conference, or congress, and individuals who would like to organize a treatise or large review are invited to communicate directly with one of the Special Issues Editors for further information: Dr. C. H. Calisher, Fort Collins, Colorado, or Dr. H.-D. Klenk, Marburg, FRG.

Subscription Information:
1993. Vols. 128–133 (4 issues each): DM 1.944,–, öS 13.608,–, plus carriage charges

 # New by Springer-Verlag

Thomas Boller and
Frederick Meins, Jr. (eds.)

Genes Involved in Plant Defense

(Plant Gene Research)

1992. 34 figures. Approx. 350 pages.
Cloth DM 218,–, öS 1526,–
ISBN 3-211-82312-3

The co-evolution of plants and microbes has led to an elaborate system of genes involved in recognition, attack and defense. This volume deals with these genes and regulation of their expression. The topic is treated broadly ranging from considerations of population genetics to the identification of defense-related genes and their regulation. The book provides a concise, authoritative review of latest developments in a rapidly developing and active field of agronomic importance.

Contents:

Resistance and Susceptibility Genes of Plants

M. S. Wolfe and C. Gessler: The Use of Resistance Genes in Breeding Epidemiological Considerations. – P. J. G. M. de Wit: Functional Models to Explain Gene-for-Gene Relationships in Plant-Pathogen Interactions.

Virulence and Avirulence Genes of Pathogens

M. A. Djordjevic, B. G. Rolfe, and W. Lewis-Henderson: An Analysis of Host Range Specificity Genes of *Rhizobium* as a Model System for Virulence Genes in Phytobacteria. – N. T. Keen and W. O. Dawson: Pathogen Avirulence Genes and Elicitors of Plant Defense. – F. Banuett and I. Herskowitz: Pathogenicity Determinants in the Smut Fungi of Cereals. – W. Schäfer, D. Stahl, and E. Mönke: Identification of Fungal Genes Involved in Plant Pathogenesis and Host Range.

Perception of Pathogens and Signal Transduction

S. C. Winans: Interactions Between *Agrobacterium tumefaciens* and its Host Plant Cells. – J. Ebel and D. Scheel: Elicitor Recognition and Signal Transduction.

Plant Genes Induced in the Defense Reaction

J. R. Cutt and D. F. Klessig: Pathogenesis-Related Proteins. – F. Meins, Jr., J.-M. Neuhaus, Ch. Sperisen, and J. Ryals: The Primary Structure of Plant Pathogenesis-Related Glucanohydrolases and Their Genes. – F. Garcia-Olmedo, M. J. Carmona, J. J. Lopez-Fando, J. A. Fernandez, A. Castagnaro, A. Molina, C. Hernandez-Lucas, and P. Carbonero †: Characterization and Analysis of Thionin Genes. – J. L. Dangl: Regulatory Elements Controlling Developmental and Stress-Induced Expression of Phenylpropanoid Genes. – M. H. Walter: Regulation of Lignification in Defense.

 # Springer-Verlag Wien New York

Sachsenplatz 4–6, P.O. Box 89, A-1201 Wien · Heidelberger Platz 3, D-1000 Berlin 33
175 Fifth Avenue, New York, NY 10010, USA · 37-3, Hongo 3-chome, Bunkyo-ku, Tokyo 113, Japan